Lecture Notes in Computer Science 8646

Commenced Publication in 1973
Founding and Former Series Editors:
Gerhard Goos, Juris Hartmanis, and Jan van Leeu

Ladjel Bellatreche   Mukesh K. Mohania (Eds.)

Data Warehousing and Knowledge Discovery

16th International Conference, DaWaK 2014
Munich, Germany, September 2-4, 2014
Proceedings

 Springer

Volume Editors

Ladjel Bellatreche
LIAS/ISAE-ENSMA
Téléport 2
1 avenue Clément Ader
BP 40109
86961 Futuroscope Chasseneuil Cedex, France
E-mail: ladjel.bellatreche@ensma.fr

Mukesh K. Mohania
IBM Research - India
4, Block-C
Institutional Area
Vasant Kunj
New Delhi 110070, India
E-mail: mkmukesh@in.ibm.com

ISSN 0302-9743   e-ISSN 1611-3349
ISBN 978-3-319-10159-0   e-ISBN 978-3-319-10160-6
DOI 10.1007/978-3-319-10160-6
Springer Cham Heidelberg New York Dordrecht London

Library of Congress Control Number: 2014945728

LNCS Sublibrary: SL 3 – Information Systems and Application,
incl. Internet/Web and HCI

Typesetting: Camera-ready by author, data conversion by Scientific Publishing Services, Chennai, India

Printed on acid-free paper

Springer is part of Springer Science+Business Media (www.springer.com)

Preface

Data warehousing and knowledge discovery technologies have been growing over time and they are now a practical need in every major small and large company. The maturity of these technologies encourages these companies to integrate heterogeneous, autonomous and evolving data issued from traditional (databases) and advanced sources such as sensors and social networks into a single large database to enable advanced querying, analysis, and recommendation. The data warehouse design passes through a well-established lifecycle that includes: conceptual modeling, ETL, logical modeling, deployment, and physical modeling. Knowledge discovery complements this lifecycle to offering exploitation capabilities of the warehouse. With the diversity of deployment platforms motivated by HPC (High Processing Computing), offering the process of retrieval and knowledge discovery from this huge amount of heterogeneous complex data builds the litmus-test for the research in the area.

During the past years, the International Conference on Data Warehousing and Knowledge Discovery (DaWaK) has become one of the most important international scientific events to bring together researchers, developers, and practitioners to discuss latest research issues and experiences in developing and deploying data warehousing and knowledge discovery systems, applications, and solutions. DaWaK is in the top 20 of the google scholar ranking related to Data Mining & Analysis: `http://scholar.google.com/citations?view_op=top_venues&hl=fr&vq=eng_datamininganalysis`. This year's conference (DaWaK 2014), builds on this tradition of facilitating the cross-disciplinary exchange of ideas, experience, and potential research directions. DaWaK 2014 seeks to introduce innovative principles, methods, models, algorithms and solutions, industrial products, and experiences to challenging problems faced in the development of data warehousing, knowledge discovery, data mining applications, and the emerging area of HPC.

This year we received 109 papers and the Program Committee finally selected 34 full papers and 8 short papers, making an acceptance rate of 31%. The accepted papers cover a number of broad research areas on both theoretical and practical aspects of data warehouse and knowledge discovery. In the area of data warehousing, the topics covered included the modeling and ETL, ontologies, real-time data warehouses, query optimization, map reduce paradigm, storage models, scalability, distributed and parallel processing and data warehouses and data mining applications integration, recommendation and personalization, multidimensional analysis of text documents, and data warehousing for real world applications such as health, bio-informatics, telecommunication, etc. In the areas of data mining and knowledge discovery, the topics included stream data analysis and mining, traditional data mining techniques topics such as frequent item sets, clustering, association, classification ranking and application of data mining technologies to real world problems, and fuzzy mining, skyline, etc.

It is especially notable to see that some papers covered emerging real world applications as bioinformatics, social network, telecommunication, brain analysis, etc.

This year we have three special issues for well-known journals: Knowledge and Information Systems: An International Journal, Springer, Journal of Concurrency and Computation: Practice and Experience, Wiley and Transactions on Large-scale Data- and Knowledge Centered Systems - TLDKS, Springer.

We would like to thank all authors for submitting their research paper in DaWaK 2014. We express our gratitude to all the Program Committee members and the external reviews, who reviewed the papers very profoundly and in a timely manner. Finally, we would like to thank Mrs Gabriela Wagner for her endless help and support.

See you all in Munich. Hope you enjoy the technical program, meeting and interacting with research colleagues, and of course, a beautiful city Munich.

July 2014   Ladjel Bellatreche
                                                        Mukesh K. Mohania

Organization

Program Committee Co-chairs

Ladjel Bellatreche   LIAS/ISAE-ENSMA, France
Mukesh Mohania   IBM Research, India

Program Committee

Alberto Abelló	Universitat Politecnica de Catalunya, Spain
Rajeev Agrawal	North Carolina A&T State University, USA
Mohammed Al-Kateb	Teradata Labs, USA
Torben Bach Pedersen	Aalborg University, Denmark
Elena Baralis	Politecnico di Torino, Italy
Sadok Ben Yahia	Faculty of Sciences of Tunis, France
Petr Berka	University of Economics in Prague, Czech Republic
Jorge Bernardino	ISEC - Polytechnic Institute of Coimbra, Portugal
Vasudha Bhatnagar	Delhi University, India
Manish Bhide	IBM Software Lab, India
Kamel Boukhalfa	USTHB, Algeria
Omar Boussaid	University of Lyon, France
Stephane Bressan	National University of Singapore, Singapore
Erik Buchmann	Karlsruhe Institute of Technology, Germany
Frans Coenen	The University of Liverpool, UK
Bruno Cremilleux	Université de Caen, France
Alfredo Cuzzocrea	ICAR-CNR and University of Calabria, Italy
Karen Davis	University of Cincinnati, USA
Alin Dobra	University of Florida, USA
Dejing Dou	University of Oregon, USA
Curtis Dyreson	Utah State University, USA
Markus Endres	University of Augsburg, Germany
Vladimir Estivill-Castro	Griffith University, Australia
Christie Ezeife	University of Windsor, Canada
Filippo Furfaro	University of Calabria, Italy
Pedro Furtado	Universidade de Coimbra, Portugal
Carlos Garcia-Alvarado	Pivotal Inc., USA
Matteo Golfarelli	DISI - University of Bologna, Italy
Sergio Greco	University of Calabria, Italy
Neelima Gupta	University of Delhi, India

Frank Hoppner	Ostfalia University of Applied Sciences, Germany
Domingo-Ferrer Josep	University Rovira i Virgili, Spain
Selma Khouri	National High School of Computer Science, Algeria
Sang-Wook Kim	Hanyang University, South Korea
Christian Koncilia	University of Klagenfurt, Austria
Jens Lechtenboerger	Westfalische Wilhelms-Universität Münster, Germany
Wolfgang Lehner	Dresden University of Technology, Germany
Carson K. Leung	The University of Manitoba, Canada
Sofian Maabout	University of Bordeaux, France
Patrick Marcel	Université François Rabelais Tours, France
Anirban Mondal	Xerox Research, India
Apostolos Papadopoulos	Aristotle University, Greece
Jeffrey Parsons	Memorial University of Newfoundland, Canada
Dhaval Patel	Indian Institute of Technology Roorkee, India
Lu Qin	University of Technology in Sydney, Australia
Sriram Raghavan	IBM Research - India
Stefano Rizzi	University of Bologna, Italy
Maria Luisa Sapino	Università degli Studi di Torino, Italy
Kai-Uwe Sattler	Ilmenau University of Technology, Germany
Neeraj Sharma	India IBM Labs, India
Alkis Simitsis	HP Labs, USA
Domenico Talia	University of Calabria, Italy
David Taniar	Monash University, Australia
Olivier Teste	IRIT, University of Toulouse, France
Dimitri Theodoratos	New Jersey Institute of Technology, USA
A Min Tjoa	Vienna University of Technology, Austria
Panos Vassiliadis	University of Ioannina, Greece
Robert Wrembel	Poznan University of Technology, Poland
Wolfram Wöß	Johannes Kepler University Linz, Austria
Karine Zeitouni	Université de Versailles Saint-Quentin-en-Yvelines, France
Bin Zhou	University of Maryland in Baltimore County, USA

External Reviewers

Christian Thomsen	Aalborg University, Denmark
Xike Xie	Aalborg University, Denmark
Anamika Gupta	CBS College, University of Delhi, India
Rakhi Saxena	Deshbandhu College, University of Delhi, India
Florian Wenzel	University of Augsburg, Germany
Patrick Roocks	University of Augsburg, Germany

Kiki Maulana Adhinugraha	Monash University, Australia
Enrico Gallinucci	University of Bologna, Italy
Lorenzo Baldacci	University of Bologna, Italy
Simone Graziani	University of Bologna, Italy
Sameep Mehta	IBM Research, India
Aditya Telang	IBM Research, India
Vinay Kolar	IBM Research, India
Kalapriya Kannan	IBM Research, India
Francesco Parisi	University of Calabria, Italy
Bettina Fazzinga	University of Calabria, Italy
Richard Kyle MacKinnon	University of Manitoba, Canada
Fan Jiang	University of Manitoba, Canada
Peter Braun	University of Manitoba, Canada
Julien Aligon	Université François Rabelais Tours, France
Arnaud Soulet	Université François Rabelais Tours, France
Ritu Chaturvedi	School of Computer Science, University of Windsor, Canada
Nhat-Hai Phan	University of Oregon, USA
Fernando Gutierrez	University of Oregon, USA
Francois Rioult	Universite de Caen, France
Cem Aksoy	New Jersey Institute of Technology, USA
Ananya Dass	New Jersey Institute of Technology, USA
Aggeliki Dimitriou	National Technical University of Athens, Greece
Xiaoying Wu	Wuhan University, China
Petar Jovanovic	Universitat Politecnica de Catalunya, Spain
Vasileios Theodorou	Universitat Politecnica de Catalunya, Spain
Yehia Taher	University of Versailles-Saint-Quentin, France
Luca Cagliero	Politecnico di Torino, Italy
Alessandro Fiori	IRCC, Institute for Cancer Research at Candiolo, Italy
Paolo Garza	Politecnico di Torino, Italy
Cristinan Molinaro	University of Calabria, Italy
Andrea Pugliese	University of Calabria, Italy
Irina Trubitsyna	University of Calabria, Italy
Christina Feilmayr	Johannes Kepler University Linz, Austria
Thomas Leitner	Johannes Kepler University Linz, Austria
Khalissa Derbal	USTHB University, Algeria
Ibtissem Frihi	USTHB University, Algeria
Imene Guellil	ESI High Scholl, Algeria
Ronan Tournier	IRIT, University Paul Sabatier, France
Zoé Faget	LIAS, Poitiers, France
Rima Bouchakri	ESI, Algeria
Stéphane Jean	LIAS, Poitiers, France
Brice Chardin	LIAS, Poitiers, France
Ahcène Boukorca	LIAS, Poitiers, France

A Secure Data Sharing and Query Processing Framework via Federation of Cloud Computing (Keynote)

Sanjay K. Madria

Department of Computer Science
Missouri University of Science and Technology, Rolla, MO, USA
madrias@mst.edu

Abstract. Due to cost-efficiency and less hands-on management, big data owners are outsourcing their data to the cloud, which can provide access to the data as a service. However, by outsourcing their data to the cloud, the data owners lose control over their data, as the cloud provider becomes a third party service provider. At first, encrypting the data by the owner and then exporting it to the cloud seems to be a good approach. However, there is a potential efficiency problem with the outsourced encrypted data when the data owner revokes some of the users' access privileges. An existing solution to this problem is based on symmetric key encryption scheme but it is not secure when a revoked user rejoins the system with different access privileges to the same data record. In this talk, I will discuss an efficient and Secure Data Sharing (SDS) framework using a combination of homomorphic encryption and proxy re-encryption schemes that prevents the leakage of unauthorized data when a revoked user rejoins the system. I will also discuss the modifications to our underlying SDS framework and present a new solution based on the data distribution technique to prevent the information leakage in the case of collusion between a revoked user and the cloud service provider. A comparison of the proposed solution with existing methods will be discussed. Furthermore, I will outline how the existing work can be utilized in our proposed framework to support secure query processing for big data analytics. I will provide a detailed security as well as experimental analysis of the proposed framework on Amazon EC2 and highlight its practical use.

Biography: Sanjay Kumar Madria received his Ph.D. in Computer Science from Indian Institute of Technology, Delhi, India in 1995. He is a full professor in the Department of Computer Science at the Missouri University of Science and Technology (formerly, University of Missouri-Rolla, USA) and site director, NSF I/UCRC center on Net-Centric Software Systems. He has published over 200 Journal and conference papers in the areas of mobile data management, Sensor computing, and cyber security and trust management. He won three best papers awards including IEEE MDM 2011 and IEEE MDM 2012. He is the co-author of a book published by Springer in Nov 2003. He serves as steering committee members in IEEE SRDS and IEEE MDM among others and has served in International conferences as a general co-chair (IEEE MDM,

IEEE SRDS and others), and presented tutorials/talks in the areas of mobile data management and sensor computing at various venues. His research is supported by several grants from federal sources such as NSF, DOE, AFRL, ARL, ARO, NIST and industries like Boeing, Unique*Soft, etc. He has also been awarded JSPS (Japanese Society for Promotion of Science) visiting scientist fellowship in 2006 and ASEE (American Society of Engineering Education) fellowship at AFRL from 2008 to 2012. In 2012-13, he was awarded NRC Fellowship by National Academies. He has received faculty excellence research awards in 2007, 2009, 2011 and 2013 from his university for excellence in research. He served as an IEEE Distinguished Speaker, and currently, he is an ACM Distinguished Speaker, and IEEE Senior Member and Golden Core awardee.

Table of Contents

Uncertainty

Preferences and Recommendation

Query Performance and HPC

Cube and OLAP

Optimization

Classification

Social Networks and Recommendation Systems

Knowledge Data Discovery

Industrial Applications

Mining and Processing Data Stream

Mining and Similarity

An Approach on ETL Attached Data Quality Management

Christian Lettner[1], Reinhard Stumptner[1], and Karl-Heinz Bokesch[2]

[1] Software Competence Center Hagenberg GmbH,
Softwarepark 21, 4232 Hagenberg, Austria
[2] Upper Austrian Health Insurance, Gruberstrasse 77, 4021 Linz, Austria
{christian.lettner,reinhard.stumptner}@scch.at,
karl-heinz.bokesch@ooegkk.at

Abstract. This contribution introduces an approach on ETL attached Data Quality Management by means of an autonomous Data Quality Monitoring System. The Data Quality Monitor can be attached (via light-weight connectors) to already implemented ETL processes and allows to quantify data quality and to suggest measures if the quality of a particular data package falls below a certain limit for instance. Furthermore, the long-term vision of this approach is to correct corrupted data (semi-)automatically according to user-defined Data Quality Rules. The Data Quality Monitor can be attached to an ETL process by defining "snapshot points", where data samples which should be validated are collected and by introducing "approval points", where an ETL process can be interrupted in case of corrupted input data. As the Data Quality Monitor is an autonomous module which is attached to instead of embedded into ETL processes, this approach supports the division of work between ETL developers and special data quality engineers.

Keywords: Data Quality, Data Warehouse, ETL, Rules, Metadata.

1 Introduction

Recently the amount of information or data collected by companies rapidly increases. At the same time, due to the enormous amounts of data, it becomes harder to keep an overview of the quality of certain data. It can easily become inconsistent, incomplete, inaccurate, duplicated, and so on. The following questions play a major role in connection with data quality. Are data complete and do have the proper meaning? Are data values consistent throughout an entire organization? Do data values fall within a defined domain? Are data values accurate? Are data duplicated?

As a consequence of low data quality there often are wrong data mining or analysis results, wrong figures in reports etc. and finally this has a remarkable impact on the daily business of an organization. To tackle these issues, the authors propose ETL (Extract Transform Load) attached data quality management which is to check the quality of new data packages before loading them into

L. Bellatreche and M.K. Mohania (Eds.): DaWaK 2014, LNCS 8646, pp. 1–8, 2014.

production databases or data warehouse systems. These new data can be profiled, matched with reference data, cleansed relying on declared rules and even corrected. The main idea of this methodology is to keep data-stores as clean as possible and to prevent the further processing of corrupted data. So data quality can be already checked in the data flows of the ETL packages.

ETL processes are used to move data from one or more source databases to target databases. During this process, data pass different transformation steps, which usually are called mappings. Having big amounts of data or complex mappings, executing ETL processes may take a very long time. In such cases it is advisable, to check the correctness of data before data loading has finished during the ETL process, before and after critical steps. If it is necessary to restart an ETL process because of corrupted data and if a restart is possible, a lot of time may be lost.

As already mentioned, the authors propose an ETL attached Data Quality Management system which should check the quality of data packages in the course of ETL execution which is shown in Fig. 1. In the ETL process so-called "snapshot points" are defined, where all necessary information for data validation is collected. These can be data samples for matching with reference data, for applying rules or for calculating and comparing certain statistics. When an ETL process reaches an "approval point", all declared constraints are evaluated (automatically or by a user) and if fulfilled, the process can continue. If certain constraints are not fulfilled, the process stops at the according approval point and loading of faulty data into a destination data store can be prevented for instance.

2   Related Work

The approach presented in this work focuses on measuring data quality to finally increase the overall quality of the corporate data. Nevertheless, faulty ETL operations are another reason for low data quality. [1] addresses faulty ETL operations by introducing a template-based ETL development environment which promises to generally increase the quality of ETL processes. This approach is extended by [2], demanding domain knowledge specified by business users to be used as a source of information for platform-independent ETL template generation.

A survey on declarative approaches to measure data quality is given in [3]. Declarative approaches provide logic based descriptions, called dependencies, of what is expected to be contained within a database instance. The advantage of declarative approaches is that these explicitly specify what a result should look like. Compared to imperative approaches, there is no need to specify how the data cleaning process is actually performed. Formally, a database instance D is said to be consistent with respect to a set of logical dependencies $\sum$, if it satisfies all dependencies within $\sum$. Logical dependencies are expressed using predicate logic and include functional dependencies, inclusion dependencies, or if the dependency is only valid for a subset of data, conditional functional dependencies and conditional inclusion dependencies. A prototype implementation of a data quality system utilizing conditional functional dependencies is presented in [4]. Conditional and functional dependencies may also be specified using the approach presented in this paper as shown in Sec. 4.

In [5] data quality rules for staging and data warehouse layers are defined using meta data tables. Based on these meta data tables, a generator creates data quality modules that implement the rules using SQL. To increase performance, the generator groups all rules that can be checked at the same time into one data quality module, so only one table scan is necessary for each group instead of one table scan for each rule. As shown in [3], attribute value constraints, foreign key constraints, etc. can be expressed as a combination of functional and inclusion dependences. As the approach presented in this paper is able to specify conditional and functional dependencies, all data quality rule types presented in that work [5] may also be defined in the frame of the ETL attached Data Quality Management approach. Nevertheless, the approach presented in [5] misses an explicit separation of data processing and data quality logic, as proposed in this work.

As an example from industry, SQL Server Data Quality Services (DQS) [6] introduced with Microsoft SQL Server 2012 represent a compact environment for data cleansing and mapping. This knowledge-driven data quality product consists of functions for describing data structures (attributes, data types etc.), defining matching rules, providing reference data, and so on.

However, a main difference to the presented approach is that data quality operations (e.g. handling corrupted data) in [6] have to be defined within the ETL processes and do not work (fully) autonomously.

The authors of [8] present NADEEF, "an extensible, generalized and easy-to-deploy data cleaning platform". The programming interface of NADEEF allows specifying different types of data quality rules, which describe expected data corruption and how to repair it if applicable. In contrast to that, the approach in this contribution looks at the data quality problem at a higher level as data cleansing is just a part of data quality management. This contribution aims at introducing a methodology or a process for data quality management including an approach for integrating these new methods into existing data processing infrastructures with minimum efforts. Furthermore, there is a difference between these approaches concerning the point of time of checking data quality. As in big data processing infrastructures data usually will be further processed as soon as these become available, it is important to correct data (or to prevent loading corrupted data) as soon as possible. Nevertheless, NADEEF could be an interesting extension of ETL-attached Data Quality Management in terms of a sophisticated "data cleansing module".

3 ETL Attached Data Quality Management

In this work data quality rules are evaluated using a separate Data Quality Monitor as depicted in Fig. 1. The data quality monitor is connected to the

ETL process using snapshot points and approval points, which are placed in the control flow, i.e. between the mappings of the ETL process. The result of the data quality rule evaluation is sent to the approval point, which, depending on the evaluation result, continues or interrupts the ETL process.

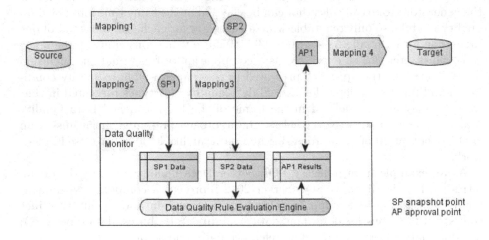

Fig. 1. Data quality monitor using snapshot points and approval points

Snapshot points represent locations within the ETL process where data is collected for data quality rule evaluation. Data are copied into a special data repository of the Data Quality Monitor or accessed on the fly using views. A materialized copy of snapshot point data is necessary, if it is changed or deleted in the course of ETL process execution before the data quality rules are evaluated, or if the snapshot point data should be used as reference data for future rule checking.

Approval points represent locations within the ETL process where data quality rules are evaluated and actions based on the result of the check are triggered. Possible actions are: (i) if the check is successful, the ETL process continuous automatically, (ii) if the check fails, a notification may be sent to a person in charge, a manual approval will be demanded or the ETL process is aborted at all. Which of these actions have to be triggered is specified using upper and lower bounds for the check result.

Snapshot points and approval points are implemented as *lightweight* ETL components that are placed in the control flow of an ETL process and act as connection points between ETL processes and the Data Quality Monitor. This way a strict separation of the data processing logic, represented by ETL processes, from the data quality processing logic, represented by the data quality monitor is established. This enables a separation of roles when implementing and maintaining these two data engineering areas.

Compared to implementing data quality checks on demand as part of the ETL process and finally becoming blurred into the data processing logic, the conceptual breakup into snapshot points and approval points allows dependencies and temporal synchronization between them to be explicitly modelled. Even more important, in the opinion of the authors, such an approach is absolutely needed for implementing separated data processing and data quality roles. It provides a borderline between roles and allows data quality rules to be implemented as an orchestration of snapshot points and approval points. This creates explicit and easy to maintain data quality checking processes.

To test the approach presented in this work, a prototypical Data Quality Rule Evaluation Engine based on SQL queries has been implemented. Each data quality rule is specified using six different types of custom expressions that will be composed to a query using the following SQL template. Custom expressions are depicted in angle brackets. We would like to point out that the approach presented in this work is open, in terms of encouraging the use of other rule evaluation engines, especially ones that are based on declarative approaches as discussed in Sec. 2.

SQL template used to implement data quality rules

```
select
  <Test-Check> result,
  <Test-Granularity> test_datarow_id
from <Test-Snapshot point>
where
  <Test-Condition>
group by <Test-Granularity>
```

    <Test-Snapshotpoint> contains the data to be checked.
    <Test-Check> defines a boolean quality check.
    <Test-Granularity> defines the granularity for the data quality check.
    <Test-Condition> filters the test snapshot points' data rows.

4   Results

4.1   ETL Tool Integration

Fig. 2 shows the implementation of the ETL process depicted in Fig. 1 using Kettle [9]. The transformations to be executed in parallel are modeled in a separate sub job, indicated as a box within the figure. Snapshot points (SP) and approval points (AP) are added to the ETL job by placing them into the control flow and setting the snapshot point and approval point identifier parameter respectively.

Fig. 3 shows the implementation of the snapshot point component in Kettle [9]. It must be placed into the ETL processes whenever a data snapshot should be created. Based on the snapshot point identifier specified by the data quality engineer and passed as a parameter, the snapshot point component (i) loads

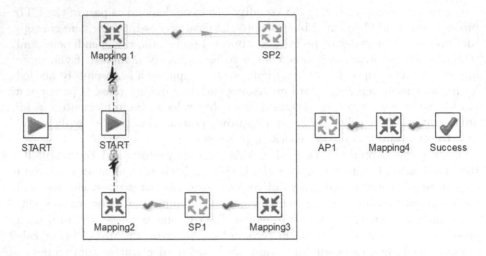

Fig. 2. Snapshot and approval points placed in a Kettle job

Fig. 3. Snapshot point component implemented in Kettle

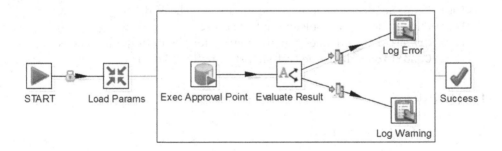

Fig. 4. Approval point component implemented in Kettle

configuration parameters, (ii) notifies the Data Quality Monitor that the execution of the respective snapshot point has been started, (iii) selects and executes the data load package registered for the provided snapshot point identifier, and finally (iv) notifies the Data Quality Monitor that the execution of the snapshot point has been finished.

Fig. 4 shows the implementation of the approval point component. Again, based on the approval point identifier that must be specified by the data quality engineer, it (i) loads configuration parameters, and (ii) evaluates all data quality rules assigned to this approval point in a sub job, again indicated as a box within the figure.

5   Conclusion

The approach on ETL attached Data Quality Management presented in this contribution promises a flexible tool-set for data quality monitoring. The basic idea is that a Data Quality Monitor can be connected to (already implemented) ETL processes, extract data, which should be validated, evaluate user-defined data quality rules and finally decide, if an ETL process should be stopped or if data is clean enough to being loaded into the according destination. These rules are fully flexible (written in SQL) and can refer to snapshot data from the ETL process and to reference data which is also extracted from the same ETL process or which is already available in the data quality systems internal repository. The Data Quality Monitor is attached to an ETL process by defining snapshot points, where data samples which should be validated are collected and by defining approval points, where, depending on the result of data quality rule evaluation, the ETL process can be halted automatically or by a user in case of not fulfilling certain constraints. So, if certain constraints are not fulfilled in case of faulty data packages and the process stops at an approval point, then loading of these faulty data into a data destination can be prevented and global data quality can be assured. All data quality rules or constraints can be declared independently from any ETL process. Consequently, this approach supports the division of work between ETL developers and data quality engineers which often is in accord with prevalent corporate structures. There is no need for data quality engineers to know about ETL syntax and components and ETL developers do not need to take care about rules or constraints for data quality validation. They can centrally specify rules, collect reference data, etc. and make this knowledge available to all ETL developers. This is one of the main advantages of the presented approach on ETL attached Data Quality Management when comparing it with other Data Quality Management systems.

Future investigation and enhancements will focus on increasing the level of sophistication of the data quality validation approach (e.g. improved data profiling methodologies) and on the ETL connection with supporting automated correction of corrupted data for instance. To get an improved measure of the quality of a certain data package, data profiles should be investigated as a main data quality criteria. Such profiles should be created based on statistics like the distribution of certain data, certain measures like averages, correlations, etc. and data spectra for instance. In many cases new data packages have very similar characteristics in terms of distribution, frequencies or other measures as the base data which new data should be added to. These measures should enable a data quality engineer to assess the data quality of a data package also independent from particular data quality rules. This could be presented by a dashboard giving an overview of the profiles of incoming data packages and of contents of the according data destination, visualizing the described measures and consequently support a user deciding if the quality of a data package should be good enough to load it into the data destinations or if measures concerning apparent data quality issues have to be taken.

Acknowledgements. This work has been supported by the COMET-Program of the Austrian Research Promotion Agency (FFG).

References

1. Stumptner, R., Freudenthaler, B., Krenn, M.: BIAccelerator – A template-based approach for rapid ETL development. In: Chen, L., Felfernig, A., Liu, J., Raś, Z.W. (eds.) ISMIS 2012. LNCS, vol. 7661, pp. 435–444. Springer, Heidelberg (2012)
2. Lettner, C., Zwick, M.: A data analysis framework for high-variety product lines in the industrial manufacturing domain. To appear in Proceedings of the 16th International Conference on Enterprise Information Systems, Lisbon, Portugal (2014)
3. Bertossi, L., Bravo, L.: Generic and declarative approaches to data quality management. In: Handbook of Data Quality, pp. 181–211. Springer (2013)
4. Fan, W., Geerts, F., Jia, X.: Semandaq: A data quality system based on conditional functional dependencies. Proc. VLDB Endow. 1(2), 1460–1463 (2008)
5. Rodic, J., Baranovic, M.: Generating data quality rules and integration into etl process. In: Proceedings of the ACM Twelfth International Workshop on Data Warehousing and OLAP, DOLAP 2009, pp. 65–72. ACM, New York (2009)
6. Microsoft: Data quality services, sql server 2012 books online, http://msdn.microsoft.com/en-us/library/ff877925.aspx (online; accessed January 27, 2014)
7. Farinha, J., Trigueiros, M.J., Belo, O.: Using inheritance in a metadata based approach to data quality assessment. In: Proceedings of the First International Workshop on Model Driven Service Engineering and Data Quality and Security. MoSE+DQS 2009, pp. 1–8. ACM, New York (2009)
8. Dallachiesa, M., Ebaid, A., Eldawy, A., Elmagarmid, A., Ilyas, I.F., Ouzzani, M., Tang, N.: Nadeef: A commodity data cleaning system. In: Proceedings of the 2013 ACM SIGMOD International Conference on Management of Data, SIGMOD 2013, pp. 541–552. ACM, New York (2013)
9. Kettle, http://community.pentaho.com/projects/data-integration/ (online; accessed February 05, 2014)
10. PostgreSQL, http://www.postgresql.org/ (online; accessed February 05, 2014)
11. Celko, J.: Joe Celko's SQL for Smarties: Advanced SQL Programming. The Morgan Kaufmann Series in Data Management Systems. Elsevier Science (2010)

Quality Measures for ETL Processes

Vasileios Theodorou[1], Alberto Abelló[1], and Wolfgang Lehner[2]

[1] Universitat Politècnica de Catalunya, Barcelona, Spain
{vasileios,aabello}@essi.upc.edu
[2] Technische Universität Dresden, Dresden, Germany
wolfgang.lehner@tu-dresden.de

Abstract. ETL processes play an increasingly important role for the support of modern business operations. These business processes are centred around artifacts with high variability and diverse lifecycles, which correspond to key business entities. The apparent complexity of these activities has been examined through the prism of Business Process Management, mainly focusing on functional requirements and performance optimization. However, the quality dimension has not yet been thoroughly investigated and there is a need for a more human-centric approach to bring them closer to business-users requirements. In this paper we take a first step towards this direction by defining a sound model for ETL process quality characteristics and quantitative measures for each characteristic, based on existing literature. Our model shows dependencies among quality characteristics and can provide the basis for subsequent analysis using Goal Modeling techniques.

Keywords: ETL, business process, quality measures.

1 Introduction

Business Intelligence nowadays involves identifying, extracting, and analysing large amount of business data coming from diverse, distributed sources. In order to facilitate decision-makers, complex IT-systems are assigned with the task of integrating heterogeneous data deriving from operational activities and loading of the processed data to data warehouses, in a process known as Extraction Transformation Loading (ETL). This integration requires the execution of real-time, automated, data-centric business processes in a variety of workflow-based tasks. The main challenge is how to turn the integration process design, which has been traditionally predefined for periodic off-line mode execution, into a dynamic, continuous operation that can sufficiently meet end-user needs.

During the past years, there has been considerable research regarding the optimization of ETL flows in terms of functionality and performance [26, 7]. Moreover, in an attempt to manage the complexity of ETL processes on a conceptual level that reflects organizational operations, tools and models from the area of Business Process Management (BPM) have been proposed [29, 3]. However, the dimension of process quality [25] has not yet been adequately examined in a systematic manner. Unlike other business processes, important quality factors for ETL process design are tightly coupled to information quality while

L. Bellatreche and M.K. Mohania (Eds.): DaWaK 2014, LNCS 8646, pp. 9–22, 2014.

depending on the interoperability of distributed engines. Added to that, there is increasing need for process automation in order to become more cost-effective [28] and therefore there needs to be a common ground between business users and IT that would allow the first to express quality concerns in a high level language, which would automatically be translated to design choices.

In this paper we take a first step towards quality-aware ETL process design automation by *defining a set of ETL process quality characteristics and the relationships between them, as well as by providing quantitative measures for each characteristic.* For this purpose, we conduct a systematic literature review, extract the relevant quality aspects that have been proposed in literature and adapt them for our case. Subsequently, we produce a model that represents ETL process quality characteristics and the dependencies among them. In addition, we gather from existing literature metrics for monitoring all of these characteristics and quantitatively evaluating ETL processes. Our model can provide the basis for subsequent analysis that will use Goal Modeling techniques [19] to reason and make design decisions for specific use cases, using as input only the user-defined importance of each quality characteristic.

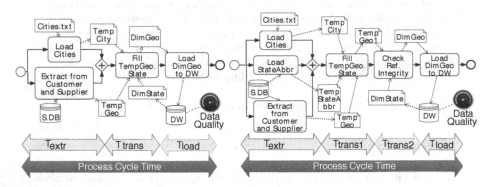

Fig. 1. Example alternative ETL process design with same functionality

We illustrate how our model works through a running example, borrowing the use case from the work of El Akkaoui and Zimanyi [11] . The use case is an ETL process that extracts geographical data from a source database (S.DB) and after processing, loads a dimension table into a data warehouse (DW). The tables extracted from the source database are *Customer* and *Supplier*, which contain information about *City*, *State*, *Country* and *ZipCode*. However, the attribute *State* might be missing from some records and therefore a flat text file (Cities.txt) containing *City*, *State*, and *Country* tuples is also used. After *State* entries have been filled, the table *DimGeo* is loaded to the data warehouse, with attributes *City*, *StateKey* and *ZipCode*, where the *StateKey* for each state is derived from *DimState*, which is another dimension table in the data warehouse. This process is modelled using the Business Process Model and Notation (BPMN[1]) and two alternative designs can be seen in Fig. 1.

---

[1] http://www.bpmn.org

The paper is organized as follows. Section 2 presents related work regarding quality characteristics for design evaluation. In Section 3 we present the extraction of our model from related work. The definitions, measures and dependencies among characteristics are presented in Section 4 and Section 5 where we distinguish between characteristics with construct implications and those only for design evaluation, respectively. Finally, we provide our conclusions and future work in Section 6.

2 Related Work

The significance of quality characteristics for the design and evaluation of ETL processes has recently gained attention. Simitsis et al. [28] recognise the importance of considering not only process functionality but also quality metrics throughout a systematic ETL process design. Thus, they define a set of quality characteristics specific to ETL processes that they refer to as *QoX metrics* and provide guidelines for reasoning about the degree of their satisfaction over alternative designs and the tradeoffs among them. A more recent work that has also considered the ideas from [28] is the work from Pavlov [24]. Based on well-known standards for software quality, the author maps software quality attributes to ETL specific parameters which he calls *QoX factors*. He defines these factors in ETL context and reasons about the impact that the different ETL subsystems might have on each characteristic.

Focusing on Information Quality, Naumann [23] provides a comprehensive list of criteria for the evaluation of Information Systems for data integration. In the same area, Dustdar et al. [10] identify most important challenges for Data Integration and highlight quality concerns in distributed, heterogeneous environments. Likewise, Jarke et al. [15] identify the various stakeholders in Data Warehousing activities and the differences in their roles as well as the importance of reasoning among alternative quality concerns and how that affects design choices.

In the last years, there has been an effort in the area of Business Process Management to quantify process quality characteristics and to empirically validate the use of well-defined metrics for the evaluation of specific quality characteristics. In this respect, García et al. [13] propose a framework for managing, modeling and evaluating software processes; define and experimentally validate a set of measures to assess, among others understandability and modifiability of process models. Similar empirical validation is provided by Sánchez-González et al. [25], who relate understandability and modifiability to innate characteristics of business process models.

Our approach differs from the above-mentioned ones in that we specifically focus on the process perspective of ETL processes. Instead of providing some characteristics as examples like in [28], we propose a comprehensive list of quality characteristics and we adjust them for our case. In addition, for each of these characteristics we provide quantitative metrics that are backed by literature.

Group	Characteristic	Barbacci et al. [5]	Simitsis et al. [28]	Jarke et al. [15]	Pavlov [24]	Naumann [23]	Dustdar et al. [10]	Kimball[2]
Characteristics with construct implications	**data quality**	-	data characteristics	quality dimensions	-	relevancy, reputation	data quality	data quality
	data accuracy			data accuracy		accuracy	accuracy	
	data completeness			data completeness		completeness	completeness	
	data freshness		freshness	data freshness, timeliness		timeliness	timeliness	
	data consistency		consistency	data coherence, correctness, minimality		consistent representation	consistency	consistency, deduplication, data conformance
	data interpretability			interpretability		interpretability		
	performance	performance	performance	performance, software efficiency	performance efficiency		performance	paralleling & pipelining
	time efficiency	latency	latency		time behaviour	latency, response time		
	resource utilization				resource utilization			change data capture
	capacity	capacity, throughput				quality of service		types of fact tables
	modes	modes						
	cost efficiency	-	cost, affordability	-	-	value-added, price	pricing	-
	upstream overhead	-	overhead of source systems	-	upstream overhead			
	security	security	-	security		security	security	security
	confidentiality	confidentiality						compliance management
	integrity	integrity						
	availability	availability	availability	availability			availability	availability
	auditability		auditability		auditability			lineage & dependency
	traceability		traceability	traceability	traceability	documentation	provenance	self-documenting
	reliability	reliability	reliability	reliability	reliability		reliability	reliability
	process availability	availability			availability	availability		
	fault tolerance	fault tolerance			fault tolerance			
	robustness	integrity	robustness	responsiveness	robustness			
	recoverability		recoverability		recoverability			recoverability, problem escalation
Characteristics for design evaluation	**adaptability**	-	scalability	-	adaptability	-	-	scalability
	scalability	scalability	scalability		scalability			
	flexibility	flexibility	flexibility					
	reusability			portability	reusability			
	usability	-		usability	-	concise representation, understandability	-	visibility
	understandability							understanding source data
	manageability	maintainability	maintainability	maintainability	modularity, analyzability			manageability
	maintainability				maintainability			maintainability
	testability			validation	testability	verifiability		
n/a		safety		accessibility, usefulness, believability		customer support, believability, objectivity, amount of data	licencing	

Fig. 2. ETL Process Characteristics

3   Extracting Quality Characteristics

Our model mainly derives from a systematic literature review that we conducted, following the guidelines reported by Kitchenham et al. [18]. The research questions addressed by this study are the following:

RQ1) What ETL process quality characteristics have been addressed?

RQ2) What is the definition for each quality characteristic?

Our search process used an automated keyword search of SpringerLink [2], ACM Digital Library [3], ScienceDirect [4] and IEEE Xplore [5]. The search strings were the following:

- (quality attributes OR quality characteristics OR qox) AND ("etl" OR "extraction transformation loading") AND ("information technology" OR "business intelligence")
- (quality attributes OR quality characteristics OR qox) AND ("data integration" OR "information systems integration" OR "data warehouses") AND ("quality aware" OR "quality driven")

The inclusion criterion for the studies was that they should identify a wide range of quality characteristics for data integration processes and thus only studies that mentioned at least 10 different quality characteristics were included. The quality characteristics could refer to any stage of the process as well as to the quality of the target repositories as a result of the process. One exclusion criterion was that studies should be written in English. Moreover, whenever multiple studies from same researcher(s) and line of work were identified, our approach was to include only the most relevant or the most recent study.

The result of our selection process was a final set of 5 studies. Nevertheless, in an attempt to improve the completeness of our sources, we also considered the ETL subsystems as defined in [2] for an industry perspective on the area, as well as standards from the field of software quality. Regarding software quality, our approach was to study the work by Barbacci et al. [5] and include in our model all the attributes relevant to ETL processes, with the required definition adjustments. This way we reviewed a commonly accepted, generic taxonomy of software quality attributes, while at the same time avoiding the adherence to more recent, strictly defined standards for practical industrial use, which we are nevertheless aware of [4]. The complete list from the resulting 7 sources, covering the most important characteristics from a process perspective that are included in our model can be seen in Fig. 2.

Data quality is a prime quality chracteristic of ETL processes. Its significance is recognized by all the approaches presented in our selected sources, except for Pavlov [24] and Barbacci et al. [5]. since the factors in their analyses derive directly or indirectly from generic software quality attributes. Our model was

---

[2] http://link.springer.com

[3] http://dl.acm.org

[4] http://www.sciencedirect.com/

[5] http://ieeexplore.ieee.org

enriched with a more clear perspective of data quality in Information Systems and a practical view of how quality criteria can lead to design decisions, after reviewing the work by Naumann [23].

Performance, was given attention by all presented approaches , which was expected since time behaviour and resource efficiency are the main aspects that have traditionally been examined as optimization objectives. On the other hand, the works of Pavlov [24] and Simitsis et al. [28] were the only approaches to include the important characteristic of *upstream overhead*. However, [24] does not include *security*, which is discussed in the rest of the works. The same is true for *auditability*, which is absent from the work of Barbacci et al. [5] but found in all other works. Reliability on the other hand, is recognized as a crucial quality factor by all approaches. As expected, the more abstract quality characteristics *adaptability* and *usability* are less commonly found in the sources, in contrast with *manageability* which is found in all approaches except for Dustdar et al. [10], who do not discuss about intangible characteristics.

Although we include *cost efficiency* in Fig. 2, in the remainder of this paper this characteristic is not examined as the rest. The reason is that we view our quality-based analysis in a similar perspective as Kazman et al. [17] , according to which any quality attribute can be improved by spending more resources and it is a matter of weighting the benefits of this improvement to the required cost that can lead to rational decisions. In addition, we regarded *safety* as non-relevant for the case of ETL processes, since these processes are computer-executable, non-critical and hence the occurrence of accidents or mishaps is not a concern. Similarly, we considered that the characteristics of *accessibility*, *usefulness*, *customer support*, *believability*, *amount of data* and *objectivity* found in [15] and [23] are not relevant for our case, as they refer to the quality of source or target repositories, yet do not depend on the ETL process. Likewise, *licencing* [10] refers to concrete tools and platforms while our ETL quality analysis is platform independent.

Through our study we identified that there are two different types of characteristics — characteristics that can actively drive the generation of patterns in the ETL process design and characteristics that cannot explicitly indicate the use of specific design patterns, but can still be measured and affect the evaluation of and the selection among alternative designs. In the remainder of this paper we refer to the first category as *characteristics with construct implications* and to the second as *characteristics for design evaluation*.

4   Process Characteristics with Construct Implications

In this section, we present our model for characteristics with construct implications. The proposed list of characteristics and measures can be extended or narrowed down to match the requirements for specific use cases.

4.1   Characteristics and Measures

In this subsection, we provide a definition for each characteristic as well as candidate metrics under each definition, based on existing approaches that we discovered coming from literature and practice in the areas of Data Warehousing and Software Engineering. For each metric there is a definition and a symbol, either (+) or (−) denoting whether the maximization or minimization of the metric is desirable, respectively.

1. *data quality*: the fitness for use of the data produced as the outcome of the ETL process. It includes:
   (a) *data accuracy*: percentage of data without data errors.
       M1: % of correct values [6] (+)
       M2: % of delivered accurate tuples [6] (+)
   (b) *data completeness*: degree of absence of null values and missing values.
       M1: % of tuples that should be present at their appropriate storage but they are not [27, 6] (−)
       M2: % of non-null values [6] (+)
   (c) *data freshness*: indicator of how recent data is with respect to time elapsed since last update of the target repository from the data source.
       M1: Instant when data are stored in the system - Instant when data are updated in the real world [6] (−)
       M2: Request time - Time of last update [6] (−)
       M3: 1 / (1 - age * Frequency of updates) [6] (−)
   (d) *data consistency*: degree to which each user sees a consistent view of the data and data integrity is maintained throughout transactions and across data sources.
       M1: % of tuples that violate business rules [27, 6] (−)
       M2: % of duplicates [6] (−)
   (e) *data interpretability*: degree to which users can understand data that they get.
       M1: # of tuples with interpretable data (documentation for important values) [6] (+)
       M2: Score from User Survey (Questionnaire) [6] (+)
2. *performance*: the performance of the ETL process as it is implemented on a system, relative to the amount of resources utilized and the timeliness of the service delivered. It includes:
   (a) *time efficiency*: the degree of low response times, low processing times and high throughput rates.
       M1: Process cycle time [21] (−)
       M2: Average latency per tuple in regular execution [27] (−)
       M3: Min/Max/Average number of blocking operations [27] (−)
   (b) *resource utilization*: the amounts and types of resources used by the ETL process.
       M1: CPU load, in percentage of utilization [21] (−)
       M2: Memory load, in percentage of utilization [21] (−)

(c) *capacity*: the demand that can be placed on the system while continuing to meet time and throughput requirements.

M1:   Throughput of regular workflow execution [27] (+)

(d) *modes*: the support for different modes of the ETL process based on demand and changing requirements, for example batch processing, real-time event-based processing, etc.

M1:   Number of supported modes / Number of all possible modes (+)

3. *upstream overhead*: the degree of additional load that the process causes to the data sources on top of their normal operations.

M1:   Min/Max/Average timeline of memory consumed by the ETL process at the source system [27] (−)

4. *security*: the protection of information during data processes and transactions. It includes:

(a) *confidentiality*: the degree to which data and processes are protected from unauthorized disclosure.

M1:   % of mobile computers and devices that perform all cryptographic operations using FIPS 140-2 cryptographic modules [9] (+)

M2:   % of systems (workstations, laptops, servers) with latest antispyware signatures [1] (+)

M3:   % of remote access points used to gain unauthorized access [9] (−)

M4:   % of users with access to shared accounts [9] (−)

(b) *integrity*: the degree to which data and processes are protected from unauthorized modification.

M1:   % of systems (workstations, laptops, servers) with latest antivirus signatures [1] (+)

(c) *reliability*: the degree to which the ETL process can maintain a specified level of performance for a specified period of time. It includes:

i. *availability*: the degree to which information, communication channels, the system and its security mechanisms are available when needed and functioning correctly.

M1:   Mean time between failures (MTBF) [27] (+)

M2:   Uptime of ETL process [27] (+)

ii. *fault tolerance*: the degree to which the process operates as intended despite the presence of faults.

M1:   Score representing asynchronous resumption support [27] (+)

iii. *robustness*: the degree to which the process operates as intended despite unpredictable or malicious input.

M1:   Number of replicated processes [27] (+)

iv. *recoverability*: the degree to which the process can recover the data directly affected in case of interruption or failure.

M1:   Number of recovery points used [27] (+)

M2:   % of successfully resumed workflow executions [27] (+)

M3:   Mean time to repair (MTTR) [27] (−)

(d) *auditability*: the ability of the ETL process to provide data and business rule transparency. It includes:

i. *traceability*: the ability to trace the history of the ETL process execution steps and the quality of documented information about runtime. M1: % of KPIs that can be followed, discovered or ascertained by end users [20] (+)

Referring to our running example from Fig. 1, the difference between the first and the second design is that the latter includes an additional task for loading state's abbreviations (e.g., CA for California). This way less records would remain without state only because abbreviation would not be recognized. The second design additionally includes one task for checking the referential integrity constraint between the two dimension tables of the data warehouse. Consequently, the *Data Quality* for the second design is improved compared to the first one, in terms of *data completeness* and *data consistency*. This could be demonstrated by the measures *% of non-null values* and *% of tuples that violate business rules*, respectively.

4.2 Characteristics Relationships

In the same direction as [28] and [5] we also recognise that ETL process characteristics are not independent of each other and each time a decision has to be made, the alternative options might affect different characteristics differently, but that this is not realized in completely ad hoc ways. On the contrary, we argue that there is an inherent relationship between characteristics and it can be depicted in a qualitative model that can be instantiated per case and facilitate reasoning and automation.

Our model for the dependencies among characteristics with construct implications can be seen in Fig. 3 . In this model we include all the characteristics with construct implications that we have identified and defined in Sec. 3.

Our model consists of first-level characteristics and in some cases second- or even third-level sub-characteristics and can be read in a cause-and-effect fashion, i.e., improving one characteristic leads to improvement or deterioration of another characteristic. We should notice that although traditionally availability

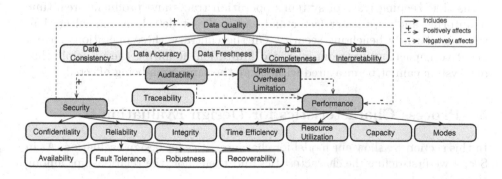

Fig. 3. Dependencies among process characteristics with construct implications

is classified directly under security, for our case availability is in fact a subgoal of reliability. In other words, reliability requires not only the satisfaction of availability but also maintaining specified SLAs for the ETL process and that justifies our decision to place availability under reliability and reliability under security.

Coming back to our running example from Fig. 1, it is clear that the second design would require more time and more computational resources than the first one in order to perform the additional tasks. The measures of *Process execution time* and *CPU load measured in percentage of utilization* would have higher values indicating worse *time efficiency* and *resource utilization*. Thus, improved *Data Quality* would have to be considered at the price of decreased *Performance* and whether or not the decision to select the second design would be optimal, would depend on the importance of each of these characteristics for the end-user.

As can be seen in Fig. 3 the improvement of any other characteristic negatively affects performance. That is reasonable since such improvements would require the addition of extra complexity to the ETL process, diverging from the optimal simplicity that favours performance. Improving Data Quality would require additional checks, more frequent refreshments, additional data processing and so on, thus utilizing more resources and imposing a heavier load on the system. In the same manner, improving security would require more complex authentication, authorization and accounting (AAA) mechanisms, encryption, additional recovery points, etc., similarly having negative impact on performance. Likewise, improving auditability would require additional processes for logging, monitoring as well as more resources to constantly provide real-time access to such information to end-users. In a similar fashion, promoting upstream overhead limitation would demand locks and scheduling to minimize impact of ETL processes on competing resources and therefore time and throughput limitations.

On the other hand, improving security positively affects data quality since data becomes more protected against ignorant users and attackers, making it more difficult for data and system processes to be altered, destroyed or corrupted. Therefore, data integrity becomes easier to maintain. In addition, improved system availability and robustness leads to improved data quality in the sense that processes for data refreshing, data cleaning and so on remain undisrupted.

Regarding the impact that improving auditability has to security, it is obvious that keeping track of system's operation traces and producing real-time monitoring analytics foster faster and easier threat detection and mitigation, thus significantly benefiting security. On the contrary, these operations have a negative impact on upstream overhead limitation, following the principle that one system cannot be measured without at the same time being affected.

5   Process Characteristics for Design Evaluation

In this section we show our model for characteristics for design evaluation. As in Sec. 4 we first define the characteristics and then show the relationships among them.

5.1   Characteristics and Measures

Following the same approach as with characteristics with construct implications, in this subsection, we provide a definition for each characteristic for design evaluation, as well as proposed metrics deriving from literature.

1. *adaptability*: the degree to which ETL process can effectively and efficiently be adapted for different operational or usage environments. It includes:
   (a) *scalability*: the ability of the ETL process to handle a growing demand, regarding both the size and complexity of input data and the number of concurrent process users.
      M1: Ratio of system's productivity figures at two different scale factors, where productivity figure = throughput * QoS/ cost [16] (+)
      M2: # of Work Products of the process model [13] (−)
   (b) *flexibility*: the ability of the ETL flow to provide alternative options and dynamically adjust to environmental changes (e.g., by automatically switching endpoints).
      M1: # of precedence dependences between activities [13] (−)
   (c) *reusability*: the degree to which components of the ETL process can be used for operations of other processes.
      M1: # of dependences between activities with locality (e.g., in the same package) [12] (+)
      M2: # of dependences between activities without locality (e.g., from different packages) [12] (−)
      The following measures are valid in the case where there are statistical data about the various modules (e.g., transformation or mapping operations) of the ETL process:
      M3: % of reused low level operations in the ETL process [12] (+)
      M4: Average of how many times low level operations in the ETL process have been reused per specified time frame [12] (+)
2. *usability*: the ease of use and configuration of the implemented ETL process on the system. It includes:
   (a) *understandability*: the clearness and self-descriptiveness of the ETL process model for (non-technical) end users.
      M1: # of activities of the software process model [13] (−)
      M2: # of precedence dependences between activities [13] (−)
3. *manageability*: the easiness of monitoring, analyzing, testing and tuning the implemented ETL process.
   (a) *maintainability*: the degree of effectiveness and efficiency with which the ETL process can be modified to implement any future changes.
      M1: Length of process workflow's longest path [27] (−)
      M2: # of relationships among workflow's components [27] (−)
      M3: Cohesion of process workflow (viewed as a directed graph) [8] (+)
      M4: Coupling of process workflow (viewed as a directed graph) [8] (−)
      M5: # of input and output flows in the process model [22] (−)
      M6: # of output elements in the process model [22] (−)

M7:   # of merge elements in the process model [22] (−)
M8:   # of input and output elements in the process model [22] (−)
(b) *testability*: the degree to which the process can be tested for feasibility, functional correctness and performance prediction.
M1:   Cyclomatic Complexity of the ETL process workflow [14] (−)

Regarding our running example from Fig. 1, we can clearly see how the second design is less usable since it is less understandable. This is not only an intuitive impression from looking at a more complex process model but can also be measured using the measures of *# of activities of the process model*, which is greater for the second design.

5.2   Characteristics Relationships

In Fig. 4 we show the dependencies among characteristics for design evaluation. Increased usability favours manageability because a more concise, self-descriptive system is easier to operate and maintain. Similarly, adaptability positively affects usability, since an easily configured system is easier to use and does not require specialized skill-set from the end user. On the other hand, adaptability can be achieved with more complex systems and therefore it negatively affects manageability. This negative relationship might appear counter-intuitive, but it should be noted that our view of adaptability does not refer to autonomic behaviour, which would possibly provide self-management capabilities. Instead, we regard manageability from an operator's perspective where control is desirable and the addition of unpredictable, "hidden" mechanisms would make the process more difficult to test and maintain. Regarding the apparent conflict between the negative direct relationship among Adaptability and Manageability and the transitive positive affection of Adaptability–Usability–Manageability, this can be explained by the different effect of each influence, which can be considered as differing weights on the edges of the Digraph.

Going back to our running example from Fig. 1, it is apparent that the first design is easier to manage, since it is easier to maintain. This can be verified using any of the metrics for maintainability defined in this section. Thus, we can see how *usability* positively impacts *manageability*.

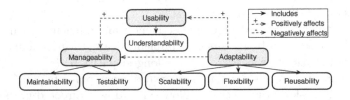

Fig. 4. Dependencies among characteristics for design evaluation

6   Summary and Outlook

The automation of ETL processes seems as a promising direction in order to effectively face emerging challenges in Business Intelligence . Although information systems are developed by professionals with technical expertise, it is important to orientate the design of underlying processes in an end-user perspective that reflects business requirements. In this paper, we have proposed a model for ETL process quality characteristics that constructively absorbs concepts from the fields of Data Warehousing, ETL, Data Integration and Software Engineering. One important aspect about our model is that for each and every characteristic, there has been suggested measurable indicators that derive solely from existing literature. Our model includes in a high level the relationships between different characteristics and can indicate how improvement of one characteristic by the application of design modifications can affect others.

Our vision is that our defined models can be used as a palette for techniques that will automate the task of selecting among alternative designs. Future work will target the development of a framework that will use this model as a stepping stone to provide automatic pattern generation and evaluation for ETL processes, keeping quality criteria at the center of our analysis.

Acknowledgements. This research has been funded by the European Commission through the Erasmus Mundus Joint Doctorate "Information Technologies for Business Intelligence - Doctoral College" (IT4BI-DC). This work has also been partly supported by the Spanish Ministerio de Ciencia e Innovación under project TIN2011-24747.

References

[1] KPI library, http://kpilibrary.com (cited January 2014)
[2] The subsystems of ETL revisited,
    http://www.informationweek.com/software/information-management/
    kimball-university-the-subsystems-of-etl-revisited/d/d-id/1060550
    (cited January 2014)
[3] El Akkaoui, Z., Mazón, J.-N., Vaisman, A., Zimányi, E.: BPMN-based conceptual modeling of ETL processes. In: Cuzzocrea, A., Dayal, U. (eds.) DaWaK 2012. LNCS, vol. 7448, pp. 1–14. Springer, Heidelberg (2012)
[4] Al-Qutaish, R.: An investigation of the weaknesses of the ISO 9126 Intl. Standard. In: ICCEE, pp. 275–279 (2009)
[5] Barbacci, M., Klein, M., Longstaff, T., Weinstock, C.: Quality Attributes. Tech. rep., Carnegie Mellon University, Pittsburgh, Pennsylvania (1995)
[6] Batini, C., Cappiello, C., Francalanci, C., Maurino, A.: Methodologies for data quality assessment and improvement. ACM Comput. Surv. 41(3), 1–52 (2009)
[7] Böhm, M., Wloka, U., Habich, D., Lehner, W.: GCIP: Exploiting the generation and optimization of integration processes. In: EDBT, pp. 1128–1131. ACM (2009)
[8] Briand, L., Morasca, S., Basili, V.: Property-based software engineering measurement. IEEE Trans. on Soft. Eng. 22(1), 68–86 (1996)

[9] Chew, E., Swanson, M., Stine, K.M., Bartol, N., Brown, A., Robinson, W.: Performance Measurement Guide for Information Security. Tech. rep. (2008)

[10] Dustdar, S., Pichler, R., Savenkov, V., Truong, H.L.: Quality-aware service-oriented data integration: Requirements, state of the art and open challenges. SIGMOD 41(1), 11–19 (2012)

[11] El Akkaoui, Z., Zimanyi, E.: Defining ETL worfklows using BPMN and BPEL. In: DOLAP, pp. 41–48. ACM (2009)

[12] Frakes, W., Terry, C.: Software reuse: Metrics and models. ACM Comput. Surv. 28(2), 415–435 (1996)

[13] García, F., Piattini, M., Ruiz, F., Canfora, G., Visaggio, C.A.: FMESP: Framework for the modeling and evaluation of software processes. In: QUTE-SWAP, pp. 5–13. ACM (2004)

[14] Gill, G., Kemerer, C.: Cyclomatic complexity density and software maintenance productivity. IEEE Trans. on Soft. Eng. 17(12), 1284–1288 (1991)

[15] Jarke, M., Lenzerini, M., Vassiliou, Y., Vassiliadis, P.: Fundamentals of Data Warehouses. Springer (2003)

[16] Jogalekar, P., Woodside, M.: Evaluating the scalability of distributed systems. IEEE Trans. on Parallel and Distributed Systems 11(6), 589–603 (2000)

[17] Kazman, R., Asundi, J., Klein, M.: Quantifying the costs and benefits of architectural decisions. In: ICSE, pp. 297–306 (2001)

[18] Kitchenham, B., Pearl Brereton, O., Budgen, D., Turner, M., Bailey, J., Linkman, S.: Systematic literature reviews in software engineering - a systematic literature review. Inf. Softw. Technol. 51(1), 7–15 (2009)

[19] van Lamsweerde, A.: Goal-oriented requirements engineering: a guided tour. In: Requirements Engineering, pp. 249–262 (2001)

[20] Leite, J., Cappelli, C.: Software transparency. Business and Information Systems Engineering 2(3), 127–139 (2010)

[21] Majchrzak, T.A., Jansen, T., Kuchen, H.: Efficiency evaluation of open source ETL tools. In: SAC, pp. 287–294. ACM, New York (2011)

[22] Muñoz, L., Mazón, J.N., Trujillo, J.: Measures for ETL processes models in data warehouses. In: MoSE+DQS, pp. 33–36. ACM (2009)

[23] Naumann, F.: Quality-Driven Query Answering for Integrated Information Systems. LNCS, vol. 2261. Springer, Heidelberg (2002)

[24] Pavlov, I.: A QoX model for ETL subsystems: Theoretical and industry perspectives. In: CompSysTech, pp. 15–21. ACM (2013)

[25] Sánchez-González, L., García, F., Ruiz, F., Mendling, J.: Quality indicators for business process models from a gateway complexity perspective. Inf. Softw. Technol. 54(11), 1159–1174 (2012)

[26] Simitsis, A., Vassiliadis, P., Sellis, T.: Optimizing ETL processes in data warehouses. In: ICDE, pp. 564–575 (2005)

[27] Simitsis, A., Vassiliadis, P., Dayal, U., Karagiannis, A., Tziovara, V.: Benchmarking ETL Workflows. In: Nambiar, R., Poess, M. (eds.) TPCTC 2009. LNCS, vol. 5895, pp. 199–220. Springer, Heidelberg (2009)

[28] Simitsis, A., Wilkinson, K., Castellanos, M., Dayal, U.: QoX-driven ETL design: Reducing the cost of ETL consulting engagements. In: SIGMOD, pp. 953–960. ACM, New York (2009b)

[29] Wilkinson, K., Simitsis, A., Castellanos, M., Dayal, U.: Leveraging business process models for ETL design. In: Parsons, J., Saeki, M., Shoval, P., Woo, C., Wand, Y. (eds.) ER 2010. LNCS, vol. 6412, pp. 15–30. Springer, Heidelberg (2010)

A Logical Model for Multiversion Data Warehouses

Waqas Ahmed[1,*], Esteban Zimányi[1], and Robert Wrembel[2,**]

[1] Dept. of Computer & Decision Engineering (CoDE),
Université Libre de Bruxelles, Belgium
{waqas.ahmed,ezimanyi}@ulb.ac.be
[2] Institute of Computing Science,
Poznań University of Technology, Poland
robert.wrembel@cs.put.poznan.pl

Abstract. Data warehouse systems integrate data from heterogeneous sources. These sources are autonomous in nature and change independently of a data warehouse. Owing to changes in data sources, the content and the schema of a data warehouse may need to be changed for accurate decision making. Slowly changing dimensions and temporal data warehouses are the available solutions to manage changes in the content of the data warehouse. Multiversion data warehouses are capable of managing changes in the content and the structure simultaneously however, they are relatively complex and not easy to implement. In this paper, we present a logical model of a multiversion data warehouse which is capable of handling schema changes independently of changes in the content. We also introduce a new hybrid table version approach to implement the multiversion data warehouse.

1 Introduction

A data warehouse (DW) is a repository of historical, subject-oriented, and heterogeneous data that is integrated from external data sources (EDSs). An inherent feature of EDSs is that they are not static and their schema may change as a result of adaptation of new technologies, changes in the modeled reality, or changes in the business requirements. As a result of schema changes in EDSs, DWs may become obsolete and thus need to be redesigned. Often, after applying schema changes, users demand to preserve the existing content and the schema in a DW. The Multiversion data warehouse (MVDW) is an available solution to manage the schema changes in a DW. This solution is based on the schema versioning approach where every change produces a new DW version and the old content and the schema are also kept available.

The content of a DW changes as a result of periodic loading of new data into it. One particular scenario that requires consideration is changes in the states of

* This research is funded by the Erasmus Mundus Joint Doctorate IT4BI-DC.
** The Polish National Science Center (NCN), grant No. 2011/01/B/ST6/05169.

L. Bellatreche and M.K. Mohania (Eds.): DaWaK 2014, LNCS 8646, pp. 23–34, 2014.

existing dimension members. For analysis purposes, it may be important for the user to keep the history of changes in the states of members. Slowly changing dimensions [8] and temporal data warehouses [7] are the solutions to manage changes in the content of a DW. These solutions maintain the content history by associating timestamps with the values. Though these solutions partially solve the problem of managing DW content changes, they are unable to separate different DW states that describe different real world scenarios.

Most of the available MVDW proposals try to solve the issues of content and schema changes simultaneously which makes it complicated to understand and implement these proposals. Some solutions for MVDW present metamodels to maintain metadata supporting the life cycle of a DW. These metamodels store DW versions in separate physical structures. Creation and maintenance of these separate data structures makes a MVDW complex and may also negatively impact the query performance. If the user only needs to manage the schema changes in a DW then a simplified model can serve this purpose.

We envision a MVDW in which content and structure change functionality can be implemented as independent plug-ins. In this paper we propose a model of the MVDW that supports structural changes in dimensions and facts. The proposed model is relatively simple and helps to identify the desired features of schema versioning. We also introduce a new hybrid table version (HTV) approach to implement the MVDW. This paper is organized as follows: Section 2 discusses the approaches related to handling the evolution of DWs. Section 3 presents a running example and defines the requirements for the MVDW. Section 4 presents a model of the MVDW while Sect. 5 proposes an approach to implement it. Finally, Sect. 6 concludes the paper by providing a summary and considerations for future research.

2   Related Work

In [7], the proposals for managing DW evolutions are classified into three broad categories: schema modification, schema evolution, and schema versioning. The systems that allow *schema modification* support changes in a data source schema but as a result of these changes, the existing data may become unavailable. *Schema evolution* is supported when a system has capability of accommodating schema changes while preserving existing data. *Schema versioning* is a mechanism through which systems store data using multiple schema versions.

Most of the research related to schema versioning deals with the issues of managing schema and content changes at the same time and therefore, presents solutions to manage schema versioning for temporal databases. Brahmia et al. [3] presented one such schema version management mechanism for multi-temporal databases. The proposed approach however, does not provide a generic model to manage schema versioning and is only specific to relational databases. The model presented in [5] also supports changes in both the content and structure of data. This model supports structural changes to the dimension members only and does not explain how to map the members from one version to another.

In [1], the authors presented a formal model of a MVDW. This model supports the real schema versions, which represent the changes in the real world, and also provides the capability of deriving alternative schema versions. The alternative schema versions can be used for what-if analysis . In [12], the authors presented a logical model for the implementation of a MVDW and discussed various constraints to maintain data integrity across DW versions. They used the work presented in [13] to query data from multiple versions of a DW. [10] is a prototype implementation which manages schema versions by creating graph based metamodels. An augmented schema [6] is also created to perform data transformations among different schema versions.

The aforementioned approaches to MVDWs track the history of content and schema changes and can query multiple DW versions but they maintain DW versions in separate structures. The creation and maintenance of these structures is relatively complex. Further, the approaches to handling DW evolution either manage changes in the content only [9], changes in the schema only [2], or changes in the content and schema simultaneously [12]. Data warehouse versioning approaches support both changes in the content and the schema at the same time but none of the existing versioning approaches deals with issues of schema and content evolution independently of each other.

Three approaches may be used to convert a database structure after applying schema changes: single table version (STV), multiple table version (MTV), and partial multiple table version (PMTV) [11]. A disadvantage of the STV approach is the null space overhead: null values are introduced as a result of both the addition and the deletion of attributes. There is no null space overhead when using the MTV approach as the deleted attributes are dropped from the new version. However, multiple table versions increase the maintenance overhead. Also, running a query that spans multiple table versions requires data adaptations and table joins which negatively affect the query performance. The PMTV approach also requires joins to construct the complete schema of the table. Joins are costly operations and may have a negative impact on the system performance.

3 Multiversion Data Warehouses

A multiversion data warehouse is a collection of its versions. Each *DW version* is composed of a schema version and an instance version. A *schema version* describes the structure of the data within a time period. The *instance version* represents the data that is stored using a particular schema version. We assume that at a given time instant, only one DW version is current and used to store data. A new version can be derived by applying changes to the current version only. To represent the period for which a version was used to store data, each DW version is assigned a closed-open time interval [9], represented by begin application time (BAT) and end application time (EAT). The EAT for the current version is set to UC (until changed). It is possible to create alternative schema versions [12] using the model presented in Sect. 4 but for simplicity's sake, we do not consider the branching versioning model. Figure 1 shows an example of multiple DW versions.

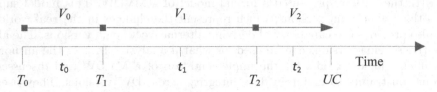

Fig. 1. Multiple versions of a data warehouse

Suppose that the initial version V_0 was created at time T_0. The BAT and EAT for V_0 are set to T_0 and UC, respectively. Figure 2a shows the DW schema in version V_0 which contains three dimensions: Time, Product, and Geography. The Geography dimension has a hierarchy, which consists of three levels: Store, City, and Region. Fact Sales relates these dimensions and contains measures Quantity and Amount. The granularity of dimension Geography is at level Store. Consider that the following schema changes were applied to the DW. (1) At time T_1, attribute Manager was added to and attribute Area was deleted from level Store. As a result, version $V_1[T_1, UC)$ was created and the EAT of V_0 was changed from UC to T_1. Figure 2b shows the schema of level Store in V_1. (2) At time T_2, level Store was deleted and new level State was added between levels City and Region. This schema modification changed the granularity of dimension Geography and thenceforth all fact members were assigned to a city. Moreover, level City rolled-up to level State. Since V_1 was the current version at that time, version $V_2[T_2, UC)$ was derived from V_1 and the EAT of V_1 was changed from UC to $T1$. Figure 2c shows the schema of V_2.

Changes in dimensions and/or facts create a new DW version. The possible changes to a dimension include (1) adding a new attribute to a level, (2) deleting an attribute from a level, (3) changing the domain of an attribute, (4) adding a level to a hierarchy, and (5) deleting a level from a hierarchy. The schema of a fact changes when (1) a new dimension is added, (2) a dimension is deleted, (3) a measure is added, (4) a measure is deleted, and (5) the domain of a measure is changed. In certain cases, a schema change in the dimension also requires changes in the fact. For example, adding a new level into a hierarchy at the lowest granularity will also require a new attribute to be added to the fact which will link the new level to the fact. Similarly, in case of removing the lowest level from a hierarchy, an attribute is removed from the fact to unlink the deleted level and a new attribute is added into it to link the next level of the hierarchy. Thus, effects of adding and deleting the lowest level of a hierarchy on the DW schema are similar to adding and deleting a new dimension to the DW.

It is worth mentioning that since our versioning model deals with the structural changes and not with the temporal evolution of facts and dimension members, at any instant t, all facts and dimension members in the MVDW are valid. To summarize the requirements of a multiversion data warehouse, we can say that (1) *in a MVDW, new data is loaded using current version only*; (2) *a MVDW must retain all the data loaded into it throughout its lifespan*; and (3) *all the data stored in the MVDW must be viewable using any MVDW schema version*.

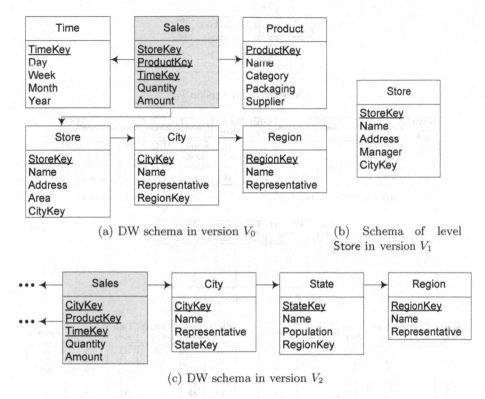

(a) DW schema in version V_0

(b) Schema of level Store in version V_1

(c) DW schema in version V_2

Fig. 2. Multiple schema versions of the running example

The first requirement is straightforward. We elaborate the second and third requirements with the help of examples. The initial state of level Store in version V_0 at an instant t_0 in $[T_0, T_1)$ is shown in Fig. 3a. As a result of the first schema change, version V_1 is derived from V_0. Figure 3b shows the schema of level Store in V_1. The shaded column represents the deleted attribute. The second requirement is that when deriving version V_1 from V_0, for the existing store members, the DW must retain the values of the deleted attribute Area. Suppose that the user decides to add a new store member s4 at an instant t_1 in $[T_1, T_2)$. The third requirement states that this newly added member must also be available in version V_0. So, when accessing the store members using version V_0, the user should be able to access all the members stored in V_0 and V_1. As attribute Manager does not exist in V_0, it is not available in V_0 at t_1. This situation is shown in Fig. 3c where shaded and crossed out cells represent the unavailable attribute. Similarly, attribute Area does not exist in V_1 therefore, the value of attribute Area for s4 is not available in V_0.

As a result of the second schema change, version V_2 is derived from V_1. In V_2 both the dimensions and the fact are affected. Figure 3d shows the state of fact Sales in V_0. After second schema change, the fact members are assigned to level City and that is why attribute StoreKey is deleted and CityKey is added to the new version of fact Sales. Since in previous versions level Store rolled-up to level

Store Key	Name	Address	Area	City Key
s1	Store1	ABC	20	c1
s2	Store2	DEF	30	c1
s3	Store3	HIJ	50	c2

(a) Store at t_0 in version V_0

Store Key	Name	Address	~~Area~~	Manager	City Key
s1	Store1	ABC	~~20~~		c1
s2	Store2	DEF	~~30~~		c1
s3	Store3	HIJ	~~50~~		c2
s4	Store4	KLM		John	c2

(b) Store at t_1 in version V_1

Store Key	Name	Address	Area	~~Manager~~	City Key
s1	Store1	ABC	20		c1
s2	Store2	DEF	30		c1
s3	Store3	HIJ	50		c2
s4	Store4	KLM		~~John~~	c2

(c) Store at t_1 in version V_0

Store Key	Product Key	Time Key	Quantity	Amount
s1	p1	t1	5	20
s2	p1	t1	3	15
s3	p2	t2	2	18
s4	p2	t3	3	9

(d) Sales at t_0 in version V_0

~~Store Key~~	City Key	Product Key	Time Key	Quantity	Amount
~~s1~~		~~p1~~	~~t1~~	~~5~~	~~20~~
~~s2~~		~~p1~~	~~t1~~	~~3~~	~~15~~
	c1	p1	t1	8	35
~~s3~~	c2	p2	t2	2	18
~~s4~~	c2	p2	t3	3	9
	c3	p1	t4	2	25

(e) Sales at t_2 in version V_2

Store Key	~~City Key~~	Product Key	Time Key	Quantity	Amount
s1		p1	t1	5	20
s2		p1	t1	3	15
	~~c1~~	~~p1~~	~~t1~~	~~8~~	~~35~~
s3	~~c2~~	p2	t2	2	18
s4	~~c2~~	p2	t3	3	9
	~~c3~~	p1	t4	2	25

(f) Sales at t_2 in version V_0

City Key	Name	Representative	RegionKey
c1	Brussels	John	r1
c2	Liège	Doe	r2
c3	Charleroi	Ahmed	r2

(g) City at t_0 in version V_0

City Key	Name	Representative	State Key	~~Region Key~~
c1	Brussels	John	e1	~~r1~~
c2	Liège	Doe	e2	~~r2~~
c3	Charleroi	Ahmed	e3	~~r2~~

(h) City at t_2 in version V_2

City Key	Name	Representative	~~State Key~~	Region Key
c1	Brussels	John	~~e1~~	r1
c2	Liège	Doe	~~e2~~	r2
c3	Charleroi	Ahmed	~~e3~~	r2

(i) City at t_2 in version V_0

Fig. 3. Contents of the DW in multiple versions of the running example

City, it is possible to obtain the values of CityKey for the existing fact members. While doing so, the fact members belonging to the stores that are located in the same city should be combined provided that the other key attributes are the same. One such example is shown in Fig. 3e where the first two fact members belong to stores s1 and s2. These stores are located in the same city c1 and the values of the other dimensions are the same, that is why the fact members associated to these stores are combined and represented as a single member in the new version of fact Sales. If the user tries to access fact Sales in V_0 at an instant t_2 in $[T_1, T_2)$, the value of StoreKey for the last fact member will not be available as this attribute is not present in the new version of fact Sales. This situation is depicted in Fig. 3f.

In V_2, a new level State is also added between levels City and Region. Figure 3g shows level City in V_0. Figure 3h shows that the effect of adding a level into the DW schema is similar to adding attribute StateKey to level City and deleting attribute RegionKey from it. If the members of level City are accessed in V_0 at time t_2 then the information about the regions will not be available. Figure 3i depicts this scenario.

Suppose at some point the user decides to add a new dimension Supplier to the DW. This addition will require a new version of fact Sales because the new facts must be linked to Supplier as well. It is worth mentioning that the supplier information will not be available for the existing facts and they will roll-up to unknown supplier.

4   A Multiversion Data Warehouse Model

In this section, we first introduce the formal definition of a data warehouse model and then we extend the definition for a multiversion data warehouse.

Definition 1 (Multidimensional schema). Multidimensional schema S has a name and is composed of (1) the set of dimensions $\mathcal{D} = \{D_1, \ldots, D_n\}$ and (2) the set of facts $\mathcal{F} = \{F_1, \ldots, F_n\}$.

Dimension $D_i \in \mathcal{D}$, $i = 1, \ldots, n$, has a name and is composed of (1) the set of levels $\mathcal{L} = \{L_1, \ldots, L_n, All\}$, (2) aggregation relation $\mathcal{R}$, and (3) the set of hierarchies $\mathcal{H} = \{H_1, \ldots, H_n\}$. Level $L_1 \in \mathcal{L}$ is called the *base level* of the dimension and every dimension has a unique level All. The dimension names are unique in $\mathcal{D}$.

Level $L_j \in \mathcal{L}$, $j = 1, \ldots, n$, is defined by its schema $L_j(A_1 : T_1, \ldots, A_n : T_n)$, where L_j is the level name and it is unique in $\mathcal{L}$. Each attribute A_k, $k = 1, \ldots, n$, is defined over domain T_k and attribute name A_k is unique in L_j. Level $All \in D_i$ does not have any attribute.

Aggregation relation $\mathcal{R}$ is a partial order binary relation on $\mathcal{L} \in D_i, i = 1, \ldots, n$, and contains ordered pairs of form $\langle L_p, L_q \rangle$, where L_p and L_q are levels belonging to $\mathcal{L}$. $\mathcal{R}^*$ denotes the transitive closure of $\mathcal{R}$ such that if $\langle L_1, L_2 \rangle$ and $\langle L_2, L_3 \rangle$ also belong to $\mathcal{R}$, then $\langle L_1, L_3 \rangle$ belongs to $\mathcal{R}^*$. Any level $L_j \in \mathcal{L}$ is, directly or transitively, reachable in $\mathcal{R}^*$ from the base level and any level $L_j \in \mathcal{L}$ reaches in $\mathcal{R}^*$, directly or transitively, top level All.

Hierarchy $H_m \in \mathcal{H} \subseteq \mathcal{R}^*$, $m = 1, \ldots, n$, and has a unique name. Each hierarchy $H_m \in \mathcal{H}$ begins from the base level and has top level All.

Fact $F_i \subset \mathcal{F}$, $i = 1, \ldots, n$, is defined by its schema $F_i(R_1 : L_1, \ldots, R_m : L_m, M_1 : T_1, \ldots, M_n : T_n)$, where R_s, $s = 1, \ldots, m$, is a role name, L_s is a base level of a dimension $D_i \in \mathcal{D}$ and each measure M_t, $t = 1, \ldots, n$, is defined over domain T_t. Role name R_s and measure name M_t are unique in F_i.   □

Definition 2 (Multidimensional instance). Multidimensional instance I is composed of dimension instance and fact instance.

An instance of dimension $D_i \in \mathcal{D}$ is as follows: For each level $L_j \in D_i$, the set of members $M_{L_j} = \{m_1, \ldots, m_n\}$ where member $m_k \in M_{L_j}$, $k = 1, \ldots, n$, is uniquely identifiable. Level All has a special member all. For each $L_j \in D_i$ with schema $L_j(A_1 : T_1, \ldots, A_n : T_n)$, a subset of $M_{L_j} \times T_1 \times \ldots \times T_n$. For each pair of level names $\langle L_p, L_q \rangle$ in $\mathcal{R} \in D_i$, a partial function $Roll\_up_{L_p}^{L_q}$ from M_{L_p} to M_{L_q}.

An instance of fact F_i, which is defined by its schema $F_i(R_1 : L_1, \ldots, R_m : L_m, M_1 : T_1, \ldots, M_n : T_n)$, is a subset of $M_{L_1} \times \ldots \times M_{L_m} \times T_1 \times \ldots \times T_n$. □

Definition 3 (Multiversion multidimensional schema). *Multiversion multidimensional schema $\mathcal{S}_{mv}$ has a name and is composed of (1) the set of multiversion dimensions $\mathcal{D}^v = \{\mathcal{D}_1^v, \ldots, \mathcal{D}_n^v\}$, (2) the set of multiversion facts $\mathcal{F}^v = \{\mathcal{F}_1^v, \ldots, \mathcal{F}_n^v\}$, and (3) the set of schema versions $\mathcal{S}^v = \{\mathcal{S}_1, \ldots, \mathcal{S}_n\}$.*

Multiversion dimension $\mathcal{D}_i^v \in \mathcal{D}^v, i = 0, \ldots, n$, defines the set $\{D_i^{v_1}, \ldots, D_i^{v_n}\}$ of versions of dimension D_i. Dimension version $D_i^{v_j} \in \mathcal{D}_i^v, j = 0, \ldots, n$, is a dimension as defined in Def. 1. The dimension names for all $D_i^{v} \in \mathcal{D}_i^v$ are the same.

Multiversion fact $\mathcal{F}_m^v \in \mathcal{F}^v, m = 1, \ldots, n$, defines the set $\{F_m^{v_1}, \ldots, F_m^{v_n}\}$ of versions of fact F_m. Fact version $F_m^{v_j}, j = 0, \ldots, n$, is a fact as defined in Def. 1. The fact names for all $F_m^{v_j} \in \mathcal{F}_m^v$ are the same.

Schema version $\mathcal{S}_l \in \mathcal{S}^v, l = 0, \ldots, n$, has an associated time interval $T_l = [B_l, E_l)$ and is a multidimensional schema as defined in Def. 1, that is, it is composed of a set of dimensions $\mathcal{D} = \{D_1, \ldots, D_n\}$ and a set of facts $\mathcal{F} = \{F_1, \ldots, F_n\}$, where each $D_i \in \mathcal{D}$ is a dimension version $D_i^{v_j} \in \mathcal{D}_i^v$ and each $F_m \in \mathcal{F}$ is a fact version $F_m^{v_j} \in \mathcal{F}_m^v$. Only one version of dimension $\mathcal{D}_i^v$ and fact $\mathcal{F}_m^v$ can exist in $\mathcal{D}$ and in $\mathcal{F}$, respectively. The time intervals associated to all schema versions in $\mathcal{S}^v$ are disjoint, contiguous, and their union cover the time interval since the creation of the first version until now. □

Example 1. The initial schema of the multiversion data warehouse in Fig. 2a can be represented as follows:

$\mathcal{D}^v = \{\mathcal{D}_G^v, \mathcal{D}_T^v, \mathcal{D}_P^v\}$, $\mathcal{F}^v = \{\mathcal{F}_S^v\}$, $\mathcal{S}^v = \{\mathcal{S}_0\}$, where G, T, P, and S denote Geography, Time, Product, and Sales, respectively.

$\mathcal{D}_G^v = \{\mathcal{D}_G^{v0}\}$, $\mathcal{D}_T^v = \{\mathcal{D}_T^{v0}\}$, $\mathcal{D}_P^v = \{\mathcal{D}_P^{v0}\}$, $\mathcal{F}_S^v = \{\mathcal{F}_S^{v0}\}$, and $\mathcal{S}_0 = \{\{\mathcal{D}_G^{v0}, \mathcal{D}_T^{v0}, \mathcal{D}_P^{v0}\}, \{\mathcal{F}_S^{v0}\}\}$. For brevity, we omit the schema definitions of the levels and the fact.

As a result of the first schema change, a new version of the Geography dimension is derived from the previous version and the other dimensions and the fact remain unchanged. Thus, the MVDW schema is modified as follows: $\mathcal{D}_G^v = \{\mathcal{D}_G^{v0}, \mathcal{D}_G^{v1}\}$, $\mathcal{S}^v = \{\mathcal{S}_0, \mathcal{S}_1\}$, $\mathcal{S}_1 = \{\{\mathcal{D}_G^{v1}, \mathcal{D}_T^{v0}, \mathcal{D}_P^{v0}\}, \{\mathcal{F}_S^{v0}\}\}$.

Finally, new versions of the dimension Geography and fact Sales are derived as a result of the second schema change. The resulting schema of the MVDW is modified as follows: $\mathcal{D}_G^v = \{\mathcal{D}_G^{v0}, \mathcal{D}_G^{v1}, \mathcal{D}_G^{v2}\}$, $\mathcal{F}_S^v = \{\mathcal{F}_S^{v0}, \mathcal{F}_S^{v1}\}$, $\mathcal{S}^v = \{\mathcal{S}_0, \mathcal{S}_1, \mathcal{S}_2\}$, $\mathcal{S}_2 = \{\{\mathcal{D}_G^{v2}, \mathcal{D}_T^{v0}, \mathcal{D}_P^{v0}\}, \{\mathcal{F}_S^{v1}\}\}$. □

Definition 4 (Multiversion multidimensional global schema). The multiversion multidimensional global schema is a multidimensional schema as defined in Def. 1 and it is constructed as follows:

In the global schema of dimension $D_i \in \mathcal{D}s$, $\mathcal{L} \in D_i$ is the union of all the levels existing in all versions of D_i; the schema of level $L_i \in \mathcal{L}$ is the union of all the attributes of L_i from all versions of D_i in which L_i is present; aggregation relation $\mathcal{R}$ is the union of the aggregation relations from all version of D_i, and $\mathcal{H} \in D_i$ is empty. Since the global schema is for system use only, there is no need to maintain hierarchies in it.

The global schema of fact $F_i \in \mathcal{F}$ consists of the set of base levels $\mathcal{B}$ and the set of measures $\mathcal{C}$, where $\mathcal{B}$ is the union of all the base levels in all versions of F_i and C is the union of all the measures in all versions of F_i.   □

The global schema, defined in Def. 4, is a traditional multidimensional schema and its instance is obtained by using Def. 2. A multiversion multidimensional instance is actually an instance of the global schema. Figure 3b, including the shaded column Area, shows the global instance of level Store. The MVDW also contains a transformation function $\mathcal{T}(\mathcal{S}_i)$ which transforms the global instance into the instance of schema $\mathcal{S}_i \in \mathcal{S}^v$, as defined in Def. 3. This transformation function can be implemented as view definitions. Figure 6b shows how the global instances of Store can be transformed into the instance of Store in version V_1.

5   Implementation of the Multiversion DW

We discuss next how the single table version (STV) and the multiple table version (MTV) approaches can be used to implement the MVDW. Then, we present a new hybrid table versioning approach to implement the MVDW.

In the STV approach, the newly added attributes are appended to the existing ones and the deleted attributes are not dropped from the table. A default or null value is stored for the deleted or the unavailable attributes. Figure 4 shows the effect of the schema changes on the relational implementation of a DW that uses the STV approach. Figure 4a shows the state of the DW after the first schema change where attribute Manager is added and Area is deleted from table Store. Records s1, s2, and s3 have null values for attribute Manager because its value is unknown for these records. As attribute Area has been deleted, all newly added records such as s4, will have null values for it. These null values may incur a space overhead in case of huge amount of data. Some DBMSs partially resolve the issue of null space overhead by offering specific features but the implementation of these features has its own limitations[1].

In the MTV approach, each change in the schema of a table produces a new version of the table. Figure 5a shows the new version of table Store which is created as a result of the first schema change. This version includes the newly added attribute Manager and excludes the deleted attribute Area. The new version of table Store results in a new version of table Sales because Sales uses attribute

---

[1] http://technet.microsoft.com/en-us/library/cc280604.aspx

Store Key	Name	Address	Area	City	Manager
s1	Store1	ABC	20	c1	null
s2	Store2	DEF	30	c1	null
s3	Store3	HIJ	50	c2	null
s4	Store4	KLM	null	c2	John

(a) Store in V_2

Store Key	City Key	Product Key	Time Key	Quantity	Amount
s1	c1	p1	t1	5	20
s2	c2	p1	t1	3	15
s3	c2	p2	t2	2	18
s4	c2	p2	t3	3	9
null	c3	p1	t4	2	25

(b) Sales in V_2

City Key	Name	Representative	State Key	Region Key
c1	Brussels	John	s1	r1
c2	Liège	Doe	s2	r2
c3	Charleroi	Ahmed	s3	r2

(c) City in V_2

Fig. 4. State of the data warehouse using the STV approach

Store Key	Name	Address	City	Manager
s4	Store4	KLM	c2	John

(a) Store in V_2

City Key	Product Key	Time Key	Quantity	Amount
c3	p1	t4	2	25

(b) Sales in V_2

City Key	Name	Representative	State Key
c1	Brussels	John	s1
c2	Liège	Doe	s2
c3	Charleroi	Ahmed	s3

(c) City in V_2

Fig. 5. State of the data warehouse using the MTV approach

StoreKey of Store as a foreign key. A new foreign key constraint is required to associate the new version of Store with fact table Sales. This is possible by creating a new version of table Sales and using StoreKey attribute of the new version of Store in it. As a result of the second schema change, level Store is deleted from dimension Geography and a version of table Sales is created because henceforth, the facts are assigned to level City. Figures 5b and 5c show tables Sales and City created as a result of the second schema change.

An advantage of the MTV approach over the STV one is that it does not require the null values to be stored for the dropped columns thus it prevents the storage space overhead. The disadvantages of this approach are that the data belonging to a table can be accessed either by creating materialized views or performing joins. The materialized views introduce the problem of view maintenance whereas, depending upon the data size in the DW and the number of existing versions, the join operations may become a performance overhead.

We propose a new Hybrid Table Version (HTV) approach for implementing the MVDW. Usually, the dimension tables in a DW have fewer records as compared to the number of records in the fact tables. We propose for changes in dimension schema, a single table version for each dimension level throughout

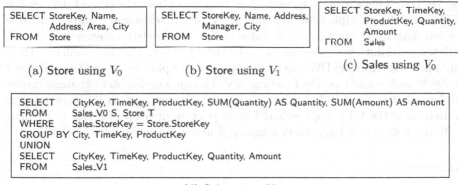

SELECT StoreKey, Name, Address, Area, City FROM   Store

(a) Store using V_0

SELECT StoreKey, Name, Address, Manager, City FROM   Store

(b) Store using V_1

SELECT StoreKey, TimeKey, ProductKey, Quantity, Amount FROM Sales

(c) Sales using V_0

SELECT   CityKey, TimeKey, ProductKey, SUM(Quantity) AS Quantity, SUM(Amount) AS Amount FROM   Sales_V0 S, Store T WHERE   Sales.StoreKey = Store.StoreKey GROUP BY City, TimeKey, ProductKey UNION SELECT   CityKey, TimeKey, ProductKey, Quantity, Amount FROM   Sales_V1

(d) Sales using V_2

Fig. 6. Derivation of version instances from the global instance using views

the lifespan of the DW. This table version is defined as the union of all the attributes that have ever been defined for the dimension level. If the attributes are added or deleted from the level, they are treated in the same way as they are treated in the STV approach. Since, the data is loaded more frequently into fact tables and they contain more records than dimension tables, it is more advantageous to create a new table versions for every change in the schema of the facts. We are aware that the creation of a new structure for every fact version may negatively impact the query performance but indexing techniques [4] can be used to address the issue of efficiency. The HTV approach avoids the null space overhead in case of fact tables and limits the number of joins by managing dimension versions in a single table. In this way, the HTV approach combines the advantages of both the STV and MTV approaches. For brevity, we do not show the the state of the DW after schema changes using HTV approach but the state of levels Store and City can be envisioned as shown in Figs. 4a and 4c, respectively and the fact Sales is represented by Figs. 3d and 5b.

The data from the multiple versions of the MVDW can be accessed by defining a set of views. Whenever a new version of a dimension or fact is created, a view definition is also created to access the existing members using this newly created version. To access the new members using the existing versions, existing view definitions need to be modified. For example, the members of level Store can be accessed in versions V_0 and V_1 using the views defined in Figs. 6a and 6b, respectively. Similarly, the views defined in Figs. 6c and 6d return the fact members in versions V_0 and V_2, respectively. As a result of second schema change, the granularity of dimension Geography was changed. The view in Fig. 6d aggregates the existing sales facts to display them at the granularity of level City.

6   Conclusions

In this paper, we presented (1) a logical model of a multiversion data warehouse (MVDW), and (2) the hybrid table version (HTV) approach to implement the MVDW. This approach combines the benefits of both the single table version (STV) and the multiple table version (MTV) approaches and creates new table versions only for the fact tables. As future work, we plan to combine our approach with temporal data warehouses so that the history of both the changes in the structure and content of the DW can be maintained. We plan to extend the presented model in such a way that the functionality of schema and content changes can be implemented as independent plug-ins. We are also working on the experimental evaluation of the HTV approach and its impact on query performance. Further, we have plans to develop a query language and data structures for MVDWs.

References

1. Bebel, B., Eder, J., Koncilia, C., Morzy, T., Wrembel, R.: Creation and management of versions in multiversion data warehouse. In: Proc. of ACM SAC, pp. 717–723. ACM (2004)
2. Blaschka, M., Sapia, C., Höfling, G.: On schema evolution in multidimensional databases. In: Mohania, M., Tjoa, A.M. (eds.) DaWaK 1999. LNCS, vol. 1676, pp. 153–164. Springer, Heidelberg (1999)
3. Brahmia, Z., Mkaouar, M., Chakhar, S., Bouaziz, R.: Efficient management of schema versioning in multi-temporal databases. International Arab Journal of Information Technology 9(6), 544–552 (2012)
4. Chmiel, J.: Indexing multiversion data warehouse: From ROWID-Based multiversion join index to bitmap-based multiversion join index. In: Grundspenkis, J., Kirikova, M., Manolopoulos, Y., Novickis, L. (eds.) ADBIS 2009. LNCS, vol. 5968, pp. 71–78. Springer, Heidelberg (2010)
5. Eder, J., Koncilia, C., Morzy, T.: The COMET metamodel for temporal data warehouses. In: Pidduck, A.B., Mylopoulos, J., Woo, C.C., Ozsu, M.T. (eds.) CAiSE 2002. LNCS, vol. 2348, pp. 83–99. Springer, Heidelberg (2002)
6. Golfarelli, M., Lechtenbörger, J., Rizzi, S., Vossen, G.: Schema versioning in data warehouses: Enabling cross-version querying via schema augmentation. Data & Knowledge Engineering 59(2), 435–459 (2006)
7. Golfarelli, M., Rizzi, S.: A survey on temporal data warehousing. International Journal of Data Warehousing and Mining 5(1), 1–17 (2009)
8. Kimball, R., Ross, M.: The Data Warehouse Toolkit: The Definitive Guide to Dimensional Modeling, 3rd edn. John Wiley & Sons (2013)
9. Malinowski, E., Zimányi, E.: A conceptual model for temporal data warehouses and its transformation to the ER and the object-relational models. Data & Knowledge Engineering 64(1), 101–133 (2008)
10. Rizzi, S., Golfarelli, M.: X-time: Schema versioning and cross-version querying in data warehouses. In: Proc. of ICDE, pp. 1471–1472. IEEE (2007)
11. Wei, H.-C., Elmasri, R.: Schema versioning and database conversion techniques for bi-temporal databases. Annals of Mathematics and Artificial Intelligence 30(1-4), 23–52 (2000)
12. Wrembel, R., Bębel, B.: Metadata management in a multiversion data warehouse. In: Spaccapietra, S., et al. (eds.) Journal on Data Semantics VIII. LNCS, vol. 4380, pp. 118–157. Springer, Heidelberg (2007)
13. Wrembel, R., Morzy, T.: Managing and querying versions of multiversion data warehouse. In: Ioannidis, Y., et al. (eds.) EDBT 2006. LNCS, vol. 3896, pp. 1121–1124. Springer, Heidelberg (2006)

OntoWarehousing – Multidimensional Design Supported by a Foundational Ontology: A Temporal Perspective

João Moreira, Kelli Cordeiro, Maria Luiza Campos, and Marcos Borges

Post Graduation Program in Informatics,
Computer Science Department (PPGI/DCC/iNCE),
Federal University of Rio de Janeiro (UFRJ), Rio de Janeiro, RJ, Brazil
{joao.moreira,kelli,mluiza,mborges}@ppgi.ufrj.br

Abstract. The choice on information representation is extremely important to fulfil analysis requirements, making the modelling task fundamental in the Business Intelligence (BI) lifecycle. The semantic expressiveness in multidimensional (MD) design is an issue that has been studied for some years now. Nevertheless, the lack of conceptualization constructs from real world phenomena in MD design is still a challenge. This paper presents an ontological approach for the derivation of MD schemas, using categories from a foundational ontology (FO) to analyse the data source domains as a well-founded ontology. The approach is exemplified through a real scenario of the Brazilian electrical system, supporting the joint exploration of electrical disturbances data and their possible repercussion on the news.

Keywords: Business Intelligence, Multidimensional Design, Conceptual Modelling, Foundational Ontologies.

1   Introduction

The MD design task is a fundamental core phase in the Data Warehousing (DW) lifecycle [6] and BI solutions, requiring an engineering process to capture semantics from business entities and their relations. A problem in this context is the difficulty in choosing the correct representations to express the concepts in a MD model, considering identity principles, restrictions, dependencies and business rules. Representing conceptualization constructs from real world phenomena is still an issue in MD design where the lack of semantic expressiveness in conceptual models may compromise the accuracy of business analyses or even limit its scope and comprehensiveness.

The hybrid MD modelling is a balanced union of analysis-driven and supply-driven methods, using them through an interactive and iterative conciliating process. Kimball´s approach [6] can be considered as analysis-driven, where the MD design process uses as input the business case, bus matrix and detailed business requirements. One of its outputs is the conceptual MD model. Another approach is the Advanced DW [7], which adds spatial and temporal meta-concepts to the ER language. To enforce the semantics in MD design, Romero et al. [11] propose an approach based on end-user requirements elicitation, representing the data sources as domain

L. Bellatreche and M.K. Mohania (Eds.): DaWaK 2014, LNCS 8646, pp. 35–44, 2014.

ontology. From this, MD concepts (facts, measures, dimensions, hierarchies and attributes) are identified based on functional dependencies. GEM approach [12] operationalizes this process by representing business requirements as SQL queries in XML files, associating them to the domain ontology by annotations. It uses first order logic for functional dependencies formalization because "if we take into account the expressiveness of the algebras, they might be as semantically rich as conceptual models are" [1]. The MD formalization through Descriptive Logic (DL) is an initiative introduced in DWQ project [5]. In this direction, a FO can enhance the semantic power in domain representations for MD design [10]. FO provide fundamental categories and a systematic strategy to analyse a universe of discourse, exploring characteristics like rigidity, identity, dependencies and unicity to define these categories.

The improvement of the present paper over the state of the art is the introduction of a novel approach to support MD design based on a FO. We propose a systematic automation of the hybrid approach, where the domain ontology is built based on the Unified Foundation Ontology (UFO) [4] conceptualization, increasing its expressiveness. Thus, MD concepts are derived from the domain ontology by a set of rules: the first derives facts and measures from the events mereology (part-whole relations); the second derives potential dimensions and hierarchies from the existential dependences of events from objects participations; and the third suggests a fact pattern for the events temporal relations, based on Allen's operators [3]. An application example is presented on the Brazilian electrical system scenario, relating disturbances and unstructured data from published news.

2   Ontologies and Their Role on BI/DW Solutions

Application of ontologies in DW has been increasing in the recent years. In [2] it is discussed how the Semantic Web (SW) technologies can be useful in nonconventional BI scenarios, discussing their advantages and disadvantages. SW languages (RDF and OWL), founded in DL, can be used to define rules to be processed by reasoning services, assisting aggregations. Annotations can be applied in the ETL conceptual data flows to deal with semantic heterogeneity. In MD modeling, ontologies and reasoning are applied to discover dimension hierarchies from sub-domains and to integrate concepts from different ontologies. Most of the ontological approaches for semantic DW, such as [9], are based on analysing functional dependencies of the data sources.

The domain ontology is a representation of a conceptualization on an universe of discourse, while a FO is a philosophically well-founded and domain independent high-level category system [9].   The application of FOs categories on what is called ontological analysis is a way to better understand and classify the concepts and relations from a domain. Pardillo et al. [10] discuss the usage of ontologies specifically for the MD design, presenting a set of shortcomings for the MD modelling and how each can be addressed. The adoption of FO is mentioned for semantic-aware summarizability of measure aggregations. Considering MD schemas quality it is suggested that "a FO (...) serves us for validating MD models by a representation mapping where ontological concepts are mapped into language constructs".

2.1   Unified Foundational Ontology (UFO)

The UFO [4] is a framework for modelling domain ontologies and it has been evolving for the last years, applied in several scenarios [3]. It is a system of categories developed to support conceptual modelling, incorporating principals from formal ontology, philosophical logic, linguistics and cognitive psychology. Derived from other FO languages, it is divided in three parts, for structural, dynamic and social aspects: UFO-A, UFO-B and UFO-C, respectively. The formal characterization is based on axioms in modal logic. For a comprehensive understanding of UFO, refer to [4]. Only the categories mentioned in this paper are defined in this section. UFO-A comprises the core ontology (**Endurants**), representing entities that persist in time while keeping their identity, such as `cars, balls`. **Substantial** is the highest-level meta-concept for representing entities which carry the identity principle and rigidity property, being specialized as **Sortal** or **Mixin**. The former aggregates individuals with the same identity principle, whilst the later classifies entities that represent individuals with different identity principles, e.g. `insurable item` (`car insurance, health insurance`). UFO-B is the ontology of **Perdurants** [3] which represent entities composed of temporal parts. They happen and extend in time, accumulating temporal parts. The main category on UFO-B is **Event**, the focus of this work, but it includes other important concepts, such as **Participation** and **Situation**. UFO-C represents social aspects, not explored in this work.

The UFO top-level concept is **Thing** which can be seen as the most abstract concept, the root of this theory of **Universals** and **Individuals** categories. The first is commonly known as entities, relationships and properties, whilst the second represents their instances. **Kind** provides an identity principle to its instances in all possible worlds, such as a `person`. **Roles,** on the other hand, are anti-rigid **Sortals**, i.e. they only instantiate individuals in certain circumstances, as a consequence of some extrinsic (relational) property: e.g. `student` is someone who is `enrolled` on an `educational institution`. **Relator** represents a mediator between individuals. UFO typically considers two types of relations: **material** and **formal relations**. A **material** relation depends on an intermediate concept, a **Relator**, whilst a **formal** is valid only by the existence of the concepts connected, an existential dependency relation (e.g. **characterization** and **mediation**). The **componentOf** is one of the four **meronymic** relations, which conceptualizes part-whole relations.

The **Perdurant** denotes the **Events** which happen on the timeline, composed from temporal parts that extend in time. They are represented by possible transformations from one portion of the reality to another, i.e. can affect the reality by changing its characteristics [3]. UFO-B proposes a definition for the **Events** mereology, where **Events** can be classified as **Atomic or Complex Events**. Note that, in [7], the **Event** concept definition is different. There, the state definition is the same as the **Event** definition in UFO, which is framed by **Begin** and **End Time Points**. Moreover, in [7], the **Event** definition corresponds to an **Atomic Event** in UFO, as it is framed by a **Time Interval** where its **Begin** and **End Time Points** are the same. An **Event** is a meta-concept which is existentially dependent on the **Participation** of **Substantial** classes and it is composed by them. This is an important part of UFO-B because it presents a connection from **Perdurants** to **Endurants**, a **formal** relation, the **participationOf**. It is also defined the **Time Interval Relation** between two **Events**, a **formal** relation that is represented by Allen's operators, such as **before**, **meets** and **overlaps**. An example of **Time Interval Relation** is in the relation of promotion entity (first **Event**) to a sale entity (second **Event**) by the **before** relation. If we need to consider how it occurs, it is represented as promotion happen **before** sale **Event**.

OntoUML, an extension of UML, was built to represent UFO [4] concepts. The admissibility of some sates of affair in domains depends on factual knowledge, which only the human cognition capability can validate. Therefore, as in model-driven approaches, a verification and validation process is required. It is made by visual simulation of possible worlds, by automatically generating examples and counterexamples. This approach is available through the OLED software, a tool which provides domain ontology verification, a set of anti-patterns detection (common design errors), an OCL constraints parser and visualization through Alloy Analyser.

3   OntoWarehousing

UFO deals with temporal aspects (UFO-B), which is an important characteristic of MD design, having intuitive relation to MD concepts such as the fact as a business **Event**. In addition, it provides a top-level ontology well-founded in metaphysics, axiomatized in DL. Therefore, in our approach, we adopted UFO and ontological analysis to support the development of domain ontology used for MD schemas derivation, based on a set of rules. In the first set of rules, we concentrated on analysing events, their temporal aspects and participations. We follow hybrid approach: first, the domain representation is semantically enriched using UFO; then, applying the set of mapping rules, some alternative elements for the MD schema are identified; and, finally, based on analytical requirements, the appropriate elements are chosen.

How to classify domain ontology based on UFO has been extensively covered in many papers [3] and particularly on the original work [4]. One must assume that a well-founded domain ontology is available, marked with UFO's stereotypes and complemented by the corresponding restrictions. MD elements are mapped from the domain ontology elements: **Events** and their mereology are mapped to *Facts* and *Measures*. **Participants** in **Events** determine possible perspectives of analysis (*Dimensions* and *Hierarchies*). **Time Interval Relations** between **Events** could derive a *Fact* pattern for the analysis of those temporal relations.

1. Events as *Facts*

The *Facts* are representations of business events. By identifying the domain **Events** and their mereology, i.e. how **Complex** and **Atomic Events** are related in a taxonomy, we can derive and suggest possible *Facts* and *Measures* to the MD modeller. To do so, all **Complex Events** and their parts should be listed, as other sub **Events** (parts of the whole **Event**). The numeric values represented by **qualities** in **Events** can support useful aggregations, so they can be set as *Measures* of a *Fact*.

R1.a)   If *e* is a **Complex Event**, then *e* can be derived as a *Fact*;

R1.b)   If *f* is a *Fact* defined by **Event** *e*, then the numeric **quality** value of *e* can be derived as *Measures* of *f*;

R1.c)   If *f* is a *Fact* from **Event** *e* and *e'* is an **Event** which has mereologic relation (**partOf**) to *e* and there exists **quality** values *qs* of *e'*, then *e'* can be derived as a *Fact* or *qs* as *Measures* in the *Fact f*.

Notice that, as **Events** can bear **qualities**, in second axiom the *Measure* derived from an **Event** can be fully evaluated by its particular **qualities** (their attributes). In other words, checking the numeric **qualities** of the **Event** (or its parts), they can be described as *Measures*. For example, the sale **Event** can be composed by other **Events**, such as the product request and the payment. The payment tax is a **quality** from the payment **Event** and can be designed as a *Measure* from both cases: if it is chosen the sale **Event** or the payment **Event** as a *Fact*, having different aggregation constraints.

2. Objects Participations as *Dimensions* and *Hierarchies*

Dimensions and their *Hierarchies* constitute perspectives of analysis according to different levels of details over a business **Event**, suggested as a *MD Fact* by 3.1. **Substantial** objects can be **Participants** in an **Event** (**Participation**). An **Event** is existentially (ontologically) dependent on other **Objects** and a **Complex Event** is represented as a sum of **Object's Participations**. Therefore, **Object's Participations** in **Events** represented in the domain ontology can be viewed as possible perspective of analysis for the *Fact*, i.e. *Dimensions*, *Attributes* and *Hierarchies*. Typical examples are product with the associated category, forming the product *Hierarchy*; spatial regions classified by country, state, city; supplier (companies); vendor; equipment. Notice that we can derive possible *Dimensions* and their *Hierarchies* by analyzing the relations between its meta-concepts, such as the entities classified as **RoleMixin Participant** of the **Participation**, described as following:

R2.a)   If *f* is a *Fact* defined by **Event** *e*, *o* is an **Object** (**Substance**) which is mediated by **formal** relation (**participationOf**) to a **Participation** *p*, has mereologic relation (**part Of** *e*), then *o* can be derived as a *Dimension* of *Fact f*;

R2.b)   If *f* is a *Fact* defined by **Event** *e*, *d* is a *Dimension* defined by **Object** *o* which is **Participant** of *e*, and *o* has relationships to other **Substances**, then each **Substance** *s* of *S* can be suggested as *Hierarchies* of the *Fact f* through *Dimension d*.

For example, the sale **Event** is existentially dependent from the client **Participation**. So, the client **Participation** in a sale is a **mediation relation** between the client **Participant** (a **RoleMixin**) and the sale **Event**. Therefore, we can represent client as a *Dimension* in a *MD schema*.

3. Time Interval Relations between Events as *Fact*

The temporal properties of **Events** are represented by a **Quality Structure** composed of **Time Intervals**, linear ordered in **Time Points**. Each **Event** must be framed (associated) by a **Time Interval** that is defined by its **Begin** and **End Time Points** values. The **Temporal Relations** between **Events** in the domain ontology design are represented by the **Allen's Operators**, i.e. **before, meets, overlaps, starts, during,**

finishes and **equals formal relations**. Those operators are represented as **Relations** and join the **Time Intervals** of two **Events** by their **Time Points**. The use of those representations by the modeller indicates that the **Time Interval Relations** are important issues to the business needs. Therefore, analysis over them can be suggested through a *Fact* pattern, which relates two **Events** by their temporal parts. The mapping rule checks for the related **Events** in the domain ontology and is formalized as:

R3.a)   If *e1*, *e2* are **Events** and they have one or more **Time Interval Relations** between them, then it can be suggested as a *Fact f* representing the **Time Interval Relations** between the **Events** *e1* and *e2*, where *e1* is represented by *Dimension d1* and *e2* represented as *Dimension d2*. Each *Dimension* (*d1* and *d2*) has a *Time Hierarchy* for the **Begin** and **End Time Points**;

R3.b)   If a *Fact f* is represented by the **Time Interval Relations** from the **Events** *e1* and *e2*, having *e1* represented by *Dimension d1* and *e2* represented as *Dimension d2*, then the fundamental **Participants** of the **Events** can be suggested as *Attributes* or *Hierarchies* for *d1* and *d2*.

A temporal *Fact* standard is established, supporting the analysis of all possible **Temporal Relation** between the two **Events** being observed. The **Events Interval Times** are represented by *Time Dimensions* for each **Event** with typical time *Hierarchies*, such as *Year*, *Semester*, *Month* and *Day* for the **Begin** and **End Time Points**. Each **Event** (*Dimension*) also has its **Participants** designed as (shared or not shared) *Dimensions*, *Attributes* or *Hierarchies*. Complementary, exploring UFO **Time Interval Relations** and their intrinsic constraints, it is possible to derive data loading rules for ETL through the **Begin Time Point** and the **End Time Point** relations between the events. The **Allen´s operators** play this role because they provide the specific constraints for loading the data into the *Fact* and in the evaluation of the *Temporal Measures*. For example, the *Measure* can represent how long the **Event** *e1* **overlaps** the **Event** *e2* or how long **before** *e1* occurred until *e2*.

Those set of mappings are represented in Figure 1 through colours equivalences. Figure 1.a presents a subset of UFO meta-concepts and the derived MD schema is represented in Figure 1.c. Just as a reference, a MD meta-model is depicted in Figure 1.b, also using the colour for the corresponding concepts.

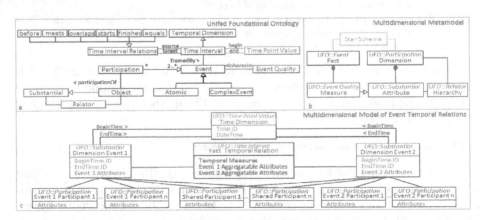

Fig. 1. Mapping rules represented from UFO to MD concepts

4   An Application Example

To evaluate the feasibility of the approach it was experimented in a BI/DW solution and the resulted MD model was verified by domain experts. The application was the context of the Electric System National Operator (ONS, www.ons.org.br), the company that operates the Integrated National System (SIN) – the set of facilities that supply electricity for Brazil – ensuring its security, continuity and economy optimization. The system parts are subjected to failures of various natures, which can cause shutdowns of one or more devices in the transmission grid, interrupting the power supply of a geographical area, popularly known as blackout. Those occurrences are known as electrical disturbances and may be caused by atmospheric electrical discharges, fires, human failures, among others. A disturbance is an occurrence in SIN characterized by forced shutdown of one or more of its components, causing any of the consequences: loss of load, shutdown of the components, equipment damage or violation of operating limits. To support the disturbances analysis, ONS has an analytical information solution based on BI/DW architecture (Disturbances BI). It consolidates data from transactional systems, which support operational processes, in a MD schema. One of its measures, the load cut level, is directly related to blackouts. Due to the negative consequences of a blackout to the populations' lives, the Brazilian press gives great focus to the subject, often citing the ONS when such a situation occurs, affecting its corporate image. A daily summary of all news pertaining to the electricity sector is provided in Intranet home page, named Clippings.

The current information systems present the data in an isolated way and manual work is necessary for a joint analysis over historic data, often making the desired results unfeasible. Therefore, our approach is applied in the MD design for joint analysis over disturbances and news, integrating news articles in a DW such as in [8]. The integration process of textual information into typical BI/DW solutions is still an open topic and because of space concerns cannot be discussed here. To support this process, Information Retrieval (IR) and Natural Language Processing (NLP) techniques are used, such as named entity recognition and morphologic analysis. A set of documentation was used in the analysis-driven approach. For the supply-driven the data sources were checked. The domain ontology was represented through EA/OntoUML tool, having a cyclic verification and validation activity performed through OLED/Alloy Analyser tool, increasing its quality.

The main concepts of the domain ontology are illustrated in Figure 2.a, where a Disturbance is a **Complex Event**. The **Participations** of Equipment as the origin of the occurrence is represented by the Source Equipment **Participation** class, stereotyped as **Participation**. It has Load Cut Value and Restoration Time as attributes and it is a part of the Disturbance **Event** (**memberOf**). Also, this **Participation** may vary depending on the Equipment Type, which is represented by the disjoint specialization for Transmission Line **Participation** and Power Transformer **Participation**. The **RoleMixin** Transmission Line **Participant** (specialization of Transmission Line **Kind**) is linked by the **participationOf**. The same was applied to Power Transformer class. The News Publication was classified as a **Complex Event**, having the News Article Text **Participation** as a mereological relation (**partOf**). The News Article Text (the document) is composed by Terms, being owned by a Press Company (**<ownerOf**) and written by an Author (**writtenBy>**). The **before Time Interval Relation** between a News Publication and a Disturbance is also represented. It means that for each Disturbance there is a **before** relation to a News Publication with 0..* cardinality. We could test the rules execution through a prototype that derives MD elements on this domain, resulting in the MD schema represented in Figure 2b.

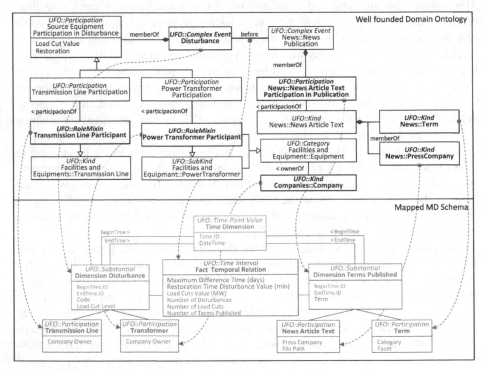

Fig. 2. Well founded domain ontology mapped to MD Schema from rules execution

The Disturbances and News Publication were firstly derived as *Facts* following rule R1.a. R1.b resulted in the Load Cut Value and Restoration Time *Measures* derived from the Source Equipment **Participation qualities** defined as *Attributes* of the class. The **memberOf** relation characterizes that those *Measures* can be represented in the Disturbance *Fact*. The Transmission Line and Power Transformer that **participate on** the Disturbance were derived as *Dimensions*, as they form the link from the Equipment **Event Participation** to the **Substantial** concepts (R2.a). The Owner Company *Hierarchy* for each of these were derived from the **Participants** relations (R2.b). The *MD Schema* for the **Time Interval Relation (before)** between Disturbance and News Publication was derived from R3.a. It also derived the **Events** Disturbance and Terms Published as *Dimensions*. R3.b provided the main **Participants** from those **Events** as *Dimension Attributes* or *Hierarchies*. The Transmission Line *Hierarchy* of the Disturbance *Dimension* was defined as having Company Owner at the first *Level*, such as Power Transformer. The Terms Published *Dimension* also contains *Hierarchies* set, such as the Term Category and Press Companies. Almost all the rules were used in the derivation process and from R3 it was possible to set the ETL constraints for **Time Interval Relation** *Fact* data load: the Disturbance **End Time** value should be lower than the **Begin Time** value of a News Publication.

A database was implemented reflecting the MD schema. The News Articles were downloaded through a web crawler from 2011 to 2013. The existent conventional ETL process was evolved to load the Disturbance *Dimension* and its *Hierarchies*. Thereafter, a complementary ETL process loads the News Article Publication *Dimension* and its *Hierarchies* from the unstructured data source. A third ETL process runs the *Linkage* process, loading the temporal relation *Fact*. A SQL query was used to relate them, following the constraints mentioned for **before** relationship, ensuring that a News Article Publication can be related to a Disturbance if it occurred previously. It was set a delta time to relate them, that is, a Disturbance is related somehow with a News Article if that news is published until ten days after the occurrence. It was firstly defined by a domain expert and confirmed by the initial analysis, which evidenced an increase on the number of News Publications after a severe Disturbance had occurred, which returns to its regular publication rate after ten days, on average. An important issue about this choice is that: the higher this value is set, the number of rows in the fact grows exponentially. The MD schema built and loaded in relational database was transformed in a data cube and get available to the users by an OLAP tool (Tableau).

The impact of blackouts in ONS corporate image can be analysed in this environment. By selecting the Total Number of Disturbances (36,778) and the Total Number of Terms occurrences (210,373,632), the calculated measure Terms by Disturbances presents the average of 5,720 Term occurrences per Disturbance. When navigating through Disturbance *Dimension* by selecting Load Cut Level *Attribute* and analysing the Terms by Disturbances *Measure*, it is possible to see a direct relation between the severity of the Disturbances and the number of News Published. The more severe are the Disturbances, more Terms are published. When Load Cut Level is minor than 49MW the average is 4,794 Terms/Disturbances; between 50MW and 99MW is 6,486 Terms/Disturbances; greater then 99MW is 7,006 Terms/Disturbances.

Comparing the rules introduced here with Kimball's approach, we could formalize some of his informal design guidelines, such the business event as a *Fact*. The main benefit on the application of the approach was the increase of the choices for the MD modeler to design a MD schema. However, the problem about it is the assumption that the domain ontology is built upon UFO concepts, once it is not so common yet.

5   Conclusions

In this paper we introduced the application of FO in MD design for the derivation of MD concepts. By referring to UFO categories, the semantic expressiveness of MD models is increased, as more real world meaning is associated to the representation constructs. We presented a subset of derivation rules, more specifically those related to the temporal exploration of events. The first rule deals with the event, its mereological structure and its qualities mapped to facts and measures. The second rule supports the creation of dimensions, hierarchies and attributes by analysing the participation of substantials in events. The third rule derives a fact pattern from the time interval relations (Allen's operators), associating events, where each event is represented as a dimension related to the fact. Those are part of a more comprehensive research effort exploring UFO categories to derive common MD conceptualizations. Besides the FO application, we considered unstructured data in the example.

For now, we have formalized the rules only textually, but they can be full axiomatized in DL, serving as a basis for automatic reasoning in the derivation process. Therefore, the full potential of reasoning cannot be performed yet. Moreover, there is a full set of concepts in UFO to be analysed and mapped to MD concepts, such as situations and objects dispositions. The links between facts and dimensions represented as part-whole relationships needs to be explored. To prove the approach effectiveness a set of metric for evaluating MD model quality can also be used. Nevertheless, this work can lead to new initiatives in the modelling and semantics externalization of BI/DW solutions research topic and may cause potential new results. In the limit, having information systems designed using a well-founded approach, i.e. represented by a FO, and linked to their data sources, it may be possible to automatically derive analytical information systems based in BI/DW architecture by applying a full set of mapping rules like those introduced in this work.

References

1. Abelló, A.: YAM: A Multidimensional Conceptual Model. PhD Thesis (2002)
2. Berlanga, R., Romero, O., Simitsis, A., Nebot, V., Pedersen, T., Abelló, A.: SW Technologies for BI. In: BI App and the Web: Models, Systems, and Technologies (2011)
3. Guizzardi, G., Wagner, G., de Almeida Falbo, R., Guizzardi, R.S.S., Almeida, J.P.A.: Towards Ontological Foundations for the Conceptual Modeling of Events. In: Ng, W., Storey, V.C., Trujillo, J.C. (eds.) ER 2013. LNCS, vol. 8217, pp. 327–341. Springer, Heidelberg (2013)
4. Guizzardi, G.: Ontological Foundations for Structural Conceptual Models. Phd thesis (2005)
5. Jarke, M., Lenzerini, M., Vassiliou, Y., Vassiliadis, P.: Fundamentals of Data Warehouses. Springer Book (2003)
6. Kimball, R., Ross, M.: The Data Warehouse Toolkit: The definitve Guide to Dimensional Modeling. Wiley (2013) ISBN-13: 978-1118530801
7. Malinowski, E., Zimányi, E.: Advanced Data Warehouse Design: From Conventional to Spatial and Temporal Applications. Springer (2009) ISBN 978-3-540-74405-4
8. Martínez, J., Berlanga, R., Aramburu, M., Pedersen, T.: Contextualizing data warehouses with documents. Decision Support Systems 45(1), 77–94 (2008)
9. Nebot, V., Berlanga, R., Martínez, J., Aramburu, M., Pedersen, T.: MD Integrated Ontologies: A Framework for Designing Semantic DW. J. Data Semantics 13, 1–36 (2009)
10. Pardillo, J., Mazón, J.-N.: Using Ontologies for the Design of Data Warehouses. International Journal of Database Management Systems (2011)
11. Romero, O., Abelló, A.: A framework for multidimensional design of data warehouses from ontologies. Data & Knowledge Engineering 69(11), 1138–1157 (2010)
12. Romero, O., Simitsis, A., Abelló, A.: GEM: Requirement-Driven generation of ETL and Multidimensional Conceptual Designs. In: Cuzzocrea, A., Dayal, U. (eds.) DaWaK 2011. LNCS, vol. 6862, pp. 80–95. Springer, Heidelberg (2011)

Modeling and Querying Data Warehouses
on the Semantic Web Using QB4OLAP

Lorena Etcheverry[1], Alejandro Vaisman[2], and Esteban Zimányi[3]

[1] Universidad de la República, Uruguay
lorenae@fing.edu.uy
[2] Instituto Tecnológico de Buenos Aires, Argentina
avaisman@itba.edu.ar
[3] Université Libre de Bruxelles, Belgium
ezimanyi@ulb.ac.be

Abstract. The web is changing the way in which data warehouses are designed and exploited. Nowadays, for many data analysis tasks, data contained in a conventional data warehouse may not suffice, and external data sources, like the web, can provide useful multidimensional information. Also, large repositories of semantically annotated data are becoming available on the web, opening new opportunities for enhancing current decision-support systems. Representation of multidimensional data via semantic web standards is crucial to achieve such goal. In this paper we extend the QB4OLAP RDF vocabulary to represent balanced, recursive, and ragged hierarchies. We also present a set of rules to obtain a QB4OLAP representation of a conceptual multidimensional model, and a procedure to populate the result from a relational implementation of the multidimensional model. We conclude the paper showing how complex real-world OLAP queries expressed in SPARQL can be posed to the resulting QB4OLAP model.

1 Introduction

The web is changing the way in which data warehouses (DW) are designed, used, and exploited [4]. For some data analysis tasks (like worldwide price evolution of some product), the data contained in a conventional data warehouse may not suffice. The web can provide useful multidimensional information, although usually too volatile to be permanently stored [1]. Further, the advent of initiatives such as Open Data[1] and Open Government promotes publishing multidimensional data using standards and non-proprietary formats[2]. Also, the Linked Data paradigm allows sharing and reusing data on the web by means of semantic web (SW) standards [8]. Domain ontologies expressed in RDF[3], or in languages built on top of RDF like RDF-S or OWL, define a common terminology for the concepts involved in a particular domain. In spite of the above, although in the last decade several open-source BI platforms have emerged, they still do not provide an open format to publish and share cubes among organizations [7], and the most popular commercial Business Intelligence (BI) tools are still proprietary.

---

[1] https://okfn.org/opendata/
[2] http://opengovdata.org/
[3] http://www.w3.org/TR/rdf-concepts/

L. Bellatreche and M.K. Mohania (Eds.): DaWaK 2014, LNCS 8646, pp. 45–56, 2014.
© Springer International Publishing Switzerland 2014

Usually, in a relational DW representation, a conceptual model is implemented at the logical level as a collection of tables organized in specialized structures, basically star and snowflake schemas, which relate a fact table to several dimension tables through foreign keys. In a semantic web DW scenario, the logical model becomes the RDF data model. The first proposal of an RDF vocabulary to cover many of the multidimensional model components was the QB4OLAP vocabulary [5,6]. In this paper, after a brief introduction to semantic web concepts (Section 2) and related work (Section 3), we present an extension of QB4OLAP that supports the most used model characteristics, like balanced, recursive, ragged, and many-to-many hierarchies (Section 4). We then propose a mechanism to translate a conceptual multidimensional model into a logical RDF model using the QB4OLAP vocabulary (Section 5), and show how we can transform an existent relational implementation of a DW into QB4OLAP via an R2RML mapping. Finally, we show how QB4OLAP cubes can be queried using SPARQL (Section 6), and discuss open challenges and future work (Section 7) .

2   Preliminary Concepts

RDF and SPARQL. The Resource Description Framework (*RDF*) allows expressing assertions over resources identified by an Internationalized Resource Identifier (IRI) as triples of the form *subject - predicate - object*, where *subject* are always resources, and *predicate* and *object* could be resources or strings. *Blank nodes* are used to represent anonymous resources or resources without an IRI, typically with a structural function, e.g., to group a set of statements. Data values in RDF are called *literals* and can only be *objects*. A set of RDF triples can be seen as a directed graph where *subject* and *object* are nodes, and *predicates* are arcs. Usually, triples representing schema and instance data coexist in RDF datasets. A set of reserved words defined in RDF Schema (called the RDF-S vocabulary) is used to define classes, properties, and hierarchical relationships. Many formats for RDF serialization exist. In this paper we use Turtle [4].

SPARQL 1.1[5] is the current W3C standard query language for RDF. The query evaluation mechanism of SPARQL is based on subgraph matching: RDF triples are interpreted as nodes and edges of directed graphs, and the query graph is matched to the data graph, instantiating the variables in the query graph definition. The selection criteria is expressed as a graph pattern in the WHERE clause, composed by *basic graph patterns (BGP)*. The '.' operator represents the conjunction of graph patterns. Relevant to our study, SPARQL supports aggregate functions and the GROUP BY clause.

R2RML[6] is a language for expressing mappings from relational databases to RDF datasets, allowing representing relational data in RDF using a customized structure and vocabulary. Both, R2RML mapping documents (written in Turtle syntax) and mapping results, are RDF graphs. The main object of an R2RML mapping is the *triples map*, a collection of triples composed of a *logical table*, a *subject map*, and one or more *predicate object maps*. A logical table is either a base table or a view (using the predicate

---

[4] http://www.w3.org/TeamSubmission/turtle/
[5] http://www.w3.org/TR/sparql11-query/
[6] http://www.w3.org/TR/r2rml/

rr:tableName), or an SQL query (using the predicate rr:sqlQuery). A predicate object map is composed of a predicate map and an object map. Subject maps, predicate maps, and object maps are either constants (rr:constant), column-based maps (rr:column), or template-based maps (rr:template). Templates use column names as placeholders. Foreign keys are handled referencing object maps, which use the subjects of another triples map as the objects generated by a predicate-object map. A set of R2RML mappings can either be used to generate a static set of triples that represent the underlying relational data (*data materialization*) or to provide a non-materialized RDF view of the relational data (*on-demand mapping*).

3   Related Work

At least two approaches are found concerning OLAP analysis of SW data. In a nutshell, the first one consists on extracting multidimensional data from the SW and loading them into traditional OLAP repositories. The second one consists in performing OLAP-like analysis directly over SW data, e.g., over multidimensional data represented in RDF.

Along the first line of research are the works by Nebot et al. [14] and Kämpgen et al. [9]. The former proposes a semi-automatic method for on-demand extracting semantic data into a multidimensional database, where data will be exploited. Kämpgen et al. follow a similar approach, although restricted to a particular kind of SW data. They explore the extraction of statistical data, published using the RDF Data Cube vocabulary (QB)[7], into a multidimensional database. Thus, these two approaches allow using the existent knowledge in this area, and reusing available tools, at the expense of requiring the existence of a local DW to store the semantic web data extracted. This constraint clashes with the autonomous and high volatile nature of web data sources.

The second line of research tries to overcome this restriction, exploring data models and tools that allow publishing and analyzing multidimensional data directly over the SW. The aim of this models is to support concepts like *self-service BI*, *situational BI* [12], and *on-demand BI*, in order to take advantage of web data to enrich decision-making processes. Abello et al. [1] envision a framework to support self-service BI, based on the notion of *fusion cubes*, i.e., multidimensional cubes that can be dynamically extended both in their schema and their instances. Beheshti et al. [2] propose to extend SPARQL to express OLAP queries. However, we believe that not working with standard languages and established data models limits the applicability of the approach. With a different approach, Kämpgen et al. [10,11] proposed an OLAP data model on top of QB and other related vocabularies, and a mechanism to implement OLAP operators over these extended cubes, using SPARQL queries. This approach inherits the drawbacks of QB, that is, multidimensional modeling possibilities are limited.

4   RDF Representation of Multidimensional Data

The QB4OLAP vocabulary[8] extends the QB vocabulary mentioned in Section 3, to enhance the support to the multidimensional model, overcoming several limitations of

---

[7] http://www.w3.org/TR/vocab-data-cube/
[8] http://purl.org/qb4olap/cubes

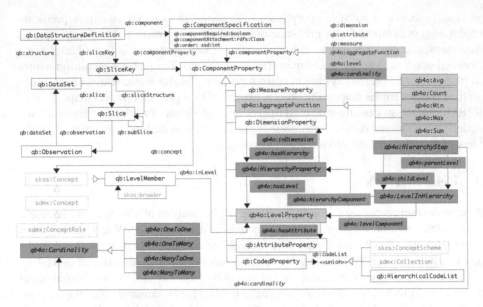

Fig. 1. A new version of the QB4OLAP vocabulary

QB [6]. Unlike QB, QB4OLAP allows implementing the main OLAP operations, such as rollup, slice, and dice, using standard SPARQL queries. Two different kinds of sets of RDF triples are needed to represent a data cube in QB4OLAP: (i) the *cube schema*, and (ii) the *cube instances*.The former defines the structure of the cube, in terms of dimension levels and measures, but also defines the hierarchies within dimensions and the parent-child relationships between levels. These metadata can then be used to automatically produce SPARQL queries that implement OLAP operations [6], which is not possible in QB. Cube instances are sets of triples that represent level members, facts and measured values. Several cube instances may share the same cube schema. Figure 1 depicts the QB4OLAP vocabulary. We can see that QB4OLAP embeds QB, allowing data cubes published using QB to be represented using QB4OLAP without affecting existing applications. Original QB terms are prefixed with qb:. Capitalized terms represent RDF classes and noncapitalized terms represent RDF properties. Classes in external vocabularies are depicted in light gray font. QB4OLAP classes and properties are prefixed with qb4o.

The previous version of QB4OLAP had some limitations regarding the representation of dimension hierarchies. For example, it did not allow more than one hierarchy per dimension, or to represent cardinalities in the relationships between level members. To overcome these limitations we have introduced new classes and properties, depicted in Fig. 1 in dark gray background and bold italics; existent QB4OLAP classes and properties are depicted in light gray background.

A data structure definition (DSD) specifies the schema of a data set of the class qb:DataSet. The DSD can be shared among different data sets, and has properties for representing dimensions, dimension levels, measures, and attributes, called

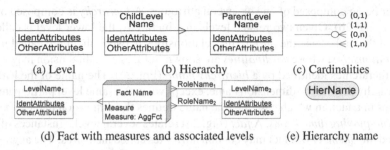

Fig. 2. Notation of the MultiDim model

qb:dimension, qb:measure, qb4o:level, and qb:attribute, respectively. Observations (facts) represent points in a multidimensional space. An observation is linked to a value in each dimension level of the DSD using instances of qb4o:LevelProperty. This class models dimension levels. Instances of the class qb4o:LevelMember represent level members, and relations between them are expressed using the skos:broader property. Dimension hierarchies are defined via the class qb4o:HierarchyProperty. The relationship between dimensions and hierarchies is represented via the property qb4o:hasHierarchy and its inverse qb4o:inDimension. A level may belong to different hierarchies, and in each hierarchy it may have a different parent level. Also, the relationships between level members may have different cardinalities (e.g. one-to-many, many-to-many, etc.). The qb4o:LevelInHierarchy class represents pairs of hierarchies and levels, and properties are provided to relate a pair with its components: qb4o:hierarchyComponent and qb4o:levelComponent. A parent-child relationship between two levels is represented using the class qb4o:HierarchyStep and the properties qb4o:childLevel and qb4o:parentLevel. The cardinality of this relationship is represented via the qb4o:cardinality property and members of the qb4o:Cardinality class. This property can also be used to represent the cardinality of the relationship between a fact and a level. QB4OLAP also allows to define level attributes via the qb4o:hasAttribute property and aggregate functions via the qb4o:AggregateFunction class. The association between measures and aggregate functions is represented using the property qb4o:aggregateFunction. This property, together with the concept of component sets, allows a given measure to be associated with different aggregate functions in different cubes.

5   QB4OLAP Implementation of Multidimensional Data Cubes

We next show that we can translate both, data cube schema and instances, into an RDF representation using QB4OLAP. To represent the cube schema we will use the MultiDim model [15,13], whose main components are depicted in Fig. 2. Of course, any conceptual model could be used instead. A *schema* is composed of a set of dimensions and a set of facts. A *dimension* is composed of either one *level*, or one or more hierarchies. Instances of a level are called *members*. A level has a set of *attributes* that describe the characteristics of their members (Fig. 2a), and one or more *identifiers*, each identifier being composed of one or several attributes. A *hierarchy* is composed of a set of levels (Fig. 2b). Given two related levels in a hierarchy, the lower level is called the *child* and the higher one the *parent*; the relationships between them are called *parent-child relationships*, whose *cardinalities* are shown in Fig. 2c. A dimension may contain several hierarchies identified by a *hierarchy name* (Fig. 2e). The name of the leaf level in a hierarchy defines the dimension name, except when the same level participates several times in a fact, in which case the role name defines the dimension name. These are called *role-playing dimensions*. A *fact* (Fig. 2d) relates several levels. Instances of a fact are called *fact members*. A fact may contain attributes called *measures*. The aggregation function associated to a measure can be specified next to the measure name (Fig. 2d), the default being the SUM function.

5.1   QB4OLAP Implementation of a Cube Schema

We first present an algorithm to obtain a QB4OLAP representation of a cube schema, from a conceptual schema. We assume that we have a conceptual schema that represents a cube C, with a fact F composed of a set M of measures, and a set D of dimensions. Each dimension $d \in D$ is composed of a set L of levels, organized in hierarchies $h \in H$. Each level $l \in L$ is described by a set of attributes A. Figure 3 depicts a simplified version of the *MultiDim* representation of the well-known Northwind DW, which we will use as our running example. The algorithm comprises seven steps described next. We call CS_{RDF} the RDF graph that represents the cube schema, which is built incrementally.

Step 1 (Dimensions). For each dimension $d \in D$, $CS_{RDF} = CS_{RDF} \cup \{t\}$, where t is a triple stating that there exists a resource d_{RDF} of type qb:DimensionProperty. Triples indicating the name of each dimension can be added using property rdfs:label.

The triples below show how some dimensions in Fig. 3 are represented (@en indicates that the names are in English, and nw: is a prefix for the cube schema graph).

```
nw:employeeDim a qb:DimensionProperty ; rdfs:label "Employee Dimension"@en .
nw:orderDateDim a qb:DimensionProperty ; rdfs:label "OrderDate Dimension"@en .
```

Step 2 (Hierarchies). For each hierarchy $h \in H$, $CS_{RDF} = CS_{RDF} \cup \{t\}$, where t is a triple stating that h_{RDF} is a resource with type qb4o:HierarchyProperty. Triples indicating the name of each hierarchy can be added using property rdfs:label.

Applying Step 2 to the hierarchies in the Employee dimension we obtain:

```
nw:supervision a qb4o:HierarchyProperty ; rdfs:label "Employee Supervision Hierarchy"@en .
nw:territories a qb4o:HierarchyProperty ; rdfs:label "Employee Territories Hierarchy"@en .
```

Step 3 (Levels and Attributes). For each level $l \in L$, $CS_{RDF} = CS_{RDF} \cup \{t\}$, where t is a triple stating that l_{RDF} is a resource with type qb4o:LevelProperty. For each attribute $a \in A$, add to CS_{RDF} a triple stating that there exists a resource a_{RDF} with type qb4o:AttributeProperty. Finally, add triples relating a level l_{RDF} with its corresponding attribute a_{RDF}, using the property qb4o:hasAttribute. Triples indicating the names of levels and attributes can be added using property rdfs:label.

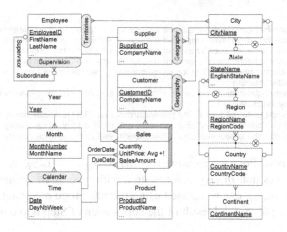

Fig. 3. A simplified version of the conceptual schema of the NorthwindDW

Applying Step 3 to level **Employee** in the **Employee** dimension we obtain:

```
nw:employee a qb4o:LevelProperty ; rdfs:label "Employee Level"@en ;
    qb4o:hasAttribute nw:firstName ; qb4o:hasAttribute nw:lastName .
nw:firstName a qb:AttributeProperty ; rdfs:label "First Name"@en .
nw:lastName a qb:AttributeProperty ; rdfs:label "Last Name"@en .
```

Step 4 (Dimension-Hierarchy Relationships). For each $h \in H$ in $d \in D$, relate d_{RDF} in CS_{RDF} to h_{RDF}, and h_{RDF} to d_{RDF}. Then, $CS_{RDF} = CS_{RDF} \cup \{d_{RDF}$ qb4o:hasHierarchy $h_{RDF}\} \cup \{h_{RDF}$ qb4o:inDimension $d_{RDF}\}$.

Applying Step 4 to the **Employee** dimension and its hierarchies we obtain:

```
nw:employeeDim qb4o:hasHierarchy nw:Supervision ; qb4o:hasHierarchy nw:territories .
nw:supervision qb4o:inDimension nw:employeeDim. nw:territories qb4o:inDimension nw:employeeDim.
```

Step 5 (Hierarchy Structure). For each hierarchy $h \in H$ composed of a level $l \in L$, relate h_{RDF} in CS_{RDF} to l_{RDF} as $CS_{RDF} = CS_{RDF} \cup \{h_{RDF}$ qb4o:hasLevel $l_{RDF}\}$. Also, create a blank node lh_{RDF} of type qb4o:LevelInHierarchy that represents the pair (l_{RDF}, h_{RDF}) and add triples as $CS_{RDF} = CS_{RDF} \cup \{lh_{RDF}$ qb4o:hierarchyComponent $h_{RDF}\} \cup \{lh_{RDF}$ qb4o:levelComponent $l_{RDF}\}$. Let (l, l') be a pair of levels in the C, such that $l, l' \in h$, and $parentLevel(l, h) = l'$ with cardinality car. Also, let l_{RDF}, l'_{RDF}, and h_{RDF} be the representations of l, l' and h in CS_{RDF}, and lh_{RDF}, lh'_{RDF} be the resources of type qb4o:LevelInHierarchy that represent the pairs (l_{RDF}, h_{RDF}) and (l'_{RDF}, h_{RDF}) respectively. Then add to CS_{RDF} a blank node hs_{RDF} of type qb4o:HierarchyStep, and the triples hs_{RDF} qb4o:childLevel lh_{RDF}, hs_{RDF} qb4o:parentLevel lh'_{RDF}. Finally, add a triple hs_{RDF} qb4o:cardinality car_{RDF}, car_{RDF} being the relationship's cardinality.

A part of the **Employee** dimension structure obtained is shown below. Note the relationship _:ih3; qb4o:parentLevel _:ih6 supporting a ragged hierarchy.

```
nw:supervision qb4o:hasLevel nw:employee .
nw:territories qb4o:hasLevel nw:employee, nw:city, nw:state, nw:country, nw:continent .
# nw:supervision levels
_:ih1 a qb4o:LevelInHierarchy ; qb4o:levelComponent nw:employee ; qb4o:hierarchyComponent nw:supervision .
```

```
# nw:territories levels
_:ih2 a qb4o:LevelInHierarchy ; qb4o:levelComponent nw:employee ; qb4o:hierarchyComponent nw:territories .
_:ih3 a qb4o:LevelInHierarchy ; qb4o:levelComponent nw:city ; qb4o:hierarchyComponent nw:territories .
_:ih4 a qb4o:LevelInHierarchy ; qb4o:levelComponent nw:state ; qb4o:hierarchyComponent nw:territories .
_:ih5 a qb4o:LevelInHierarchy ; qb4o:levelComponent nw:region ; qb4o:hierarchyComponent nw:territories .
_:ih6 a qb4o:LevelInHierarchy ; qb4o:levelComponent nw:country ; qb4o:hierarchyComponent nw:territories .
# nw:supervision hierarchy structure
_:pl1 a qb4o:HierarchyStep ; qb4o:childLevel _:ih1 ; qb4o:parentLevel _:ih1 ; qb4o:cardinality qb4o:OneToMany .
# nw:territories hierarchy structure
_:pl2 a qb4o:HierarchyStep ; qb4o:childLevel _:ih2 ; qb4o:parentLevel _:ih3 ; qb4o:cardinality qb4o:ManyToMany .
_:pl3 a qb4o:HierarchyStep ; qb4o:childLevel _:ih3 ; qb4o:parentLevel _:ih4 ; qb4o:cardinality qb4o:OneToMany .
_:pl4 a qb4o:HierarchyStep ; qb4o:childLevel _:ih3 ; qb4o:parentLevel _:ih6 ; qb4o:cardinality qb4o:OneToMany .
```

Step 6 (Measures). For each measure $m \in M$, $CS_{RDF} = CS_{RDF} \cup \{t\}$, such that t is a triple that states that m_{RDF} is a resource with type qb4o:MeasureProperty. The range of each m_{RDF} can be defined using the rdfs:range predicate.

The following triples are the result of the application of Step 6 to our example.

```
nw:quantity a qb:MeasureProperty ; rdfs:label "Quantity"@en ; rdfs:range xsd:integer .
nw:unitPrice a qb:MeasureProperty ; rdfs:label "UnitPrice"@en ; rdfs:range xsd:decimal .
nw:salesAmount a qb:MeasureProperty ; rdfs:label "SalesAmount"@en ; rdfs:range xsd:decimal .
```

Step 7 (Cube). For each fact F, $CS_{RDF} = CS_{RDF} \cup \{t\}$, such that t is a triple stating that c_{RDF} is a resource with type qb:DataStructureDefinition. For each measure $m \in M$, $CS_{RDF} = CS_{RDF} \cup \{c_{RDF}$ qb:component [qb:measure m_{RDF}; qb4o:aggregateFunction f_{RDF}]$\}$, where f_{RDF} is an aggregation function. Also, for each of the levels $l \in L$ related to a fact F in the conceptual schema, $CS_{RDF} = CS_{RDF} \cup \{c_{RDF}$ qb:component [qb:level l_{RDF}; qb4o:cardinality car_{RDF}]$\}$, where car_{RDF} represents the cardinality of the relationship between facts and level members and is one of cardinality restrictions defined in QB4OLAP (qb4o:OneToOne, qb4o:OneToMany, qb4o:ManyToOne, qb4o:ManyToMany).

The following triples are the result of the application of Step 7.

```
# Cube definition (Data structure)
nw:Northwind a qb:DataStructureDefinition ;
# Lowest level for each dimension in the cube
qb:component [qb4o:level nw:employee ; qb4o:cardinality qb4o:ManyToOne] ;
qb:component [qb4o:level nw:orderDate ; qb4o:cardinality qb4o:ManyToOne] ;
qb:component [qb4o:level nw:dueDate ; qb4o:cardinality qb4o:ManyToOne] ;
qb:component [qb4o:level nw:supplier ; qb4o:cardinality qb4o:ManyToOne] ;
qb:component [qb4o:level nw:customer ; qb4o:cardinality qb4o:ManyToOne] ;
# Measures in the cube
qb:component [qb:measure nw:quantity ; qb4o:aggregateFunction qb4o:sum] ;
qb:component [qb:measure nw:unitPrice ; qb4o:aggregateFunction qb4o:avg] ;
qb:component [qb:measure nw:salesAmount ; qb4o:aggregateFunction qb4o:sum] .
```

5.2   QB4OLAP Implementation of Cube Instances

We now show a procedure to obtain a QB4OLAP implementation from a relational cube instance, starting from: (a) The conceptual schema of a cube C; (b) CS_{RDF}, the RDF representation of the schema of C; (c) the relational implementation of the cube C, which we denote CI_{ROLAP}. The procedure produces an R2RML mapping file CI_{RDF} that generates an RDF representation of the data stored in CI_{ROLAP}, using the schema CS_{RDF}. A collection of IRI-safe strings P are used to generate unique level members

IRIs with R2RML rr:template, such that each level $l_i \in L$ has its corresponding $p_i \in P$. Analogously, f is used to generate unique IRIs for fact instances. The procedure comprises two parts: (1) Define mappings to generate level members; (2) Define mappings to generate facts (observations). We define a procedure for each relational representation of a multidimensional model construct as follows. We assume the reader is familiar with the usual kinds of dimension hierarchies.

Step 1 (Balanced hierarchies.) These hierarchies can be represented as *snowflake schemas* or as *star schemas*. If $h \in H$ is a balanced hierarchy composed of a set of levels L, represented as a **snowflake schema**, there exists a set of tables $T_h \in CI_{ROLAP}$, where each table $t_i \in T_h$ represents a level in $l_i \in L$, and contains a key attribute pk_i and one attribute a_i for each level attribute $at_i \in l_i$. For each pair $l_i, l_{i+1} \in h$, represented as $t_i, t_{i+1} \in T$, such that $parentLevel(l_i, h) = l_{i+1}$, there exists a foreign key attribute $fk_i \in t_i$ referencing $pk_{i+1} \in t_{i+1}$.

To generate the instances of a *balanced hierarchy represented as a snowflake schema*, for each level $l_i \in h$, $CI_{RDF} = CI_{RDF} \cup \{t\}$, where t is an R2RML rr:TripleMap that generates the members of l_i. The components of t are: the rr:logicalTable t_i, a rr:subjectMap which is an IRI built using the rr:template $p_i\{pk_i\}$, and one or more rr:predicateObjectMap that express: (1) To which level $l_{i_{RDF}} \in CS_{RDF}$ the members generated by t belong; (2) The value of each attribute $a_{RDF} \in l_{i_{RDF}}$, which is obtained from the attributes in t_i, specified using rr:column; (3) The associated members in other levels l_j, using skos:broader and the rr:template $p_j\{fk_j\}$.

The R2RML mapping that generates the members in level Region in Fig. 3 is:

```
<#TriplesMapRegion > a rr:TriplesMap ;
rr:logicalTable [ rr:tableName "State" ] ;
rr:subjectMap [ rr:termType rr:IRI ;
    rr:template "http://www.fing.edu.uy/inco/cubes/instances/northwind/Region#{RegionCode}" ; ] ;
rr:predicateObjectMap [ rr:predicate qb4o:inLevel ; rr:object nw:region ; ] ;
rr:predicateObjectMap [ rr:predicate nw:regionCode ; rr:objectMap [ rr:column "RegionCode" ] ; ] ;
rr:predicateObjectMap [ rr:predicate nw:regionName ; rr:objectMap [ rr:column "RegionName" ] ; ] ;
rr:predicateObjectMap [ rr:predicate skos:broader;
    rr:objectMap [ rr:termType rr:IRI ;
        rr:template "http://www.fing.edu.uy/inco/cubes/instances/northwind/Country#{CountryKey}" ] ; ] .
```

If $h \in H$ is a balanced hierarchy composed of a set of levels L, represented as a **star schema**, there exists a table $t_h \in CI_{ROLAP}$ representing all levels in $l_i \in L$. For each l_i there exists an attribute $pk_i \in t_h$ which identifies each level member and for each level attribute $at_i \in l_i$ there exists an attribute $a_i \in t_h$.

The mapping for a *balanced hierarchy represented as a star schema* is similar to the one in Step 1, except that the rr:logicalTable is the same for all levels. The R2RML mapping that produces the members in levels Month and Year is:

```
<#TriplesMapMonth> a rr:TriplesMap ;
rr:logicalTable [ rr:tableName "Time" ] ;
rr:subjectMap [ rr:termType rr:IRI ; rr:template "http://www.fing.edu.uy/instances/nw/Month#{MonthName}{Year}" ; ] ;
rr:predicateObjectMap [ rr:predicate qb4o:inLevel ; rr:object nw:month; ] ;
rr:predicateObjectMap [ rr:predicate nw:monthNumber ; rr:objectMap [ rr:column "MonthNumber" ] ; ] ;
rr:predicateObjectMap [ rr:predicate nw:monthName ; rr:objectMap [ rr:column "MonthName" ] ; ] ;
rr:predicateObjectMap [ rr:predicate skos:broader;
    rr:objectMap [ rr:termType rr:IRI ; rr:template "http://www.fing.edu.uy/instances/nw/Year#{Year}" ] ;] .
<#TriplesMapYear > a rr:TriplesMap ;
rr:logicalTable [ rr:tableName "Time" ] ;
rr:subjectMap [ rr:termType rr:IRI ; rr:template "http://www.fing.edu.uy/instances/nw/Year#{Year}" ; ] ;
rr:predicateObjectMap [ rr:predicate qb4o:inLevel ; rr:object nw:year ; ] ;
rr:predicateObjectMap [ rr:predicate nw:yearNumber ; rr:objectMap [ rr:column "Year" ] ; ] .
```

Step 2 (Parent-child (Recursive) hierarchies.) These are represented as a table containing all attributes in a level, and a foreign key to the same table, relating child members to their parent. If $h \in H$ is a parent-child hierarchy, composed of a pair of levels $l_i, l_{i+1} \in h$ such that $parentLevel(l_i, h) = l_{i+1}$, there exists a table $t_h \in CI_{ROLAP}$ which contains a key attribute pk_i that identifies the members of l_i and an attribute $fk_i \in t_h$, that identifies the members of l_{i+1} and is a foreign key referencing $pk_i \in t_h$.

The mapping for level members in a parent-child hierarchy is similar to the one presented for a star representation of a balanced hierarchy, since all hierarchy levels are populated from the same `rr:logicalTable`. For the Supervision hierarchy we have:

```
<#TriplesMapEmployee> a rr:TriplesMap ;
rr:logicalTable [ rr:tableName "Employee" ] ;
rr:subjectMap [ rr:termType rr:IRI ; rr:template "http://www.fing.edu.uy/instances/nw/Employee#{EmployeeKey}" ; ];
rr:predicateObjectMap [ rr:predicate qb4o:inLevel ; rr:object nw:employee ; ] ;
rr:predicateObjectMap [ rr:predicate nw:firstName ; rr:objectMap [ rr:column "FirstName" ] ; ];
rr:predicateObjectMap [ rr:predicate nw:lastName ; rr:objectMap [ rr:column "LastName" ] ; ] ;
rr:predicateObjectMap [ rr:predicate skos:broader;
    rr:objectMap [ rr:termType rr:IRI ;
        rr:template "http://www.fing.edu.uy/instances/nw/Employee#{SupervisorKey}" ] ; ].
```

Step 3 (Nonstrict hierarchies). Here, each level is represented as a separate table and a *bridge table* is used to represent the many-to-many relationship between level members. If $h \in H$ is a nonstrict hierarchy, composed of a set of levels L, there exists a set of tables $T_h \in CI_{ROLAP}$, one table $t_i \in T_h$ with a key attribute pk_i, for each level $l_i \in L$. For each pair of levels $l_i, l_{i+1} \in h$, represented as $t_i, t_{i+1} \in T$, such that $parentLevel(l_i, h) = l_{i+1}$ and members of l_i have exactly one associated member in l_{i+1}, the mapping is the same as for the snowflake representation of balanced hierarchies. If members of l_i have more than one associated member in l_{i+1}, there exists a bridge table $b_i \in T$ that contains two attributes fk_i, fk_{i+1} referencing $pk_i \in t_i$ and $pk_{i+1} \in t_{i+1}$ respectively. Thus, each pair of levels is populated by three `rr:TriplesMap`: two of them generate level members, while the third uses the bridge table as `rr:logicalTable` to generate parent-child relationships between level members.

The R2RML mapping that generates the parent-child relationship between members in the Employees and City levels, in the Territories hierarchy is:

```
<#TriplesMapTerritories>
rr:logicalTable [ rr:tableName "Territories" ] ;
    rr:subjectMap [ rr:template "http://www.fing.edu.uy/instances/nw/Employee#{EmployeeKey}" ; ] ;
rr:predicateObjectMap [ rr:predicate skos:broader ;
    rr:objectMap [ rr:termType rr:IRI ; rr:template "http://www.fing.edu.uy/instances/nw/City#{CityKey}" ] ; ] .
```

Step 4 (Facts.) For each fact F, $CI_{RDF} = CI_{RDF} \cup \{t\}$; t is an R2RML `rr:TripleMap` that generates fact instances (observations). The components of t are as follows: one `rr:logicalTable`, one `rr:subjectMap`, which is an IRI built using the rr:template $f\{F\_KEY\}$, one `rr:predicateObjectMap` stating the dataset the observation belongs to, one `rr:predicateObjectMap` for each level related to the fact, and one `rr:predicateObjectMap` for each measure. F_KEY provides a unique value for each fact, and can be obtained from a fact table column, or concatenating the keys of all the level members that participate in the fact.

The R2RML mapping that generates the members in the Sales facts is as follows. Note the representation of the role-playing dimensions, and the key in rr:template.

```
<#TriplesMapSales> a rr:TriplesMap ;
rr:logicalTable [ rr:tableName "Sales" ] ;
rr:subjectMap [ rr:termType rr:IRI ;
    rr:template "http://www.fing.edu.uy/instances/nw/Sale#{OrderNo}_{OrderLineNo}"; rr:class qb:Observation ; ] ;
rr:predicateObjectMap [ rr:predicate qb:dataSet ; rr:object nwi:dataset1 ; ] ;
rr:predicateObjectMap [ rr:predicate nw:customer ;
    rr:objectMap [ rr:termType rr:IRI ;
        rr:template "http://www.fing.edu.uy/instances/nw/Customer#{CustomerKey}" ] ; ] ;
rr:predicateObjectMap [ rr:predicate nw:employee ;
    rr:objectMap [ rr:termType rr:IRI ;
        rr:template "http://www.fing.edu.uy/instances/nw/Employee#{EmployeeKey}" ] ; ];
rr:predicateObjectMap [ rr:predicate nw:orderDate ;
    rr:objectMap [ rr:termType rr:IRI ; rr:template "http://www.fing.edu.uy/instances/nw/Time#{OrderDateKey}" ] ; ];
rr:predicateObjectMap [ rr:predicate nw:dueDate ;
    rr:objectMap [ rr:termType rr:IRI ; rr:template "http://www.fing.edu.uy/instances/nw/Time#{DueDateKey}" ] ; ];
rr:predicateObjectMap [ rr:predicate nw:supplier ;
    rr:objectMap [ rr:termType rr:IRI ; rr:template "http://www.fing.edu.uy/instances/nw/Supplier#{SupplierKey}" ] ; ] ;
rr:predicateObjectMap [ rr:predicate nw:quantity ; rr:objectMap [ rr:column "Quantity" ] ; ] ;
rr:predicateObjectMap [ rr:predicate nw:unitPrice ; rr:objectMap [ rr:column "UnitPrice" ] ; ] ;
rr:predicateObjectMap [ rr:predicate nw:salesAmount ; rr:objectMap [ rr:column "SalesAmount" ] ; ] .
```

6   Querying a QB4OLAP Cube

The representation produced by the procedures described in Section 5 supports expressing in SPARQL commonly-used real-world OLAP queries. We next give the intuition
of this, showing two typical OLAP queries expressed in MDX, the *de facto* standard
language for OLAP, and its equivalent SPARQL query. We assume the reader is familiar with MDX. We start with the query *"Three best-selling employees"*, which reads in
MDX:

```
SELECT Measures.[Sales Amount] ON COLUMNS,
    TOPCOUNT(Employee.[Full Name].CHILDREN, 3,Measures.[Sales Amount]) ON ROWS
FROM Sales
```

The query above reads in SPARQL(LIMIT 3 keeps the first three results):

```
SELECT ?fName ?lName (SUM(?sales) AS ?totalSales)
WHERE { ?o qb:dataSet nwi:dataset1 ; nw:employee ?emp ; nw:salesAmount ?sales .
    ?emp qb4o:inLevel nw:employee ; nw:firstName ?fName ;nw:lastName ?lName . }
GROUP BY ?fName ?lName
ORDER BY DESC (?totalSales) LIMIT 3
```

Consider now: *"Total sales and average monthly sales by employee and year"*

```
WITH MEMBER Measures.[Avg Monthly Sales] AS
AVG(DESCENDANTS([Order Date].Calendar.CURRENTMEMBER,
    [Order Date].Calendar.Month),Measures.[Sales Amount]),FORMAT_STRING = '$###,##0.00'
SELECT {Measures.[Sales Amount], Measures.[Avg Monthly Sales] } ON COLUMNS,
Employee.[Full Name].CHILDREN * [Order Date].Calendar.Year.MEMBERS ON ROWS FROM Sales
```

Below we show the equivalent SPARQL query. The inner query computes the total
sales by employee and month; the outer query aggregates this result to the Year level,
and computes the total yearly sales and the average monthly sales.

```
SELECT ?fName ?lName ?yearNo (SUM(?monthlySales)) (AVG(?monthlySales) )
WHERE { {
    SELECT ?fName ?lName ?month (SUM(?sales) AS ?monthlySales)
    # Montly sales by employee
    WHERE { ?o qb:dataSet nwi:dataset1 ; nw:employee ?emp ;
        nw:orderDate ?odate ; nw:salesAmount ?sales .
        ?emp qb4o:inLevel nw:employee , nw:firstName ?fName ; nw:lastName ?lName .
        ?odate qb4o:inLevel nw:orderDate ; skos:broader ?month . ?month qb4o:inLevel nw:month . }
    GROUP BY ?fName ?lName ?month }
    ?month skos:broader ?year . ?year qb4o:inLevel nw:year ; nw:yearNumber ?yearNo . }
GROUP BY ?fName ?lName ?yearNo ORDER BY ?fName ?lName ?yearNo
```

7   Discussion and Open Challenges

In this paper we focused in studying modeling issues, and in showing that writing real-world OLAP queries in SPARQL based on an appropriate model is a plausible approach. We presented a new version of QB4OLAP that allows representing most of the concepts in the multidimensional model, and a procedure to obtain a QB4OLAP representation of such model. We also proposed a set of steps that produce a QB4OLAP representation of a relational implementation of a data cube. It would be possible to automatically generate queries like the ones presented in Section 6, starting from a high-level algebra like the one proposed in [3], and this is the approach we will follow in future work. We will also study mechanisms for obtaining QB4OLAP data cubes using other data sources, not necessarily multidimensional ones.

References

1. Abelló, A., Darmont, J., Etcheverry, L., Golfarelli, M., Mazón, J.N., Naumann, F., Pedersen, T.B., Rizzi, S., Trujillo, J., Vassiliadis, P., Vossen, G.: Fusion cubes: Towards Self-Service Business Intelligence. IJDWM 9(2), 66–88 (2013)
2. Beheshti, S.-M.-R., Benatallah, B., Motahari-Nezhad, H.R., Allahbakhsh, M.: A Framework and a Language for On-Line Analytical Processing on Graphs. In: Wang, X.S., Cruz, I., Delis, A., Huang, G. (eds.) WISE 2012. LNCS, vol. 7651, pp. 213–227. Springer, Heidelberg (2012)
3. Ciferri, C., Ciferri, R., Gómez, L., Schneider, M., Vaisman, A., Zimányi, E.: Cube Algebra: A Generic User-Centric Model and Query Language for OLAP Cubes. IJDWM 9(2), 39–65 (2013)
4. Cohen, J., Dolan, B., Dunlap, M., Hellerstein, J.M., Welton, C.: Mad skills: New Analysis Practices for Big Data. PVLDB 2(2), 1481–1492 (2009)
5. Etcheverry, L., Vaisman, A.A.: Enhancing OLAP Analysis with Web Cubes. In: Simperl, E., Cimiano, P., Polleres, A., Corcho, O., Presutti, V. (eds.) ESWC 2012. LNCS, vol. 7295, pp. 469–483. Springer, Heidelberg (2012)
6. Etcheverry, L., Vaisman, A.: QB4OLAP: A Vocabulary for OLAP Cubes on the Semantic Web. In: Proc. of COLD 2012. CEUR-WS.org, Boston (November 2012)
7. Golfarelli, M.: Open source BI platforms: A functional and architectural comparison. In: Pedersen, T.B., Mohania, M.K., Tjoa, A.M. (eds.) DaWaK 2009. LNCS, vol. 5691, pp. 287–297. Springer, Heidelberg (2009)
8. Heath, T., Bizer, C.: Linked Data: Evolving the Web into a Global Data Space. Morgan & Claypool Publishers (2011)
9. Kämpgen, B., Harth, A.: Transforming statistical linked data for use in OLAP systems. In: Proceedings of the 7th International Conference on Semantic Systems, I-Semantics 2011, pp. 33–40. ACM, New York (2011)
10. Kämpgen, B., O'Riain, S., Harth, A.: Interacting with Statistical Linked Data via OLAP Operations. In: ESWC Workshops, Heraklion, Crete, Greece (May 2012)
11. Kämpgen, B., Harth, A.: No size fits all – running the star schema benchmark with SPARQL and RDF aggregate views. In: Cimiano, P., Corcho, O., Presutti, V., Hollink, L., Rudolph, S. (eds.) ESWC 2013. LNCS, vol. 7882, pp. 290–304. Springer, Heidelberg (2013)
12. Löser, A., Hueske, F., Markl, V.: Situational Business Intelligence. In: Castellanos, M., Dayal, U., Sellis, T. (eds.) BIRTE 2008. LNBIP, vol. 27, pp. 1–11. Springer, Heidelberg (2009)
13. Malinowski, E., Zimányi, E.: Advanced Data Warehouse Design: From Conventional to Spatial and Temporal Applications. Springer (2008)
14. Nebot, V., Llavori, R.B.: Building data warehouses with semantic web data. Decision Support Systems 52(4), 853–868 (2011)
15. Vaisman, A., Zimányi, E.: Data Warehouse Systems: Design and Implementation. Springer (2014)

Extending Drill-Down through Semantic Reasoning on Indicator Formulas

Claudia Diamantini, Domenico Potena, and Emanuele Storti

Dipartimento di Ingegneria dell'Informazione,
Università Politecnica delle Marche,
via Brecce Bianche, 60131 Ancona, Italy
{c.diamantini,d.potena,e.storti}@univpm.it

Abstract. Performance indicators are calculated by composition of more basic pieces of information, and/or aggregated along a number of different dimensions. The multidimensional model is not able to take into account the compound nature of an indicator. In this work, we propose a semantic multidimensional model in which indicators are formally described together with the mathematical formulas needed for their computation. By exploiting the formal representation of formulas an extended drill-down operator is defined, which is capable to expand an indicator into its components, enabling a novel mode of data exploration. Effectiveness and efficiency are briefly discussed on a prototype introduced as a proof-of concept.

1 Introduction

Performance measurement is the subject of extensive interdisciplinary research on information systems, organizational modeling and operation, decision support systems and computer science. Much work is devoted to categorize reference performance measures, or indicators [1, 2]. Strategic support information systems exploit results on data warehouse architectures. The multidimensional model has been introduced to suitably represent Performance Indicators (PI) and to enable flexible analyses by means of OLAP operators, facilitating managers in visualization, communication and reporting of PIs. Nevertheless, design and management of PIs are still hard. Among the major obstacles: (1) differences between the business view and the technical view of PIs, (2) information overload syndrome: managers are inclined to ask for more indicators than those actually needed [3], (3) interpretation of the meaning of indicators and their values. To some extent these obstacles relate to the fact that indicators are complex data with an aggregate and/or compound nature, as their values are calculated by applying some *formulas* defined over other indicators, or by aggregating raw data values, or both. Unawareness of the dependencies among indicators leads people to treat indicators as independent pieces of information, and is a cause of the information overload syndrome. Similarly, many disputes during managerial meetings come from a lack of a common understanding of indicators. To give an example, somebody states that the amount of investment in higher education

L. Bellatreche and M.K. Mohania (Eds.): DaWaK 2014, LNCS 8646, pp. 57–68, 2014.
© Springer International Publishing Switzerland 2014

in Italy is too low (far below the EU average), while somebody else states it is too high and should be lowered. Digging into this apparent contradiction, one discovers that both evaluate the amount of investment as the ratio between "total expenditure" and "student population", but whereas the former defines the "student population" as people officially enrolled in a course, the latter subtracts students who do not actually take exams. In order to fully grasp the meaning of the indicator an analysis of the way it is calculated is necessary. Furthermore, a correct interpretation of the value and trend of the indicator is unachievable without analysing its components.

Although the complex nature of indicators is well-known, it is not fully captured in existing models. The multidimensional model takes into account the aggregative aspect, defining a data cube as a multi-level, multidimensional database with aggregate data at multiple granularities [4]. The definition of powerful OLAP operators like drill-down directly comes from this model. Semantic representations of the multidimensional model have been recently proposed [4–8] mainly with the aim to reduce the gap between the high-level business view of indicators and the technical view of data cubes, to simplify and to automatize the main steps in design and analysis.

The compound nature of indicators is far less explored. Proposals in [2, 9–14] include in the representations of indicators' properties some notion of formula in order to support the automatic generation of customized data marts, the calculation of indicators [9, 10, 13], or the interoperability of heterogeneous and autonomous data warehouses [12]. In the above proposals, formula representation does not rely on logic-based languages, hence reasoning is limited to formula evaluation by ad-hoc modules. No inference mechanism and formula manipulation is enabled. Formal, logic-based representations of dependencies among indicators are proposed in [2,14]. These properties are represented by logical predicates (e.g. *isCalculated* [14], *correlated* [2]) and reasoning allows to infer implicit dependencies among indicators useful for organization modeling, design, as well as reuse, exchange and alignment of business knowledge. An ontological representation of indicator's formulas is proposed in [11] in order to exchange business calculation definitions and to infer their availability on a given data mart through semantic reasoning, with strong similarities with our previous work [15].

In the present paper we propose to extend the data cube model with the description of the structure of an indicator given in terms of its algebraic formula. Relations between the aggregation function and the formula are taken into account. As the traditional multidimensional model leads to the definition of the drill-down operator, so the novel structure enables the definition of a novel operator we call *indicator drill-down*. Like the usual drill-down, this indicator increases the detail of a measure of the data cube, but instead of disaggregating along the levels of dimensions, it expands an indicator into its components. The two notions of drill-down are integrated thus allowing a novel two-dimensional way of exploring data cubes. To the best of our knowledge this is the first time that such an extended drill-down is considered. As a further contribution, we introduce a first-order logic theory for the representation of indicators' formulas and

Table 1. An excerpt of the enterprise glossary

Indicator	Description	Aggr	Formula
AvgCostsExtIdeas	Average costs of ideas produced by external users through the crowdsourcing platform	n/a	$\dfrac{CrowdInv}{NumExternalIdeas}$
CrowdInv	Total Investments for crowdsourcing activities: management of the platform, promoting and facilitating the participation,...	SUM	
IdeasAcceptedEng	The number of ideas accepted for engineering. These ideas are intended for future production	SUM	
IdeasProposed	Number of (internal/external) ideas proposed	SUM	$NumInternalIdeas+$ $NumExternalIdeas$
IdeaYield	Ratio of proposed ideas to ideas that have entered the engineering phase and will be developed	n/a	$\dfrac{IdeasAcceptedEng}{IdeasProposed}$
NumExternalIdeas	Number of ideas generated from the external stakeholders	SUM	
NumInternalIdeas	Number of ideas generated from enterprise's employees	SUM	

manipulation based on equation resolution. The set of predicates defines a knowledge base for reference, domain-independent indicators specification. Other predicates are introduced to define members' roll-up along dimensional hierarchies, as well as to state the set of indicators actually implemented in a given data mart, thus providing the specification for a certain application domain. The theory enables formula manipulation thus providing the full drill-down functionalities, allowing to expand an indicator into its components even if some are not explicitly stored in the data mart. Besides enabling the definition of a novel operator, the proposed logic representation extends the state of the art in the following ways: (1) we are not limited to the reference indicator specification, since equivalent definitions can be inferred, hence the evaluation of the formula can follow several paths, (2) indicators that are not explicitly stored in the data mart can be calculated by exploiting their relationships with other indicators represented in the knowledge base, (3) relevant relationships among indicators, like (inverse) dependency, (inverse) correlation, causality, influence can be inferred, while they have to be explicitly introduced with other approaches [2,13,14].

The rest of the paper is organized as follows: Section 2 introduces the case study that will be used as an illustrative example through the paper. Section 3 presents the proposed model, then Section 4 discusses its application to the definition of the extended drill-down operator. In Section 5 the proposal is evaluated. Finally Section 6 draws some conclusion and discusses future work.

2   Case Study

The present work is conceived within the EU Project BIVEE[1]. In this Section we present a case study that is based on the data mart (DM) used by one of the end-users of the project, and will be used as an illustrative example through the paper. In particular, we refer to an enterprise that develops innovative solutions,

---

[1] http://bivee.eu

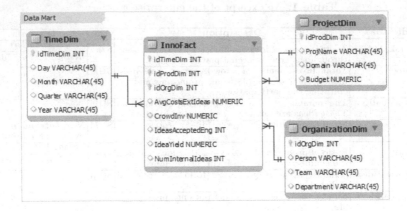

Fig. 1. The Innovation data mart

from metrology to robotics domains, to satisfy specific requests of customers. The enterprise adopts the open innovation paradigm [16], where innovation extends beyond the boundaries of the enterprise, involving both partners and customers as sources of innovative ideas. To this end, the enterprise hosts and manages a crowdsourcing platform. Finally, in the enterprise the work is organized on a project basis. Figure 1 shows the data mart used by the enterprise to analyse innovation projects. The data mart provides 5 indicators (i.e. measures in data warehouse terminology) and 3 dimensions, which represent different perspectives of analysis: time, organization and project. The attribute *budget* in the project dimension is not a level, but an informative attribute of a project. The description of measures is given in Table 1. In particular note that together with a textual description, the mathematical formula to calculate the indicator is provided for some indicators. When the formula does not exist, the indicator is said *atomic*, and is independent on other indicators. Otherwise the indicator is said *compound*, and the operands occurring in its formula are called *dependencies*. Note that compound indicators can be defined in terms of other compound indicators, producing a tree of dependencies, although it cannot be fully grasped in an informal glossary like the one presented. Also, aggregation functions are specified when applicable. When an aggregation function exists, it means that it is applied to the result of the formula in order to aggregate over dimensions. According to the case study and for the sake of simplicity, hereafter it is assumed that the same aggregation function applies to each dimension.

3   Semantic Multidimensional Model

The semantic multidimensional model is based on a first-order logic representation of indicators' formulas and their properties as well as of a multidimensional structure. Hence we first introduce the basic notions related to the multidimensional model.

Definition 1. *(Cube schema)* A cube schema S is a pair $\langle \mathcal{D}, I \rangle$, where I is a set of indicators $\{ind_1, \ldots, ind_m\}$ and $\mathcal{D}$ is the set of dimensions for I.

A dimension $D_i \in \mathcal{D}$ is the hierarchy of levels $L_1^{D_i} \preceq \ldots \preceq L_l^{D_i}$ along which measures are aggregated. The partial order $\preceq$ is such that if $L_1^D \preceq L_2^D$ then L_1^D rolls up to L_2^D (and L_2^D drills down to L_1^D). The domain of a level $L_j^{D_i}$ is a set of members $\{m_{j_1}, \ldots, m_{j_n}\}$ and will be denoted by $\alpha(L_j^{D_i})$, e.g. $\alpha(\text{Department}) = \{\text{RnD, RforI, ElectricDept, MechanicDept}, \ldots\}$. Referring to the Organization dimension of the case study, the following hierarchy holds: Person $\preceq$ Team $\preceq$ Department. In general the definition allows also for multiple hierarchies.

Definition 2. *(Cube element)* A cube element ce for the cube schema $S = \langle \mathcal{D}, I \rangle$ is the tuple $\langle m_1, \ldots, m_n, v_1, \ldots, v_m \rangle$ where each member belongs to a level of a dimension in $\mathcal{D}$ (i.e., $\forall m_i \; \exists L_j^{D_i}$ such that $m_i \in \alpha(L_j^{D_i})$), and $\{v_1, \ldots, v_m\}$ are values for I.

In the following, we will not assume *completeness* of data cubes, i.e. cubes that include a cube element for any possible combination of members.

Central to our model is the notion of indicator: while in the standard cube schema definition (1) indicators are just labels, in the proposed model the structure of an indicator is taken into account.

Definition 3. *(Indicator)* An indicator $ind \in I$ is defined by the pair $\langle aggr, f \rangle$, where:

- $aggr \in \{\text{SUM, MIN, MAX, AVG, VAR, COUNT, NONE}, \ldots\}$ is an aggregation function that represents how to group values of the indicator;
- $f(ind_1, \ldots, ind_n)$ is the formula of the indicator, i.e. a mathematical expression that describes the way ind is computed in terms of other indicators $(ind_1, \ldots, ind_n) \in I$.

The label NONE is used to denote the absence of aggregation function, e.g. $IdeaYield = \langle NONE, IdeasAcceptedEng/IdeasProposed \rangle$. According to widely accepted models (e.g. [17]), aggregation is categorized in distributive, algebraic or holistic. Indicators with a distributive aggregator can be directly computed on the basis of values at the next lower level of granularity (e.g., SUM, MIN, MAX). Algebraic aggregator cannot be computed by means of values at next lower level unless a set of other indicators are also provided, which transform the algebraic indicator in a distributive one; a classical example is given by the average aggregator (AVG). Indicators described by holistic functions can never be computed using values at next lower level (e.g., MEDIAN and MODE).

A formula is said to be *additive* if it includes only summation and differences of indicators, e.g. $f(ind_x, ind_y) = ind_x + ind_y$. Additivity is a relevant property since indicators with additive formulas and distributive aggregation functions (e.g., SUM) define a special subclass of indicators, for which holds that:

$$aggr(f(ind_1, \ldots, ind_n)) = f(aggr(ind_1), \ldots, aggr(ind_n)).$$ This is not true in the general case, e.g. $AVG(x/y) \neq AVG(x)/AVG(y)$.

Indicators and their properties are represented by first-order logic predicates. We refer to Horn Logic Programming, and specifically to Prolog, as the representation language. $\text{formula}(ind, f)$ is a fact representing the formula related to an indicator. In the predicate ind is an indicator label, while f is an expression including algebraic operators like sum, difference, product, power, and operands are indicators' labels. Additivity is expressed by the predicate $\text{isAdditive}(ind)$. The set of formulas and its properties define a reference knowledge base for indicators specification. A further predicate $\text{hasInd}(c, ind)$ allows to state the presence of the indicator ind in the schema of the cube c. For what concerns the multidimensional structure, similarly to [18], the predicate $\text{rollup}(X, Y)$ is introduced to assert that a member X is mapped to the member Y of the next higher level to perform the roll-up operation.

Given the set of facts, rules are devised to implement reasoning functionalities. For instance the following rule implements the transitive closure of roll-up:

$$\text{partOf}(X, Y) : - \text{rollup}(X, Y).$$
$$\text{partOf}(X, Y) : - \text{rollup}(X, Z), \text{partOf}(Z, Y).$$

While formulas are represented as facts, manipulation of mathematical expressions is performed by specific predicates from PRESS (PRolog Equation Solving System), which is a formalization of algebra in Logic Programming for solving equations. Such predicates implement axioms of a first-order mathematical theory, and can manipulate an equation to achieve a specific syntactic effect (e.g., to reduce the number of occurrence of a given variable in an equation) through a set of rewriting rules. PRESS works in the domain of R-elementary equations, that is on equations involving polynomials, and exponential, logarithmic, trigonometric, hyperbolic, inverse trigonometric and inverse hyperbolic functions over the real numbers, although all indicators found in the analysis of real-world scenarios until now have linear formulas. PRESS is demonstrated to always find a solution for linear equations.

Due to lack of space, we cannot go into the details of the rule system. We just enlighten that the use of PRESS here allows to derive a new formula for an indicator that is not explicitly given. Among all possible inferred formulas, we are able to individuate the subset that can be actually calculated on the given data cube by the predicate $\text{hasInd}(c, \_)$. In the following we will refer to this high-level reasoning functionality as $\mathcal{F}(ind, C)$. Referring to the cube DM of the case study, the following formulas are derived by PRESS:

$$IdeasProposed = NumInternalIdeas + NumExternalIdeas;$$
$$IdeasProposed = NumInternalIdeas + \frac{CrowdInv}{AvgCostsExtIdeas};$$
$$IdeasProposed = \frac{IdeasAcceptedEng}{IdeaYield}.$$

$\mathcal{F}(IdeasProposed, DM)$ returns only the last two since, although the first one is the definition of the indicator provided in the knowledge base, it cannot be calculated in this way on DM due to the lack of $NumExternalIdeas$.

The formal representation and manipulation of the structure of a formula enables advanced functionalities like the definition of an extended drill-down operator, described in the next Section.

4   Extended Drill-Down

The present Section discusses how to exploit reasoning capabilities over indicator's formula by introducing a novel *indicator* drill-down operator. Like the usual drill-down, it increases the detail of a measure of the data cube, but instead of disaggregating along the levels of dimensions, it expands an indicator into its components. As the traditional drill-down is enabled by the notion of dimension's hierarchies, so the indicator drill-down arises from introducing the indicator's structure in the model. Furthermore, by reasoning on the logic representation proposed, the operator is able to extract values even for indicators not explicitly stored in the cube. We hasten to note that the rules defined to this end must work jointly on the structure of an indicator and on the structure of its dimensions. This integration of the notion of indicator in the multidimensional model is what enables the definition of the extended drill-down as the composition of the classic drill-down and of the indicator drill-down defined as follows.

Definition 4. *(Indicator drill-down)*
Given a schema $S=\langle \mathcal{D}, \{ind_1,\ldots,ind_i,\ldots,ind_n\}\rangle$ and an indicator ind_i with a formula $f_{ind_i} = f(ind_{i_1},\ldots,ind_{i_k})$, the indicator drill-down on ind_i is a function that maps a cube with schema S in a cube with schema S' such that:

- $S' = \langle \mathcal{D}, I'\rangle, I' = (I \setminus \{ind_i\}) \cup \{ind_{i_1},\ldots,ind_{i_k}\}$;
- instances are cube elements $ce = \langle m_1,\ldots,m_h, v_1,\ldots,v_{i_1},\ldots,v_{i_k},\ldots,v_n\rangle$, such that v_j is the value of the j-th indicator, m_p is the member of Dimension D_p and $v_i = f(v_{i_1},\ldots,v_{i_k})$.

Operationally, this means to access the definition of ind_i, extract the dependencies from its formula, and extract values for dependencies in order to build the new cube. This can be expressed as the rewriting of the multidimensional query generating the cube S.

Definition 5. *(Multi-dimensional Query)* A multi-dimensional query MDQ on a cube C with schema $\langle \mathcal{D}, I\rangle$ is a tuple $\langle \delta, \{ind_1,\ldots,ind_m\}, W, K, \sigma\rangle$, where:

- δ is a boolean value introduced here to make explicit how the query is performed. If $\delta = false$ then indicator values are materialized in the cube, otherwise they are virtual, hence we assume they are calculated by aggregation of values at the lowest levels of dimensional hierarchies;
- $\{ind_1,\ldots,ind_m\}$ is the set of requested indicators;
- W is the set of levels $\{L^{D_1}, ..., L^{D_n}\}$ on which to aggregate, such that $L^{D_i} \in D_i$ and $\{D_1,\ldots,D_n\} \subseteq \mathcal{D}$;
- K is the collection of sets $K_h = \{m_{h_1}, ..., m_{h_k}\}$, $h := 1, ..., n$, of members on which to filter, such that each m_{h_j} belongs to $\alpha(L^{D_h})$. K_h can be an empty set. In this case all members of the corresponding level are considered;
- σ is an optional boolean condition on indicators' values.

$\{ind_1, \ldots, ind_m\}$ are the elements of the target list, W is the desired roll-up level (or group-by components) for each dimension, while K allows slice and dice (suitable selections of the cube portion). While K works on members, the filter σ defines a condition on other elements of the DM: both descriptive attributes of dimensional schema (e.g. $Budget > 50K$) and values of indicators (e.g. $NumberInternalIdeas < NumberExternalIdeas$).

The result of a MDQ is a subset of the original cube where cube elements are $ce = \langle m_1, \ldots, m_n, v_1, \ldots, v_m \rangle$, where $\langle m_1, \ldots, m_n \rangle \in K_1 \times \ldots \times K_n$ and v_i is a value of the indicator ind_i. Given the notion of query, the drill-down can be seen as a rewriting of the original query $MDQ = \langle \delta, I, W, K, \sigma \rangle$ as $MDQ' = \langle \delta, I', W, K, \sigma \rangle$. Rewriting an indicator as its direct dependencies produces a correct query only if the data cube has been designed to store the set of indicators $\{ind_{i_1}, \ldots, ind_{i_k}\}$. The rewriting rules allowing to correctly specify the indicator drill-down are discussed in the following. They depend on the typology of formula and aggregation function. For the sake of simplicity we consider multidimensional queries with only one indicator in the target list.

Indicator Drill-Down Rule. Let $MDQ = \langle \delta, \{ind\}, W, K, \sigma \rangle$ be a query over the cube C with schema $\langle D, I \rangle$, where $ind = \langle aggr, f(ind_1, \ldots, ind_k) \rangle$. The indicator drill-down of MDQ is $MDQ' = \langle \delta, \{ind_1, \ldots, ind_k\}, W, K, \sigma \rangle$ where $\forall ind_i$ either $ind_i \in I$ or one of the following equivalence rules applies.

Equivalence Rules. Let $MDQ = \langle false, \{ind\}, W, K, \sigma \rangle$ be a query over the cube C with schema $\langle D, I \rangle$, where $ind = \langle aggr, f \rangle$

- if ($aggr$ is distributive AND f is additive) OR ($aggr = NONE$):
  $MDQ = \langle false, g, W, K, \sigma \rangle$, $g \in \mathcal{F}(ind, C)$
- else: $MDQ = \langle true, g, W, K, \sigma \rangle$, $g \in \mathcal{F}(ind, C)$

The equivalence rules make use of the inference mechanism represented by $\mathcal{F}(ind, C)$, which defines any formula equivalent to f that can be inferred and is computed by indicators of the cube. The rule described in the else case captures the fact that for general aggregation functions the correct value of an indicator at a given level can be only obtained by calculating the formula at the lowest level of granularity (given that $\delta = true$), and then applying the aggregation on the resulting values. The first rule accounts for the commutativity property stated in the previous section that allows to apply the formula directly on the requested aggregation levels. This rule can be easily extended to algebraic aggregators given the well-known relation with distributive aggregators. For instance, in the case of the AVG the query becomes $MDQ = \langle \delta, \{\frac{ind'}{CountM}\}, W, K, \sigma \rangle$, where $ind' = \langle SUM, f \rangle$ and $CountM$ is a special function which returns the number of members m_0 of the lowest level such that $\text{partOf}(m_0, m), m \in K$.

5   Evaluation

A prototype of the system has been implemented as a proof-of-concept. This is part of a system offering additional services developed within the BIVEE

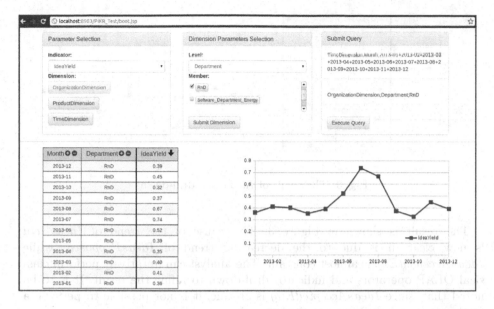

Fig. 2. Interface of the system for extended drill-down

Project [19, 20]. As logic programming system we refer to XSB[2], which extends conventional Prolog systems with an operational semantics based on tabling, i.e., a mechanism for storing intermediate results and avoiding to prove sub-goals more than once. XSB also provides interfaces to Java, through which service interfaces are written and calls to Prolog rules are managed, and MDQs are finally translated in SQL. We refer to MySQL to store all the cube elements that are in the enterprise's data mart, without adding pre-aggregations.

Figure 2 shows the interface with the query specification form and the visualization of results. The query is aimed to analyse the monthly trend of $IdeaYield$ in 2013 for the RnD Department: $\langle false, \{IdeaYield\}, \{Month, Department\}, \{\{2013-01, \ldots, 2013-12\}, \{RnD\}\}, \{\}\rangle$. The result is shown both as a table and as a chart. Symbols near to the labels of levels enable classical drill-down/roll-up operators, while the arrow near $IdeaYield$ enables the indicator drill-down. The chart enlightens a peak in July 2013. In order to understand the reason for such a variability, the analyst performs an indicator drill-down on $IdeaYield$. The indicator has been chosen since it allows to demonstrate both rewriting rules and formula inference.

The operator rewrites the query by replacing $IdeaYield$ with its dependencies as in the knowledge base (see Table 1), namely $IdeasAcceptedEng$ and $IdeasProposed$. Since the latter is not in the DM, the equivalence rule is used and the query becomes: $\langle false, \{IdeasAcceptedEng, (NumInternalIdeas + \frac{CrowdInv}{AvgCostsExtIdeas})\}, \{Month, Department\}, \{\{2013-01, \ldots\}, \{RnD\}\}, \{\}\rangle$.

---

[2] http://xsb.sourceforge.net/

Month ◐◑	Department ◐◑	IdeasAcceptedEng	IdeasProposed ▼
2013-12	RnD	16	41
2013-11	RnD	17	38
2013-10	RnD	12	37
2013-09	RnD	13	35
2013-08	RnD	14	21
2013-07	RnD	14	19
2013-06	RnD	13	25
2013-05	RnD	14	36
2013-04	RnD	13	37
2013-03	RnD	16	40
2013-02	RnD	18	44
2013-01	RnD	14	39

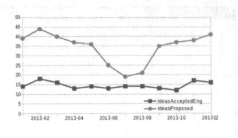

Fig. 3. The result of *IdeaYield* drill-down

The result is shown in Figure 3. The cause of the trend of *IdeaYield* is now clear: it is due to the decreasing trend of *IdeasProposed*, while *IdeasAcceptedEng* is almost constant. The analyst can iteratively perform classical OLAP operators and indicator drill-down to refine the result. It is to be noted that, since *IdeasAcceptedEng* is atomic, it is not possible to perform a further indicator drill-down on it; hence the arrow icon is not shown.

In order to evaluate the cost of the novel operator, we observe that the main steps required to perform an indicator drill-down on *ind* are: (1a) searching the formula of *ind* in the knowledge base, or (1b) inferring any valid formula for *ind* (i.e. $\mathcal{F}(ind, C)$) and (2) executing the query over the data mart, until a rewriting succeeds in data retrieval. Steps (1a) and (1b) concern the rewriting of the query, and their costs depend on the number of indicators in the knowledge base, and on the structure of their formulas. The cost of step (2) depends on the cardinality of data, their schema and the adopted management system. It is noteworthy that these parameters do not affect the cost of other steps. Here, we discuss the costs due to query rewriting, which is the cost added by the proposed operator to the classical execution of a query.

Since step (1a) has negligible cost compared to (1b), the complexity of the query rewriting is comparable to that of inferring $\mathcal{F}(ind, C)$ that, in the worst case, corresponds to all possible rewritings of the original formula by traversing all the dependencies' paths. In order to provide an evaluation of these costs we give the average execution time of $\mathcal{F}(ind, C)$ over each *ind* in the knowledge base, for a real and a synthetic scenario[3].

In the real-world scenario, the knowledge base representing the business domains of the BIVEE project is characterized by 356 indicators. The dependency tree has on average 2.67 operands per indicator, height 5 for the root node (i.e. the number of layers of indicators in the tree) and average height 3.14. In this situation, the average execution time of $\mathcal{F}(ind, C)$ is 219ms. We recall that 100ms is about the limit for making the user feel that the system is reacting instantaneously.

---

[3] Experiments have been carried on an Intel Xeon CPU 3.60GHz with 3.50GB memory, running Windows Server 2003 SP 2.

A synthetic knowledge base has been generated to perform more extensive tests. In particular, we have generated 10 different random trees with height 5, where each indicator is calculated on the basis of 4 random indicators of the lower layer; the number of operands in the formula is fixed. This kind of tree has 1365 different indicators. The average execution time of $\mathcal{F}(ind, C)$ mediated over the 10 trees is 431ms, with 449ms as maximum value. We believe these execution times are perfectly in line with the notion of On Line Analytical Processing, also in the view of possible optimizations of the system.

6 Conclusions

The paper proposed an extension of the data cube model to take into account the structure of an indicator given in terms of a formula. The extension allows to introduce a novel drill-down operator able to increase the detail of a measure of the data cube along the tree of indicators dependencies. Relations between the aggregation function and the formula are taken into consideration, so that the novel and the classic drill-down can be integrated. The logic representation adopted is a powerful way to reason over the tree of indicators' dependencies to calculate measures not explicitly provided in a data mart through formula rewriting. This approach can extend existing DB management systems, as queries can be rewritten either in SQL and executed on relational database, like the prototype shown, or in MDX queries on traditional OLAP systems. The evaluation of a prototype on real and synthetic scenarios enlightens the effectiveness and efficiency of the approach. For the sake of simplicity, the presentation assumed that the data mart schema adopts the knowledge base terminology to define measures, but the model can be simply extended with mapping predicates to relax this assumption. Although only indicators with the analytic expressions managed by PRESS can be represented, this does not limit the model applicability since other kinds of indicators (e.g. qualitative indicators) can be introduced as atomic. We plan to study extensions of the theory towards more complex expressions manipulation. Other extensions regard the representation of relational algebra expressions as indicator's formulas, and of different aggregation functions for different indicator's dimensions.

Acknowledgments. This work has been partly funded by the European Commission through the FoF-ICT Project BIVEE (No. 285746). The authors wish to thank project partners for providing data useful for the case study, and Haotian Zhang for the implementation of the interface.

References

1. Kaplan, R.S., Norton, D.P.: The Balanced Scorecard: Measures that Drive Performance. Harvard Business Review 70, 71–79 (1992)
2. Popova, V., Sharpanskykh, A.: Modeling organizational performance indicators. Information Systems 35, 505–527 (2010)

3. Ackoff, R.L.: Management misinformation systems. Management Science 14 (1967)
4. Lakshmanan, L.V.S., Pei, J., Zhao, Y.: Efficacious data cube exploration by semantic summarization and compression. In: VLDB, pp. 1125–1128 (2003)
5. Neumayr, B., Anderlik, S., Schrefl, M.: Towards Ontology-based OLAP: Datalog-based Reasoning over Multidimensional Ontologies. In: Proc. of the Fifteenth International Workshop on Data Warehousing and OLAP, pp. 41–48 (2012)
6. Niemi, T., Toivonen, S., Niinimäki, M., Nummenmaa, J.: Ontologies with semantic web/grid in data integration for olap. Int. J. Sem. Web Inf. Syst. 3, 25–49 (2007)
7. Huang, S.M., Chou, T.H., Seng, J.L.: Data warehouse enhancement: A semantic cube model approach. Information Sciences 177, 2238–2254 (2007)
8. Priebe, T., Pernul, G.: Ontology-Based Integration of OLAP and Information Retrieval. In: Proc. of DEXA Workshops, pp. 610–614 (2003)
9. Pedrinaci, C., Domingue, J.: Ontology-based metrics computation for business process analysis. In: Proc. of the 4th International Workshop on Semantic Business Process Management, pp. 43–50 (2009)
10. Xie, G., Yang, Y., Liu, S., Qiu, Z., Pan, Y., Zhou, X.: EIAW: Towards a Business-Friendly Data Warehouse Using Semantic Web Technologies. In: Aberer, K., et al. (eds.) ISWC/ASWC 2007. LNCS, vol. 4825, pp. 857–870. Springer, Heidelberg (2007)
11. Kehlenbeck, M., Breitner, M.H.: Ontology-based exchange and immediate application of business calculation definitions for online analytical processing. In: Pedersen, T.B., Mohania, M.K., Tjoa, A.M. (eds.) DaWaK 2009. LNCS, vol. 5691, pp. 298–311. Springer, Heidelberg (2009)
12. Golfarelli, M., Mandreoli, F., Penzo, W., Rizzi, S., Turricchia, E.: OLAP Query Reformulation in Peer-to-peer Data Warehousing. Inf. Sys. 37, 393–411 (2012)
13. Horkoff, J., Barone, D., Jiang, L., Yu, E., Amyot, D., Borgida, A., Mylopoulos, J.: Strategic business modeling: representation and reasoning. Software & Systems Modeling (2012)
14. del-Río-Ortega, A., Resinas, M., Ruiz-Cortés, A.: Defining process performance indicators: An ontological approach. In: Meersman, R., Dillon, T.S., Herrero, P. (eds.) OTM 2010. LNCS, vol. 6426, pp. 555–572. Springer, Heidelberg (2010)
15. Diamantini, C., Potena, D.: Semantic enrichment of strategic datacubes. In: Proc. of the ACM 11th International Workshop on Data Warehousing and OLAP, DOLAP 2008, pp. 81–88 (2008)
16. Chesbrough, H.: Open Innovation: The New Imperative for Creating and Profiting from Technology. Harvard Business Press, Boston (2003)
17. Gray, J., Chaudhuri, S., Bosworth, A., Layman, A., Reichart, D., Venkatrao, M., Pellow, F., Pirahesh, H.: Data cube: A relational aggregation operator generalizing group-by, cross-tab, and sub-totals. Data Min. Knowl. Discov. 1, 29–53 (1997)
18. Neumayr, B., Schrefl, M.: Multi-level conceptual modeling and OWL. In: Heuser, C.A., Pernul, G. (eds.) ER 2009. LNCS, vol. 5833, pp. 189–199. Springer, Heidelberg (2009)
19. Diamantini, C., Potena, D., Storti, E.: A logic-based formalization of KPIs for virtual enterprises. In: Franch, X., Soffer, P. (eds.) CAiSE Workshops 2013. LNBIP, vol. 148, pp. 274–285. Springer, Heidelberg (2013)
20. Diamantini, C., Potena, D., Proietti, M., Smith, F., Storti, E., Taglino, F.: A semantic framework for knowledge management in virtual innovation factories. International Journal of Information System Modeling and Design 4, 70–92 (2013)

An OLAP-Based Approach to Modeling and Querying Granular Temporal Trends

Alberto Sabaini[1], Esteban Zimányi[2], and Carlo Combi[1]

[1] Department of Computer Science, University of Verona, Italy
{alberto.sabaini,carlo.combi}@univr.it
[2] Department of Computer and Decision Engineering,
Université Libre de Bruxelles, Belgium
ezimanyi@ulb.ac.be

Abstract. Data warehouses contain valuable information for decision-making purposes, they can be queried and visualised with Online Analytical Processing (OLAP) tools. They contain time-related information and thus representing and reasoning on temporal data is important both to guarantee the efficacy and the quality of decision-making processes, and to detect any emergency situation as soon as possible. Several proposals deal with temporal data models and query languages for data warehouses, allowing one to use different time granularities both when storing and when querying data. In this paper we focus on two aspects pertaining to temporal data in data warehouses, namely, *temporal patterns* and *temporal granularities*. We first motivate the need for discovering granular trends in an OLAP context. Then, we propose a model for analyzing granular temporal trends in time series by taking advantage of the hierarchical structure of the time dimension.

1 Introduction

Data warehouses contain valuable information for decision-making purposes. They can be queried and visualised with Online Analytical Processing (OLAP) tools. Data warehouses are different from usual databases, as they contain aggregated data described by a multidimensional data model. They contain time-related information and thus representing and reasoning on temporal data is important both to guarantee the efficacy and the quality of decision-making processes, and to detect any emergency situation as soon as possible. Several proposals deal with temporal data models and query languages for data warehouses, allowing one to use different time granularities when storing and querying data [2,5,7].

In this paper, we focus on two general aspects of these data, namely, temporal patterns and temporal granularites. Temporal patterns provide a significant source of information for decision-makers. They represent specific sequences of data values relevant to the considered domain. Typically, temporal patterns are made of some basic *temporal trends* (i.e., increase and decrease). As an example, negative trends of articles sold could induce a manager to promote or to lower

L. Bellatreche and M.K. Mohania (Eds.): DaWaK 2014, LNCS 8646, pp. 69–77, 2014.

their prices. On the other hand, temporal granularities are used for representing time in the modelled reality. They can be used for describing time units at a finer or coarser level of detail. Applications have to deal with various granularities even for the same domain, and they need to be carefully considered both when representing, storing, and querying temporal data. As an example, purchases may be registered in minutes or hours, while suppliers' deliveries may be scheduled in weeks or months.

In the following, we address the problem of finding *granular temporal trends* in data warehouses by proposing two new OLAP operators, namely, *Trend* and *TrendAggregator*. The first operator exploits the hierarchical structure of dimensions in order to find trends of possibly aggregated data at some granularity level. It allows users to use standard OLAP operations for retrieving time series at the desired level of detail, as a mean to find trends that yield decision-making, e.g., "Display for each product, the positive trends in days for each quarter of 2013." The second operator allows users to further investigate the discovered trends, represented by means of a multidimensional model. This operator aggregates trends on the dimension hierarchies, e.g., "Display for each city, the longest trends between days of 2013."

This paper is organized as follows. In Sect. 2 we describe the notion of *granular temporal trends* introduced in [1]. Then, we briefly discuss some main contributions from the literature in the area of trend analysis on time series. In Sect. 3 we describe the novel trend operators for time series in OLAP and the aggregation of the discovered trends. Lastly, in Sect. 4 we draw some conclusions.

2   Related Work

In this section we start by discussing a logic-based taxonomy for the description of granular temporal trends, presented in [1], on which our work is based. The authors presented a graphical representation of the dimensions related to granular trends, called *Trend Dimension Tree*. It systematically describes the temporal features related to a granular trend. The main feature we are going to consider is the *granular type* one: it allows one to distinguish two kinds of trends, called *intragranule* and *intergranule*, respectively. In intragranular trends, temporal properties expressed by means of a trend must be satisfied inside a given granule; in intergranular trends, properties must be satisfied over different granules.

Now we briefly discuss some contributions from the literature regarding time series and temporal trends. In [8], the author introduced trend dependencies, which allow one to express significant temporal trends, e.g., *Salaries of employees should never decrease across months*. The time dimension is captured by trend dependencies through the concept of time accessibility relation, which can express also time granularities in a simple and elegant way. Trend dependencies can compare attribute values by some comparison operator.

The authors of [4] developed a method for extracting trends by means of neural networks, used as a tool to discover all existing hidden trends in several different types of crimes in US cities.

The authors of [6] argue that traditional temporal data models are mainly focused on describing temporal data based on versioning of objects, tuples, or attributes. Time series data are frequently found in real-world applications, such as sales, economics, and scientific data, but they are not effectively managed by this approach. They propose a temporal query language that treats both version-based and time series data in an uniform manner.

3 Granular Temporal Trends in OLAP

In this section, we recall the definition of the multidimensional model and describe how it may be queried by means of Online Analytical Processing (OLAP) tools. We discuss how OLAP queries may be interpreted as *time series*. At last, we define trends and we describe two new operators for discovering granular temporal trends on time series.

A multidimensional model is based on the notion of dimensions, measures, and facts. Each dimension is organized in a hierarchy of levels, corresponding to data domains at different granularities. A multidimensional schema describes numerical facts defined with respect to a particular combination of levels and quantified by measures. An example of a multidimensional schema is depicted in Fig. 1.

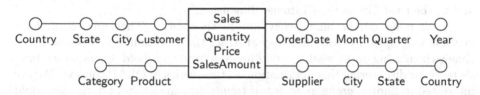

Fig. 1. An example of a multidimensional fact with four dimensions and three measures represented by means of the dimensional fact model (DFM) [3]

Data may be retrieved from instances of multidimensional schemas by means of OLAP queries. They consist of three basic analytical operations: roll-up for aggregating data (hence, showing data at a coarser level), drill-down for shown-ing data at a finer level, and slicing and dicing for filtering data. Consider the following OLAP query:

Example 1. Display for each Day the total Quantity of items sold. This query may be represented in a more compact way as:

Sales(OrderDate:Day, Qty)

The schema name is Sales, all the omitted atemporal dimensions are aggregated to the All level, the level Day has been selected for the time dimension OrderDate, and the selected measure is Qty.

Now consider the following OLAP query:

Example 2. Display for each City and for each Day the total Quantity of items sold.

Sales(Customer:City, OrderDate:Day, Qty)

The City level of the Customer dimension has been selected, the omitted atemporal dimensions are aggregated to the All level, and the time dimension OrderDate is still at the Day level.

The results of both queries may be interpreted as *time series*. In an OLAP context, they are ordered sets of triples $\langle \bar{d}, t, v \rangle$, where $\bar{d}$ represents the grouping values for the atemporal dimensions, t is a time point at a particular granularity, and v is its possibly aggregated value. The grouping component $\bar{d}$ contains all the level members for each atemporal dimension considered in a query. It resembles the grouping clause of SQL queries, since it identifies one time series among the selected ones. A fact aggregated to the All level in each dimension is represented by a time series in which $\bar{d}$ is empty.

The result of the query in Example 1 may be interpreted as triples $\langle \emptyset, \text{Day}, value \rangle$, ordered according to the time dimension OrderDate. *value* is the value of the measure Qty, aggregated on each dimension to the level All. Values are not aggregated on the selected temporal dimension OrderDate, since the level Day corresponds to the bottom one.

On the other hand, the result of the query in Example 2 may be represented as triples $\langle \text{City}, \text{Day}, value \rangle$, ordered according to the time dimension OrderDate. *value* refers to the measure Qty, aggregated on each dimension to the level All, and to the level City in the Customer dimension.

Trends point out peculiar behaviours in time series. They are essential for decision making-processes, since they abstract the evolution of a measure across time as increasing or decreasing according to a fixed threshold. Users need analysis tools for discovering this information on time series in OLAP context. We are interested in finding *granular temporal trends* on time series of data that could possibly be aggregated. According to the classification presented in [1], there are eight different characterizations of granular trend. We are going to consider two of them, namely, *intragranular trend* and *intergranular trend*, depicted in Fig. 2.

In the first case, one is interested in finding time points within one granule of the specified granularity in which a trend holds. An example of a query is "Find positive trends of items sold for each product category between days in months." In the second case, one is interested in finding trends between some time points within a granule and a representative in the consecutive one. An example of a query is "Find positive trends that hold in at least 90% of days for each city across months." In both cases a base level and a grouping level have to be selected on the temporal dimension for discovering granular temporal trends. Time points will be considered at the base level and granules will be considered at the grouping level (e.g., Day as base level and Quarter as grouping level). Intragranular trends express a relationship between members of the base level, while intergranular trends express a relationship between members of the grouping level.

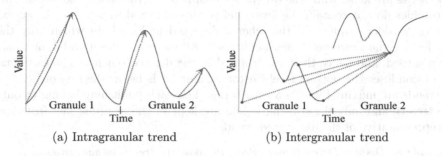

(a) Intragranular trend   (b) Intergranular trend

Fig. 2. Trend types: (a) *intragranular* case (b) *intergranular*

In the following, we propose a new OLAP operator, *Trend*, that discovers *intragranular* or *intergranular* trends on time series of possibly be aggregated data. It takes as input six parameters: *Trend (GranuleType, TrendType, Base-Level, GroupingLevel, Threshold, [Percentage])*. (1) *GranuleType* may be either *intragranular* or *intergranular*. (2) *TrendType* may be either ↑ for increasing or ↓ for decreasing. (3) *BaseLevel* and *GroupingLevel* are the levels of the temporal dimension to be used for discovering the trends. (4) *Threshold* is a numerical number that quantifies the minimum difference between two time points. (5) *Percentage* is a numerical value between 0 and 100, that can be specified only in the *intergranular* case: it is the minimum of time points in a granule that needs to participate in the trend. The trends discovered by the operator can be represented a multidimensional way, allowing users to perform more in-depth analysis. To this purpose, we also propose a new aggregation operator, namely, *TrendAggregator*, that summarizes the previously discovered trends.

3.1 Intragranular *Trend* Operator

An *intragranular trend* expresses a relationship between time points in the same granule within a time series. Two (or more) consecutive time points are considered a trend if they belong to the same granule, and the specified pattern (increase, or decrease) holds according to a threshold.

The *Trend* operator analyses OLAP queries results interpreted as time series. An *intragranular trend* is a subset of time points of a time series. Given three consecutive time points t_i, t_i, t_{i+1} within the same grouping granule, t_i belongs to a trend if the pattern between t_i and t_{i+1} holds. The trend starts in t_i if the pattern does not hold between t_{i-1} and t_i.

An intragranular trend is characterized by its starting point, its measure value at the trend start, its length expressed in number of time points, and its total steepness. The latter is the sum of the differences between all values in it.

The result of the *Trend* operator is represented as a multidimensional instance of a schema Fact', derived from its source schema Fact. It contains two new measures, the trend Length and its Steepness. The dimensions used for grouping time series are kept, while the others are dropped. Furthermore, the dimensions hierarchies do not contain the levels below the one used in a query. As an example, consider a query on the schema depicted in Fig. 1, in which only the City level of the Customer dimension is used. All the other atemporal dimensions are not used, and hence they are dropped in the derived schema. The Customer dimension loses the bottom level Customer, since it is below the City one.

Trends are maximal within a time series (i.e., each point may belong to only one trend), and only the starting points are kept in the resulting fact instance, as representative of the discovered trends.

Example 3. Display for customer cities, the positive trends of quantities of items sold between days in months of 2013.

$$\text{SalesIncrease} \leftarrow \text{Trend}_{Intragranular,\uparrow,\text{Day},\text{Month, 0}}(\text{Sales}(\text{Customer:City,}$$
$$\text{OrderDate:Day, OrderDate:Year}=2013, \text{Qty}))$$

First, data contained in the Sales cube are aggregated to the City level in the Customer dimension, and to the All one in the other dimensions. Then, the aggregated data are sliced, and only the orders placed in 2013 are considered.

Each day t_i is compared with the next one t_{i+1} to check if it belongs to a positive trend. Table 1 shows an example of a time series and the discovered trends. Specific days do not appear within the table, indicating that no sales occurred that day. Thus, July 22nd and 31st are considered subsequent days within the month granule. According to that time series, two trends have been discovered. The first one starts on July 21st, lasts for 10 days, and has a total steepness of 156. The second one starts on August 1st, it lasts 10 days, and it has a steepness of 20.

The resulting trends are stored in a new fact named SalesIncrease. Its schema consists of the dimension Customer, from the City level and above, and the OrderDate dimension that starts from the Day level, as shown in Fig. 3.

Table 1. Example of a time series regarding quantities of items sold in 2013 and the discovered trends

Date	Quantity		OrderDate	Quantity	Length	Steepness
2013-07-21	9		2013-07-21	9	10	156
2013-07-22	17		2013-08-01	250	10	20
2013-07-31	165		...	...	...	...
2013-08-01	250		...	...	...	...
2013-08-10	270		...	...	...	...

3.2   Intergranular Trend4BI Operator

An *intergranular trend* expresses a relationship between time granules. This kind of trend requires that a given pattern holds in some of the consecutive granules

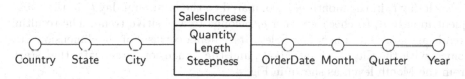

Fig. 3. Example of application of *Trend* on the schema depicted in Fig. 1, for finding intragranular trends. The derived schema consists of the dimension Customer, from the City level and above, and the OrderDate dimension.

of the specified grouping granularity level. Two consecutive granules g_1, g_2 are considered a trend if the specified pattern holds between their time points: the pattern must hold among a specified percentage of time points in g_1, ranging from 0% to 100%, and a representative in g_2. A different representative is chosen depending on the trend type. In the case of a positive trend, all the points of g_1 will be compared to the minimum of g_2 (in case of negative trend, the representative is the maximum). Figure 2 shows a positive trend between almost all the time points in g_1 and the minimum of g_2. As an example, two months belong to a positive increasing trend with threshold equal to 30 for the measure and 100% threshold for the subgranules, if the difference between the measure within all the days in the first month and the minimum in the second month is equal or greater than the 30.

The *Trend* operator analyses OLAP queries interpreted as time series. A trend is a subset of grouping time granules of a series. Each time point t_i in the first granule g_i is compared with the representative t_j in the consecutive granule, for assessing if g_i belongs to a trend or not. The starting point of a trend is the granule for which the pattern does not hold with the previous one. An intergranular trend is characterized by its starting point, the length of the trend (expressed in number of granules), and the total steepness. The steepness between two consecutive granules, is the maximal difference between all points in the first granule, and the representative in the second one. The result of the *Trend* operator may be represented by the multidimensional schema Fact′, derived from its source schema Fact. It contains two new measures, the trend Length and its Steepness. The dimensions used for grouping time series are kept, while the others are dropped. Furthermore, the dimensions hierarchies do not contain the levels below the one used in a query.

Example 4. Display for customer cities, the positive trends of quantities of items sold across months with a 100% threshold. The set of time series to be analyzed is the result of the following query:

SalesIncrease ← $Trend_{Intergranular,\uparrow,Day,Month,0,100}$(Sales(Customer:City,
            Supplier:All, Product:All, OrderDate:Day, Qty))

First, data contained in the Sales cube is aggregated to the City level in the Customer dimension, and to the All in the other dimensions. The result is interpreted as a set of time series, one for each product city.

Each day t_i in the month g_i is compared to the minimum day t_j in the subsequent month g_j, to check whether g_i belongs to a positive trend. The resulting trends are stored in a new fact Sales'. Its schema consists of the dimension Customer, with the City level and above, and the dimension OrderDate, that starts from the Month level, as shown in Fig. 4.

Fig. 4. Example of application of the *Trend* operator on the schema depicted in Fig. 1, for finding intergranular trends. The derived schema consists of the dimension Customer, from the City level and above, and the OrderDate dimension, from the Month and above.

3.3   Trend Aggregator

The trends discovered by the *Trend* operator are represented in a multidimensional way, where only the time dimension and the atemporal dimensions used for grouping purposes are kept, as shown in Figures 3 and 4. The derived schema allows users to continue their analysis by means of OLAP operators. Thus, aggregating the discovered trends is a problem that needs to be addressed: they are characterized by two measures, the length and the steepness, which are strictly linked. Usual aggregation operators would consider them separately, leading to erroneous results. Consider the example in Table 2, in which several trends have been discovered. The temporal dimension may be aggregated to the Quarter

Table 2. Example of discovered trends. The *TrendAggregator* operator may be applied for aggregating them to the Quarter level on the time dimension. The highlighted trend is the result of this aggregation.

OrderDate	Length	Steepness
2013-07-21	*10*	*156*
2013-08-01	10	20
2013-08-15	3	190

level, in order to discover its longest trend. The standard *max* operator would return 10 as maximum for the Length measure, and 190 for the Steepness. This could mislead the user, since that trend does not exists. An ad hoc operator for preserving this link is needed.

TrendAggregator allows one to perform a *prioritized maximum or minimum aggregation*: it consists of choosing the longest trend, with the maximal or minimal steepness, depending on the user's choice. In the example shown in Table 2, the trend aggregated to the Quarter level is the first one, since the maximum Length is 10, and the maximal Steepness associated to it is 156.

4 Conclusions and Future Work

Users need to perform more in-depth analysis for enhancing their decision-making processes. Temporal patterns provide a significant source of information for decision-makers, but they are usually made of some basic temporal trends. To address this problem, we merged two general aspects of temporal data, namely, temporal patterns and temporal granularities, with analysis on aggregated data. We proposed two new operators, we called, *Trend* and *TrendAggregator*, for finding *granular temporal trends* in Online Analytical Processing context. *Trend* points out peculiar behaviour in data, by exploiting the hierarchical structure of dimensions in order to find trends of possibly aggregated data at some granularity levels. *TrendAggregator* allows users to further investigate and summarize the discovered trends, represented by means of a multidimensional model.

We are extending our approach to consider time series with multiple values associated to each time point. We are also working on how to temporally summarize the discovered trends. We are interested in finding temporal trends in medical data, and in particular in reports of unexpected adverse reactions induced by drug administrations.

References

1. Combi, C., Pozzi, G., Rossato, R.: Querying temporal clinical databases on granular trends. Journal of Biomedical Informatics 45(2), 273–291 (2012)
2. Eder, J., Koncilia, C., Morzy, T.: The COMET metamodel for temporal data warehouses. In: Pidduck, A.B., Mylopoulos, J., Woo, C.C., Ozsu, M.T. (eds.) CAiSE 2002. LNCS, vol. 2348, pp. 83–99. Springer, Heidelberg (2002)
3. Golfarelli, M., Maio, D., Rizzi, S.: The dimensional fact model: A conceptual model for data warehouses. Int. J. Cooperative Inf. Syst. 7(2-3), 215–247 (1998)
4. Kaikhah, K., Doddameti, S.: Discovering trends in large datasets using neural networks. Appl. Intell. 24(1), 51–60 (2006)
5. Khatri, V., Ram, S., Snodgrass, R.T., Terenziani, P.: Capturing telic/atelic temporal data semantics: Generalizing conventional conceptual models. IEEE Trans. Knowl. Data Eng. 26(3), 528–548 (2014)
6. Lee, J.Y., Elmasri, R.: An EER-based conceptual model and query language for time-series data. In: Ling, T.-W., Ram, S., Li Lee, M. (eds.) ER 1998. LNCS, vol. 1507, pp. 21–34. Springer, Heidelberg (1998)
7. Malinowski, E., Zimányi, E.: A conceptual model for temporal data warehouses and its transformation to the ER and the object-relational models. Data Knowl. Eng. 64(1), 101–133 (2008)
8. Wijsen, J.: Reasoning about qualitative trends in databases. Inf. Syst. 23(7), 463–487 (1998)

Real-Time Data Warehousing:
A Rewrite/Merge Approach

Alfredo Cuzzocrea[1], Nickerson Ferreira[2], and Pedro Furtado[2]

[1] ICAR-CNR and University of Calabria, Italy
[2] University of Coimbra, Portugal
cuzzocrea@si.deis.unical.it, {nickerson,pnf}@dei.uc.pt

Abstract. This paper focuses on *Real-Time Data Warehousing systems*, a relevant class of *Data Warehouses* where the main requirement consists in executing classical data warehousing operations (e.g., loading, aggregation, indexing, OLAP query answering, and so forth) under *real-time constraints*. This makes classical DW architectures not suitable to this goal, and puts the basis for a novel research area which has tight relationship with emerging *Cloud architectures*. Inspired by this motivation, in this paper we proposed a novel framework for supporting *Real-Time Data Warehousing* which makes use of a *rewrite/merge approach*. We also provide an extensive experimental campaign that confirms the benefits deriving from our framework.

1 Introduction

Data Warehouses (e.g., [11]) are more and more demanding for high-performance which allow them to deal with *real-time paradigms* (e.g., [16]), which may turn to be extremely useful in next-generation *Big Data research*. Indeed, there exists a plethora of emerging applications where *Real-Time Data Warehousing* plays a leading role, such as: *sensor networks*, *real-time business intelligence*, *real-time Cloud applications*, and so forth.

The traditional data warehouse architecture model assumes that new data loading occurs only at certain times, when the warehouse is taken offline, and the data is integrated during a more or less lengthy time interval. This offline procedure is required for three main reasons: there should be no interference between the loading process and the query sessions running on the data warehouse. Therefore, there is no significant slowdown. Looking at data formats, a data warehouse is typically a set of interconnected *data marts*, schemas (*stars*), with constraints (e.g., foreign keys, not null constraints, primary keys), lots of indexes (e.g., *b-trees*, *bitmap indexes*), *materialized views* and other summary or derived data, which are created to speedup query answering. From the point of view of data integration, constraints and indexes considerably slow the process down, as well as the refreshing of all those structures with the new data. The appropriate solution for these problems in traditional warehouses is to take the whole warehouse offline, disable/drop the constraints and indexes that cause loading slowdown, load the whole data and refresh the datasets, and then rebuild the auxiliary structures and constraints (e.g., [10, 12,13,14]).

L. Bellatreche and M.K. Mohania (Eds.): DaWaK 2014, LNCS 8646, pp. 78–88, 2014.

It would be convenient if, instead of using completely different architecture and database engine solutions, it would be possible to transform a warehouse to make it real-time, by adding some mechanisms to it. We explore this possibility in this paper, propose an approach and show that it achieves the desired objectives.

Our approach is based on a *real-time integration component* that is added to the traditional data warehouse and provides real-time capabilities without modifications to the existing setup. We concentrate on the warehouse itself, which includes loading and refreshing. This assumes that extraction and transformation phases of the ETL processes are real-time-capable, or modified to be so. In short, the frequency of extraction should increase to provide the desired freshness (e.g., instead of once every day it could be once every 5 minutes), using *Change-Data-Capture* approaches and producing mini-batches, and transformation procedures should be guaranteed to run efficiently for the small mini-batches.

2   A Rewrite/Merge Approach for Supporting Real-Time Data Warehousing

The purpose of the Real-time Warehousing is to allow data to be integrated in much shorter time than the traditional periodic loading intervals. The requirements are that it be an add-on to the traditional architecture, requiring no or few modifications to the existing infrastructure. It is based on two components: the static and the dynamic data warehouse components.

Data warehouses have multiple structures, including star schemas, multiple indexes, summary tables. Loading and refreshing all this redundant data slows the process down significantly if there are queries running simultaneously. If a schema has many indexes, insertion of new data will be slowed due to index updates. But dropping indexes before loading the data and rebuilding them after the loads is a heavy process if it is done online, with queries running simultaneously. In our architecture those problems are eliminated by completely separating the light real-time data integration component, the *D-DW*, from the component holding the bulk of the data and all the indexing and summarizing elements on the huge less recent data, the *S-DW*.

The S-DW follows the typical data warehouse organization and holds the bulk of the data. First of all, it is expected to have multiple star schemas. Indexes such as B-trees (I) or Bitmap indexes (BI) (possibly many) are expected there, to speedup accesses. The cost of reorganizing those indexes can be quite high. Another existing element in the S-DW are any number of summary tables, which summarize the data to allow very fast answers to queries that may access those them instead of base data.

Only recent data will live in the D-DW (e.g., data from the current day), resulting in a comparatively small amount of data. Queries over the D-DW are always answered fast in absolute terms, since the quantity of data that needs to be processed is small. This is an important point, since it also means that this component can work efficiently even without indexes and aggregated views, at least comparatively to the S-DW. The fact that it dispenses indexes and materialized views altogether means that loading of mini-batches can be online, with extreme efficiency, in real-time and with no or very little noticeable performance degradation for the queries. This is shown in Fig. 1.

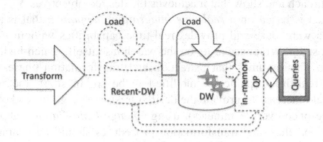

Fig. 1. Real-Time Warehousing

Comparing the S-DW to the D-DW, the base star schemas have the same fact and dimension tables. Materialized aggregates (*MV*) are replaced by non-materialized views (*V*), there are no constraints (e.g., no explicit links between facts and dimensions), and both B-tree and Bitmap indexes are absent. These differences and the size of the data (the D-DW contains a much smaller amount of data than the S-DW) makes the system load and run much faster. There is no need to update indexes or views.

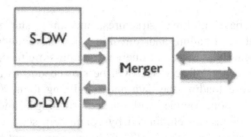

Fig. 2. Merger Component

2.1   The Merger Component

Since in this design the data is spread over two different components, and those should reside in different nodes for faster processing, the approach features a merger, which submits the query in each of the two independent components simultaneously (S-DW and D-DW), which send the results back to the merger. This merger then computes the final answer. The merger is simply another database instance, in our prototype implementation it was an in-memory database for faster merging operation (we used H2). Fig. 2 illustrates the merger component.

Query processing in the merger is very similar to the typical parallel query processing approaches. We describe the approach using a simple SQL-based example. In Fig. 3, a query is broken down into two queries and the results are merged using an UNION operator.

```
select sum(l_revenue), d_year, p_brand
from (select l_revenue, d_year, p_brand
          from lineorder, date_dim, part, supplier
            where
            l_orderdate = d_datekey and
            l_partkey = p_partkey and
            l_suppkey = s_suppkey and
            p_brand = 'MFGR#2221' and
            s_region = 'Nevada')
UNION ALL
(select l_revenue, d_year, p_brand
          from lineorder@ddw, date_dim@ddw, part@ddw,
supplier@ddw
            where
            l_orderdate = d_datekey and
            l_partkey = p_partkey and
            l_suppkey = s_suppkey and
            p_brand = 'MFGR#2221' and
            s_region = 'Nevada')
group by d_year, p_brand
order by d_year, p_brand;
```

Fig. 3. Merging SQL Statement

2.2   Alternative Implementations

The dynamic data warehouse component (D-DW) can also be implemented in the same instance as the S-DW, as temporary tables. The problem then is that this alternative does not eliminate the simultaneity of querying and loading activity, therefore the performance is still degraded.

Fig. 4 illustrates indexes (triangles), a table (big left rectangle), and a time-partitioned table (right rectangles). While in the left organization a whole index has to be dropped and rebuilt when loading data, using the right organization only the index for the last partition that is the one loaded needs to be rebuilt. This saves a large amount of time.

Fig. 4. Indexing vs Partitioning

3   Experimental Assessment and Analysis

The following experiments show that it is possible to add the RWC component to achieve real-time. We test the real-time loading limitations of the DW for the setup scenario, compare it with the real-time loading capability when the RWC is added. We also show that query response times do not suffer significantly. The experimental setup and results, taken from [7], are based on the SSB (*Star Schema Benchmark*) [15], augmented to include: A) the Orders star, representing sales, and the corresponding dimension tables (*Time_dim*, *Date_dim* and *Customer*); B) a star representing *LineOrder*, which contains order items, with 5 dimension tables

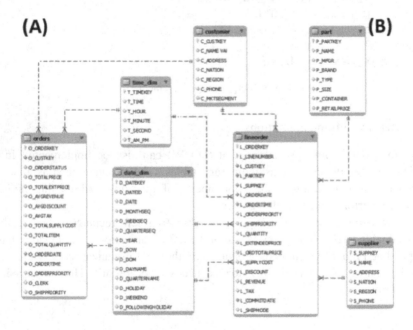

Fig. 5. RT-Benchmark Schema

(*Time_dim*, *Date_dim*, *Customer*, *Supplier* and *Part*). Fig. 5 shows the star schemas. Aggregated views were also added, measuring sales by *Supplier*, *Part* and *Customer* over granularities (hour, day, month and year). The schema was indexes - 8 B-tree indexes for the star schemas (4+4), and 6 bitmap indexes. The size of the data warehouse was 30 GB. The same query workload of the original SSB benchmark was used, consisting of 13 queries that target data with different granularity [15]. Experiments included loading time, from log files in the operating system. Additional details on the modified benchmark (SSB-RT) are available in [7]. The database Server used in the experiments was an Intel(R) Core(TM) i5 3.40GHz with 16GB of RAM memory, running Oracle version 11g.

The SSB-RT experimental setup includes the ETL. The extracted data was simulated by means of TPC-H *dbgen* generated orders and *lineitem* log files (for facts), and *customer*, *part* and *supplier* (for dimensions). The transformation process started by reading those files, and included selection of columns to load, translation of coded values (5 fields), encoding of *lineorder*, *orderpriority* and *ship-priority* fields, computation of the total price from *lineorder* prices, generation of surrogate keys, generation of data and time rows and surrogate keys for those. Details of the SSB-RT are available in [7].

3.1   Impact of Query Sessions on ETL(R)

This section assesses how much query sessions impact the ETL process, that is, the throughput of the ETLR process. The setup places 10 simultaneous query sessions

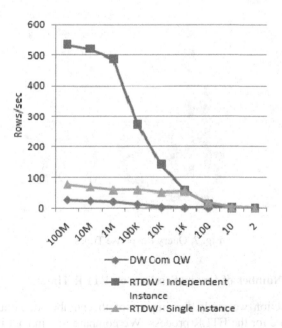

Fig. 6. Throughput of ETLR

always running queries that they choose randomly form the SSB-RT query set. The throughput is measured as the number of rows of the batch that our system is able to process per second. Three alternatives are compared: traditional DW (S-DW), same-instance D-DW, and the solution of independent instance D-DW and merging in a third component. Fig. 6 shows the results.

3.2   Impact of Load on Queries

In this experiment we consider the best approach (IINM - In Memory DB) and study the impact of the real-time loading over all queries, in terms of execution time. Standard deviation of the results was always less than 9%. Fig. 7 compares the results of running the queries against a traditional warehouse, with no loading, with those of running them against a real-time warehouse without loading and those of running it while loading data simultaneously.

It is possible to see that the impact of the merge and the impact of the loading are both minimal, thanks to the fact that loading happens on the D-DW with no indexing and light queries only (since it has smaller amounts of data), and merging is quite fast, in-memory. In 10 queries the impact was less than 2%. The only more significant impacts were in 3 queries that had a very small response time, however the absolute response times for those queries are still in the order of few seconds or milliseconds. This means that our solution enables very efficient real-time integration of data without impacting query response times.

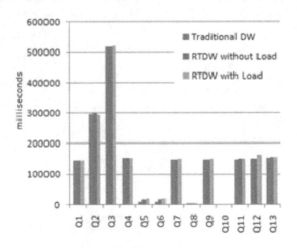

Fig. 7. Query Response Times

3.3   Impact of Number of Query Sessions on ETLR Times

Finally, in this section we study the impact of the number of simultaneous query sessions on the time for the ETLR process. We compare that impact if we try to load in a traditional warehouse architecture with loading over our optimized architecture.

For the traditional DW we had to test with only 10 rows, because as the number of sessions increases, the capacity of the system to integrate degrades exponentially. If we had chosen many more rows to integrate, the time needed would be enormous. So, while for the traditional DW system we integrate only 10 rows, for our RTDW approach we integrate 100k rows. Fig. 8 shows the results.

We can see that, even with a much bigger mini-batch, our proposed architecture is able to integrate at a high speed, taking much less time than the traditional architecture. With these results, we conclude that our solution is able to support multiple query sessions and simultaneous data integration, even for large data sets.

3.4   Minimizing D-DW to S-DW Loading Times

Fig. 9 shows all the components of the loading and refreshing overheads for 10M rows. From Fig. 9, it is clear that the largest overhead was index rebuilding if we had to rebuild the whole indexes for the 30GB dataset (two columns on the right). If we have time-interval partitions of 1GB (two columns on the left), index rebuild times are reduced drastically. This means that if we partition the warehouse by time, it is possible to have short offline periods by loading reasonably small amounts of data and having small index rebuild times.

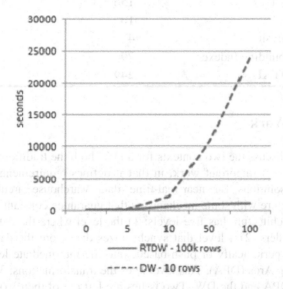

Fig. 8. ETLR Execution Time versus Number of Simultaneous Query Sessions

Next we experimented with loading 10K rows with 1GB time-interval partitions, and measured the time taken on each part of the procedure, including ETL. Table 1 shows the results. The whole loading procedure took less than 4 minutes. The DW data warehouse only has to be offline for 4 minutes every time 10000 rows need to be loaded, and can answer to queries with total freshness (capacity to access real-time data if necessary), thanks to the real time component that was added.

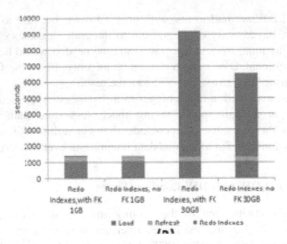

Fig. 9. Loading 10 M rows and Refreshing 1GB partitions versus 30 GB partitions

Table 1. Time Taken to Load 100k rows

Breakdown of time to load 100k rows (in secs)	
E + T	156
Load	10
Refresh	4
Rebuilding Indexes	79
TOTAL	**249**

4   Related Work

In [1] the authors discuss the two contexts for a DW, both the traditional and the real-time one. Theirs is an important work, in that it defines requirements and provides some indicative solutions for near-real-time data warehouse architecture. They suggest an architecture with some mechanisms that guarantee constant refresh of data in the DW. The architecture has five levels: 1) the level where the data extraction is done into data holders; 2) a level that synchronizes data from the data holders and transfers the data, periodically or push-based, into the intermediate level, known as the Data Processing Area (DPA). This level does the transformations; 3) the level that synchronizes the DPA and the DW. Two issues are left out of their work: the authors do not consider, during the loading and refreshing phase, the existence of indexes, materialized summaries and data marts. The performance problems introduced by the simultaneity between online querying and continuous data loading, and the difficulties in real-time loading introduced by having indexes and summary tables are not considered or evaluated.

The authors of [2,3,4] all propose a similar, temporary-tables based, approach to deal with real-time. For instance, in [3] the authors add time-interval granularity partitions, e.g. one for the last hour and one for the last day. The main limitation of those solutions is that there is not an adequate decoupling between the data loading and querying services, since they are done in the same database instance in the same machine.

In [5] the authors propose to build a tool based on Data Streams that allows data freshness and continuous integration, besides other criteria also evaluated by the authors. However, these stream-based solutions return approximate results, due to the large cost that would be incurred to have exact results using continuous streams.

The authors of [6] suggest a solution based in SOA (Service Oriented Architecture). This approach does data extraction initially, through a web service, and stores the results in caches. There are various levels of cache concerning different update levels, for instance, 5, 10, 30 and 60 minutes. After the data passes through all caches, it is stored in the DW. This solution also has a component that joins information from the caches. This proposal modifies the architecture and the structures of the data warehouse.

Finally, there are also commercial systems that propose real-time and freshness in DW, such as the ones in [8,9].

5   Conclusions and Future Work

In this paper we described an approach to transform a data warehouse into a real-time-capable one. The real-time warehouse is able to integrate data at the rate that may be needed. We have shown that this can be done transparently to the user and existing data warehouse by adding a Real-time Warehouse component, and small middleware component that rewrites client queries into queries to two components and merges results in a single result to be returned to clients. We also discussed how to minimize D-DW to S-DW loading times. Experimental results have shown the validity of the proposed approach.

Future work is mainly oriented to improve the efficiency of the proposed framework via data compression paradigms (e.g., [17,18]).

References

1. Vassiliadis, P., Simitsis, A.: Near Real Time ETL. In: New Trends in Data Warehousing and Data Analysis. Annals of Information Systems, vol. 3, pp. 1–31 (2009)
2. Jain, T., Rajasree, S., Saluja, S.: Refreshing Datawarehouse in Near Real-Time. International Journal of Computer Applications 46(18), 24–29 (2012)
3. Zuters, J.: Near Real-Time Data Warehousing with Multi-stage Trickle and Flip. In: Grabis, J., Kirikova, M. (eds.) BIR 2011. LNBIP, vol. 90, pp. 73–82. Springer, Heidelberg (2011)
4. Santos, R.J., Bernardino, J.: Real-Time Data Warehouse Loading Methodology. In: Proceedings of ACM IDEAS, pp. 49–58 (2008)

5. Nguyen, M., Tjoav, A.M.: Zero-Latency Data Warehousing for Heterogeneous Data Sources and Continuous Data Streams. In: Proceedings of iiWAS (2003)
6. Zhu, Y., An, L., Liu, S.: Data Updating and Query in Real-time Data Warehouse System. In: Proceedings of IEEE CSSE, pp. 1295–1297 (2008)
7. Ferreira, N.: Realtime Warehouses: Architecture and Evaluation. MSc Thesis, U. Coimbra (June 2013)
8. Vertica, http://www.vertica.com/the-analytics-platform/real-time-loading-querying/
9. Oracle, Best Practices for Real-time Data Warehousing. White Paper (2012)
10. Zhu, Y., An, L., Liu, S.: Data Updating and Query in Real-time Data Warehouse System. In: Proceedings of IEEE CSSE (2008)
11. Chaudhuri, S., Dayal, U.: An Overview of Data Warehousing and OLAP Technology. SIGMOD Record 26(1), 65–74 (1997)
12. Shi, J., Bao, Y., Leng, F., Yu, G.: Study on Log-based Change Data Capture and Handling Mechanism in Real-time Data Warehouse. In: Proceedings of IEEE CSSE, pp. 478–481 (2008)
13. Ram, P., Do, L.: Extracting Delta for Incremental Data Warehouse Maintenance. In: Proceedings of IEEE ICDE, pp. 220–229 (2000)
14. Furtado, P.: Efficiently Processing Query-Intensive Databases over a Non-Dedicated Local Network. In: Proceedings of IEEE IPDPS, p. 72 (2005)
15. O'Neil, P., O'Neil, E., Chen, X., Revilak, S.: The Star Schema Benchmark and Augmented Fact Table Indexing. In: Nambiar, R., Poess, M. (eds.) TPCTC 2009. LNCS, vol. 5895, pp. 237–252. Springer, Heidelberg (2009)
16. Cuzzocrea, A.: A Framework for Modeling and Supporting Data Transformation Services over Data and Knowledge Grids with Real-Time Bound Constraints. Concurrency and Computation: Practice and Experience 23(5), 436–457 (2011)
17. Cuzzocrea, A.: Providing probabilistically-bounded approximate answers to non-holistic aggregate range queries in OLAP. In: Proceedings of ACM DOLAP, pp. 97–106 (2005)
18. Cuzzocrea, A., Serafino, P.: LCS-Hist: taming massive high-dimensional data cube compression. In: Proceedings of ACM EDBT, pp. 768–779 (2009)

Towards Next Generation BI Systems:
The Analytical Metadata Challenge

Jovan Varga[1], Oscar Romero[1], Torben Bach Pedersen[2],
and Christian Thomsen[2]

[1] Universitat Politècnica de Catalunya, BarcelonaTech, Barcelona, Spain
{jvarga,oromero}@essi.upc.edu
[2] Aalborg Universitet, Aalborg, Denmark
{tbp,chr}@cs.aau.dk

Abstract. Next generation Business Intelligence (BI) systems require integration of heterogeneous data sources and a strong user-centric orientation. Both needs entail machine-processable metadata to enable automation and allow end users to gain access to relevant data for their decision making processes. Although evidently needed, there is no clear picture about the necessary metadata artifacts, especially considering user support requirements. Therefore, we propose a comprehensive metadata framework to support the user assistance activities and their automation in the context of next generation BI systems. This framework is based on the findings of a survey of current user-centric approaches mainly focusing on query recommendation assistance. Finally, we discuss the benefits of the framework and present the plans for future work.

Keywords: BI 2.0, metadata, user support.

1 Introduction

Next generation BI systems (BI 2.0 systems) shift the focus to the user and claim for a strong user-centric orientation. Through automatic user support functionalities, BI 2.0 systems must enable the user to perform data analysis tasks without fully relying on IT professionals designing/maintaining/evolving the system. Ideally, the end user should be as autonomous as possible and the system should replace the designer by providing maximal feedback with minimal efforts. This is even more important if we consider heterogeneous data sources. However, this scenario is yet far from being a reality. A research perspective on this new scenario can be found in [1]. As discussed in the paper, BI 2.0 systems should provide a global unified view of different data sources. To address new requirements, such as dynamic exploration of relevant data sources at the right-time, it is outlined that automatic information exploration and integration is a must. These characteristics raise the need for semantic-aware systems and machine-processable metadata.

Metadata in BI 2.0 systems are needed to support query formulation, relevant source discovery, data integration, data quality, data presentation, user guidance, pattern detection, mappings of business and technical terms, visualization, and

L. Bellatreche and M.K. Mohania (Eds.): DaWaK 2014, LNCS 8646, pp. 89–101, 2014.

any other automatable task that are to be provided by the system. However, current approaches address specific metadata needs in an ad-hoc manner and using customized solutions. A unified global view of the metadata artifacts needed to support the user is yet missing. In this paper we perform a survey to identify what user assistance functionalities should be supported and by means of which metadata artifacts their automation should be enabled. Identifying such metadata is mandatory to enable their systematic gathering, organization, and exploration.

Contributions. We propose a comprehensive metadata framework that supports user assistance activities and their automation in the context of BI 2.0 systems. Specifically, we describe the additional process of the user-system interaction so that the system does not only answer queries but supports the user during the interaction. We identify the main user assistance activities to be supported and the metadata artifacts to be gathered, modeled, and processed to enable the automation of such tasks. These results are based on a survey of specific approaches devoted to provide user assistance on activities like querying and visualization. Finally, we categorize the metadata artifacts to support the user assistance and their processing to enable automation.

The rest of the paper is organized as follows. Section 2 discusses the related work. Section 3 presents a survey of user-centric approaches. Then, Section 4 defines the metadata framework according to our findings and finally, Section 5 concludes the paper and provides directions for future work.

2   Related Work

The main challenges of user assistance are highlighted in [16]. This paper outlines the increasing need for system support for user analytical tasks in the settings of fast growing, large-scale, shared-data environments. To assist the user with query completion, query correction, and query recommendations, the authors propose meta querying paradigms for advanced query management and discuss the challenge of query representation and modeling. This challenge motivated our research for metadata capturing queries and other related metadata artifacts for user assistance purposes.

Indeed, as discussed in [10], metadata are important for data warehousing users. The article gives a perspective about the metadata benefits for end users' understanding and usage of data warehouse and BI systems, especially for inexperienced ones. The authors present an end user metadata taxonomy consisting of *definitional, data quality, navigational,* and *lineage* categories. This is a first attempt to characterize end user metadata in BI systems that we aim at extending for a BI 2.0 context.

With the expansion of the Web, analytical data requirements have exceeded traditional data warehouse settings and now entail the incorporation of new and/or external data sources with unstructured or semi-structured data. In this environment, user assistance and its automation are even a greater need but also a challenge. By analyzing the architectural solutions provided in BI 2.0

systems, in the next paragraphs we discuss that metadata are to be one of the key resources for such task.

Some architectural solutions have been recently presented for BI 2.0 (e.g., [1] and [20]). These systems focus on supporting source discovery, data integration, and user guidance for large and often unstructured data sets. However, they do not provide much details about specific metadata artifacts.

[21] proposes the creation of a knowledge base to support data quality-awareness for the user assistance. This knowledge base is to be represented by means of metadata. Furthermore, in the BI Software-as-a-Service deployment model presented in [9], one of the essential business intelligence services is a meta-data service. It defines the business information to support information exchange and sharing among all other services. For metadata handling, these two approaches refer to a specific metadata framework [25].

Finally, the vision of next-generation visual analytics services is presented in [23]. The challenges of visualization and data cleaning, data enrichment and data integration naturally match the goals of next generation BI systems. The authors analyze these challenges for structured data sets but they note that unstructured and semi-structured data represent even greater challenges. As discussed, these tasks raise the need for a common formalism. The metadata are to address such requirements.

Overall, BI 2.0 systems focus on end-to-end architectural solutions and typically pay little attention to describe how they deal with metadata. Indeed, most of these systems just mention the crucial role of metadata for the system overall success. As mentioned, [9] and [21] suggest the usage of Common Warehouse Metamodel (CWM) [25]. Nevertheless, CWM is a standard for interchange of warehouse metadata that provides means for describing data warehouse concepts but the support provided is incomplete for the BI 2.0 metadata artifacts discussed in the following sections (which we refer to as analytical metadata). In this line, the Business Intelligence Markup Language (BIML) Framework [19] presents the automation achieved by using metadata for tasks like data integration, but it does not cover the user assistance perspective. Hence, in order to gain insight on the needed artifacts, in the next section we focus on approaches addressing user assistance tasks, such as query recommendation, and describe in more detail their metadata needs and management.

3   A Survey of User-Centric Approaches

As mentioned in the previous section, we subsequently discuss the approaches providing user support functionalities (typically, query recommendation), and primarily focusing on the metadata artifacts used and their exploitation. Thus, we focus on *analytical metadata* meant to support the user analytical tasks.

3.1   Methodology

As our work is motivated by [16], the search for relevant references started with it and the papers citing it / cited by it. We iteratively followed the references found looking for relevant approaches (on journals and conferences) and detecting the most relevant authors in this field. This search was complemented by the keyword searches on the main research related engines. Soon, the area of query recommendation proved to be the most related one and we repeated and refined our searches for relevant papers on this topic. During our search, our primary focus was to detect metadata artifacts used for the user assistance. Therefore, we selected those papers proposing concrete solutions (implementations and/or theoretical foundations for the user assistance) providing enough details about the detection and definition of metadata artifacts. Due to the space limitation, we present a subset of papers representing the whole set of papers found.

3.2　Classification of the Surveyed Approaches

In the typical user-system interaction, the user poses a query, the system processes the query and returns the query result to the user. Throughout this process the user often needs assistance. In the reviewed approaches, we encountered various forms of user support. Figure 1 describes the additional process flows triggered to provide such support and refers to the main *metadata artifacts* collected and exploited.

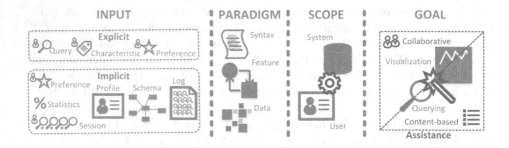

Fig. 1. Assistance Process

The system first gathers the needed **input** to be processed in order to achieve the ultimate **goal** of the user support (e.g., query recommendation). We further classify this process according to the level of abstraction or **paradigm** used to process the input data (syntactic, feature-based or data-based approaches) and the process **scope** (profiling the user, the system or both). Table 1 shows how the reviewed approaches were classified.

The **input** includes *metadata artifacts* that are defined in Table 2. As illustrated in Figure 1, we distinguish between *explicit* and *implicit* input. *Explicit* input represents data currently produced by the user that triggers the user assistance process of the system, whereas *implicit* input refers to input that was previously gathered (and stored in the metadata repository), either coming from the system or the user, as well as further metadata inferred from both. *Explicit*

Table 1. Classification of the reviewed approaches

Approach	Explicit Input	Implicit Input	Paradigm	Scope	Goal	Assistance Techniques
SQL QueRIE Recommendations [4]	Query	Query Log, User Session	Syntax, Data	User Profiling	Querying	Collaborative
Similarity Measures for OLAP Sessions [5]	Query	User Session, Schema	Syntax, Feature (Schema)	/	Querying	Content-based
Predicting Your Next OLAP Query Based on Recent Analytical Sessions [6]	Query	Query Log, Schema	Syntax, Feature (Schema)	System Profiling	Querying	Content-based
A Personalization Framework for OLAP Queries [7]	Query, Preferences	User Profile	Feature (Preferences)	User Profiling	Visualization	Content-based
Query Recommendations for Interactive Database Exploration [8]	Query	Query Log, User Session	Data	User Profiling	Querying	Collaborative
Expressing OLAP Preferences [12]	Query, Preferences	Schema	Feature (Preferences)	User Profiling	Querying	Content-based
myOLAP: An Approach to Express and Evaluate OLAP Preferences [13]	Query, Preferences	Schema	Feature (Preferences)	User Profiling	Querying	Content-based
SnipSuggest: Context-Aware Autocompletion for SQL [17]	Query	Query Log	Syntax	System Profiling	Querying	Content-based
The Meta-Morphing Model Used in TARGIT BI Suite [22]	Query	Statistics, Preferences, Schema	Feature (Preferences, Statistics)	User Profiling	Querying, Visualization	Content-based
"You May Also Like" Results in Relational Databases [30]	Query	Query Log, Schema	Data, Feature (Schema)	User Profiling, System Profiling	Querying	Content-based, Collaborative
Recommending Join Queries via Query Log Analysis [31]	Query	Query Log	Syntax	System Profiling	Querying	Content-based
A Framework for Recommending OLAP Queries [11]	Query	Query Log, User Session	Syntax	System Profiling	Querying	Content-based
Meet Charles, Big Data Query Advisor [28]	Query	Statistics	Feature (SDL, Statistics)	System Profiling	Querying	Content-based

input detected in our survey are queries, preferences, and user characteristics. *Implicit* input are previously logged queries, typically stored in the sequence they were posed (i.e., sessions), stored user profiles, automatically inferred user preferences, detected system statistics, and the system schema.

Queries are the main input considered in all surveyed approaches. Processing queries is primarily performed at three abstraction levels or **paradigms**: at the *syntax* level, at the *data* level, or modeled according to a certain *feature*. According to [16], *syntactic* processing happens when the syntactic structure of the query is the main facet to be explored (e.g., to combine fragments and propose new queries). For example, [6] presents a framework for recommending OLAP queries based on a probabilistic model of the user's behavior, which is computed by means of query similarities at the syntactic level. *Data-based* processing describes the query in terms of the data it retrieves. For example, in [8] the user queries are characterized by the retrieved tuples. Alternatively, *feature-based* processing models and stores the query in terms of a certain *feature*. *Features* encountered in the reviewed approaches are schema information, user preferences (and visual constraints), statistics, and ad-hoc languages to capture semantic fragments from the queries (e.g., the Segmentation

Table 2. Analytical Metadata Artifacts

Metadata Artifact	Definition	Example
Query	The user inquiry for certain data, disregarding the form it takes	What is the total quantity per product and location?
Preferences	The result set selection and/or representation prioritization	Preferred results of sales where amount is in between 1000 and 5000 range; Preferred representation is a pie chart
User characteristics	Explicitly stated data characterizing the user	Job position, department, office location
Query log	The list of all queries ever posed	{Query 1, query 2, query 3, query 4, ..., query N}
User session	Automatically detected sequence of queries posed by the user when analyzing or searching for certain data	<Query 1, query 3, query 7>
User profile	The set of user characteristics and preferences	Characteristics: User id '1', job position 'manager' Preferences: Preferred monthly over quarter overview
Statistics	Automatically detected data usage indicators	Product id 'P' searched in 23% (12345) of cases
System schema	The data model of the system	Dimension Tables: ProductDimension, DateDimension, LocationDimension Fact Tables: SalesFact

Description Language in [28]). Interesting conclusions can be drawn for each paradigm. *Syntactic* approaches lack semantics and suffer from several drawbacks. First, differently formulated queries returning the same data cannot be identified as equivalent. Second, the natural interconnection between a series of queries in a single analytical session is lost (or cannot be easily represented). Third, the data granularity produced cannot be detected (as in general, a pure syntactic approach is performed). All in all, exploitation is limited due to the usage of recorded syntactic artifacts only. *Data-based* approaches characterize queries according to the data they retrieve, which entails similar deficiencies due to the lack of semantics gathered. Similar queries returning disjoint results due to some filtering conditions or data aggregation cannot be identified. Also, their interconnection cannot be easily represented. Nevertheless, the main deficiency of this paradigm is its questionable feasibility in the context of BI 2.0 systems, which typically consider Big Data settings with large amounts of data.

None of the two previous paradigms are powerful and flexible enough to fully capture the intention of the user and detect similar queries in a broader sense. This is the main goal behind the last paradigm, which opens new possibilities for addressing this challenge. The concept of a *feature* focuses on modeling the input query to gain additional semantics representing meaningful information that previous paradigms miss. Several current approaches can be classified according to these terms. For example, [5], [6], and [30] represent the queries in terms of the schema. Moreover, [22] proposes the use of the multidimensional model to both model the system schema and the queries posed. For recommendations, it uses recorded user actions and preferences or predefined settings. However, the recommendation potential based on multidimensional semantics such as hierarchical dimension organization is not fully exploited. Finally, other approaches such as [12,13] introduce the means for the user to explicitly express her preferences when querying the data. However, the user is assumed to manually express the preferences.

The process **scope** defines the profiling need for user assistance purposes. We consider two scope types, *user profiling* that correlates (input) metadata artifacts with the user and *system profiling* that creates a general set of (input) metadata artifacts about the system. While metadata generated in both cases are then used for multiple user assistance purposes, most approaches focus on *user profiling* and few pay attention to *system profiling*, which opens new interesting possibilities, such as self-tuning systems.

The ultimate **goal** and the final step in the assistance process is the concrete user assistance produced. There are multiple forms of user assistance related to the different phases of user-system interaction. We generalize them into two major categories. The first category is *querying* assistance which covers various forms of user support when querying a database. The most typical querying assistance is query recommendation, but other tasks such as query completion, result selection, result recommendations, and join recommendations are identified in our survey. The second major category of user assistance is *visualization*. This assistance refers how to represent the query output in the most satisfactory way for the user. Although undoubtedly crucial for BI 2.0, little attention has been paid to this issue.

An orthogonal aspect to the two categories just discussed are the **assistance techniques** used to provide support. We typically talk about *collaborative* techniques, which entail assistance generation based on the metadata gathered for similar users, and *content-based* techniques, which only exploit the metadata related to a certain user for providing support.

Finally, since many ideas for database recommendation systems come from web solutions we aim at completing our survey by briefly positioning web recommender systems (e.g., see [2]) in terms of our classification. For generating personalized recommendations, web recommendation approaches typically rely on *user profiles*. They process user *queries* at the *data* level, i.e., according to the results retrieved, and profile users and, to some extent, systems. On this base, they provide *content-based, collaborative*, or hybrid recommendations. Additionally, as suggested in [2] and thoroughly elaborated in [3], current web recommendation systems should take into account context information for generating context-aware recommendations. In terms of our classification, the context is generally covered by either *user characteristics* and *user preferences*, or by data itself (e.g., in BI, the time/location information are typically covered by appropriate dimensions). Relevantly, web-based recommender systems strongly rely on *user characteristics* (mostly ignored in the database field) and profiles.

As result, by analyzing the described *user assistance process* (Figure 1), we identified currently used *metadata artifacts* (Table 2), outlined the importance of a *feature-based* approach for gathering and modeling metadata artifacts, remarked that *system profiling* can be used for more than just user assistance and highlighted the importance of *user characteristics* and profiling metadata.

4   The Analytical Metadata Challenge

The survey presented currently used metadata artifacts, their handling, and exploitation potential. Nevertheless, in the context of BI 2.0 systems there is a need for the automation of user assistance activities. Therefore, in this section, we propose a comprehensive framework to address this challenge. First, we present the metadata artifacts to be gathered in the Analytical Metadata (AM) repository and then we discuss how to gather, model, and process them in order to automate their management.

4.1   Analytical Metadata

AM are the set of metadata artifacts entailed by BI 2.0 systems to support the user decision making process. To clarify these artifacts we extend the end user metadata taxonomy from [10] as illustrated in Figure 2. The original taxonomy includes *definitional, data quality, navigational,* and *lineage* metadata categories that are business oriented and do not refer to technical artifacts. We define the technical interpretation of these categories, add a new category, and classify concrete metadata artifacts accordingly. The *definitional* category defines the integration schema, user characteristics, and a vocabulary of business terminology. The *data quality* category describes data set characteristics. The *navigational* category keeps evidence about how the user explores and navigates data. The *lineage* category captures the origin of data including data sources, transformations, and mappings. The *ratings* category covers metadata artifacts about user interests and data usage statistics.

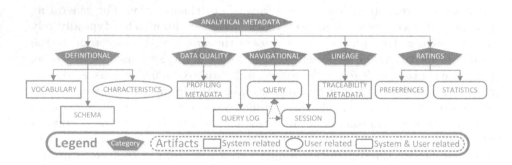

Fig. 2. Analytical Metadata Taxonomy

The schema, characteristics, query, session, query log, preferences, and statistics artifacts were introduced in Section 3. In this section, we classify these artifacts and introduce new ones to enhance automation. The metadata artifacts may refer to the system (e.g., schema), the user (e.g., characteristics) or both (e.g., a query defines user interests and refers to the schema structure).

Definitional. The *vocabulary* and *schema* artifacts are fundamental for all the other categories, i.e., all other metadata artifacts should be defined in terms of them. *Vocabulary* defines business terms, their relationships, and their mappings to the integration schema. Its primary role is to act as a reference terminology where to map all gathered metadata artifacts. It can efficiently be represented with an ontology [29] that is machine-processable and enables the automatic reasoning needed for the automation of user support. Next, in the context of BI systems, we propose the *schema* to be represented by means of the multidimensional (MD) model [18]. As discussed in [26], the MD model is mature and well-founded and has key applicability in data warehousing, on-line analytical processing (OLAP), and increasingly in data mining. It captures analytical perspectives by means of facts and dimensions. In this context, it defines necessary constraints and is covered by the MD algebra [27] that together determine potential user actions. For example, if a user analyzes data on a Month level, just based on the *schema* she can be suggested to change the granularity to the Day or Year level even if no one performed this analysis before. Lastly, the *user characteristics* artifact is borrowed from web recommender systems and defines the user by capturing explicitly asserted information that cannot be automatically detected (e.g., job position, age, etc.). It is typically stored as unstructured data and if defined in terms of the *vocabulary* it can be used for metadata processing, e.g., as parameter for recommending algorithms, pattern detection, etc.

Data Quality. To tackle *data quality* from a technical point of view, we propose metadata profiling processes (e.g., as explained in [24]) to gain insight into data. *Profiling metadata* characterize data sets like value range, number of values, number of unique values, sparsity, and similar metrics. This way data can be automatically annotated so that domain experts are provided with quality evidences for the data used. For example, inaccurate analytical results might come from data sparsity, i.e., a non-representative data sample.

Navigational. In compliance with the MD representation suggested for the *schema*, we propose the MD model as modeling feature (see previous section for further details) used to capture *queries*, *logs*, and *sessions*. Moreover, to better capture the user intentions, we propose to represent *queries* as ETL flows. As elaborated in [15], a query can be represented as a directed graph of operators. In turn, each operator is characterized as its input and output schema, which should follow the MD principles. This solution is more generic than a typical query definition and enables representation of more complex transformations (e.g., *rollup*), supports lineage, and can be represented as a graph that facilitates manipulations in comparison to declarative queries. Although more powerful, managing such a complex representation is more demanding (e.g., computing similarities or containment between ETL flows) and remains as an open challenge.

Lineage. For *lineage*, we propose the *traceability metadata* artifact that must capture the information about data sources, transformations performed when migrating data from the sources, and mappings to the integration *schema* (e.g., see [14, 15]). This way the system may provide the user with explanations about how an analytical value is computed and from what sources.

Ratings. The *user preferences* artifact can be manually stated, e.g., using a preference algebra [13]. Although this option might suit advanced users, whenever possible, it is preferable to automatically detect them by appropriate processing techniques over other metadata artifacts. For example, we may infer from the *queries* gathered that the user systematically applies some filtering predicates when navigating the data. Finally, the *statistics* artifact represents data usage indicators that can be considered as a kind of query profiling, i.e., to keep evidence about what data are more explored, as well as more complex indicators (e.g., the most popular combinations of fact and dimension tables [22]).

4.2 Automation and Processing

For the efficient management and storage of the AM we alternatively categorize its artifacts into the categories illustrated in Figure 3. These categories elaborate on how to gather each artifact, its level of processing automation, and exploitation purposes. This categorization must be used to guide the AM storing organization.

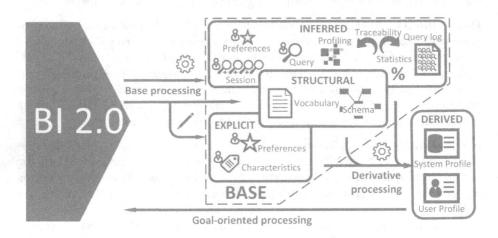

Fig. 3. Analytical Metadata Processing Categories

In the categorization, we talk about **structural, explicit, inferred**, and **derived** metadata. As discussed earlier, the *schema* and *vocabulary* artifacts are fundamental for all the other artifacts and for this reason we refer to them as structural artifacts. Typically, the system designers define and maintain these artifacts. Explicit artifacts (i.e., *user preferences* and *characteristics*) are those explicitly stated by the user and not automatically detected by the system. Contrary, inferred artifacts are automatically gathered by the system and thus, they can be automatically detected and stored without the explicit help of the user. This category mainly refers to *query logs* and *sessions*, but also to automatically detected *preferences*, data usage *statistics*, as well as the *profiling* and *traceability* metadata artifacts. Inferred artifacts are feature-based, i.e., either modeled according to the MD *schema* or in terms of the definitional *vocabulary*. The structural, explicit, and inferred categories jointly represent the minimal set of information about user actions and interests to be gathered for automating user assistance within the system. For this reason, we refer to these three categories as the **base** metadata, which determine the exploitation possibilities of the AM.

The remaining **derived** metadata category results from processing base metadata according to certain exploitation purposes (typically user and/or system profiling). The produced *user/system profiles* are materialized pieces of derived metadata typically aimed at improving the performance of the algorithms used to provide user assistance. For example, a collaborative recommending system requires to compute similar users. Computing similar users must be performed from base metadata artifacts and it can be previously materialized as derived metadata (i.e., user profiling) in order to improve the response time of the recommending system. Note that system profiling opens new possibilities for system self-tuning capabilities.

Automation implies processing flows (denoted as arrows in Figure 3) to populate and exploit the AM artifacts. Our goal here is not to define concrete algorithms but to point out the metadata management and processing flows to be considered when implementing the metadata repository. Consequently, we talk about **base**, **derivative**, and **goal-oriented processing**. The base processing populates the base metadata (i.e., the structural, explicit, and inferred artifacts) by interacting with the BI system. Specifically, the explicit metadata is stated by the user, whereas inferred artifacts are automatically detected and gathered from the system. Structural metadata is typically maintained by the system administrator. Contrary, the derivative processing populates the derived metadata from the base metadata gathered (i.e., derivative processing happens within the AM repository). Finally, goal-oriented processing exploits both the base and derived metadata in order to give concrete user assistance. It represents the purpose of the AM storage and exploitation. The aims of goal-oriented processing are, for example, query recommendations or visualization techniques, whereas final products are recommended queries, graphs/charts, and potentially self-tuning database actions for database optimizations (e.g., data usage *statistics* can serve to trigger indexing of some frequently used attribute).

All in all, due to the expected large volumes of metadata in these systems and the demanding processing capabilities described, the AM repository should be implemented in a dedicated subsystem.

5   Conclusions and Future Work

We have presented a comprehensive metadata framework to support user assistance and its automation in the context of BI 2.0 systems. The framework is based on the findings of a survey of existing user-centric approaches where we describe the user assistance process and identify assistance activities and metadata artifacts needed. It proposes an AM repository by categorizing the metadata artifacts to support the user assistance and their processing to enable automation. By introducing the subsets of automatically inferred and derived metadata artifacts with corresponding processing techniques, our framework motivates and directs the automation of the user assistance process which is one of the key requirements of BI 2.0 systems. As metadata artifacts are described on a high abstraction level, the framework is a base to support user assistance features over BI 2.0 heterogeneous data sources. Moreover, since AM also capture the information about the system usage, they can serve for other purposes typically overlooked, such as system self-tuning and optimization.

In our future work, we plan to define the metamodel of AM and provide an implementation of the AM support for query recommendation.

Acknowledgments. This research has been funded by the European Commission through the Erasmus Mundus Joint Doctorate "Information Technologies for Business Intelligence - Doctoral College" (IT4BI-DC) and it has been partly supported by the Spanish Ministerio de Ciencia e Innovación under project TIN2011-24747.

References

1. Abelló, A., Darmont, J., Etcheverry, L., Golfarelli, M., Mazón, J.N., Naumann, F., Pedersen, T.B., Rizzi, S., Trujillo, J., Vassiliadis, P., Vossen, G.: Fusion Cubes: Towards Self-Service Business Intelligence. IJDWM 9(2), 66–88 (2013)
2. Adomavicius, G., Tuzhilin, A.: Toward the Next Generation of Recommender Systems: A Survey of the State-of-the-Art and Possible Extensions. IEEE TKDE 17(6), 734–749 (2005)
3. Adomavicius, G., Tuzhilin, A.: Context-Aware Recommender Systems. In: Recommender Systems Handbook, pp. 217–253. Springer (2011)
4. Akbarnejad, J., Chatzopoulou, G., Eirinaki, M., Koshy, S., Mittal, S., On, D., Polyzotis, N., Varman, J.S.V.: SQL QueRIE Recommendations. PVLDB 3(2), 1597–1600 (2010)
5. Aligon, J., Golfarelli, M., Marcel, P., Rizzi, S., Turricchia, E.: Similarity Measures for OLAP Sessions. In: KAIS, pp. 1–27 (2013)
6. Aufaure, M.-A., Kuchmann-Beauger, N., Marcel, P., Rizzi, S., Vanrompay, Y.: Predicting Your Next OLAP Query Based on Recent Analytical Sessions. In: Bellatreche, L., Mohania, M.K. (eds.) DaWaK 2013. LNCS, vol. 8057, pp. 134–145. Springer, Heidelberg (2013)
7. Bellatreche, L., Giacometti, A., Marcel, P., Mouloudi, H., Laurent, D.: A Personalization Framework for OLAP Queries. In: DOLAP, pp. 9–18 (2005)
8. Chatzopoulou, G., Eirinaki, M., Polyzotis, N.: Query Recommendations for Interactive Database Exploration. In: Winslett, M. (ed.) SSDBM 2009. LNCS, vol. 5566, pp. 3–18. Springer, Heidelberg (2009)
9. Essaidi, M.: ODBIS: Towards a Platform for On-Demand Business Intelligence Services. In: EDBT/ICDT Workshops (2010)

10. Foshay, N., Mukherjee, A., Taylor, A.: Does Data Warehouse End-User Metadata Add Value? Commun. ACM 50(11), 70–77 (2007)
11. Giacometti, A., Marcel, P., Negre, E.: A Framework for Recommending OLAP Queries. In: DOLAP, pp. 73–80 (2008)
12. Golfarelli, M., Rizzi, S.: Expressing OLAP Preferences. In: Winslett, M. (ed.) SSDBM 2009. LNCS, vol. 5566, pp. 83–91. Springer, Heidelberg (2009)
13. Golfarelli, M., Rizzi, S., Biondi, P.: myOLAP: An Approach to Express and Evaluate OLAP Preferences. IEEE TKDE 23(7), 1050–1064 (2011)
14. Jovanovic, P., Romero, O., Simitsis, A., Abelló, A., Mayorova, D.: A Requirement-Driven Approach to the Design and Evolution of Data Warehouses. IS (2014)
15. Jovanovic, P., Romero, O., Simitsis, A., Abelló, A.: Integrating ETL Processes from Information Requirements. In: Cuzzocrea, A., Dayal, U. (eds.) DaWaK 2012. LNCS, vol. 7448, pp. 65–80. Springer, Heidelberg (2012)
16. Khoussainova, N., Balazinska, M., Gatterbauer, W., Kwon, Y., Suciu, D.: A Case for A Collaborative Query Management System. In: CIDR (2009)
17. Khoussainova, N., Kwon, Y., Balazinska, M., Suciu, D.: SnipSuggest: Context-Aware Autocompletion for SQL. PVLDB 4(1), 22–33 (2010)
18. Kimball, R., Reeves, L., Ross, M., Thornthwaite, W.: The Data Warehouse Lifecycle Toolkit: Expert Methods for Designing, Developing, and Deploying Data Warehouses. John Wiley & Sons, Inc. (1998)
19. Leonard, A., Masson, M., Mitchell, T., Moss, J., Ufford, M.: Business Intelligence Markup Language. In: SQL Server 2012 Integration Services Design Patterns, pp. 301–326. Springer (2012)
20. Löser, A., Hueske, F., Markl, V.: Situational Business Intelligence. In: Castellanos, M., Dayal, U., Sellis, T. (eds.) BIRTE 2008. LNBIP, vol. 27, pp. 1–11. Springer, Heidelberg (2009)
21. Mazón, J.N., Zubcoff, J.J., Garrigós, I., Espinosa, R., Rodríguez, R.: Open Business Intelligence: On the Importance of Data Quality Awareness in User-Friendly Data Mining. In: EDBT/ICDT Workshops, pp. 144–147 (2012)
22. Middelfart, M., Pedersen, T.B.: The Meta-Morphing Model Used in TARGIT BI Suite. In: De Troyer, O., Bauzer Medeiros, C., Billen, R., Hallot, P., Simitsis, A., Van Mingroot, H. (eds.) ER 2011 Workshops. LNCS, vol. 6999, pp. 364–370. Springer, Heidelberg (2011)
23. Morton, K., Balazinska, M., Grossman, D., Mackinlay, J.D.: Support the Data Enthusiast: Challenges for Next-Generation Data-Analysis Systems. PVLDB 7(6), 453–456 (2014)
24. Naumann, F.: Data Profiling Revisited. SIGMOD Record 32(4) (2013)
25. Object Management Group: Common Warehouse Metamodel Specification 1.1, http://www.omg.org/spec/CWM/1.1/PDF/ (last accessed January 2014)
26. Pedersen, T.B.: Multidimensional Modeling. In: Encyclopedia of Database Systems, pp. 1777–1784. Springer US (2009)
27. Romero, O., Abelló, A.: On the Need of a Reference Algebra for OLAP. In: Song, I.-Y., Eder, J., Nguyen, T.M. (eds.) DaWaK 2007. LNCS, vol. 4654, pp. 99–110. Springer, Heidelberg (2007)
28. Sellam, T., Kersten, M.L.: Meet Charles, Big Data Query Advisor. In: CIDR (2013)
29. Skoutas, D., Simitsis, A.: Ontology-Based Conceptual Design of ETL Processes for Both Structured and Semi-Structured Data. IJSWIS 3(4), 1–24 (2007)
30. Stefanidis, K., Drosou, M., Pitoura, E.: "You May Also Like" Results in Relational Databases. In: PersDB Workshop (2009)
31. Yang, X., Procopiuc, C.M., Srivastava, D.: Recommending Join Queries via Query Log Analysis. In: ICDE, pp. 964–975 (2009)

Mining Fuzzy Contextual Preferences

Sandra de Amo and Juliete A. Ramos Costa

Federal University of Uberlândia,
Faculty of Computer Science,
Uberlândia, Brazil
deamo@ufu.br, juliete@mestrado.ufu.br

Abstract. Recent research work on preference mining has focused on
the development of methods for mining a preference model from preference data following a *crisp* pairwise representation. In this representation, the user has two options regarding a pair of objects u and v: either
he/she prefers u to v or v to u. In this article, we propose *FuzzyPrefMiner*,
a method for extracting *fuzzy contextual* preference models from *fuzzy*
preference data characterized by the fact that, given two objects u, v the
user has a spectrum of options according to his *degree of preference* on
u and v. Accordingly, the mined preference model is *fuzzy*, in the sense
that it is capable to predict, given two new objects u and v, the degree of preference the user would assign to these objects. The efficiency
of FuzzyPrefMiner is analysed through a series of experiments on real
datasets.

1 Introduction

In recent years the topic of Preference Mining has been attracting the attention
of the data mining research community due its important role in the development
of efficient content-based recommendation systems [4, 13]. A preference mining
algorithm aims at extracting an accurate *preference model* from a sample dataset
of preferences provided by the user.

The formalisms commonly used for preference representation in the preference
mining techniques found in the literature are *score functions* [3, 6, 7] (the user
provides a rating for each object in the sample dataset), pairwise comparison of
alternatives [1, 2, 9, 10, 12] (for each pair of objects in the sample database, the
user informs which one he/she prefers) and ranking of alternatives [8, 11] (the
user provides a sequence of objets in decreasing order of preference).

Classifiers versus Preference Mining Techniques. Most of the preference mining
techniques used in content-based recommendation systems are simply traditional
classification techniques: the training set is constituted by user ratings on items,
ratings are viewed as *classes*, and a classifier is used to infer the rating a user
would assign to a new item. However, depending on the application, the use
of a classifier to predict user ratings is not suitable. For instance, when a user
evaluates a movie, the given rating is normally related not only to the intrinsic
characteristics of the movie, but also to the ratings already given by the user

L. Bellatreche and M.K. Mohania (Eds.): DaWaK 2014, LNCS 8646, pp. 102–114, 2014.

on other movies. So, the important information here is not the exact rating given to movies, but how the user *compares* pairwise alternatives. So, for some applications, specific preference mining techniques for extracting a preference model from a set of pairs of alternatives are more suitable.

In our previous work [2] we proposed the algorithm CPrefMiner for mining a *crisp* contextual preference model from a set of preference samples represented as pairs of alternatives. The model is said to be *crisp* since the input data provided by the user is based on *yes/no* alternatives: either he/she prefers object u to object v (denoted by $u \succ v$ for short) or vice-versa and the output model allows to predict if a new object w_1 is preferred to a new object w_2 or vice-versa. The mined preference model is contextual, constituted by a set of *contextual preference rules* ([15]) of the form *IF <context> THEN I prefer 'this' to 'that'* . Example of such rules is: *For films directed by Spielberg (context) I prefer 'Action' to 'Comedy'*. The preference model extracted by CPrefMiner is capable to infer an *order relation on objects* (a task a classifier would not be able to achieve) that is, it is capable to infer *transitivity* relatioships (if $u \succ v$ and $v \succ w$ then $u \succ w$ besides satisfying *irreflexibility* (if $u \succ v$ then $v \not\succ u$).

Preference mining techniques that use input data following a pairwise representation normally execute a pre-processing step for transforming user ratings into pairs. For instance, if the user rating for u is 5 and for v is 1, this information is transformed into the pair (u, v). We claim that in this transformation, a lot of information is lost. Indeed, if the ratings were 5 and 4 for u and v respectively, then the same pair (u, v) would be obtained. However, in the first situation, the user prefers object u to v with more *intensity* than in the second situation.

In this paper we propose the method FuzzyPrefMiner to mine *fuzzy* contextual preference model from a set of preferences samples represented as triples (u, v, n) where u and v are objects evaluated by a user U and n is the *degree of preference* $(0 \leq n \leq 1)$ from u to v provided by this user. An extensive set of experiments comparing CPrefMiner and FuzzyPrefMiner validates this hypothesis: FuzzyPrefMiner is far more accurate then CPrefMiner to predict user preferences. Besides, the preference knowledge extracted by FuzzyPrefMiner is more refined then the one provided by CPrefMiner, since it is capable to predict not only if $u \succ v$ but also the degree of preference of u with respect to v.

Fuzzy preference modeling and reasoning has been extensively studied in the last decade [5,14,16]. However, in the best of our knowledge this kind of preference model has not yet been explored in preference mining research.

2 Background on Crisp Contextual Preference Mining [2]

A *Crisp Preference Database* (or *CPD* for short) over a schema $R(A_1, ..., A_n)$ is a finite set $\mathcal{P} \subseteq \mathrm{Tup}(R) \times \mathrm{Tup}(R)$ which is *consistent*, that is, if $(u, v) \in \mathcal{P}$ then $(v, u) \notin \mathcal{P}$. Here, $\mathrm{Tup}(R)$ denotes the set all tuples over relational schema R. The pair (u, v) represents the fact that the user prefers *the tuple u to the tuple v* $(u \succ v)$. In the crisp scenario, a user has two options concerning objects u and v: either $u \succ v$ or $v \succ u$.

Fig. 1(Ib) illustrates a preference database, representing a sample of alternatives extracted from the original information (ratings) provided by the user about his/her preferences (Fig. 1(Ia)).

The problem of mining *crisp* preferences, as illustrated in Figure 1(I), consists in extracting a *crisp preference model* (Fig.1(Ic)) from a *crisp* preference database (Fig.1(Ib)) provided by the user. A crisp preference model is a target function designed to predict, given two tuples t_1 and t_2, which one is preferred by the user. The preference model we considered in our previous research [2] as well as in the present work is specified by a *Bayesian Preference Network* (BPN), consisting in (1) a directed acyclic graph G whose nodes are attributes and the edges stand for attribute dependency and (2) a mapping θ that associates to each node of G a finite set of conditional probabilities. Fig. 1(II) illustrates a c-BPN **PNet$_1$** over the relational schema $R(A, B, C, D)$. Notice that the preference on values for attribute B depends on the context C: if $C = c1$, the probability that value b_1 is preferred to value b_2 for the attribute B is 60%. For that reason, we call our model (the BPN) a *contextual* preference model. In order to emphasize the fact that the BPN is extracted from a crisp preference database, we denote it by c-BPN.

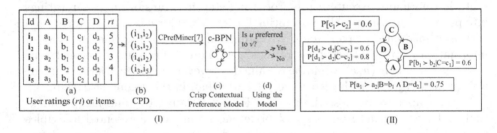

Fig. 1. (I) The Crisp Contextual Preference Mining Process; (II) a c-BPN **PNet$_1$**

How a c-BPN Is Used to Predict the User Preference on New Items.
A c-BPN should be capable to infer a *preference ordering* on tuples (Fig. 1(Id)). The following example illustrates how this ordering is effectively calculated. For more details on the theoretical background behind this ordering, please see [15].

Example 1. Let us consider the c-BPN **PNet$_1$** depicted in Fig. 1(II). In order to compare two **new** tuples $u_1 = (a_1, b_1, c_1, d_1)$ and $u_2 = (a_2, b_2, c_1, d_2)$, we proceed as follows: (1) Let $\Delta(u_1, u_2)$ be the set of attributes for which u_1 and u_2 differ. In this example, $\Delta(u_1, u_2) = \{A, B, D\}$; (2) Let $\min(\Delta(u_1, u_2)) \subseteq \Delta(u_1, u_2)$ such that the attributes in $\min(\Delta)$ have no ancestors in Δ (according to graph G underlying the c-BPN **PNet$_1$**). In this example $\min(\Delta(u_1, u_2)) = \{D, B\}$. The necessary and sufficient conditions for u_1 to be preferred to u_2 are: $u_1[D] > u_2[D]$ and $u_1[B] > u_2[B]$; (3) Compute the following probabilities: $p_1 =$ probability that $u_1 > u_2 = P[d_1 > d_2 | C = c_1] * P[b_1 > b_2 | C = c_1] = 0.6 * 0.6 = 0.36$; $p_2 =$ probability that $u_2 > u_1 = P[d_2 > d_1 | C = c_1] * P[b_2 > b_1 | C = c_1] = 0.4 * 0.4 = 0.16$. In order to compare u_1 and u_2 we select the higher between p_1 and p_2. In this example, $p_1 > p_2$ and so, we infer that u_1 is preferred to u_2. If $p_1 = p_2$, a random choice decides if u_1 is preferred to u_2 or vice-versa. So, a BPN is capable to compare any pair of tuples: either it *effectively* infer the preference ordering without randomness or it makes a random choice.

How to Measure the Efficiency of a c-BPN. Let M, N_r, N_a and N_c denote respectively the cardinality of the following sets: (1) $\mathcal{P}$; (2) set of pairs in $\mathcal{P}$ compatible with the preference ordering effectively inferred by **PNet**; (3) set of pairs in $\mathcal{P}$ whose ordering has been randomly correctly inferred and (4) set of pairs in $\mathcal{P}$ that have been effectively compared by **PNet** (correctly or not). The following measures are used to evaluate the quality of the mined preference model:

(1) **Accuracy**(acc): defined by $acc(\textbf{PNet},\mathcal{P})=\frac{N_r+N_a}{M}$.

(2) ***Recall(rec)***: defined by $rec(\textbf{PNet},\mathcal{P})=\frac{N_r}{M}$.

(3) ***Randomness Rate(rr)***: defined by $rr(\textbf{PNet},\mathcal{P})=\frac{N_a}{M}$.

(4) ***Precision(prec)***: defined by $prec(\textbf{PNet},\mathcal{P})=\frac{N_r}{N_c}$.

(5) ***Comparability Rate(cr)***: defined by $cr(\textbf{PNet},\mathcal{P})=\frac{N_c}{M}$.

Notice that these five measures are related by the following equations $acc = rec + rr$ and $rec = prec * cr$.

The Problem of Mining Crisp Contextual Preferences. This problem consists in, given a crisp preference database, return a BPN having good quality with respect to $acc, prec, rec, cr$ and rr measures. It has been introduced in [2] as well as the algorithm CPrefMiner designed to solve it.

3   Fuzzy Contextual Preference Mining

A *Fuzzy Preference Relation* over a set of objects $X = \{x_1, ..., x_n\}$ is a $n \times n$ matrix P with $P_{ij} \in [0, 1]$ representing the *degree of preference (dp)* on objects x_i and x_j. Fuzzy preference relations have been introduced in [5] and should satisfy certain conditions (see [5] for details), the most important being the *reciprocity* property $P_{ij} = 1 - P_{ji}$ (which entails that $P_{ii} = 0.5$). One way of obtaining fuzzy preference relations from a set of scores provided by the user is to consider $P_{ij} = f(\frac{r_i}{r_j})$, where r_i is the rating u assigned to object x_i and f is a function belonging to the family $\{f_n : n \geq 1\}$, $f_n(x) = \frac{x^n}{x^n+1}$. Fig. 2(b) illustrates the fuzzy preference relation corresponding to the item-rating database of Fig. 2(a).

A *Fuzzy Preference Database (FDP)* over a relation schema $R(A_1, ..., A_n)$ is a finite set $\mathcal{P} \subseteq \text{Tup}(R) \times \text{Tup}(R) \times [0,1]$ which is *consistent*, that is if (u, v, n) and $(v, u, m) \in \mathcal{P}$ then $m = 1 - n$. The triple (u, v, n), represents the fact that the user prefers *the tuple u to the tuple v* with a degree of preference n. In the fuzzy scenario, a user has a whole spectrum of options concerning objects u and v, by varying his/her degree of preference concerning these objects. Fig. 2(c) illustrates an FDP, corresponding to a *subset* of the information contained in the matrix depicted in Fig. 2(b).

The problem of mining *fuzzy* preferences, as illustrated in Figure 2, consists in extracting a *fuzzy preference model* from an *FDP*. A fuzzy preference model

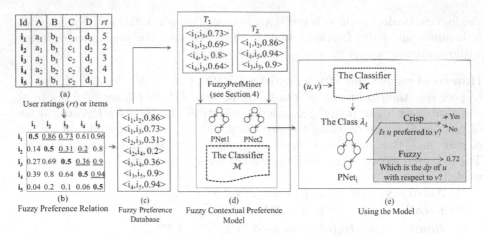

Fig. 2. Fuzzy Contextual Preference Mining Process

is a target function designed to predict, given two tuples t_1 and t_2, which is the degree of preference of t_1 with respect to t_2.

The Fuzzy Preference Model. Let $\mathcal{P}$ be an FDP and dp_{min} (resp. dp_{max}) be the smaller (resp. the larger) $dp > 0.5$ appearing in $\mathcal{P}$. Let $k \in \mathbb{N}$, $k \geq 1$.

k-Partitioning: First, let us partition the interval $I = [dp_{min}, dp_{max}]$ into k disjoints and contiguous subintervals $I_1, ..., I_k$, such that $\bigcup_{i=1}^{k} I_i = I$. Second, partition $\mathcal{P}$ into k disjoint subsets of triples $\mathcal{P}_1, ..., \mathcal{P}_k$. Each $\mathcal{P}_i$ is obtained by inserting into $\mathcal{P}_i$ all triples (u, v, n) with $n \geq 0.5$ such that (u, v, n) or $(u, v, 1 - n) \in \mathcal{P}$. Figure 2(d) illustrates this step.

Layers can be viewed as classes: Let $I_i = [a_i, b_i]$ and $\lambda_i = \frac{a_i + b_i}{2}$. Triples (u, v, n) in $\mathcal{P}_i$ are characterized by tuples u and v having an average "intensity" of preference λ_i between them.

The Fuzzy Preference Model Proposed: The fuzzy preference model we propose in this paper is a structure $\mathcal{F}_k = <\mathcal{M}, PNet_1, ..., PNet_k>$, where $\mathcal{M}$ is a classification model extracted from the set of layers $\mathcal{P}_1, ..., \mathcal{P}_k$ (considering elements in $\mathcal{P}_i$ as (u, v, n, λ_i) with an extra "class" dimension λ_i), and $PNet_i$ is a BNP extracted from layer $\mathcal{P}_i$, for all $i = 1, ..., k$.

It is important to emphasize that the BPN extracted from an FDP (denoted by f-BPN) is not the same as the c-BPN extracted from the corresponding CPD. Indeed, the f-BPN encloses more information than its crisp counterpart c-BPN.

Using $\mathcal{F}_k$ to Predict Fuzzy Preference between Tuples. Let u, v be two new tuples. First of all, we have to infer the intensity of preference between them. The classifier $\mathcal{M}$ is responsible for this task: it is executed on (u, v) and returns a class λ_i for this pair. Then the f-BPN $PNet_i$, specific for this layer, is executed on (u, v) to decide which one is preferred. This execution follows the same steps as illustrated in Example 1. If $PNet_i$ decides that u is preferred to v (resp. v is preferred to u), the final prediction is : *u is preferred to v with degree of preference* λ_i (resp. $1 - \lambda_i$).

Measuring the Quality of a Fuzzy Preference Model. Since an exact match between the predicted dp (dp_p) and the *real* one (dp_r) is quite unlikely to occur, we consider a threshold $\sigma > 0$ to measure how good is the prediction. We consider that the predicted dp_p is *correct* if the following conditions are verified: (1) $\mid dp_r - dp_p \mid \leq \sigma$ and (2) ($dp_r \geq 0.5$ and $dp_p \geq 0.5$) or ($dp_r < 0.5$ and $dp_p < 0.5$).

Example 2. Let us consider tuples u_1 and u_2 of Example 1. Let us also suppose the input FDP depicted in Fig. 2(d) with its 2-layer partition T_1 and T_2. The corresponding classes are $\lambda_1 = 0.72$ (average dp of $[0.64, 0.8]$) and $\lambda_2 = 0.9$ (average dp of $[0.86, 0.94]$). Let us suppose that $dp_r(u_1, u_2) = 0.78$ and that the classifier $\mathcal{M}$ assigns the class λ_1 to (u_1, u_2). Notice that the classifier only predicts that the insensity of preference between the two tuples is 0.72. It doesn't say anything about the *ordering* of preference between them. This task is achieved by the BNP_1 associated to partition T_1. Let us suppose that BPN_1 is the same as the one illustrated in Fig. 1(2). It compares u_1 and u_2 as described in Example 1, by calculating the probabilities $p_1 = 0.36$ and $p_2 = 0.16$. As $p_1 > p_2$ then the predicted dp is $dp_p = \lambda_1 = 0.72$. So, both dp_r and dp_r are ≥ 0.5. Let us consider parameter $\sigma = 0.05$. Since $\mid 0.78 - 0.72 \mid = 0.06 > 0.05$, we conclude that the fuzzy model wrongly predicted the dp of u_1 with respect to u_2.

The Problem of Mining Fuzzy Contextual Preferences. This problem consists in, given an FDP and $k > 0$, return a fuzzy preference model $\mathcal{F}_k$ having good quality with respect to $acc, prec, rec, cr$ and rr measures and a given threshold σ.

4   The FuzzyPrefMiner Algorithm

The FuzzyPrefMiner algorithm we propose to solve the Fuzzy Contextual Preference Mining Problem follows the same strategy of the algorithm CPrefMiner proposed in [2]. The general framework of the algorithm is showed in Alg.1. In this section we present the details only of the parts of the method that have been modified in order to treat the fuzzy information (indicated inside a rectangle, in boldface).

Algorithm 1. The FuzzyPrefMiner Algorithm

Input: An FDP $\mathcal{P}_i$ over relational schema $R(A_1, ..., A_n)$ % *corresponding to a layer in $\mathcal{P}$*;
Output: An f-BPN **PNet**

1 *Extract*($\mathcal{P}, G$);
2 **for** *each* $i = 1, ..., n$ **do**
3   *Parents*(G, A_i) $= [A_{i_1}, ..., A_{i_m}]$;
4      $\boxed{CPTable(A_i, \textbf{Parents}(A_i, G)) = [CProb_1, ..., CProb_l]}$;
5 **return** *PNet*

4.1   Learning the Network Structure

Procedure $Extract(\mathcal{P}, G)$ is responsible for learning the network topology G from the training preference database $\mathcal{P}$. G is a graph with vertices in $A_1, ..., A_n$. An edge (A_i, A_j) in G means that the preference on values for attribute A_j depends on values of attribute A_i. That is, A_i is part of the context for preferences over the attribute A_j This learning task is performed by a genetic algorithm which generates an initial population $\mathbf{P}$ of graphs with vertices in $\{A_1, ..., A_n\}$ and for each graph $G \in \mathbf{P}$ evaluates a fitness function $score(G)$. Full details on the codification of individuals and on the crossover and mutation operations are presented in [2].

Algorithm 2. *Extract* Procedure - (from [2])

Input: $\mathcal{P}$: an FDP over relational schema $R(A_1, ..., A_n)$; β: population size; ϑ: number of generations; γ: number of attribute orderings

Output: A directed graph $G = (V, E)$, with vertices $V \subseteq \{A_1, ..., A_n\}$ that fits best to the fitness function

1 **for** *each* $i = 1, ..., \gamma$ **do**
2   Randomly generate an attribute ordering;
3   Generate an initial population I_0 of random individuals;
4      $\boxed{\textbf{\textit{Evaluate individuals in }} I_0, \textbf{\textit{ i.e, calculate their fitness}}}$;
5   **for** *each* j-*th generation*, $j = 1, ..., \vartheta$ **do**
6   Select $\beta/2$ pairs of individuals from I_j;
7   Apply crossover operator for each pair, generating an offspring population I_j';
8   Evaluate individuals of I_j';
9   Apply mutation operator over individuals from I_j', and evaluate mutated ones;
10   From $I_j \cup I_j'$, select the β fittest individuals through an elitism procedure;
11   Pick up the best individual after the last generation;
12 **return** *the best individual among all* γ *GA executions*

The Fitness Function. The main idea of the fitness function is to assign a real number (a score) in $[-1, 1]$ for a candidate structure G, aiming to estimate how good it captures the dependencies between attributes in a fuzzy preference database $\mathcal{P}$. In this sense, each network arc is "punished" or "rewarded", according to the matching between each arc (X, Y) in G and the corresponding *degree of dependence* of the pair (X, Y) with respect to $\mathcal{P}$.

The *degree of dependence* of a pair of attributes (X, Y) with respect to an FDP $\mathcal{P}$ is a real number that estimates how preferences on values for the attribute Y are influenced by values for the attribute X. Its computation is described in Alg. 3. In order to facilitate the description of Alg. 3 we introduce some notations as follows: **(1)** For each $y, y' \in \mathbf{dom}(Y)$, $y \neq y'$ we denote by $T_{yy'}$ the subset of pairs $(t, t') \in \mathcal{P}$, such that $t[Y] = y \wedge t'[Y] = y'$ or $t[Y] = y' \wedge t'[Y] = y$;

Algorithm 3. The degree of dependence between a pair of attributes

Input: $\mathcal{P}$: an FDP ; (X, Y): a pair of attributes; two thresholds $\alpha_1 \geq 0$ and $\alpha_2 \geq 0$.

Output: The Degree of Dependence of (X, Y) with respect to $\mathcal{P}$

1 **for** *each pair* $(y, y') \in \boldsymbol{dom}(Y) \times \boldsymbol{dom}(Y)$, $y \neq y'$ *and* (y, y') *comparable* **do**

2   **for** *each* $x \in \boldsymbol{dom}(X)$ *where* x *is a cause for* (y, y') *being comparable* **do**

3   Let $f_1(S_{x|(y,y')}) = \max\{N, 1 - N\}$, where
$$N = \frac{\{\sum dp \in DP(S_{x|(y,y')})\} : t > t' \wedge (t[Y] = y \wedge t'[Y] = y')}{\sum dp \in DP(S_{x|(y,y')})}$$

4   Let $f_2(T_{yy'}) = \max \{f_1(S_{x|(y,y')}) : x \in \boldsymbol{dom}(X)\}$

5 Let $f_3((X, Y), \mathcal{T}) = \max\{f_2(T_{yy'}) : (y, y') \in \boldsymbol{dom}(Y) \times \boldsymbol{dom}(Y), y \neq y', (y, y')$ comparable$\}$

6 **return** $f_3((X, Y), \mathcal{T})$

---

(2) If S is a set of triples in $\mathcal{P}$, we denote by $DP(S)$ the (multi)set constituted by the dp appearing in the triples of S (repetitions are considered as different elements); **(3)** We define support$((y, y'), \mathcal{P}) = \frac{\sum dp \in DP(T_{y,y'})}{\sum dp \in DP(\mathcal{P})}$. We say that the pair $(y, y') \in \textbf{dom}(Y) \times \textbf{dom}(Y)$ is *comparable* if support$((y, y'), \mathcal{P}) \geq \alpha_1$, for a given threshold α_1, $0 \leq \alpha_1 \leq 1$; **(4)** For each $x \in \textbf{dom}(X)$, we denote by $S_{x|(y,y')}$ the subset of $T_{yy'}$ containing the triples (t, t', n) such that $t[X] = t'[X] = x$; **(5)** We define support$((x|(y, y')), \mathcal{P}) = \frac{\sum_{dp \in DP(S_{x|(y,y')})} dp}{\sum_{z \in \textbf{dom}(X), dp \in DP(S_{z|(y,y')})} dp}$; **(6)** We say that x is a *cause for* (y, y') *being comparable* if support$(S_{x|(y,y')}, \mathcal{P}) \geq \alpha_2$, for a given threshold α_2, $0 \leq \alpha_2 \leq 1$.

The fitness function $score(G)$ is calculated by $\dfrac{\sum_{X,Y} g((X, Y), G)}{n(n-1)}$ where function g is defined as follows: **(1)** If $f_3((X, Y), \mathcal{P}) \geq 0.5$ and edge $(X, Y) \in G$, then $g((X, Y), G) = f_3((X, Y), \mathcal{P})$; **(2)** If $f_3((X, Y), \mathcal{P}) \geq 0.5$ and edge $(X, Y) \notin G$, then $g((X, Y), G) = -f_3((X, Y), \mathcal{P})$; **(3)** If $f_3((X, Y), \mathcal{P}) < 0.5$ and edge $(X, Y) \notin G$, then $g((X, Y), G) = 1$; **(4)** If $f_3((X, Y), \mathcal{P}) < 0.5$ and edge $(X, Y) \in G$, then $g((X, Y), G) = 0$. More details on the motivation behind this computation, please see [2].

4.2 Calculating the Probability Tables Associated to each Vertex

Procedure $CPTable(A_i, \text{Parents}(A_i))$ returns, for each vertex A_i of the graph returned by *Extract*, a list of conditional probabilities $[CProb_1, ..., CProb_l]$. Each conditional probability $CProb_i$ is of the form $Pr[E_1|E_2]$ where E_2 is an event of the form $A_{i_1} = a_1 \wedge ... \wedge A_{i_l} = a_l$ and E_1 is event of the form $(B = b_1) \succ (B = b_2)$.

The second point in the FuzzyPrefMining algorithm where the degree of preference is relevant is in procedure *CPTable*. We calculate the maximum likelihood estimates for each conditional probability distribution of our model. The underlying intuition of this principle uses frequencies as estimates; for instance, if want to estimate $P(A = a > A = a'|B = b, C = c)$ we need to calculate $\frac{\sum n \in DP(S(a,a'|b,c))}{\sum m \in DP(S(a,a'|b,c)) + \sum m' \in DP(S(a',a|b,c))}$, where $S(a, a'|b, c) = $ set of triples (t, t', dp) such that $t[B] = t'[B] = b$ and $t[C] = t'[C] = c$ and $t[A] = a$ and $t'[A] = a'$.

5   Experimental Results

5.1   Preparing the Experiments

The datasets used in the experiments have been obtained from the GroupLens Project[1]. The original data are of the form (UserId, FilmId, rating). More details about the movies, namely Genre, Actors, Director, Year and Language have been obtained by means of a crawler which extracted this information from the IMDB website[2]. Six datasets of film evaluations, corresponding to six different users have been considered. Details about them are shown in Fig. 3(a).

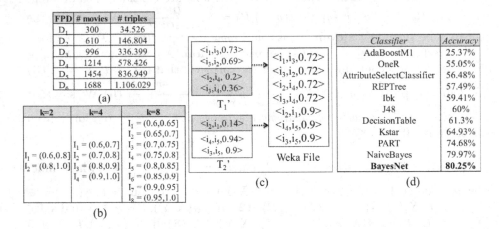

(a)

FPD	# movies	# triples
D_1	300	34.526
D_2	610	146.804
D_3	996	336.399
D_4	1214	578.426
D_5	1454	836.949
D_6	1688	1.106.029

(b)

k=2	k=4	k=8
		$I_1 = (0.6, 0.65]$
		$I_2 = (0.65, 0.7]$
	$I_1 = (0.6, 0.7]$	$I_3 = (0.7, 0.75]$
$I_1 = (0.6, 0.8]$	$I_2 = (0.7, 0.8]$	$I_4 = (0.75, 0.8]$
$I_2 = (0.8, 1.0]$	$I_3 = (0.8, 0.9]$	$I_5 = (0.8, 0.85]$
	$I_4 = (0.9, 1.0]$	$I_6 = (0.85, 0.9]$
		$I_7 = (0.9, 0.95]$
		$I_8 = (0.95, 1.0]$

(c)

$<i_1,i_3,0.73>$
$<i_3,i_2,0.69>$
$<i_2,i_4, 0.2>$
$<i_3,i_4,0.36>$
T_1'

$<i_2,i_1,0.14>$
$<i_4,i_5,0.94>$
$<i_3,i_5, 0.9>$
T_2'

$<i_1,i_3,0.72>$
$<i_3,i_2,0.72>$
$<i_2,i_4,0.72>$
$<i_3,i_4,0.72>$
$<i_2,i_1,0.9>$
$<i_4,i_5,0.9>$
$<i_3,i_5,0.9>$
Weka File

(d)

Classifier	Accuracy
AdaBoostM1	25.37%
OneR	55.05%
AttributeSelectClassifier	56.48%
REPTree	57.49%
Ibk	59.41%
J48	60%
DecisionTable	61.3%
Kstar	64.93%
PART	74.68%
NaiveBayes	79.97%
BayesNet	**80.25%**

Fig. 3. (a) The FDPs, (b) The Preference Layers, (c) Preparing the training data for classification, (d) Classifier's Accuracy

Training a Classifier. A classifier has been trained on each dataset in order to predict the "intensity of preference" existing between pairs of movies. A preliminary pre-processing is needed: Each preference layer should contain 50% of triples with $dp \geq 0.5$ and 50% with $dp < 0.5$. This pre-processing is illustrated in Fig. 3(b), where the preference layers T_1 and T_2 of Fig. 2(d) have been transformed into layers T_1' and T_2'. Due to the computation carried out by the *fuzzification* function f_2 (as explained in Section 3) we have $dp_{min} = 0.6$ and $dp_{max} = 1.0$ for each FPD considered in our experiments. The interval I =

---

[1] www.grouplens.org
[2] imdb.com

$[0.6, 1.0]$ is partionned into k layers, and triples of each layer receive the extra class attribute with value λ_i.

Ten Weka classifiers have been trained on a FDP of 31713 triples partionned into 4 layers: $T_1 = (0.6, 0.7]$, $T_2 = (0.7, 0.8]$, $T_3 = (0.8, 0.9]$ and $T_4 = (0.9, 1.0]$. The accuracy results are shown in Fig. 3(d). The bayesian classifier *BayesNet* presented the best results. Besides it executes very fast. Thus, *BayesNet* has been chosen for the remaining experiments concerning FuzzyPrefMiner evaluation.

5.2   Performance and Scalability Results

FuzzyPrefMiner was implemented in Java and all the experiments were performed on a core i7 3.20GHZ processor, with 32GB RAM, running on operation system Linux Ubuntu 12.10. The following default values have been set for some parameters involved in the algorithm: (1) *Extract* Procedure: $\beta = 50$, $\vartheta = 100$ and $\gamma = 20$. In Algorithm 3 (calculating the degree of dependence between attributes): $\alpha_1 = 0.2$ e $\alpha_2 = 0.2$. A *30-cross-validation* protocol has been considered in the experiments with FuzzyPrefMiner.

The Effect of the Number of Layers. Of course, FuzzyPrefMiner can also solve the crisp preference mining problem. Our first group of experiments aims at evaluating the performance of the f-BPN extracted by FuzzyPrefMiner, in the *crisp* scenario, by varying the number of layers considered in the partition of the FDP. The results on the accuracy, precision and recall are shown in Fig. 4(a). As expected, the more refined the partitioning (8 layers) the better are the performance results.

FPD	k=2			k=4			k=8			FPD	k=2		k=4		k=8	
	acc	rec	prec	acc	rec	prec	acc	rec	prec		cr	rr	cr	rr	cr	rr
D1	75.59%	73.23%	76.72%	82.48%	80.76%	83.67%	**87.41%**	**85.64%**	**88.80%**	D1	95.46%	2.35%	**96.53%**	**1.72%**	96.45%	1.76%
D2	74.96%	73.28%	75.84%	83.29%	81.98%	84.20%	**88.42%**	**87.17%**	**89.45%**	D2	96.62%	1.68%	97.38%	1.30%	**97.46%**	**1.25%**
D3	72.27%	70.71%	72.97%	87.49%	86.39%	88.33%	**90.09%**	**89.21%**	**90.83%**	D3	96.91%	1.56%	97.80%	1.09%	**98.21%**	**0.89%**
D4	72.84%	72.14%	73.16%	86.39%	85.71%	86.91%	**89.96%**	**89.23%**	**90.56%**	D4	98.61%	0.69%	**98.62%**	**0.68%**	98.53%	0.73%
D5	73.98%	73.10%	74.40%	86.11%	85.34%	86.66%	**89.25%**	**88.50%**	**89.85%**	D5	98.26%	0.87%	98.48%	0.76%	**98.49%**	**0.75%**
D6	73.04%	72.31%	73.37%	83.80%	82.98%	84.35%	**89.48%**	**88.70%**	**90.07%**	D6	**98.56%**	**0.73%**	98.37%	0.81%	98.48%	0.77%
	(a)										(b)					

Fig. 4. FuzzyPrefMiner performance in the *Crisp* scenario (a) *acc, rec, prec* and (b) *cr, rr*

As far as comparability and randomness rates are concerned, we notice that the results do not significantly differ with respect to the number of layers, as shown in Fig. 4(b)).

The Influence of the Degree of Preference. In the second group of experiments we compared FuzzyPrefMiner with the algorithm CPrefMiner [2], designed to solve the "pure" crisp problem discussed in Section 2. Both algorithms returns BNPs, but CPrefMiner learns its BNP from a *crisp* training set and FuzzyPrefMiner from a *fuzzy* one. Fig. 5(a) presents the results comparing CPrefMiner and FuzzyPrefMiner. A partitioning with 8 layers has been considered in this group of experiments. These results show the superiority of FuzzyPrefMiner over CPrefMiner. As expected the performance of FuzzyPrefMiner in the *fuzzy* scenario decreases with respect to its performance in the *crisp* scenario, since the fuzzy validation protocol requires predicting the degree of preference between two tuples and not only their relative preference ordering. However, even when it is validated under a more strict protocol, the results are very satisfactory and better then the results obtained by CPrefMiner. As far as measures cr and rr are concerned, the results do not differ significantly, as shown in 5(b).

FPD	CPrefMiner			FPMiner: Crisp			FPMiner: Fuzzy		
	acc	rec	prec	acc	rec	prec	acc	rec	prec
D1	81.24%	79.45%	82.34%	87.41%	85.64%	88.80%	81.60%	79.93%	82.77%
D2	80.48%	79.46%	81.18%	88.42%	87.17%	89.45%	85.90%	84.73%	86.96%
D3	83.85%	83.32%	84.32%	90.09%	89.21%	90.83%	88.33%	87.41%	89.13%
D4	81.79%	80.44%	82.71%	89.96%	89.23%	90.56%	88.16%	87.44%	88.76%
D5	81.36%	80%	82.25%	89.25%	88.50%	89.85%	86.97%	86.18%	87.65%
D6	81.29%	80.86%	81.56%	89.48%	88.70%	90.07%	86.96%	86.18%	87.58%

(a)

FPD	CPrefMiner		FPMiner: Crisp		FPMiner: Fuzzy	
	cr	rr	cr	rr	cr	rr
D1	95.50%	1.78%	96.45%	1.76%	96.47%	1.67%
D2	97.91%	1.02%	97.46%	1.25%	97.43%	1.17%
D3	98.72%	0.63%	98.21%	0.89%	98.07%	0.92%
D4	97.26%	1.35%	98.53%	0.73%	98.51%	0.72%
D5	97.29%	1.36%	98.49%	0.75%	98.31%	0.79%
D6	99.14%	0.43%	98.48%	0.77%	98.40%	0.77%

(b)

FPD	Building the Model	
	CPrefMiner	FuzzyPrefMiner
D1	1 sec 25 msec	3 sec 93 msec
D2	2 sec 933 msec	4 sec 667 msec
D3	5 sec 431 msec	8 sec 427 msec
D4	8 sec 914 msec	14 sec 117 msec
D5	13 sec 3 msec	20 sec 241 msec
D6	15 sec 442 msec	26 sec 835 msec

(c)

Using the Model	
CPrefMiner	FuzzyPrefMiner
35 msec	50 msec

(d)

Fig. 5. (a) Comparing the 3 methods w.r.t. *acc, rec* and *prec*; (b) Comparing the 3 methods w.r.t. *cr* and *rr*; (c) Comparing the time spent by each method for building the model (d) Comparing the time spent by each method for using the model on 1000 pairs of tuples.

Execution Time. Fig. 5(c) presents the time spent by CPrefMiner and FuzzyPrefMiner to build their respective models. A 8-partitioning has been considered in the pre-processing step. In the case of FuzzyPrefMiner the total time spent includes the time for pre-processing the input data for the classifier (involving I/O operations), the time spent by the classifier BayesNet to build the classification model and the time spent to mine the 8 BNPs, one for each partition. As expected, FuzzyPrefMiner is more costly than CPrefMiner, taking an average time $T_1 = 1.5 \times T_2$, where $T_2 =$ time spent by CprefMiner to build its BNP. Fig. 5(d) shows the time spent to predict the preference for 1000 pairs of tuples. Again, FuzzyPrefMiner is a little more costly than CPrefMiner, taking nearly $1.4 T_3$, where T_3 is the time spent by CPrefMiner to accomplish the same task.

6   Conclusion and Further Work

In this article we proposed to study the influence of the *degree of preference* in the performance of Preference Mining techniques specific for learning contextual preference models. We proposed a formalism for specifying Fuzzy Contextual Preference Models, constituted by a classifier and a small set of Bayesian Preference Networks. This model generalizes the Contextual Preference Model we treated in our previous work in the *crisp* scenario. We proposed the algorithm FuzzyPrefMiner to solve this problem. It is capable to treat richer input data information (the degree of preference) and also to produce more refined results.

Although the original BPN introduced in our previous work is capable to infer a strict partial order on the set of tuples, the fuzzy relation produced by FuzzyPrefMiner does not verify some transitivity properties of fuzzy relations such as weak transitivity or additive transitivity [14]. We are focusing our efforts on developing new techniques to ensure that the mined fuzzy preference model verifies such properties.

Acknowledgements. We thank the Brazilian Research Agencies CNPq and FAPEMIG for supporting this work.

References

1. de Amo, S., Diallo, M.S., Diop, C.T., Giacometti, A., Li, H.D., Soulet, A.: Mining contextual preference rules for building user profiles. In: Cuzzocrea, A., Dayal, U. (eds.) DaWaK 2012. LNCS, vol. 7448, pp. 229–242. Springer, Heidelberg (2012)
2. de Amo, S., Bueno, M.L.P., Alves, G., Silva, N.F.: Cprefminer: An algorithm for mining user contextual preferences based on bayesian networks. In: 24th IEEE International Conference on Tools with Artificial Intelligence (2012)
3. Burges, C.J.C., Shaked, T., Renshaw, E., Lazier, A., Deeds, M., Hamilton, N., Hullender, G.N.: Learning to rank using gradient descent. In: ICML, pp. 89–96 (2005)
4. Burke, R.: Hybrid recommender systems: Survey and experiments. User Modeling and User-Adapted Interaction 12(4), 331–370 (2002)
5. Chiclana, F., Herrera, F., Herrera-Viedma, E.: Integrating three representation models in fuzzy multipurpose decision making based on fuzzy preference relations. Fuzzy Sets and Systems 97(1), 33–48 (1998)
6. Cohen, W.W., Schapire, R.E., Singer, Y.: Learning to order things. J. Artif. Intell. Res. 10, 243–270 (1999)
7. Crammer, K., Singer, Y.: Pranking with ranking. In: NIPS, pp. 641–647 (2001)
8. Freund, Y., Iyer, R., Schapire, R.E., Singer, Y.: An efficient boosting algorithm for combining preferences. J. Mach. Learn. Res. 4, 933–969 (2003)
9. Holland, S., Ester, M., Kießling, W.: Preference mining: A novel approach on mining user preferences for personalized applications. In: Lavrač, N., Gamberger, D., Todorovski, L., Blockeel, H. (eds.) PKDD 2003. LNCS (LNAI), vol. 2838, pp. 204–216. Springer, Heidelberg (2003)
10. Jiang, B., Pei, J., Lin, X., Cheung, D.W., Han, J.: Mining preferences from superior and inferior examples. In: ACM SIGKDD, pp. 390–398 (2008)
11. Joachims, T.: Optimizing search engines using clickthrough data. In: KDD, pp. 133–142 (2002)
12. Koriche, F., Zanuttini, B.: Learning conditional preference networks. Artif. Intell. 174(11), 685–703 (2010)

13. Lops, P., de Gemmis, M., Semeraro, G.: Content-based recommender systems: State of the art and trends. In: Recommender Systems Handbook, pp. 73–105 (2011)
14. Ma, J., Fan, Z.P., Jiang, Y.P., Mao, J.Y., Ma, L.: A method for repairing the inconsistency of fuzzy preference relations. Fuzzy Sets and Systems 157(1), 20–33 (2006)
15. Wilson, N.: Extending cp-nets with stronger conditional preference statements. In: AAAI, pp. 735–741 (2004)
16. Xu, Y., Patnayakuni, R., Wang, H.: The ordinal consistency of a fuzzy preference relation. Information Sciences 224, 152–164 (2013)

BLIMP: A Compact Tree Structure for Uncertain Frequent Pattern Mining

Carson Kai-Sang Leung* and Richard Kyle MacKinnon

University of Manitoba, Canada
kleung@cs.umanitoba.ca

Abstract. Tree structures (e.g., UF-trees, UFP-trees) corresponding to many existing uncertain frequent pattern mining algorithms can be large. Other tree structures for handling uncertain data may achieve compactness at the expense of loose upper bounds on expected supports. To solve this problem, we propose a compact tree structure that captures uncertain data with tighter upper bounds than the aforementioned tree structures. The corresponding algorithm mines frequent patterns from this compact tree structure. Experimental results show the compactness of our tree structure and the tightness of upper bounds to expected supports provided by our uncertain frequent pattern mining algorithm.

1 Introduction and Related Works

Over the past few years, many frequent pattern mining algorithms have been proposed [7, 9, 13, 14], which include those mining uncertain data [3–5, 11, 16]. For instance, the *UF-growth algorithm* [10] is one of the uncertain frequent pattern mining algorithms. In order to compute the expected support of each pattern, paths in the corresponding UF-tree are shared only if tree nodes on the paths have the same item and same existential probability. Consequently, the UF-tree may be quite large when compared to the FP-tree [6] (for mining frequent patterns from precise data). In an attempt to make the tree compact, the *UFP-growth algorithm* [2] groups *similar* nodes (with the same item x and similar existential probability values) into a cluster. However, depending on the clustering parameter, the corresponding UFP-tree may be as large as the UF-tree (i.e., no reduction in tree size). Moreover, because UFP-growth does not store every existential probability value for an item in a cluster, it returns not only the frequent patterns but also some infrequent patterns (i.e., false positives). The *PUF-growth algorithm* [12] addresses these deficiencies by utilizing the idea of upper bounds to expected support with much more aggressive path sharing (in which paths are shared if nodes have the same item in common regardless of existential probability), to yield a final tree structure that can be as compact as the FP-tree is for precise data.

In this paper, we study the following questions: can we further tighten the upper bounds on expected support (e.g., than the PUF-tree)? Can the resulting

---

* Corresponding author.

L. Bellatreche and M.K. Mohania (Eds.): DaWaK 2014, LNCS 8646, pp. 115–123, 2014.

tree still be as compact as the FP-tree? How would frequent patterns be mined from such a tree? Would such a mining algorithm be faster than PUF-growth? Our *key contributions* of this paper are as follows:

1. a branch-level item prefixed-cap **tree (BLIMP-tree)**, which can be as compact as the original FP-tree and PUF-tree; and
2. a mining algorithm (namely, **BLIMP-growth**), which finds all frequent patterns from uncertain data.

The remainder of this paper is organized as follows. The next section presents background material. We then propose our BLIMP-tree structure and BLIMP-growth algorithm in Sections 3 and 4, respectively. Experimental results are shown in Section 5, and conclusions are given in Section 6.

2   Background

Let (i) Item be a set of m domain items and (ii) $X = \{x_1, x_2, \ldots, x_k\} \subseteq$ Item be a k-itemset (i.e., a pattern consisting of k items), where $1 \leq k \leq m$. Then, a transactional database $= \{t_1, t_2, \ldots, t_n\}$ is a set of n transactions. The projected database of X is the set of all transactions containing X. Each item x_i in a transaction $t_j = \{x_1, x_2, \ldots, x_h\} \subseteq$ Item in an uncertain database is associated with an **existential probability value** $P(x_i, t_j)$, which represents the likelihood of the presence of x_i in t_j [8]. Note that $0 < P(x_i, t_j) \leq 1$. The **expected support** $expSup(X)$ of X in the database is the sum (over all n transactions) of the product of the corresponding existential probability values of items within X when these items are independent [8]: $expSup(X) = \sum_{j=1}^{n} \left(\prod_{x \in X} P(x, t_j) \right)$. Hence, given (i) a database of uncertain data and (ii) a user-specified minimum support threshold *minsup*, the research problem of *frequent pattern mining from uncertain data* is to discover from the database a complete set of *frequent* patterns (i.e., to discover every pattern X having $expSup(X) \geq minsup$).

To mine frequent patterns from uncertain data, the *PUF-growth algorithm* [12] scans the uncertain data to build a PUF-tree for capturing the contents of transactions in the uncertain data. Specifically, each node in a PUF-tree captures (i) an item x and (ii) its *prefixed item cap*.

Definition 1. The **prefixed item cap** [12] of an item x_r in a tree path $t_j = \langle x_1, \ldots, x_r, \ldots, x_h \rangle$ representing a transaction where $1 \leq r \leq h$—denoted as $I^{Cap}(x_r, t_j)$—is defined as the product of (i) $P(x_r, t_j)$ and (ii) the highest existential probability value M_1 of items from x_1 to x_{r-1} in tree path t_j (i.e., in the *proper prefix* of x_r in t_j):

$$I^{Cap}(x_r, t_j) = \begin{cases} P(x_1, t_j) & \text{if } h = 1 \\ P(x_r, t_j) \times M_1 & \text{if } h > 1 \end{cases} \tag{1}$$

where $M_1 = \max_{1 \leq q \leq r-1} P(x_q, t_j)$. $\square$

Table 1. A transactional database of uncertain data ($minsup$=1.1)

TID	Transactions
t_1	a:0.6, b:0.1, c:0.2, f:0.8, g:0.5
t_2	a:0.5, b:0.2, c:0.1, e:0.9, g:0.6
t_3	a:0.7, b:0.2, c:0.2, f:0.9
t_4	a:0.9, b:0.1, c:0.1, e:0.8, f:0.6
t_5	b:0.9, c:0.9, d:0.4

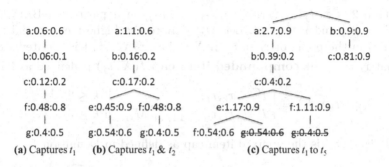

(a) Captures t_1 (b) Captures t_1 & t_2 (c) Captures t_1 to t_5

Fig. 1. BLIMP-trees for the database shown in Table 1 when $minsup$=1.1

It was proven [12] that the expected support of any k-itemset X (where $k > 2$) (which is the product of two or more probability values) must be $\leq I^{Cap}(x_r, t_j)$. However, such an upper bound may not be too tight when dealing with long patterns mined from long transactions of uncertain data. See Example 1. In many real-life situations, it is not unusual to have long patterns to be mined from long transactions of uncertain data.

Example 1. Consider a path $t_1 = \langle a$:0.6, b:0.1, c:0.2, f:0.8, g:0.5$\rangle$ in a PUF-tree capturing the contents of uncertain data shown in Table 1. If $X=\{a, b, c, f\}$, then $I^{Cap}(f, t_1) = P(f, t_1) \times M_1 = 0.8 \times 0.6 = 0.48$ because 0.6 is the highest existential probability value in the proper prefix $\langle a$:0.6, b:0.1, c:0.2$\rangle$ of f in t_1. Note that $I^{Cap}(f, t_1)$ also serves as an upper bound to the expected support of patterns $\{a, f\}, \{b, f\}, \{c, f\}, \{a, b, f\}$, $\{a, c, f\}, \{b, c, f\}$ and $\{a, b, c, f\}$. While this upper bound is tight for short patterns like $\{a, f\}$ having $P(\{a, f\}, t_1)$=0.48, it becomes loose for long patterns like $\{a, b, f\}$ having $P(\{a, b, f\}, t_1)$ =0.048 and $\{a, b, c, f\}$ having $P(\{a, b, c, f\}, t_1)$=0.0096.   □

3   Our BLIMP-tree Structure

To tighten the upper bound for all k-itemsets ($k > 2$), we propose a branch-level **item prefixed-cap tree structure (BLIMP-tree)**. The key idea is to keep track of a value calculated solely from the maximum of all existential probabilities for the single item represented by that node. Every time a frequent extension is added to the suffix item to form a k-itemset (where $k > 2$), this "blimp" value will be used. Hence, each node in a BLIMP-tree contains: (i) an item x_r,

(ii) an item cap $I^{Cap}(x_r, t_j)$ and (iii) a "blimp" value, which is the maximum existential probability of x_r in t_j. Fig. 1(c) shows the contents of a BLIMP-tree for the database in Table 1.

With this information, BLIMP-trees give a *tightened upper bound* on the expected support of an itemset by the product of $I^{Cap}(x_r, t_j)$ and the "blimp" values in the prefix of x_r. This new *compounded item cap* of any k-itemset $X = \{x_1, x_2, \ldots, x_k\}$ in a tree path $t_j = \langle x_1, x_2, \ldots, x_h \rangle$ (denoted as $I(\widehat{X, t_j})$ where $x_k = x_r$) can be defined as follows.

Definition 2. Let $t_j = \langle x_1, x_2, \ldots, x_r, \ldots, x_h \rangle$ be a path in a BLIMP-tree, where $h = |t_j|$ and $r \in [1, h]$. Let M_{x_i} denote the highest existential probability of x_i in the prefix of x_r in t_j. If $X = \{x_1, x_2, \ldots, x_k\}$ is a k-itemset in t_j such that $x_k = x_r$, its **compounded item cap** $I(\widehat{X, t_j})$ is defined as follows:

$$I(\widehat{X, t_j}) = \begin{cases} I^{Cap}(x_r, t_j) & \text{if } k \leq 2 \\ I^{Cap}(x_r, t_j) \times \prod_{i=1}^{k-2} M_{x_i} & \text{if } k \geq 3 \end{cases} \qquad (2)$$

where $I^{Cap}(x_r, t_j)$ is the prefixed item cap as defined in Definition 1. □

Example 2. Let us revisit Example 1. If $X = \{a, b, c, f\}$, then $I(\widehat{X, t_1}) = I^{Cap}(f, t_1) \times (\prod_{i=1}^{2} M_{x_i}) = (0.8 \times M_1) \times (M_{x_1} \times M_{x_2}) = (0.8 \times 0.6) \times (0.6 \times 0.1) = 0.0288$. Simiarly, if $X' = \{b, c, g\}$, then $I(\widehat{X', t_1}) = I^{Cap}(g, t_1) \times (\prod_{i=1}^{1} M_{x_i}) = (0.5 \times M_1) \times M_{x_1} = (0.5 \times 0.8) \times 0.1 = 0.04$. □

To construct a BLIMP-tree, we scan the transactional database of uncertain data to compute the expected support of every domain item. Any infrequent items are removed. Then, we scan the database a second time to insert each transaction into the BLIMP-tree. An item is inserted into the BLIMP-tree according to a predefined order. If a node containing that item already exists in the tree path, we (i) update its item cap by *summing* the current $I^{Cap}(x_r, t_j)$ with the existing item cap value and (ii) update its "blimp" value by taking the *maximum* of the current $P(x_r, t_j)$ with the existing "blimp" value. Otherwise, we create a new node with $I^{Cap}(x_r, t_j)$ and $P(x_r, t_j)$ (i.e. the initial "blimp" value). For a better understanding of BLIMP-tree construction, see Example 3.

Example 3. Consider the database in Table 1, and let *minsup*=1.1. For simplicity, items are arranged in the alphabetic order. After the first database scan, we remove infrequent domain item d. The remaining items a:2.7, b:1.4, c:1.4, e:1.7, f:2.3 & g:1.1 are frequent. With the second database scan, we insert only the frequent items of each transaction (with their respective item cap and "blimp" values). For instance, after reading transaction $t_1 = \{a$:0.6, b:0.1, c:0.2, f:0.8, g:0.5$\}$, we insert $\langle a$:0.6:0.6, b:0.06(=0.1×0.6):0.1, c:0.12(=0.2×0.6):0.2, f:0.48(=0.8×0.6):0.8, g:0.4(=0.5×0.8):0.5$\rangle$ into the BLIMP-tree as shown in Fig. 1(a). As the tree path for t_2 shares a common prefix $\langle a, b, c \rangle$ with an existing path in the BLIMP-tree created when t_1 was inserted, (i) the item cap values of those items in the common prefix (i.e., a, b and c) are added to their corresponding nodes, (ii) the existential probability values of those items are checked against the "blimp" values for their corresponding nodes, with only the maximum saved for each node, and (iii) the remainder of the transaction (i.e., a new branch for items e

and g) is inserted as a child of the last node of the prefix (i.e., as a child of c). See Fig. 1(b). Fig. 1(c) shows the BLIMP-tree after inserting all transactions and pruning those items with infrequent extensions (i.e. item g because its total item cap is less than $minsup$).   □

Observation 1. With this compact BLIMP-tree, we observed the following: (a) $expSup^{Cap}(X)$, which sums compunded item caps for tree paths containing X, serves as an upper bound on the expected support of X. (b) $expSup^{Cap}(X)$ does *not* generally satisfy the downward closure property because $expSup^{Cap}(Y)$ can be less than $expSup^{Cap}(X)$ for some proper subset Y of X. (c) For special cases where X' and its subset Y' share the same suffix item (e.g., $Y'=\{a, b, f\}$ $\subset \{a, b, c, f\}=X'$ sharing the suffix item f), $expSup^{Cap}(Y')$ for BLIMP-trees satisfies the downward closure property. (d) The number of tree nodes in a BLIMP-tree can be equal to that of an FP-tree [6] (when the BLIMP-tree is constructed using the frequency-descending order of items). (e) The compounded item cap $I(\widehat{X}, t_j)$ computed based on the existential probability value of x_k, the highest existential probability value in its prefix and the "blimp" values of its prefix items provides a *tighter* upper bound than that based on the non-compounded item cap $I^{Cap}(x_r, t_j)$ of PUF-trees because the former tightens the bound as candidates are generated during the mining process with increasing cardinality of X, whereas the latter has no such compounding effect.   □

4 The BLIMP-growth Algorithm

To mine frequent patterns (from our BLIMP-tree structure), we propose a tree-based pattern-growth mining algorithm called **BLIMP-growth**. The basic operation in BLIMP-growth is to construct a projected database for each potential frequent pattern and recursively mine its potentially frequent extensions.

Once an item x is found to be potentially frequent, the existential probability of x must contribute to the expected support computation for every pattern constructed from the $\{x\}$-projected database (denoted as DB_x). Hence, the complete set of patterns with suffix x can be mined (ref. Observation 1(c)). Let (i) X be a k-itemset (where $k > 1$) with $expSup^{Cap}(X) \geq minsup$ in the database and (ii) Y be an itemset in DB_X. Then, $expSup^{Cap}(Y \cup X)$ in the original database $\geq minsup$ if and only if $expSup^{Cap}(Y)$ in all the transactions in $DB_X \geq minsup$. Like UFP-growth [2] and PUF-growth [12], this mining process may also lead to some false positives (i.e., those itemsets that appear to be frequent but are truly infrequent) in the resulting set of frequent patterns at the end of the second database scan.

Fortunately, all these false positives will be filtered out with a third database scan. Hence, our BLIMP-growth is guaranteed to return to the user the *exact* collection of frequent patterns (i.e., *all* and *only those* frequent patterns with neither false positives nor false negatives).

5 Experimental Results

As it was shown [12] that PUF-growth outperformed many existing algorithms (e.g., UF-growth [10], UFP-growth [2] and UH-Mine [2]), we compared the performances of our BLIMP-growth algorithm with PUF-growth. We used both real life and synthetic datasets for our tests. The synthetic datasets, which are generally sparse, are generated within a domain of 1000 items by the data generator developed at IBM Almaden Research Center [1]. We also considered several real life datasets such as kosarack, mushroom and retail. We assigned a (randomly generated) existential probability value from the range (0,1] to each item in every transaction in the dataset. The name of each dataset indicates some characteristics of the dataset. For example, the dataset u100k5L_10100 contains 100K transactions with average transaction length of 5, and each item in a transaction is associated with an existential probability value that lies within a range of [10%, 100%].

All programs were written in C++ and ran in a Linux environment on an Intel Core i5-661 CPU with 3.33GHz and 7.5GB RAM. Unless otherwise specified, runtime includes CPU and I/Os for tree construction, mining, and false-positive removal. While the number of false positives generated at the end of the second database scan may vary, all algorithms (ours and others) produce the same set of truly frequent patters at the end of the mining process. The results shown in this section are based on the average of multiple runs for each case. In all experiments, *minsup* was expressed in terms of the absolute support value, and all trees were constructed using the ascending order of item value.

False Positives. Although PUF-trees and BLIMP-trees are compact (in fact, the number of nodes in the global tree can be equal to the FP-tree for both), their corresponding algorithms generate some false positives. Hence, their overall performances depend on the number of false positives generated. In this experiment, we measured the number of false positives generated by both algorithms for fixed values of *minsup* with different datasets. Figs. 2(a)–(b) shows the results when using one *minsup* value for each of the two datasets (i.e., mushroom_5060 and u100k5L_10100). BLIMP-growth was observed to greatly reduce the number of false positives when compared with PUF-growth. The primary reason of this improvement is that the upper bounds for the BLIMP-growth algorithm are much tighter than PUF-growth for higher cardinality itemsets ($k > 2$), hence less total candidates are generated and subsequently less false positives. If fact, when existential probability values were distributed over a narrow range with a higher *minsup* as shown in Fig. 2(a), BLIMP-growth generated fewer than 1% of the total false positives of PUF-growth. When the existential probability values were distributed over a wider range with a much lower *minsup*, as in Fig. 2(b), the total number of false positives in BLIMP-growth was still fewer than 10% of the total false positives of PUF-growth. As a result, BLIMP-growth had a runtime less than or equal to that of PUF-growth in every single experiment we ran.

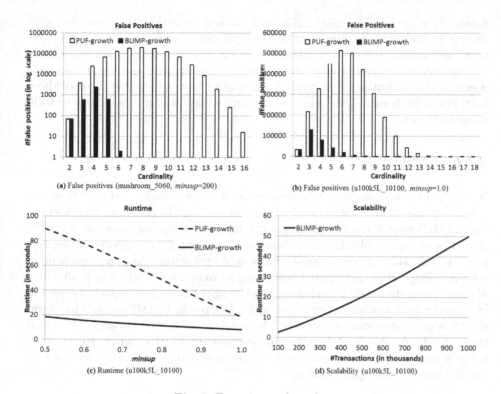

Fig. 2. Experimental results

Runtime. Recall that PUF-growth was shown [12, 15] to outperform UH-Mine and subsequently UFP-growth. Hence, we compared our BLIMP-growth algorithm with PUF-growth. Fig. 2(c) shows that, for low values of *minsup*, BLIMP-growth had shorter runtimes than PUF-growth for u100k5L_10100. The primary reason is that, even though PUF-growth finds the exact set of frequent patterns when mining an extension of X, it may suffer from the high computation cost of generating unnecessarily large numbers of candidates due to only using two values in its item cap calculation: the existential probability of the suffix item and the single highest existential probability value in the prefix of x_r in t_j. This allows large amounts of high cardinality candidates to be generated with similar expected support cap values as low cardinality candidates with the same suffix item. The use of the "blimp" values in BLIMP-growth ensures that those high cardinality candidates are never generated due to their expected support caps being much closer to the actual expected support. Moreover, for lower values of *minsup*, the number of high cardinality candidates being generated increases. In this situation, the probability is higher that the "blimp" values in each node will actually be low, tightening the upper bound even further.

Scalability. To test the scalability of BLIMP-growth, we applied the algorithm to mine frequent patterns from datasets with increasing size. The experimental results presented in Fig. 2(d) indicates that our algorithm (i) is scalable with respect to the number of transactions and (ii) can mine large volumes of uncertain data within a reasonable amount of time. The experimental results show that our algorithms effectively mine frequent patterns from uncertain data irrespective of distribution of existential probability values (whether most of them have low or high values and whether they are distributed into a narrow or wide range of values).

6   Conclusions

In this paper, we proposed a compact tree structure—called *BLIMP-tree*—for capturing important information from uncertain data. In addition, we presented the *BLIMP-growth algorithm* for mining frequent patterns from the BLIMP-tree. The algorithm obtains upper bounds on the expected supports of frequent patterns based on the *compounded item caps*. As these item caps are compounded with a *"blimp" value* (computed based on the maximum existential probability of a particular item), they further tighten the upper bound on expected supports of frequent patterns when compared to PUF-growth. BLIMP-growth has been shown to generate significantly fewer false positives than PUF-growth (e.g., 1% of the total value). Our algorithms are guaranteed to find *all* frequent patterns (with *no* false negatives). As ongoing work, we are conducting theoretical analyses on the tightness of the upper bound.

Acknowledgements. This project is partially supported by NSERC (Canada) and University of Manitoba.

References

1. Agrawal, R., Srikant, R.: Fast algorithms for mining association rules. In: VLDB 1994, pp. 487–499 (1994)
2. Aggarwal, C.C., Li, Y., Wang, J., Wang, J.: Frequent pattern mining with uncertain data. In: ACM KDD 2009, pp. 29–37 (2009)
3. Bernecker, T., Kriegel, H.-P., Renz, M., Verhein, F., Zuefle, A.: Probabilistic frequent itemset mining in uncertain databases. In: ACM KDD 2009, pp. 119–127 (2009)
4. Calders, T., Garboni, C., Goethals, B.: Approximation of frequentness probability of itemsets in uncertain data. In: IEEE ICDM 2010, pp. 749–754 (2010)
5. Calders, T., Garboni, C., Goethals, B.: Efficient pattern mining of uncertain data with sampling. In: Zaki, M.J., Yu, J.X., Ravindran, B., Pudi, V. (eds.) PAKDD 2010, Part I. LNCS (LNAI), vol. 6118, pp. 480–487. Springer, Heidelberg (2010)
6. Han, J., Pei, J., Yin, Y.: Mining frequent patterns without candidate generation. In: ACM SIGMOD 2000, pp. 1–12 (2000)
7. Jiang, F., Leung, C.K.-S.: Stream mining of frequent patterns from delayed batches of uncertain data. In: Bellatreche, L., Mohania, M.K. (eds.) DaWaK 2013. LNCS, vol. 8057, pp. 209–221. Springer, Heidelberg (2013)

8. Leung, C.K.-S.: Mining uncertain data. WIREs Data Mining and Knowledge Discovery 1(4), 316–329 (2011)
9. Leung, C.K.-S., Hao, B.: Mining of frequent itemsets from streams of uncertain data. In: IEEE ICDE 2009, pp. 1663–1670 (2009)
10. Leung, C.K.-S., Mateo, M.A.F., Brajczuk, D.A.: A tree-based approach for frequent pattern mining from uncertain data. In: Washio, T., Suzuki, E., Ting, K.M., Inokuchi, A. (eds.) PAKDD 2008. LNCS (LNAI), vol. 5012, pp. 653–661. Springer, Heidelberg (2008)
11. Leung, C.K.-S., Tanbeer, S.K.: Fast tree-based mining of frequent itemsets from uncertain data. In: Lee, S.-G., Peng, Z., Zhou, X., Moon, Y.-S., Unland, R., Yoo, J. (eds.) DASFAA 2012, Part I. LNCS, vol. 7238, pp. 272–287. Springer, Heidelberg (2012)
12. Leung, C.K.-S., Tanbeer, S.K.: PUF-tree: A compact tree structure for frequent pattern mining of uncertain data. In: Pei, J., Tseng, V.S., Cao, L., Motoda, H., Xu, G. (eds.) PAKDD 2013, Part I. LNCS (LNAI), vol. 7818, pp. 13–25. Springer, Heidelberg (2013)
13. Oguz, D., Ergenc, B.: Incremental itemset mining based on matrix Apriori algorithm. In: Cuzzocrea, A., Dayal, U. (eds.) DaWaK 2012. LNCS, vol. 7448, pp. 192–204. Springer, Heidelberg (2012)
14. Qu, J.-F., Liu, M.: A fast algorithm for frequent itemset mining using Patricia* structures. In: Cuzzocrea, A., Dayal, U. (eds.) DaWaK 2012. LNCS, vol. 7448, pp. 205–216. Springer, Heidelberg (2012)
15. Tong, Y., Chen, L., Cheng, Y., Yu, P.S.: Mining frequent itemsets over uncertain databases. PVLDB 5(11), 1650–1661 (2012)
16. Zhang, Q., Li, F., Yi, K.: Finding frequent items in probabilistic data. In: ACM SIGMOD 2008, pp. 819–832 (2008)

Discovering Statistically Significant Co-location Rules in Datasets with Extended Spatial Objects

Jundong Li[1], Osmar R. Zaïane[1], and Alvaro Osornio-Vargas[2]

[1] Department of Computing Science, University of Alberta, Edmonton, Canada
[2] Department of Paediatrics, University of Alberta, Edmonton, Canada
{jundong1,zaiane,osornio}@ualberta.ca

Abstract. Co-location rule mining is one of the tasks of spatial data mining, which focuses on the detection of sets of spatial features that show spatial associations. Most previous methods are generally based on transaction-free apriori-like algorithms which are dependent on user-defined thresholds and are designed for boolean data points. Due to the absence of a clear notion of transactions, it is nontrivial to use association rule mining techniques to tackle the co-location rule mining problem. To solve these difficulties, a transactionization approach was recently proposed; designed to mine datasets with extended spatial objects. A statistical test is used instead of global thresholds to detect significant co-location rules. One major shortcoming of this work is that it limits the size of antecedent of co-location rules up to three features, therefore, the algorithm is difficult to scale up. In this paper we introduce a new algorithm that fully exploits the property of statistical significance to detect more general co-location rules. We use our algorithm on real datasets with the National Pollutant Release Inventory (NPRI). A classifier is also proposed to help evaluate the discovered co-location rules.

Keywords: Co-location Rules, Statistically Significant, Classifier.

1 Introduction

Co-location mining, one of the canonical tasks of spatial data mining, has received increasing attention in recent years. It tries to find a set of spatial features that are frequently co-located together, i.e. in a geographic proximity. A motivating application example is the detection of possible co-location rules between chemical pollutants and cancer cases with children. Previous work [13,15,14,12] are mainly based on transaction-free algorithms with an apriori-like framework. A prevalence measure threshold is required in the property of anti-monotonicity for effective pruning, the strength of co-location rules are determined afterwards with a prevalence measure threshold. However, the support-confidence framework fails to capture the statistical dependency between spatial features. On one hand, the antecedent and consequent spatial features may be independent of each other. On the other hand, some other strong dependent co-location rules may be ignored due to a prevalence measure value. In the worst case, all detected

L. Bellatreche and M.K. Mohania (Eds.): DaWaK 2014, LNCS 8646, pp. 124–135, 2014.

co-location rules can be spurious, and strong co-location rules are totally missing. Another limitation of transaction-free apriori-like co-location mining algorithms is that they use only one distance threshold to determine the neighbourhood relationship. However, in real applications, a proper distance threshold is hard to determine. Meanwhile, with only one distance threshold, the neighbourhood relationship among spatial features cannot be fully exploited. For instance, the contaminated area around a chemical facility is affected by the amount of chemical pollutants the facility emits. It is apparently that the more amount of chemical pollutants it emits, the more neighbourhood relationships it should capture.

To solve the previous mentioned limitations of transaction-free apriori-like co-location mining algorithms, Adilmagambetov et al. [2] proposed a new transaction based framework to discover co-location rules in datasets with extended spatial objects. Buffers are built around each spatial object, the buffer zone could be the same for all spatial objects or it might be affected by some other spatial or non-spatial features, like the amount of chemical pollutants the facility emits, wind direction in this region, etc. Then, grids are imposed over the geographic space; each grid point intersecting with a set of spatial objects could be seen as a transaction. As mentioned above, the usage of support-confidence framework may result in the discovery of incorrect co-location rules and omission of strong co-location rules. Therefore, to find statistically significant co-location rules, a statistical test method is used instead of global thresholds. However, the statistical significance is not a monotonic property and it cannot be used to prune insignificant co-location rules as apriori-like algorithms. Thus in their work, they limit the size of the antecedent of a rule up to three features and test each possible candidate co-location rule to see if it passes the statistical test. The algorithm cannot scale up well for co-location rules with more than three spatial features in the antecedent, and therefore limits its use.

In this paper, we investigate how to exploit the property of statistical significance to scale it up to detect more general co-location rules. We propose a new algorithm: Co-location Mining Constrained StatApriori (CMCStatApriori) which is able to detect statistically significant co-location rules without any limitation on the rule size. CMCStatApriori is based on the work of StatApriori [8,10]. It uses the z-score to search for statistically significant co-location rules with a fixed consequent spatial feature. The results of co-location rules are hard to evaluate even for domain experts, therefore, we also propose to use a classifier to help evaluate the results of co-location rules.

The remainder of the paper is organized as follows. The overview of related work is given in Section 2. The algorithm framework is described in Section 3. Section 4 describes the experimental results and the evaluation of the results. Section 5 concludes the paper.

2 Related Work

In this section, we review some related work on co-location mining from two perspectives: the support-confidence framework and the statistical test framework.

2.1 Support-confidence Based Co-location Mining

Shekhar and Huang [13] proposed a co-location pattern mining framework which is based on neighbourhood relations and the concept of participation index. The basic concept of this method is similar to the concept of association rule mining. As an input, the framework takes a set of spatial features and a set of instances, where each instance is a vector that contains information on the instance ID, the feature type of the instance, and the location of the instance. As an output, the method returns a set of co-location rules. A co-location rule is of the form of $C_1 \rightarrow C_2(PI, cp)$, where C_1 and C_2 are a set of spatial features, PI is the prevalence measure (participation index), and cp is the conditional probability. A co-location pattern is considered as prevalent, or interesting, if for each feature of the pattern at least $PI\%$ instances of that feature form a clique with the instances of all other features of the pattern according to the neighbourhood relationship. Similar to association rule mining, only frequent $(k-1)$-patterns are used for the k-candidate generation process. Yoo and Shekhar [15] proposed a join-less algorithm which decreases the computation time of constructing neighbourhood relationship. The main idea is to find star neighbourhoods instead of calculating pairwise distances between all instances in the dataset. Huang et al. [12] continued their previous work by introducing an algorithm that finds co-location patterns with rare features. Instead of the participation index threshold, the authors proposed to use the maximal participation ratio threshold. Briefly, a co-location pattern is considered prevalent if $maxPR\%$ instances of at least one of the features in the pattern are co-located with instances of all other features, where $maxPR$ is the maximal participation ratio. Xiong et al. [14] introduced a framework for detecting patterns in datasets with extended spatial objects. Extended spatial objects are objects that are not limited to spatial points but also include lines and polygons. In the proposed buffer-based model, the candidate patterns are pruned by the coverage ratio threshold. In other words, if the area covered by the features of a candidate pattern is greater than a predefined threshold, this pattern is considered as prevalent or interesting.

2.2 Statistical Test Co-location Mining

The approaches mentioned above use thresholds on interestingness measures, which result in meaningless patterns when a low threshold is used, and a high threshold may prune interesting but rare patterns. Instead of a threshold-based approach, Barua and Sander [4] used the statistical test to mine statistically significant co-location patterns. The participation index of a pattern in the observed data is calculated as previous studies. Then for each co-location pattern the authors compute the probability p of seeing the same or greater value of prevalence measure under a null hypothesis model. The co-location pattern is considered statistically significant if $p \leq \alpha$, where α is a level of significance. Adimagambetov et al. [2] proposed a transactionization framework to find significant co-location rules on extended spatial objects. Spatial instances are transformed into transactions by buffers and grids and the expected support is used as the interesting measure. The statistical test method they used is similar to [4].

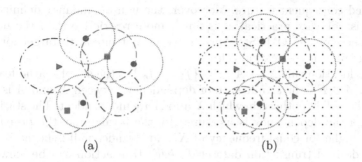

Fig. 1. Transactionization step: (a) An example of spatial dataset with point feature instances and their buffers; (b) Grids imposed over the space.

3   Algorithm Framework

3.1   Problem Definition

The objective is to discover statistically significant co-location rules between a set of antecedent spatial features and one single fixed consequent spatial feature. A real world application of this task is to detect co-location rules between chemical pollutants (antecedent) and cancer cases or other morbidities (consequent). Since we do not intend to find the causality relationships, the goal is to identify potential interesting co-location associations in order to state hypotheses for further study.

The task consists of three steps. In the initialization step, a buffer is built around each spatial object, and it defines the area affected by that object; for example, the buffer zone around an emission point shows the area polluted by a released chemical pollutant. The buffer shape is defined as circle, but it may also be affected by some other factors like wind direction. Considering the factor of wind direction, the circular buffer is transformed to elliptical. Fig. 1(a) displays an example of spatial dataset with buffers of various sizes (circular and elliptical) that are formed around spatial point objects. In the transactionization step, the transaction dataset is formed by imposing grids over all the buffer zones, as shown in Fig. 1(b). Then a transaction is defined as a set of spatial features corresponding to these objects [2]. After getting the derived transaction dataset T from the spatial dataset, we intend to detect statistically significant co-location rules in the next step.

3.2   Co-location Mining Constrained StatApriori

In this subsection, we introduce the proposed Co-location Mining Constrained StatApriori (CMCStatApriori) algorithm which is able to detect statistically significant co-location rules without any rule length limitation.

CMCStatApriori is a variation of StatApriori [8,10]; the main difference is that CMCStatApriori can efficiently detect more specific co-location rules, rules with one fixed consequent feature. Moreover, the non-redundancy definition in StatApriori is not very practical, it is much more restrictive than the normal definition. Therefore, in CMCStatApriori, we do not intend to target for non-redundant significant co-location rules.

For the co-location rule $X \to A$ ($F = \{f_1, ..., f_m\}$ is the set of spatial features and $X \subsetneq F$, $A \in F$), the significance of dependency between X and A is compared with the null hypothesis in which X and A are independent. The statistical significance of the dependency is measured by the p-value, i.e. the probability of observing higher or equal frequency of X and A under null hypothesis. Suppose in the derived transaction dataset T, each transaction can be viewed as an independent Bernoulli trial with two possible results, that $P(XA) = 1$ or $P(XA) = 0$. Thus, the statistical significance of the frequency of XA follows the binomial distribution and the p-value can be formulated as:

$$p = \sum_{i=\sigma(XA)}^{\sigma(A)} \binom{n}{i} (P(X)P(A))^i (1 - P(X)P(A))^{n-i} \tag{1}$$

where $\sigma(XA)$ is the observed frequency of XA, and n is the total number of transactions in T.

The p-value is not a monotonic property, but z-score provides an upper bound for the binomial distribution:

$$z(X \to A) = \frac{\sigma(XA) - \mu}{s} = \frac{\sqrt{nP(XA)(\gamma(XA) - 1)}}{\sqrt{\gamma(XA) - P(XA)}} \tag{2}$$

where $\mu = nP(X)P(A)$, $s = \sqrt{nP(X)P(A)(1 - P(X)P(A))}$ are the mean and standard deviation of the binomial distribution, respectively. $\gamma(XA) = \frac{P(XA)}{P(X)P(A)}$ is the lift for the co-location rule $X \to A$. It measures the strength of the dependency between X and A such that $\gamma(X \to A) > 1$ if X and A show a positive correlation. It is easy to notice that the z-score is a monotonically increasing function with the support and lift of XA: $\sigma(XA)$ and $\gamma(XA)$, therefore, it can be denoted as $z(X \to A) = f(\sigma(XA), \gamma(XA))$.

Therefore, following StatApriori [8,10], the search problem can be reformulated as searching for all statistically significant co-location rules in the form of $X \to A$ with the following requirements (the set of statistically significant co-location rules is denoted as P):

Definition 1. *Statistically significant co-location rules*

1. $X \to A$ expresses a positive correlation, i.e. $\gamma(X \to A) > 1$

2. for all $(Y \to A) \notin P, z(X \to A) > z(Y \to A)$

3. $z(X \to A) \geq z_{min}$

With this definition, the property "potentially significant" (PS) is defined as follows. It is a necessary condition to construct the set of statistically significant co-location rules.

Definition 2. *Let A be the fixed consequent feature, z_{min} is an user-defined threshold for the z-score, and upperbound(f) be an upper bound for the function f. The co-location rule $X \to A$ is defined as potentially significant, i.e. $PS(X) = 1$, iff $upperbound(z(X \to A)) \geq z_{min}$. Otherwise, the co-location rule is not considered as statistically significant.*

The property of PS displays a monotonic property in some specific situations:

Theorem 1. *Let A be the fixed consequent feature and $PS(X) = 1$, then for all $Y \subseteq X$ and $min(XA) = min(YA)$ we can get $PS(Y) = 1$, where $min(XA)$ denotes the feature with the minimum support in XA.*

The proof of Theorem 1 is straightforward, first we can see that:

$$\gamma(YA) = \frac{P(YA)}{P(Y)P(A)} \leq \frac{1}{P(Y)} \leq \frac{1}{P(min(YA))} \tag{3}$$

where $min(YA)$ denotes the feature with the smallest support among YA, the upper bound of the co-location rule $Y \to A$ now is:

$$upperbound(z(Y \to A)) = f(P(YA), \frac{1}{P(min(YA))}) \tag{4}$$

then we have:

$$upperbound(z(X \to A)) = f(P(XA), \frac{1}{P(min(XA))}) \leq$$
$$f(P(YA), \frac{1}{P(min(YA))}) = upperbound(z(Y \to A)) \tag{5}$$

for all $Y \subseteq X$ such that $min(XA) = min(YA)$. We can see that the monotonic property is kept only when the minimum feature (the feature with the minimal support) in XA and YA are the same.

With the monotonic property of PS, we can derive the algorithm that discovers the potential significant co-location rules in the same way as the general Apriori-like algorithms do, alternating between the candidate generation and candidate pruning. First, the set of antecedent features are arranged in an ascending order by their frequencies. Let the renamed features be $\{f'_1, f'_2, ..., f'_{m-1}\}$, where $P(f'_1) \leq P(f'_2) \leq ... \leq P(f'_{m-1})$. The candidate generation process is the same as that in Apriori [3], for the l-set $S_l = \{f'_{a_1}, ..., f'_{a_l}\}$ $(a_1 < a_2 < ... < a_l)$, we can generate $(l+1)$-sets $S_l \cup \{f'_{a_j}\}$, where $a_j > a_l$. After the generation of the $(l+1)$-sets $S_l \cup \{f'_{a_j}\}$, we need to check if all of its l-set "regular" parents (the parents with the same minimum support feature when combined with A as $S_l \cup \{f'_{a_j}\} \cup A$) can indicate PS co-location rules. If all of its regular parents can indicate PS co-location rules, then $S_l \cup \{f'_{a_j}\}$ is added to the candidate set for the pruning process, otherwise, $S_l \cup \{f'_{a_j}\}$ can be pruned directly. In the pruning process, the PS co-location rule $X \to A$ is kept if it meets the z_{min} threshold, otherwise, it is removed.

Algorithm 1. CMCStatApriori Algorithm

Require: Set of antecedent features $F\backslash A$, the consequent feature A, derived transaction dataset T, the threshold z_{min} for the z-score
Ensure: Set of potentially significant co-location rules P
1: $P_1 = \{f_i \in F\backslash A | PS(f_i) = 1\}$
2: $l = 1$
3: **while** $(P_l \neq \varnothing)$ **do**
4:     $C_{l+1} = GenCands(P_l, A)$
5:     $P_{l+1} = PrunCands(C_{l+1}, z_{min}, A)$
6:     $l = l + 1$
7: **end while**
8: $P = \cup_i P_i$
9: **return** P

Algorithm 2. Algorithm GenCands

Require: Potentially significant l-sets P_l, the consequent feature A.
Ensure: $(l + 1)$-candidates C_{l+1}.
1: $C_{l+1} = \emptyset$
2: **for all** $Q_i, Q_j \in P_l$ such that $|Q_i \cap Q_j| = l - 1$ **do**
3:     **if** $\forall Z \subseteq Q_i \cup Q_j$ such that $|Z| = l$ and $min(ZA) = min((Q_i \cup Q_j)A)$ and $Z \subseteq P_l$ **then**
4:        $C_{l+1}.add(Q_i \cup Q_j)$
5:     **end if**
6: **end for**
7: **return** C_{l+1}

Algorithm 3. Algorithm PruneCands

Require: l-candidates C_l, threshold z_{min}, the consequent feature A.
Ensure: Potentially significant l-sets P_l
1: $P_l = \varnothing$
2: **for all** $Q_i \in C_l$ **do**
3:     calculate $P(Q_i A)$ and the upperbound of lift $\frac{1}{P(min(Q_i A))}$
4:     **if** $f(P(Q_i A), \frac{1}{P(min(Q_i, A))}) \geq z_{min}$ **then**
5:        $P_l.add(Q_i)$
6:     **end if**
7: **end for**
8: **return** P_l

A problem of StatApriori is that for each potentially significant set C, only the best rule is derived from C. For example, if $C\backslash A \to A$ is the best rule, where $A \in C$ and the "best" indicates that the rule has the highest z-score, then no other rules in the form of $C\backslash B \to B(B \neq A)$ is output. However, in our CMCStatApriori algorithm, this kind of problem does not exist, because the PS property is for the co-location rule and the consequent feature is fixed. The detailed pseudo code of CMCStatApriori is illustrated in Algorithms 1, 2 and 3.

4   Experiments

4.1   Datasets

We conduct our experiments on two real datasets which contain pollutant emissions and information about cancer cases for children in the provinces of Alberta and Manitoba, Canada. The sources of the data are the National Pollutant Release Inventory (NPRI) [5] and the provincial cancer registries. The information on pollutants is taken for the period between 2002 and 2007 and contains the type of a chemical, location of release, and average amount of release per year. In order to get reliable results, the chemical pollutants that had been emitted from less than three facilities are excluded from the dataset. There are 47 different chemical pollutants and 1,422 chemical pollutant emission points in Alberta; 26 different chemical pollutants and 545 chemical pollutant emission points in Manitoba, several chemical pollutants might be released from the same location. The number of cancer cases are 1,254 and 520 in Alberta and Manitoba, respectively. In order to make the model more accurate, the wind speed and direction are also taken into account in these two provinces. The interpolation of wind information between wind stations is used. In Alberta, the data of 18 stations are from Environmental Canada [6] and 156 stations are from ArgoClimatic Information Service (ACIS) [1]. In Manitoba, the data of all 20 stations are all from Environment Canada [6]. We obtain the wind direction and speed in the locations of chemical facilities by making interpolations in the ArcGIS tool [7].

4.2   Experimental Settings

We are interested in co-location rules of the form of $Pol \rightarrow Cancer$, where Pol is a set of pollutant features and $Cancer$ is a cancer feature. Three different methods are compared: the co-location mining algorithm by Adilmagambetov et al. in [2] (denoted as CM), co-location mining algorithm with Kingfisher [9,11] (denoted as CMKingfisher) and the proposed CMCStatApriori method. In all of these three methods, the distance between grid points is 1km.

CM. CM needs a number of simulations to detect significant co-location rules, the number of simulations for the statistical test is set to be 99 and the level of significance α is set to be 0.05. The size of antecedent features of a candidate rule is up to three. The randomized datasets (simulations) that are used in the statistical test are generated according to the distributions of chemical pollutant emitting facilities and cancer cases. Chemical pollutant emitting facilities are not randomly distributed, and are usually located close to regions with high population density, thus, CM does not randomize the pollutant facilities all over the region, instead, it keeps locations of facilities and randomize the pollutants within these regions. For the cancer cases, most of them are located within dense "urban" regions and the rest are in "rural" regions. Therefore, the cancer cases are randomized according to the population ratio of "urban" regions to "rural" regions. In each simulation of CM, both pollutant chemicals and cancer cases are randomized.

CMKingfisher. Kingfisher [9,11] is developed to discover positive and negative dependency rules between a set of antecedent features and a single consequent feature. The algorithm is based on a branch and bound strategy to search for the best, non-redundant dependency top-K rules. Kingfisher is able to detect statistically significant positive and negative rules with any possible consequent. But we are only interested in the positive rules whose consequent is "Cancer", therefore, after getting the derived transaction dataset T, we apply Kingfisher algorithm to get the complete set of co-location rules and extract the subset of co-location rules that we are interested in. The significance level α is 0.05.

CMCStatApriori. The CMCStatApriori is the algorithm proposed in this paper. Unlike CM and CMKingfisher which use the p-value as a significance level, CMStatApriori uses the z-score which provides an upper bound for the p-value. In the experiment, the threshold of z-score is set to be 150 in the Alberta dataset. This threshold of 150 is too high in the Manitoba dataset and no co-location rules are output. Therefore, we set a lower z-score threshold of 40. Indeed, the lower the z-score threshold, the more co-location rules is generated. The parameter setting of z-score threshold of CMCStatApriori is discussed in the last subsection.

4.3   Experimental Results

Both CMKingfisher and CMCStatApriori are able to detect more general co-location rules (without limitation of size of antecedent features). However, to have a fair comparison with CM, we only list the co-location rules with up to three antecedent features. The number of rules detected by these three methods and the number of rules overlaps with CM by CMKingfisher as well as CMCStat-Apriori are listed in Table 1. It can be observed that in the dataset of Alberta, both of CMKingfisher and CMCStatApriori have a small overlap with CM rules. The situation is slightly different in the dataset of Manitoba, around 80% and 30% of detected rules by CMKingfisher and CMCStatApriori also appear in CM.

Table 1. Number of co-location generated by different methods

	Alberta		Manitoba	
	#rules	# rules in CM	rules	# rules in CM
CM	273	–	170	–
CMKingfisher	108	7	23	19
CMCStatApriori	571	5	60	16

4.4   Evaluation

Environmental pollutants are suspected to be one of the causes of cancer in children. However, there are other factors that could lead to this disease. Therefore, it is a difficult task to evaluate the detected co-location rules even for domain experts. To assist in evaluating the discovered co-location rules, we propose to use a classifier with the co-location rules as a predictive model. The results by different methods are carefully and painstakingly evaluated manually by experts in our multidisciplinary team. However, the systematic evaluation by classification provides an estimation of the best quality co-location rule set.

We consider co-location rules generated by either method as a classifier. To evaluate the discovered co-location rules, we randomly sample some grid points on the geographic space. The randomly sampled grid point has to intersect with at least one pollutant feature; it either intersects with cancer or not. For the type of grid point $(Pol_{grid}, Cancer)$ intersects with both pollutant(s) and cancer, if we can find at least one co-location rule $Pol \to Cancer$ in the classifier that correctly matches it, i.e. $Pol \subseteq Pol_{grid}$, the grid point is indicated as correctly classified. For the other type of grid point $(Pol_{grid}, \neg Cancer)$ intersects with pollutant(s) but not cancer, if there does not exist any co-location rules $Pol \to Cancer$ that match it, i.e. $Pol \not\subseteq Pol_{grid}$, the grid point is also indicated as correctly classified. Otherwise, the grid points are considered as misclassified. The ratio of correctly classified grid points to the total number of sampled grid points is output as the classification accuracy. Fig. 2 shows a toy example of the evaluation process. In the datasets of Alberta and Manitoba, we randomly sample 1000 grid points each time, repeats 100 times, and calculate the average classification accuracy for the previously mentioned three methods. Table 2 presents the evaluation results, along with the classification accuracy (ACC), the specificity (SPE) and sensitivity (SEN) are also listed. As can be observed from the classification accuracy, CMCStatApriori is better than CM and CMKingfisher. The classification accuracy is much higher in Alberta compared with Manitoba. One possible explanation is that the co-location association between chemical pollutants and children cancer cases is stronger in Alberta. Both the number of rules and the accuracy is very low in Manitoba, therefore, it is possible that chemical pollutants and children cancer cases are more likely to be independent in Manitoba. We can also notice that the specificity is much higher than the sensitivity in both datasets. High specificity means that grid points without cancer are seldom misclassified; on the other hand, low sensitivity indicates that grid points with cancer are mostly misclassified. This phenomenon may imply that the co-location associations between chemical pollutants and children cancer cases is weak. However, these assumptions still need to be carefully scrutinized.

Table 2. Evaluation of different methods using Accuracy, Specificity and Sensitivity

	Alberta			Manitoba		
	ACC	SPE	SEN	ACC	SPE	SEN
CM	83.9 ± 3.3	97.6 ± 1.6	11.4 ± 8.1	22.0 ± 4.3	55.8 ± 11.2	$\mathbf{13.4 \pm 3.7}$
CMKingfisher	69.2 ± 4.1	77.4 ± 4.1	$\mathbf{28.6 \pm 11.4}$	26.6 ± 4.6	$\mathbf{96.4 \pm 3.6}$	8.7 ± 3.0
CMCStatApriori	$\mathbf{84.7 \pm 3.4}$	$\mathbf{90.6 \pm 0.7}$	6.6 ± 0.4	$\mathbf{27.4 \pm 4.1}$	83.4 ± 7.7	12.2 ± 3.4

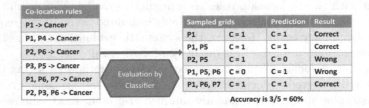

Fig. 2. Toy example of the classification evaluation

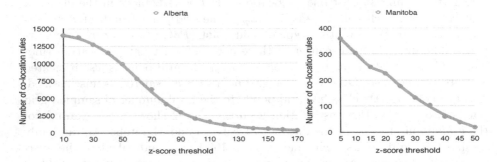

Fig. 3. Number of co-location rules on Alberta and Manitoba dataset

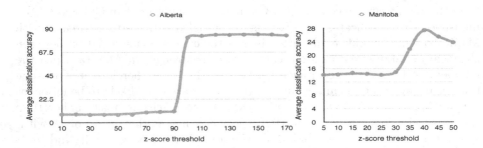

Fig. 4. Average classification accuracy of sampled grids

The only parameter in CMCStatApriori is the z_{min}. In this subsection, we discuss the effect of the parameter z_{min}. As shown in Fig. 3, the number of discovered co-location rules drops when we increase z_{min}. We were not able to find any statistically significant co-location rules when $z_{min} > 170$ in Alberta and when $z_{min} > 50$ in Manitoba. In Fig. 4, the average classification accuracy of the sampled grid points is presented. The classification performance is poor when the z-score threshold is set to be low. Besides, there exists a turning point ($z_{min} = 100$ in Alberta, $z_{min} = 30$ in Manitoba) where the accuracy improves dramatically. In the Alberta dataset, there is not much difference when z_{min} varies from 110 to 170, while in the Manitoba dataset, the performance is best when z_{min} is set to be 40.

5   Conclusion

In this paper, we propose a novel co-location mining algorithm to detect statistically significant co-location rules in datasets with extended spatial objects. By exploiting the property of statistical significance, we do not have to limit the number of antecedent features up to three in co-location rules which is a major shortcoming of previous work. Therefore, more general co-location rules can be generated and the algorithm is able to scale up well. In addition, we propose to use a classifier to help the evaluation of discovered co-location rules.

References

1. AgroClimatic Information Service (ACIS). Live alberta weather station data, http://www.agric.gov.ab.ca/app116/stationview.jsp
2. Adilmagambetov, A., Zaiane, O.R., Osornio-Vargas, A.: Discovering co-location patterns in datasets with extended spatial objects. In: Bellatreche, L., Mohania, M.K. (eds.) DaWaK 2013. LNCS, vol. 8057, pp. 84–96. Springer, Heidelberg (2013)
3. Agrawal, R., Srikant, R.: Fast algorithms for mining association rules. In: VLDB 1994, pp. 487–499 (1994)
4. Barua, S., Sander, J.: SSCP: Mining statistically significant co-location patterns. In: Pfoser, D., Tao, Y., Mouratidis, K., Nascimento, M.A., Mokbel, M., Shekhar, S., Huang, Y. (eds.) SSTD 2011. LNCS, vol. 6849, pp. 2–20. Springer, Heidelberg (2011)
5. Environment Canada. National Pollutant Release Inventory. Tracking Pollution in Canada, http://www.ec.gc.ca/inrp-npri/
6. Environment Canada. National Climate Data and Information. Canadian climate normals or averages 1971-2000, http://climate.weatheroffice.gc.ca/climate_normals/index_e.html
7. ESRI. ArcGIS Desktop: Release 10 (2011)
8. Hämäläinen, W., Nykanen, M.: Efficient discovery of statistically significant association rules. In: ICDM, pp. 203–212 (2008)
9. Hämäläinen, W.: Efficient discovery of the top-k optimal dependency rules with fisher's exact test of significance. In: ICDM, pp. 196–205 (2010)
10. Hämäläinen, W.: Statapriori: an efficient algorithm for searching statistically significant association rules. KAIS 23(3), 373–399 (2010)
11. Hämäläinen, W.: Kingfisher: an efficient algorithm for searching for both positive and negative dependency rules with statistical significance measures. KAIS 32(2), 383–414 (2012)
12. Huang, Y., Pei, J., Xiong, H.: Mining co-location patterns with rare events from spatial data sets. Geoinformatica 10(3), 239–260 (2006)
13. Shekhar, S., Huang, Y.: Discovering spatial co-location patterns: A summary of results. In: Jensen, C.S., Schneider, M., Seeger, B., Tsotras, V.J. (eds.) SSTD 2001. LNCS, vol. 2121, pp. 236–256. Springer, Heidelberg (2001)
14. Xiong, H., Shekhar, S., Huang, Y., Kumar, V., Ma, X., Yoo, J.S.: A framework for discovering co-location patterns in data sets with extended spatial objects. In: SDM (2004)
15. Yoo, J.S., Shekhar, S.: A joinless approach for mining spatial co-location patterns. TKDE 18(10), 1323–1337 (2006)

Discovering Community Preference Influence Network by Social Network Opinion Posts Mining

Tamanna S. Mumu and Christie I. Ezeife[*]

School of Computer Science, University of Windsor, Windsor, Ontario N9B 3P4, Canada
{mumut,cezeife}@uwindsor.ca

Abstract. The popularity of posts, topics, and opinions on social media websites and the influence ability of users can be discovered by analyzing the responses of users (e.g., likes/dislikes, comments, ratings). Existing web opinion mining systems such as OpinionMiner is based on opinion text similarity scoring of users' review texts and product ratings to generate database table of features, functions and opinions mined through classification to identify arriving opinions as positive or negative on user-service networks or interest networks (e.g., Amazon.com). These systems are not directly applicable to user-user networks or friendship networks (e.g., Facebook.com) since they do not consider multiple posts on multiple products, users' relationships (such as influence), and diverse posts and comments. This paper proposes a new influence network (IN) generation algorithm (Opinion Based IN:OBIN) through opinion mining of friendship networks. OBIN mines opinions using extended OpinionMiner that considers multiple posts and relationships (influences) between users. Approach used includes frequent pattern mining algorithm for determining community (positive or negative) preferences for a given product as input to standard influence maximization algorithms like CELF for target marketing.

Keywords: Social network, Influence analysis, Sentiment classification, Recommendation, Ranking, Opinion mining.

1 Introduction

People may give their opinions more often on shared posts on social network where those opinions may be positive, negative, or controversial to the shared posts. At present, considering communities extracted from social graphs and monitoring the aggregate trends and opinions discovered by these communities has shown its potential for a number of business applications such as marketing intelligence and competitive intelligence. The tasks include identification of influential posts, influential persons and services, users' opinions analysis, and community detection based on shared interest. These tasks can be performed through data mining approaches which include classification, clustering, association rule mining, sequential pattern mining.

[*] This research was supported by the Natural Science and Engineering Research Council (NSERC) of Canada under an operating grant (OGP-0194134) and a University of Windsor grant.

L. Bellatreche and M.K. Mohania (Eds.): DaWaK 2014, LNCS 8646, pp. 136–145, 2014.

Social Network Data - A social network framework is generally represented as a graph $G(V, E)$, where V is the set of nodes representing users and E is the set of edges between nodes representing a specific type of interaction between them. The edges may be directed or undirected. Bonchi et al. [3] define the links between nodes in two ways. *Explicit* e.g., users declaring explicitly their friends or connections through such actions as joining a group, liking a page, following a user or accepting a friendship request. Another type of link is *Implicit* e.g., links identified from users' activities by analyzing broad and repeated interactions between users such as voting, sharing, tagging, or commenting. Viral marketing is an approach of information spreading, where a small set of influential customers can influence greater number of customers [2]. A major and common issue in the area of existing opinion mining is to identify product popularity based on one specific feature such as sentiment of comments [6], [4], or rating on topic [8]. However, in friendship networks (e.g., facebook.com), users' opinions on a product are defined implicitly or explicitly in the networks, and unlike in interest networks (e.g., Amazon.com), influence on a product not only depends on a specific product webpage but also on the complex relationships between users connected in the networks, and this requires all kinds of implicit and explicit opinions and relationships that need to be identified and aggregated.

In standard influence maximization (IM) systems such as CELF [7], the whole social network is taken as input to find influential users as seed set for a specific product (e.g., iPhone) for target marketing [2]. The main limitation of general IM systems like CELF is that they do not consider multiple posts on multiple products as well as relationships between users, and thus, are not effectively product-specific because of the need to first search large social network data for multiple influential users/product opinions who may not be influential on a specific product (e.g., iphone). So, considering those users as influential for a product reduces the accuracy and efficiency of such general IM algorithms.

1.1   Contributions and Outline

Motivated by the issues described above, the problem we tackle is as follows:

Problem Definition – Build an influential network (IN) generation model for influence maximization of a specific product based on mining users' posts and opinions (positive or negative) on a specific product and relationships from a friendship network graph $G(V, E)$ where every edge $e_{ij} \in E$ connects nodes v_i and v_j ($v_i, v_j \in V$ and $i, j = 1, 2, \ldots, N$) and indicates v_i and v_j have relationships on a specific product. Also, measure influence acceptance score of each node v_i for a product and remove nodes that are below certain threshold before applying IM algorithm on the pruned product-specific friendship network to more effectively and efficiently compute a product-specific IM. To solve this problem, paper contributions are:

1) We propose a new influence network generation model for friendship networks by mining opinions named Opinion Based IN (OBIN) which incorporates implicit, explicit opinions and complex user-user and user-product relationships to get a ranked list of users, opinions and relationships in a Topic-Post Distribution (TPD) model.

2) Based on TPD model, we propose a new opinion mining framework named Post-Comment Polarity Miner (PCP-Miner) which is an extension of OpinionMiner [6] augmented with the Apriori [1] frequent pattern mining technique, that considers multiple posts, relationships between posts and non-tagged comments, and then generates pruned IN.

Section 2 of this paper presents related work, section 3, the proposed system, the OBIN. Section 4 the experimental results, and section 5 conclusions and future works.

2   Related Work

Recent research on Social Network have analyzed social network data to find pattern of popularity or influence in various domains such as blogging (e.g., Slashdot.org), micro-blogging (e.g., Twitter.com), bookmarking (e.g., Digg.com), co-authorship (e.g., Academia.edu), movie review (e.g., IMDb.com), and product review domains (e.g., Amazon.com). Our proposed work is motivated by [6] in product review domains; and the work of [7] and [2] in influence maximization.

Dave et al. [4] proposed an opinion extraction and mining method based on features and scoring matrices and classified review sentences as positive or negative. Hu and Liu [5] proposed a feature-based summarization (FBS) that mines product features from customers' reviews. Jin et al. [6] also worked similar as [5] and defined four entity types, eight tags and four pattern tag sets to the feature-based approach, named OpinionMiner. But they ignore opinions that contain different product information and infrequent product features. Existing influence maximization algorithms such as Cost-Effective Lazy Forward selection (CELF) [7] require that the influence probability is known and given to the algorithm as input along with a social network graph. Trust-General Threshold (TGT) model proposed by [2] works in trust network considering trust and distrust of nodes and improved [7]. Both CELF and TGT do not consider user-user relationships that can be obtained from friendship networks such as Facebook. and product specific problems. TGT also requires trust and distrust to be explicitly described.

In this paper, we propose a product-specific influence network generation model based on users' opinions on friendship network to discover community preference of relevant users by considering implicit and explicit users relationships, which shows target marketing is more focused than [7] and [2] with the benefit of computing a more accurate influential network for a product.

3   The Proposed OBIN Model

The proposed OBIN model takes a social network graph $G(V, E)$ and a product z as input and generates an influence graph $Gz(V, E)$ on product z from computed community preference where V is the relevant nodes extracted from G. Our proposed pruning strategy is based on number of nodes, likes and shares, number of positive and negative comments, and extracted user-user relationships. Fig. 1 shows the algorithm for OBIN model. OBIN has 3 main functions, TPD (Topic-Post Distribution) (lines 1-4 in Fig. 1), PCP-Miner (Post-Comment Polarity Miner) (lines 5-7 in Fig. 1), and influence network generator (line 8 in Fig. 1).

3.1   Topic – Post Distribution (TPD) Model

For a given product z , our proposed TPD extracts all nodes, posts, and comments by applying SQL queries to friendship network graph, to mine the dataset to classify relevant and irrelevant nodes of product z, determined by the number of nodes connected to influential node and number of likes on the posts denoted by $A(Approve)$, number of shares and comments on the posts denoted by $SR(Simple\ Response)$, product information as search key (e.g., iPhone screen resolution) denoted by $Term$. We compute $Approve\ A$ in two levels, (1) in the node selection step A_f represents the number of friends connected to the node, and (2) in the post selection step A_l represents the number of likes on the post.. The collective approved rating of a node V is the sum of the two approved ratings as $A(V) = A_f + A_l$.

Algorithm OBIN to generate influence network graph G_z from friendship network G
Input: Social network URL (e.g., facebook.com), product z, Approve A, Simple response SR, $Term$ in product name z
Output: Set of influential nodes V_s, influenced nodes V_t, influence graph G_z on z

1. *Execute SQL query on URL to find set of nodes on product z using Graph API*
2. *Generate nodes matrix NM with 4 attributes $< node, Term, A, SR >$*
3. *Generate relevant nodes matrix PM with 4 attributes $< node\ (V)$, Term, A, SR > by **mining** NM with three features (Term + A), (Term + SR), (A + SR), to classify relevant and irrelevant nodes using SVM classifier.*
4. *Execute SQL query to find set of posts and comments of V from PM. Store posts w in table tblPosts and comments c in table tblComments.*
5. **FOR** *each comment c in table tblComments* **DO**
    5.1. *Execute tokenization and POS-tagging process as described in section 3.2*
    5.2. *Generate FeqFT (c, FFT) matrix by identifying frequent features (FFT) in c through **Apriori** frequent pattern algorithm with minimum support 50%*
    5.3. *Identify opinion words OW for extracted FFT as described in section 3.2*
    5.4. *Determine semantic orientation OR of OW as described in section 3.2*
    5.5. *Generate OE (c, FFT, OW, OR) matrix of c*
    5.6. *Store node $V_t \in V$, who commented c, in the matrix named VTmatrix (influenced nodes matrix)*
6. **FOR** *each post W in table   tblPosts* **DO**
    6.1. *Compute the polarity score θ_z from OE matrix as $\theta_z = (\sum positive\ c - \sum negative\ c) \times 100\%$*
    6.2. *Store node $V_s \in V$, who posted W, in the matrix named VS matrix (influential nodes matrix)*
7. *Merge VT and VS matrices into influence matrix IMAT with 3 attributes $< influencial\ users(V_s), influenced\ users(V_t), Action >$ as follows:*
    7.1. **IF** *node V_t responds to node V_S* **Do** *IMAT[Action] = $\sum responses$*
    7.2. **ELSE Do** *IMAT[Action] = 0*
8. *Generate influence graph $G_z = (V, E), V \in VT, VS$ and $E = IMAT[Action]$ if there exist a relationship between VT and VS matrices (section 3.3)*

Fig. 1. Algorithm of OBIN to generate influential network from friendship network

TPD consists of three steps: relevant nodes identification (line 1 in Fig. 1), preprocessing (lines 2-3 in Fig. 1), and extraction (line 4 in Fig. 1). Identification is done by
conducting a local search in the whole social network using Graph API (to crawl social network) and SQL. Let us consider the social network as Facebook.com and for a
given product z, search using Graph API and FQL (Facebook SQL) results in a set of
relevant nodes V (users) on z as $< V, Term, Approve, Link >$ where $Link$ denotes
profile URL of a user (V). To determine if node V has any influence on z, we need
a node approval threshold Az determined by the user based on the following logic,
and this can be adjusted after various runs to get desired ranges of thresholds. The higher
the threshold, the higher the influence spread of selected relevant nodes. If $A(V) \geq Az$,
TPD extracts all the relevant posts posted by V on z as $< W, Term, A, SR >$, where W
is the published post, SR is the summation of comments and shares of W.

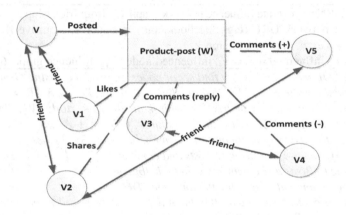

Fig. 2. An example activity in friendship network

Fig. 2 shows an example of activities in a friendship network where V posted a
post (W) on a product, and $V1$, $V2$, $V3$, $V4$, $V5$ have expressed their opinions in
different ways. For example, if we decide $Az = 50$, then TPD will extract all posts
published by V on z having ($number\ of\ likes \geq 50$). To extract all the possible
relevant posts, TPD mines the dataset with three different search features such as
search term plus number of approval ($Term + A$), term plus number of simple responses ($Term + SR$), and sum of approval and simple responses ($A + SR$), to classify
relevant and irrelevant nodes. A profile d can be denoted as a vector of N posts W_N
and node V. For N number of posts and i number of relevant nodes, we denote the
profile documents as $D: \{ d1, d2, d3, ..., di \}$, $di: \{(W_1, Vi), (W_2, Vi), ..., (W_N, Vi)\}$.
TPD keeps track of $V \times D$ (nodes by profiles) matrix, $D \times W$ (profiles by posts)
matrix, and $W \times C$ (posts by comments) matrix.

For example, if we execute the query "*SELECT id, name, category, likes, link
FROM search WHERE q = 'iPhone' AND (type = 'page' OR type = 'group')*", then
TPD will produce a set of relevant nodes on 'iPhone'. Table 1 shows an example of
extracted relevant nodes. After indexing and applying threshold, if we execute a query

such as *"SELECT post_id, message, likes.count AS A, share_count, created_time, comments.count, (comments.count+share_count) AS SR FROM stream WHERE source_id = '1' AND message !='' ORDER BY likes.count LIMIT 100"*, then TPD produces a set of 100 posts on *z* for node id = *1*. Table 2 and 3 show examples of extracted data by TPD from social network.

Table 1. Example of extracted nodes by TPD

Node ID *Vs*	Term	Approve *A*	Link
1	iphone	3116728	iphone.page
11	iphone 4	1435239	Iphone-4

Table 2. Example of extracted posts by TPD

Post ID *W*	Term	Approve *A*	Simple response *SR*
46947	Black or white	61153	11325

Table 3. Example of extracted comment data by TPD with node ID who commented

Post ID *W*	User ID *Vt*	Time	Comment *C*
46947	108936	2013-01-06	this is really cool

3.2   Post – Comment Polarity Miner (PCP – Miner)

In a friendship network, users are free to comment on any published post to express their opinions. TPD gives a ranked list of relevant nodes (*V*), posts (*W*) and comments (*C*). To determine the influential capability of a node *V*, we need to compute the polarity score (θ_z) of each post published by *V*. Proposed PCP-Miner identifies opinion comments among all the comments, identifies semantic orientation (*SO*) of the comments, and measures the polarity score (θ_z) of the posts.

Step1: In our proposed method, data cleaning includes removal of stopwords, stemming, and fuzzy matching to deal with word variations and misspelling, fuzzy duplication removal, removal of comments that are not exact replicas but exhibit slight differences in the individual data values extracted from the same node, removal of suspicious links to protect spam. The PCP-Miner then uses Tokenization and part-of-speech tagger[9] to identify phrases in the input text.

Step 2: Identify product features on which many people have expressed their opinions. In this paper, we improve existing opinion mining system OpinionMiner by identifying explicit and implicit features. To identify implicit features, we have added Apriori frequent pattern to extract frequent features. For example, in the case of "iPhone", some users may use "resolution" as a feature for "camera", some use as "screen", some use as "video call", as shown in Table 4. To identify which itemsets are product features, we use Association rule mining [1] to identify frequent itemsets, because those itemsets are likely to be product features. In our work, we define an itemset as frequent if it appears more than 50% in the review set, so as not want to lose any important comment while avoiding spam comments.

Table 4. Example of frequent features

Comment Id	Features
1	Camera {resolution, camera}
2	Picture {resolution, camera, picture}
3	Screen resolution {resolution, screen}
4	Picture {resolution, video_call, camera, picture}

The input to the Apriori algorithm is the set of nouns or noun phrases from POS-tagging, and the output itemset is product features. **Association rule mining** – In the above example, frequently occurred items such as *resolution*, *camera*, and *picture* may lead to finding association rules *resolution => camera*, which means that the comment that mentions *resolution*, may usually refer to the feature *camera*. Here, the set {*resolution, camera*}, {*resolution, camera, picture*}, {*resolution, screen*}, and {*resolution, video_call, camera, picture*} are called itemsets. Now suppose, $resolution = 30$ (the number of comments mentioning resolution), $camera = 20$ (the number of comments mentioning camera), $both = 20$ (the number of comments mentioning both resolution and camera), and $total = 100$ (the number of comments on the post). So $support = {|Rule|}/{|total|} = \frac{20}{100} = 20\%$. Rules that satisfy a minimum support, are called frequent features.

Step 3: Extract opinion words and their semantic orientations. For example, "This picture is awesome" has the word 'awesome' is the effective opinion of 'picture' and it has positive orientation. Presence of adjectives in comment text is useful for predicting whether a text is expressing opinion or not. In the opinion words extraction phase, our proposed method takes the list of tokens with corresponding POS-tags [9] from our transactional database, and search for whether it contains adjective words and/or frequent features identified by Apriori algorithm. In our proposed work, we use WordNet to get the adjective synonym (words with similar meaning) set and antonym (words with opposite meaning) set to identify the opinion expressed by the word (i.e., positive or negative opinion). To identify the semantic orientation(positive/negative) of each comment, we need to identify the semantic orientation of extracted opinion words. This is done with the help of WordNet, we store some known orientations along with the words. When we will extract any new unknown word that is not in the list of WordNet, we look for its synonym and consider its' semantic orientation as the orientation of unknown word, and store it back into the list of semantic orientations for future use. For example, we know the word "good" has positive orientation stored in the list and has a list of synonyms {great, cool, awesome, nice}. If we find a comment containing the word 'cool', we will take its orientation positive. If a comment sentence contains a set of features, then for each feature, we compute an orientation score for the feature. Positive opinion word has score $(+1)$ and negative opinion word has score (-1). All the scores are then summed up. If the final score is positive, the semantic orientation of the comment is positive otherwise negative and if the score is zero then neutral.

Step 4: We calculate the polarity score θ_z of each post as

$$\theta_z = \left(\sum positive\ responses - \sum negative\ responses\right) \times 100\% \qquad (1)$$

3.3   Social Influence Graph Generation and Community Preference

Based on polarity score, we have a ranked list of relevant nodes V, their posts W, comments C, and the set of influenced nodes who commented on the posts W. From the set of influenced nodes, we compute the influence score determined by their number of responses and then index them. Table 6 (for example) shows a list of influenced nodes. We propose an algorithm PoPGen (popularity graph generator) to generate a social network influence graph $G_z = (V, E)$ on product z using the influence matrix $IMAT$ (Table 7). PoPGen adds a node V to the vertex list according to the number of responses. For all vertices, PoPGen finds if V_i has a relation with V_j where V_i, $V_j \in V$ and $i, j \in N(number\ of\ nodes\ in\ network)$, add 1 in Table 7, for example, and $\mathbf{0}$ otherwise. The value 1 means add an edge between the vertices V_i and V_j. The generated influence graph Gz represents the community preference for a product z.

Table 5. Example data for post – user relationship and user – user relationship

Node id V_i	Post ID W	Node id V_j		User id V_1	User id V_2
1	49823667	4		3	1
2	11250901	6			

Table 6. Example data for Influence Matrix (IMAT) according to Table 6

	1	2	3	4	5	6	7
1	0	0	1	1	0	0	0
2	0	0	0	0	0	1	0

4   Experimental Evaluation

4.1   Dataset

We conducted our experiment using the users' posts and opinions in Facebook as a friendship network for iphone, iPad, Samsung Galaxy. Our TPD method automatically extracts the relevant data and stored into our generated data warehouse, called OBIN_dwh, in a temporal basis. We selected a set of datasets based on Approve (A) and Simple response (SR) as follows:

(1) Node selection – we considered $A_f \geq \mathbf{1000}$ i.e., nodes having more than $\mathbf{1000}$ friends. With this characteristic, we had $\mathbf{1178}$ nodes with $\mathbf{42,664}$ relevant and irrelevant posts. (2) Post selection – we considered $SR \geq \mathbf{20}$ and $A_l \geq \mathbf{10}$ for

posts of each node where **SR** represents number of re-shares and comments of the posts and **A** represents number of likes and the result is **3793** relevant posts.

After applying TPD and PCP-Miner, we obtained **343** influential nodes and **45, 126** influenced nodes with **47, 298** relationship edges.

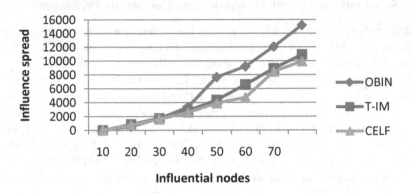

Fig. 3. Comparison of influence spread by different number of influential nodes

4.2   Performance Analysis

We evaluated proposed OBIN against accuracy of CELF and T-IM algorithms. **Recall** – is the ratio of the number of relevant nodes retrieved and the total number of relevant nodes that exist in the network, denoted by R. **Precision** – is the ratio of the number of relevant nodes retrieved and the total number of relevant and irrelevant nodes retrieved, denoted by P. **F-score** – accuracy of the proposed system = $2 \times (P \times R)/(P + R)$. Table 8 shows the accuracy measure of CELF [7] and T-IM [2] and proposed OBIN with the same dataset, and we observed that OBIN is dramatically better by retrieving more relevant nodes than CELF and T-IM.

Table 7. Comparison of discovering influential nodes by CELF, T-IM and OBIN

	CELF	T-IM	OBIN
F – score	85.4%	88.1%	95.3%

Fig 3 shows the influence spread over network by different algorithms. With small number of nodes, CELF and T-IM and proposed OBIN give almost the same performance in influence spread, but as we increase the number of nodes, OBIN performs better because, for a specific product, CELF and T-IM discover relevant nodes along with more irrelevant nodes which slow down their performances.

5   Conclusions and Future Works

This paper proposed an effective method for discovering relevant influential nodes from friendship network which enables more focused target marketing than existing influential maximization algorithms. Previous research considers opinion mining only in user-service network that are not directly applicable to user-user network i.e., friendship network, where user-user network is a more complex network that includes multiple relationships between users and users, and between users and products. The proposed OBIN miner mines opinions from complex user-user relationship network (e.g., Facebook) with multiple posts, multiple products, considering both implicit and explicit opinions. Experimental results show that the proposed technique performs better than the existing general IM methods. To handle more rapid network evolution, future work will look into using pre-updating of data such as friendship relationships to speed up the computation of the IN by the OBIN system.

References

1. Agrawal, R., Srikant, R., et al.: Fast algorithms for mining association rules. In: 20th Int. Conf. Very Large Data Bases, VLDB, pp. 487–499 (1994)
2. Ahmed, S., Ezeife, C.I.: Discovering Influential Nodes from Trust Network. In: ACM SAC International Conference, Coimbra, Portugal (2013)
3. Bonchi, F., Castillo, C., Gionis, A., Jaimes, A.: Social network analysis and mining for business applications. ACM Transaction TIST 22 (2011)
4. Dave, K., Lawrence, S., Pennock, D.M.: Mining the peanut gallery: Opinion extraction and semantic classification of product reviews. In: 12th International Conference on World Wide Web, pp. 519–528 (2003)
5. Hu, M., Liu, B.: Mining and summarizing customer reviews. In: 10th ACM SIGKDD Int. Conference on Knowl. Discov. and Data Mining, pp. 168–177 (2004)
6. Jin, W., Ho, H.H., Srihari, R.K.: OpinionMiner: a novel machine learning system for web opinion mining and extraction. In: 15th ACM SIGKDD Int. Conference on Knowl. Discov. and Data Mining, pp. 1195–1204 (2009)
7. Leskovec, J., Krause, A., Guestrin, C., Faloutsos, C., VanBriesen, J., Glance, N.: Cost-effective outbreak detection in networks. In: 13th ACM SIGKDD Int. Conference on Knowl. Discov. and Data Mining, pp. 420–429 (2007)
8. Pang, B., Lee, L., Vaithyanathan, S.: Thumbs up?: sentiment classification using machine learning techniques. In: ACL 2002 Conference on Empirical Methods in Natural Language Processing, vol. 10, pp. 79–86 (2002)
9. Marcus, M.P., Marcinkiewicz, M.A., Santorini, B.: Building a large annotated corpus of English: The Penn Treebank. Computational Linguistics, 313–330 (1993)

Computing Hierarchical Skyline Queries "On-the-Fly" in a Data Warehouse

Tassadit Bouadi[1], Marie-Odile Cordier[1], and René Quiniou[2]

[1] IRISA - University of Rennes 1
[2] IRISA - INRIA Rennes
Campus de Beaulieu, 35042 RENNES, France
{tassadit.bouadi,marie-odile.cordier}@irisa.fr, rene.quiniou@inria.fr

Abstract. Skyline queries represent a powerful tool for multidimensional data analysis and for decision aid. When the dimensions are conflicting, skyline queries return the best compromises associated with these dimensions. Many studies have focused on the extraction of skyline points in the context of multidimensional databases, but, to the best of our knowledge, none of them have investigated skyline queries, when data are structured along multiple and hierarchical dimensions. This article proposes a new method that extends skyline queries to multiple hierarchical dimensions. Our proposal, *HSky* (Hierarchical Skyline Queries) allows the user to navigate along the dimensions hierarchies (i.e. specialize / generalize) while ensuring an efficient online calculation of the associated skyline.

1 Introduction

Skyline queries represent a powerful tool for multidimensional data analysis and decision-making. When the dimensions are in conflict, the skyline queries return the best compromises on these dimensions. Skyline queries have been extensively studied [1, 2] in the database and the artificial intelligence communities. Several studies have investigated [3, 4] the problem of expressing and evaluating OLAP preferences, but few of them have addressed the problem of skyline computation when dealing with aggregated data and hierarchical dimensions in a data warehouse. The aim is to couple OLAP with skyline analysis to enable the user to select the most interesting facts from the data warehouse. Therefore, the main challenge is to compute skylines efficiently over hierarchical dimensions and over aggregated data. This problem rises several scientific and technical issues. Should the skylines be recomputed at every hierarchical level? Can the skyline of a given level be derived from the skyline at lower or higher level? Can conventional skyline algorithms be extended to cope with hierarchical dimensions?

Recent work [5, 6, 7] has considered the computation of skyline queries over aggregated data. These proposals have focused on the optimization of queries involving both *Skyline* and *Group-By* operators. But, they propose to execute the two operators sequentially without a real coupling. More interestingly, the operator *Aggregate Skyline* proposed by the authors of [8], combines the functionalities of both *Skyline* and *Group-By* operators. The definition of the dominance

L. Bellatreche and M.K. Mohania (Eds.): DaWaK 2014, LNCS 8646, pp. 146–158, 2014.
© Springer International Publishing Switzerland 2014

relation is extended to point groups. This operator enables the user to perform skyline queries on point groups in order to select the most relevant group.

The aim of this work is to propose an efficient approach simulating the effect of the OLAP operators "drill-down" and "roll-up" on the computation of skyline queries. The proposed method *HSky (Hierarchical Skyline queries)* provide the user with an interactive navigation tool that let him specialize/generalize a basic preference and its associated skyline, and derive the corresponding skylines while respecting the hierarchical structure of the involved dimensions. Properties of the hierarchical relationships between preferences associated with the different dimensions are used to design an efficient navigation schema among the preferences, while ensuring an online computation of skyline queries.

In Section 2, we introduce the basic concepts related to skyline queries and preference orders. Section 3 develops the formal aspects of our new approach *HSky* and its implementation. Section 4 gives the results of the experimental evaluation performed on synthetic datasets and highlights the relevance of the proposed solution. Section 5 concludes the paper.

2   Basic Concepts

Let $D = \{d_1, ..., d_n\}$ be an n-dimensional space, E a data set defined in space D and $p, q \in E$. $p(d_i)$ denotes the value of p on dimension d_i. A preference on the domain of a dimension d_i, denoted by $\wp_{d_i}$, is defined by a partial order $\leq_{d_i}$. This order can also be represented by the set of binary preferences (possibly infinite) $\wp_{d_i} = \{(u, v) | u \leq_{d_i} v\}$, where (u, v) is an ordered pair. $\wp = \bigcup_{i=1}^{|D|} \wp_{d_i}$ denotes the set of preferences associated to the space D. p is said to *dominate* q in D, denoted by $p <_D q$, if $\forall d_i \in D$, $p(d_i) \leq_{d_i} q(d_i)$ (i.e. p is preferred or equal to q on D) and $\exists d_i \in D$, $p(d_i) <_{d_i} q(d_i)$ (i.e. p is strictly preferred to q on some dimension d_i). For better readability, $p <_D q$ is simply noted $p < q$.

Definition 1. *(Skyline) Let $\wp$ be the preference set on D. The skyline of the dataset E on the dimension space D is the set of points that are not dominated by any point in E: $Sky(D, E)_\wp = \{p \in E | \neg(\exists q \in E, q <_D p)\}$.*

Property 1. *[9] (Monotonicity of preference extension) Let $\wp'$ and $\wp''$ be two preference sets on D. If $\wp' \subseteq \wp''$ (i.e. $\wp''$ is an extension of $\wp'$) then $Sky(D, E)_{(Z, \wp'')} \subseteq Sky(D, E)_{(Z, \wp')}$.*

Example 1. *(Running example) In this paper, we will use as running example the dataset E in Table 1. It contains 6 agricultural plots described by 3 dimensions: location (Loc), nitric pollution rate (Np) kgN/ha/year and crop yield (Yd) kg/ha. A basic preference $\wp^0_{d_i}$ is defined for each dimension d_i of the space D. The order relations $\leq_{Yd}$ and $\leq_{Np}$ are based on the order relation on natural numbers, and specify that plots with the highest crop yield (Yd) and with the lowest nitric pollution rate (Np) are preferred. The values of dimension Loc are associated with the preference order $\{Brittany <_{Loc} Epte,\ Yeres <_{Loc} Normandy\}$. The remaining values of dimension Loc are left unordered. $Sky(D, E)_{\wp^0} = \{a, b, e, c, d, f\}$.*

Table 1. A set of agricultural plots

ID plot	Loc	Np	Yd
a	Pays de la Loire (PL)	16	200
b	Yeres (YRS)	24	500
c	Vilaine (VLN)	36	100
d	Yar	30	200
e	ALL	23	400
f	Epte (EPT)	30	300

2.1   Hierarchy Formalization

In data warehouses, the domain values of dimensions are structured in hierarchies. Each dimension $d_i \in D$ is associated with a hierarchy that is represented by a directed acyclic graph whose nodes represent subsets of d_i domain values and edges represent set inclusion relationships. The most general value is at the root of the hierarchy and the leaves correspond to the most specific values.

Definition 2. *(Hierarchy) Let $H_D = \{h_{d_1}, ..., h_{d_{|D|}}\}$ be the set of hierarchies associated with the dimensions of space D where:*

- *h_{d_i} represents the hierarchical relationship, possibly empty (i.e. the graph reduced to the single node ALL), between the values of dimension $d_i \in D$,*
- *h_{d_i} is a directed acyclic graph,*
- *for each node $n_j \in h_{d_i}$, $label(n_j) = v_j$ with $v_j \in dom(d_i)$,*
- *for each value $v_j \in dom(d_i)$, $\exists n_j \in h_{d_i}$, $label(n_j) = v_j$.*

$\hat{v}_j$ denotes the set of n_j ancestor (direct or indirect) labels in the hierarchy h_{d_i} and $\check{v}_j$ denotes the set of its descendant labels.

The value ALL, the most general value in the hierarchy, is required to belong to the domain values of every d_i.

Example 2. *Figure 1 describes examples of hierarchies defined on dimensions Loc and Np. The hierarchy of dimension Loc is represented by three hierarchical levels: region (Brittany, Normandy,...), sub-region (West Brittany, North Normandy,...) and catchment (Yar, Epte,...). The domain values of Loc contains all these values associated with hierarchical levels. Similarly, the Np hierarchy introduces categorical values abstracting numerical values associated with Np.*

A hierarchy allows representing incomplete information. For example, plot e in table 1 is described by the value *ALL* on the dimension *Loc*. This means that there is no information on the exact location where plot e is located. Similarly, plot a in table 1 is described by the value *Pays de Loire* on dimension *Loc*. This means that the most detailed and most accurate value that exists about the location of plot a is its region (*Pays de Loire*). In contrast, plots b, c, d and f are described by the most precise values of the *Loc* domain.

Defining hierarchies on dimensions allows the user to generalize or specialize preferences for computing related skyline. For example, once domain values of

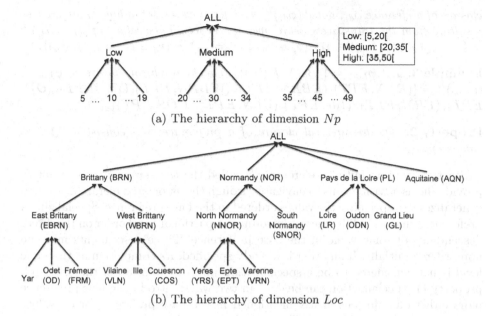

(a) The hierarchy of dimension Np

(b) The hierarchy of dimension Loc

Fig. 1

Np are partitioned in categorical values (e.g. the rates below 10 – *low* – have a low impact, the rates between 20 and 34 – *medium* – have a medium impact and the rates higher than 35 – *high* – have a strong impact), the user can express abstract preferences, and submit related abstract queries, instead of formulating preferences between individual values only (e.g. a plot with a nitric pollution rate of 16 is better than a plot with a rate of 30). Similarly, a user may be interested in more specific phenomena, e.g. green algae proliferation in coastal agricultural catchments and is likely to pay attention to specific regions of dimension Loc like Brittany, where this proliferation problem is well-known.

2.2   Hierarchical Relationships between Preferences

After defining the structure of hierarchical dimensions, we focus now on preference properties introduced by such dimensions.

Definition 3. *(Preference consistency)* Let $\wp_{d_i} = \{(v_1, v_2), ..., (v_n, v_m)\}$ be *a preference on the hierarchical dimension d_i. The preference $\wp_{d_i}$ is consistent if and only if it does not contain binary preference ordering a value and its ancestors:* $\nexists (v_k, v_j) \in \wp_{d_i},\ k \neq j,\ (v_k \in \widehat{v_j} \lor v_j \in \widehat{v_k})$.

In the rest of the paper, we only consider consistent preferences and we assimilate preferences to their hierarchical closures.

Definition 4. *(Hierarchical closure)* Let $\wp_{d_i}$ be a preference associated with the hierarchical dimension d_i of the multidimensional space D. The hierarchical *closure of preference $\wp_{d_i}$, noted $(\wp_{d_i})^H$, is defined as the set of binary preferences resulting from the transitive closure over the descendants values of $\wp_{d_i}$: $(\wp_{d_i})^H = \{(v_p, v_q) \mid \exists \ (v_n, v_m) \in \wp_{d_i}, (v_p \in \widetilde{v_n} \vee v_p = v_n) \wedge (v_q \in \widetilde{v_m} \vee v_q = v_m)\}.$*

Example 3. *Let $\wp_{Loc} = \{(BRN, EPT)\}$. The hierarchical closure of $\wp_{Loc}$ is $(\wp_{Loc})^H = \{(BRN, EPT), (EBRN, EPT), (WBRN, EPT), (Yar, EPT), (OD, EPT), (FRM, EPT), (Ille, EPT), (VLN, EPT), (COS, EPT)\}.$*

Property 2. *The hierarchical closure of a preference $\wp$ is consistent iff $\wp$ is consistent.*

Starting from an initial preference $\wp^0$, called the *base preference*, we want to provide the user with means to navigate through the associated hierarchies, i.e. to generalize or to specialize the values ordered in the base preference. Specializing a preference removes indifference: it introduces a partial or total order on the direct descendants of some value in the base preference. These descendants must be non ordered initially. If an order is already specified, its completion at the same level is not considered to be a specialization but an extension of the order (cf. property 1). Specialization can be done in two ways: introducing new preference pairs either on values explicitly mentioned in the base preference or on values ignored in the base preference. Thus, the domain values of some dimension, e.g. *Loc*, can be divided into two sets: values that can be specialized (ordered values – colored green in Figure 2 – and non ordered values – colored blue) and those that cannot be specialized (colored red in Figure 2). A border can be traced between these two kinds of values (cf. Figure 2).

Example 4. *Let $\wp_{Loc}^0$ and $\wp'_{Loc}$ be two preferences associated with the hierarchical dimension Loc. $\wp_{Loc}^0 = (\{(BRN, EPT)\})^H = \{(BRN, EPT), (BBRN, EPT), (HBRN, EPT), (Yar, EPT), (OD, EPT), (FRM, EPT), (Ille, EPT), (VLN, EPT), (COS, EPT)\}, \ \wp'_{Loc} = \wp_{Loc}^0 \bigcup \{(Yar, VLN)\}$ and $\wp''_{Loc} = \wp_{Loc}^0 \bigcup \{(LR, GL)\}$. $\wp'_{Loc}$ is a specialization of $\wp_{Loc}^0$ since the values of $\{(Yar, VLN)\}$ are descendants of BRN in h_{Loc}.*

$\wp''_{Loc}$ is a specialization of $\wp_{Loc}^0$ with respect to h_{Loc}. The ancestor of GL and LR (i.e. PL) does not belong to $\wp_{Loc}^0$. This means that the value PL is not explicitly ordered in $\wp_{Loc}^0$ with respect to the other values (i.e. BRN and NOR). Let now $\wp_{Loc}^0 = (\{(BRN, EPT)\})^H$ and $\wp'''_{Loc} = \wp_{Loc}^0 \bigcup \{(VLN, YRS)\}$. $\wp'''_{Loc}$ is an extension of $\wp_{Loc}^0$ but $\wp'''_{Loc}$ is not a specialization of $\wp_{Loc}^0$ since the value YRS mentioned in $\wp'''_{Loc}$ is not the specialization of a value mentioned in $\wp_{Loc}^0$ (i.e. values colored green or blue Figure 2).

Preference generalization, the dual operation of preference specialization, adds indifference between values ordered in the preference. Definition 3 gives the formal specification of preference specialization / generalization:

Definition 5. *(Preference specialization/generalization)*
Let $\wp_{d_i} = \{(v_1, v_2), ..., (v_n, v_m)\}$ and $\wp'_{d_i} = \{(v'_1, v'_2), ..., (v'_p, v'_q)\}$ be two preferences associated with the hierarchical dimension d_i.

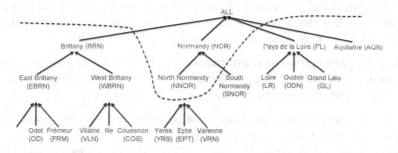

Fig. 2. Specialization/generalization implicit border

- $\wp'_{d_i}$ is a specialization of $\wp_{d_i}$ (denoted $\wp_{d_i} \sqsubset_h \wp'_{d_i}$) if:
  1. $\wp_{d_i} \subset \wp'_{d_i}$ i.e. $\wp'_{d_i}$ is an extension of $\wp_{d_i}$,
  2. $\forall (v'_i, v'_j) \in \wp'_{d_i}$:
     - either (a): $(v'_i, v'_j) \in \wp_{d_i}$,
     - or (b):
       * $\exists (v_k, v_m) \in \wp_{d_i}$, $(v'_i \in \check{v_k} \wedge v'_j \in \check{v_k}) \vee (v'_i \in \widetilde{v_m} \wedge v'_j \in \widetilde{v_m})$ or,
       * $\hat{v'_i} \cap \hat{v'_j} \neq \varnothing \wedge \forall v \in \hat{v'_i} \cap \hat{v'_j}$, v is not ordered in $\wp_{d_i}$.
       and there exists at least one pair (v'_i, v'_j) which verifies property (b).
- $\wp_{d_i}$ is a generalization of $\wp'_{d_i}$ if $\wp'_{d_i}$ is a specialization of $\wp_{d_i}$.
- $\wp_{d_i}$ and $\wp'_{d_i}$ are incomparable if there exists no specialization nor generalization relation between these values $(\neg(\wp_{d_i} \sqsubset_h \wp'_{d_i}) \wedge \neg(\wp'_{d_i} \sqsubset_h \wp_{d_i}))$.

We extend the definition of specialization/generalization to preferences expressed on several hierarchical dimensions.

Definition 6. *Let* $\wp = \bigcup_{d_i \in D} \wp_{d_i}$ *and* $\wp' = \bigcup_{d_i \in D} \wp'_{d_i}$ *be two preferences associated with dimension space* D. $\wp'$ *is a specialization of* $\wp$ *(i.e.* $\wp$ *is a generalization of* $\wp'$) *iff* $\forall d_i \in D, \wp_{d_i} \sqsubset_h \wp'_{d_i} \vee \wp_{d_i} = \wp'_{d_i} \wedge \exists d_i \in D, \wp_{d_i} \sqsubset_h \wp'_{d_i}$.

The set of specializations and generalizations of a preference associated with a set of dimensions is partially ordered with respect to the specialization / generalization relation. For example, the set of specializations and generalizations of the base preference $\wp^0$ specified on the dimensions of Table 1, induces the hierarchical structure of Figure 3. Nodes in this structure are associated with a specialization or a generalization of the base preference $\wp^0$. To improve the readibility of figures, the notation $\wp^{kl}$, where k is a level in the structure and l is a rank, is used to specify the nodes at level k. The preferences associated with the direct descendants (resp. the direct ancestors) of node N with associated preference $\wp$ are called direct specializations (resp. direct generalizations) of $\wp$.

3 Hierarchical Skyline Queries

In this section, we give some properties of hierarchical preferences and a materialization method for storing related skylines which is grounded on these properties.

Following definition 3, the specialization of a preference is also an extension of its preference, but the converse is not true. The corollary of property 1 asserts the monotonicity of preference specialization (resp. generalization).

Corollary 1. *(Hierarchical monotonicity)* *Let $\wp$ and $\wp'$ be two preferences on dimension space D. If $\wp'$ is a specialization of $\wp$ ($\wp \subset_h \wp'$) then $\wp'$ is also an extension of $\wp$ and, by property 1, $Sky(D, E)_{\wp'} \subseteq Sky(D, E)_{\wp}$.*

The hierarchical monotonicity property states that every skyline point associated with a given preference remains skyline when considering a generalization of this preference. In the sequel, the hierarchical preferences designate the set of specializations and generalizations of some base preference $\wp$.

3.1 Hierarchical Skyline Query: Algorithm *HSky*

Our goal is to minimize the number of dominance test for computing efficiently the skyline sets while navigating in the hierarchical dimensions. To do so, we characterize the skyline points that remains skyline and those that become skyline or non skyline after specializing (drill-down) or generalizing (roll-up) hierarchical preferences. We propose a compromise between (i) *materializing* all skyline points of every hierarchical preferences associated with the base preference $\wp^0$, and (ii) *compute*, for every user query, the skyline points associated with the hierarchical preferences formulated in the query.

Computation of Hierarchical Skyline Queries. In order to compute the skyline points for a user query related to some specialization $\wp'$ of preference $\wp$, we introduce the set of *hierarchical skyline* points, $HNSky(D, E)_{(\wp', \wp)}$ where $\wp$ is a direct ancestor of $\wp'$ in the specialization/generalization structure. This set gathers the skyline points that are disqualified when $\wp$ is specialized in $\wp'$ or, conversely, the skyline points introduced when $\wp'$ is generalized in $\wp$.

Definition 7. *(HNSky: Hierarchical New Skyline)* *Let $\wp$ be a preference defined on D, $Sky(D, E)_{\wp}$ its associated skyline, $\wp'$ a direct specialization of $\wp$, and $Sky(D, E)_{\wp'}$ its associated skyline. By definition: $HNSky(D, E)_{(\wp,\wp')} = \{p \in Sky(D, E)_{\wp} | p \notin Sky(D, E)_{\wp'}\}$.*

Example 5. *Let $\wp = (\{(BRN, EPT)\})^H \bigcup \wp_{Sn} \bigcup \wp_{Re}^0$ and $\wp' = \wp \bigcup \{(Yar, VLN)\}$ be two preferences defined on space D. $\wp'$ is a direct specialization of $\wp^0$. $Sky(D, E)_{\wp} = \{a, b, e, c, d, f\}$ and $Sky(D, E)_{\wp'} = \{a, b, d, e, f\}$. Consequently, $HNSky(D, E)_{(\wp,\wp')} = \{c\}$.*

The computation of $HNSky(D, E)_{(\wp,\wp')}$ does not require to compute the whole skyline associated with $\wp'$ but only to verify that the skyline points associated with $\wp$ remains skyline for $\wp'$. When specializing preference $\wp$ into $\wp'$, the skyline associated with $\wp'$ may be obtained by removing from the skyline associated with $\wp$ the points disqualified by the specialization (i.e. the set $HNSky$).

We want to build a memorization data structure $HSky$ for storing efficiently the pre-computed information. Our goal is to avoid computing and storing all the skyline points associated with every possible hierarchical preference defined on D. $HSky$ is a graph data structure whose nodes represent preferences and whose arcs connecting a child node associated with some preference $\wp'$ to a parent node associated with preference $\wp$ is labeled by $HNSky(D, E)_{(\wp, \wp')}$.

Once the base preference $\wp^0$ is chosen, the $HSky$ building process navigates in the specializations and the generalizations of its associated node:

- generate all the direct and indirect specializations and generalizations of $\wp^0$ and build the preference hierarchy. This gives the general structure of $HSky$. The node at the top of the structure denotes the empty preference $\varnothing$,
- for each arc of $HSky$, compute the set $HNSky$ that stores the disqualified points (resp. introduced) when going from a preference to a direct specialization (resp. generalization) of this preference.

In practice, these two operations are performed simultaneously. Thanks to property 1, every point in $Sky(D, E)_{\wp^0}$ belongs to the skyline associated with each parent node (generalization) of $\wp^0$. Consequently, to compute the set $HNSky$ associated with an arc going from node $\wp^0$ to one of its generalizations N, it is

---

Algorithm 1. $HSky(\wp^0, \wp, Sky(D, E)_{\wp^0})$

input : $\wp^0$: base preference associated with start node, $Sky(D, E)_{\wp^0}$: skyline associated with preference $\wp^0$, $\wp$: preference associated with target node

output: $Sky(D, E)_{\wp}$: skyline associated with $\wp$

1 $Sky \leftarrow Sky(D, E)_{\wp^0}$
2 **if** $\wp \subset_h \wp^0$ // Test whether $\wp$ is a generalization of $\wp^0$
3 **then**
4   **foreach** *parent node* $\wp^1$ *of* $\wp^0$ **do**
5   **if** $\wp = \wp^1$ **then**
6               $Sky \leftarrow Sky(D, E)_{\wp^0} \cup HNSky(D, E)_{(\wp^1, \wp^0)}$
7               $Sky(D, E)_{\wp} \leftarrow Sky$
8   **else**
9   **if** $\wp \subset_h \wp^1$ // Test whether $\wp$ is a generalization of $\wp^1$
10   **then**
11                  $Sky \leftarrow Sky(D, E)_{\wp^0} \cup HNSky(D, E)_{(\wp^1, \wp^0)}$
12                  $HSky(\wp^1, \wp, Sky)$
13   Exit

14 **if** $\wp^0 \subset_h \wp$ // Test whether $\wp$ is a specialization of $\wp^0$
15 **then**
16   **foreach** *child node* $\wp^1$ *of* $\wp^0$ **do**
17   **if** $\wp = \wp^1$ **then**
18              $Sky \leftarrow Sky(D, E)_{\wp^0} - HNSky(D, E)_{(\wp^0, \wp^1)}$
19              $Sky(D, E)_{\wp} \leftarrow Sky$
20   **else**
21   **if** $\wp^1 \subset_h \wp$ // Test whether $\wp$ is a specialization of $\wp^1$
22   **then**
23                  $Sky \leftarrow Sky(D, E)_{\wp^0} - HNSky(D, E)_{(\wp^0, \wp^1)}$
24                  $HSky(\wp^1, \wp, Sky)$
25   Exit

26 Return $Sky(D, E)_{\wp}$

---

sufficient to test the dominance of points that does not belong to the sets $Sky(D, E)_{\wp^0}$ or to the sets $HNSky$ associated with arcs targeting node N. Similarly, to compute the set $HNSky$ associated with any specialization of preference $\wp^0$, only points belonging to $Sky(D, E)_{\wp^0}$ must be tested. Figure 3 gives the structure $HSky$ of the running example.

Query Evaluation. In traditional OLAP, the *drill-down* and *roll-up* operators are applied on dimension hierarchies. In our approach, these operators are applied on hierarchical preferences associated with skyline queries. The data structure $HSky$ helps to reduce the runtime computation of skyline points associated with hierarchical preferences, which enhances interactivity. Below, the $HSky$ structure depicted in Figure 3 is used to illustrate a navigation from the base preference $\wp^0 = (\{(BRN, EPT)\})^H \bigcup \wp^0_{Sn} \bigcup \wp^0_{Re}$ and the computation of the related skylines. The skyline associated with $\wp^0$ is $Sky(D, E)_{\wp^0} = \{a, b, e, c, d, f\}$. We show how to use the structure $HSky$ for computing the skyline points associated with the preferences $\wp' = \wp^0 \bigcup \{(LR,GL),(Yar,VLN)\}$.

Skyline of a Specialized Preference. The skyline associated with preference $\wp'$, a specialization of the base preference $\wp^0$, is computed as follows. The search starts from the node associated with $\wp^0$ and explores recursively its children node depth-first, looking for the node associated with $\wp'$ (cf. algorithm 1 line 16). When the searched node is reached, $Sky(D, E)_{\wp'}$ is computed by subtracting from $Sky(D, E)_{\wp^0}$ each set $HNSky$ labeling an arc of the path going from $\wp^0$ to

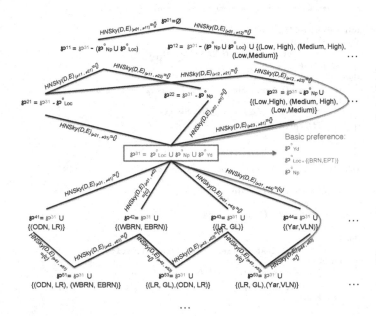

Fig. 3. The specialization / generalization data structure $HSky$

$\wp'$ (Algorithm 1, lines from 21 to 24). For the sake of efficiency, the search only explores nodes associated with some generalization of preference $\wp'$ (Algorithm 1, line 21). This pruning insures that the path going from the node associated with $\wp^0$ to the node associated with $\wp'$ has a minimal length.

Example 6. $\wp^0 = (\{(BRN, EPT)\})^H \bigcup \wp^0_{Sn} \bigcup \wp^0_{Re}$. The skyline related to $\wp^0$ is $Sky(D, E)_{\wp^0} = \{a, b, e, c, d, f\}$. Let $\wp' = \wp^0 \bigcup \{(LR,GL), (Yar,VLN)\}$ be a specialization of $\wp^0$. In Figure 3, the path from the start node $\wp^0 \equiv \wp^{31}$ to the target node $\wp' \equiv \wp^{53}$, whose skyline must be computed, is colored in red.

$$Sky(D, E)_{\wp'} = Sky(D, E)_{\wp^{31}} - (HNSky(D, E)_{(\wp^{31}, \wp^{44})} \bigcup HNSky(D, E)_{(\wp^{44}, \wp^{53})}) = \{a, b, c, e, d, f\} - (\{c\} \bigcup \{\}) = \{a, b, e, d, f\}.$$

The query evaluation of a generalization is symmetric to the specialization.

4 Experiments

In this section, we present an empirical evaluation of algorithm $HSky$ on synthetic data. $HSky$ is implemented in $JAVA$. The experiments were performed on an Intel Xeon CPU 3GHz with 16 GB de RAM under Linux. Data related to dimensions with only one hierarchical level were produced by the generator presented in [1]. Three kinds of datasets were generated: independent data, correlated data, non-correlated data. A detailed description of these datasets can be found in [1]. We only give the results concerning non-correlated data. The results on other datasets were similar, whereas the results of pre-computing and query answering are much shorter for correlated data. Data for hierarchical dimensions were generated with respect to a Zipfian distribution [10]. By default, the Zipfian parameter θ was initialized to 1 (non-correlated data). This yielded 700.000 tuples for 6 dimensions with one hierarchical level. We imposed the number of hierarchical dimensions vary from 3 to 20 and the number of hierarchical levels of these dimensions vary from 3 to 7. The base preference was chosen to yield a balanced number of specializations and generalizations. The query preference template was such that the preference represents an indirect specialization / generalization of the base preference.

To the best of our knowledge, there does not exist a work on skyline extraction within hierarchical dimensions in the literature. Thus, we have implemented algorithm $DC\text{-}H$ that computes skyline on the *Divide & Conquer* (*D&C*) principle [1]. $DC\text{-}H$ does not materialize any partial result and re-computes all the skyline sets at every hierarchical level. As *D&C* can be encoded naturally in parallel, we have parallelized $DC\text{-}H$ to improve its performance. Algorithm $HSky$ is also based on the $DC\text{-}H$ principle.

Varying the Number of Hierarchical Dimensions. In these experiments, the number of one level dimensions is set to 6 and the number of hierarchical dimensions varies from 3 to 20. Figure 4 shows that the memory size and the pre-computation runtime of $HSky$ rises with the number of hierarchical dimensions. This is related to the complexity of the data structure $HSky$ (quadratic in the number of hierarchical dimensions). Algorithm $DC\text{-}H$ does not need any memory storage, nor pre-computation runtime since it does not store any partial result.

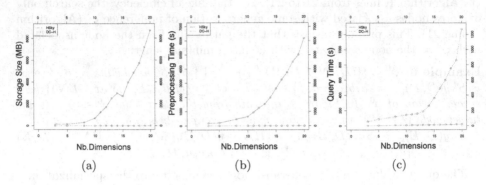

Fig. 4. Varying the number of hierarchical dimensions

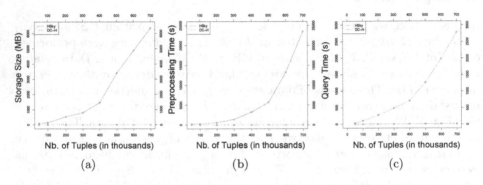

Fig. 5. Varying the size of datasets

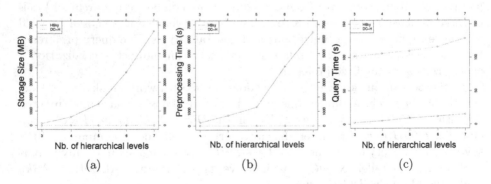

Fig. 6. Varying the number of levels in hierarchical dimension

Varying the Size of the Dataset. In these experiments, the number of tuples in the dataset varies from 50.000 to 700.000. Figure 5 shows that the memory size and the pre-computation runtime of *HSky* rises with the dataset size as well. This is due to the size of skyline sets which rises polynomially with the dataset size.

Varying the Number of Levels in Hierarchical Dimensions. In these experiments, the number of levels in hierarchical dimensions varies from 3 to 7. Figure 6 shows that the memory size and the pre-computation runtime of *HSky* rises also with the number of hierarchical levels. This is due to the volume of hierarchical dimensions which rise exponentially with the number of hierarchical levels. However, one should note that experiments were notably complex since it is not common to have to analyze dimensions with 7 hierarchical levels in real applications.

However, for each of these experiments, *HSky* outperforms *DC-H* for skyline query answering (Figures 4c, 5c and 6c). *HSky* response times are quasi-instantaneous. In fact, for each new query, *DC-H* re-computes the whole skyline whereas *HSky* deduces it using simple set operations. it should be noted that the query response time is an important criterion in the context of online analysis.

The construction of the data structure *HSky* impacts the global runtime and memory storage. To obtain some benefit, on average, from 6 to 8 navigations from the same base preference must be performed.

5 Conclusion

This article proposes a new method, called *HSky*, which extends skyline points extraction to hierarchical dimensions. Coping with hierarchies on dimensions enables to specialize or generalize user preferences when computing skylines. Hierarchical relations between preferences on dimensions were formulated and some interesting properties were explicited. These properties, e.g. the hierarchical monotonicity property, are exploited in a data structure called *HSky* for designing an efficient navigation tool along the preferences hierarchy while ensuring an online computation of skylines. The experiments underline the online performance of *HSky* compared to *DC-H* (an algorithm for computing skylines with no materialization based on [1]) when the number of queries performed during navigation is greater than some threshold (6 to 8 queries on average).

This structure could evolve with respect to the application context. For example, if the size of the structure were too high it could be possible to restrict the specializations / generalizations to values explicitly ordered in the base preference. We plan to extend the hierarchical dimensions presented in this paper to the case of dynamic preferences, as introduced in [9, 11]. In this setting, the dynamical aspect of preferences may lead to an explosion of the preferences to be materialized. A potential solution would be to restrict the preferences to the class of so-called *nth-order* preferences [9].

References

[1] Borzsonyi, S., Kossmann, D., Stocker, K.: The skyline operator. In: Proc of the 17th Int. Conf. on Data Engineering, pp. 421–430. IEEE Computer Society (2001)
[2] Raïssi, C., Pei, J., Kister, T.: Computing closed skycubes. Proc. VLDB Endow., 838–847 (2010)
[3] Golfarelli, M., Rizzi, S., Biondi, P.: myolap: An approach to express and evaluate olap preferences. IEEE Trans. Knowl. Data Eng. 23(7), 1050–1064 (2011)

[4] Golfarelli, M., Rizzi, S.: Expressing OLAP preferences. In: Winslett, M. (ed.) SSDBM 2009. LNCS, vol. 5566, pp. 83–91. Springer, Heidelberg (2009)

[5] Antony, S., Wu, P., Agrawal, D., Abbadi, A.E.: Aggregate skyline: Analysis for online users. In: Proceedings of the 2009 Ninth Annual International Symposium on Applications and the Internet, pp. 50–56. IEEE Computer Society (2009)

[6] Antony, S., Wu, P., Agrawal, D., El Abbadi, A.: Moolap: Towards multi-objective olap. In: Proceedings of the 2008 IEEE 24th International Conference on Data Engineering, pp. 1394–1396. IEEE Computer Society (2008)

[7] Jin, W., Ester, M., Hu, Z., Han, J.: The multi-relational skyline operator. In: ICDE, pp. 1276–1280 (2007)

[8] Magnani, M., Assent, I.: From stars to galaxies: skyline queries on aggregate data. In: EDBT, pp. 477–488 (2013)

[9] Bouadi, T., Cordier, M.O., Quiniou, R.: Computing skyline incrementally in response to online preference modification. T. Large-Scale Data- and Knowledge-Centered Systems 10, 34–59 (2013)

[10] Trenkler, G.: Univariate discrete distributions: N.L. Johnson, S. Kotz and A.W. Kemp, 2nd edn. John Wiley, New York (1992) ISBN 0-471-54897-9; Computational Statistics & Data Analysis, pp. 240–241 (1994)

[11] Bouadi, T., Cordier, M.O., Quiniou, R.: Incremental computation of skyline queries with dynamic preferences. In: Liddle, S.W., Schewe, K.-D., Tjoa, A.M., Zhou, X. (eds.) DEXA 2012, Part I. LNCS, vol. 7446, pp. 219–233. Springer, Heidelberg (2012)

Impact of Content Novelty on the Accuracy of a Group Recommender System*

Ludovico Boratto and Salvatore Carta

Dipartimento di Matematica e Informatica,
Università di Cagliari, Via Ospedale 72 - 09124 Cagliari, Italy
{ludovico.boratto,salvatore}@unica.it

Abstract. A group recommender system is designed for contexts in which more than a person is involved in the recommendation process. There are types of content (like movies) for which it would be advisable to recommend an item only if it has not yet been consumed by most of the group. In fact, it would be trivial and not significant to recommended an item if a great part of the group has already expressed a preference for it. This paper studies the impact of content novelty on the accuracy of a group recommender system, by introducing a constraint on the percentage of a group for which the recommended content has to be novel. A comparative analysis in terms of different values of the percentage of the group and for groups of different sizes, was validated through statistical tests, in order to evaluate when the difference in the accuracy values is significant. Experimental results, deeply analyzed and discussed, show that the recommendation of novel content significantly affects the performances only for small groups and only when content has to be novel for the majority of it.

Keywords: Group Recommendation, Content Novelty, Clustering, Accuracy.

1 Introduction

Recommender systems aim to provide information items that are expected to interest a user [1]. *Group recommendation* is a type of recommendation designed for contexts in which more than a person is involved in the recommendation process [2,3]. Group recommender systems suggest items to a group, by combining individual models that contain a user's preferences [4].

Group Recommendation and Detection of Groups. A particular application scenario in which group recommendation is useful is when the number of recommendations that can be generated by a system is limited.

---

* This work is partially funded by Regione Sardegna under project SocialGlue, through PIA - Pacchetti Integrati di Agevolazione "Industria Artigianato e Servizi" (annualità 2010).

L. Bellatreche and M.K. Mohania (Eds.): DaWaK 2014, LNCS 8646, pp. 159–170, 2014.

> *A company decides to print recommendation flyers that present suggested products. Even if the data to produce a flyer with individual recommendations for each customer is available, printing a different flyer for everyone would be technically too hard to accomplish and costs would be too high. A possible solution would be to set a number of different flyers to print, such that the printing process could be affordable in terms of costs and the recipients of the same flyer would be interested by its content.*

With respect to classic group recommendation, this type of systems adds the complexity of optimally defining groups, in order to respect the constraint on the number of recommendations that can be produced and maximize users' satisfaction. In the literature no system is able to automatically adapt to such constraints imposed by the system.

Recommendation of Novel Content. A group recommendation approach that recommends the same content previously evaluated by users would be useful for content that is always renewed and ever-changing, like news items or TV series episodes, since the user preferences for such types of content can be used to recommend items of the same type (e.g., news about the same topic or new episodes of the same series).

On the contrary, when a system produces group recommendations for types of content like movies, a new issue arises: the *novelty* of the recommended items. In fact, if an item was already evaluated by a great part of the group, the system should limit its recommendation, since users who already considered the item would be bored to reconsider it. From the user perspective, a group recommendation of an already evaluated item would be uninteresting and not significant, while from the system perspective it would be trivial to recommend items evaluated by a large part of the group. The recommendation of novel content is a key aspect that is being investigated both in the collaborative filtering [5] and the content-based [6] recommender systems literature.

Our Contributions. This paper studies the impact of content novelty on the accuracy of a group recommended system. Recommending novel content creates a trade-off that involves an improvement in the satisfaction of the users and a loss in the accuracy of the predicted ratings. Since groups of different sizes are automatically detected by the system we used, this study allows a content provider to explore such a trade-off, by controlling the level of personalization of the recommended content. To the best of our knowledge, this is the first study of this type.

The scientific contributions of our study are the following:

- this is the first time that content novelty is studied in the group recommendation literature;
- novelty is evaluated for groups of different sizes and by considering different amounts of users for which a recommended item has to be novel. This allows to evaluate how the accuracy of the system evolves when the constraints change; that is, how content novelty affects the accuracy for groups of different sizes;

 – a critical discussion of the obtained results is presented, in order to help in
   the design of a group recommender system that detects groups and allow a
   content provider to control the novelty of the recommended content.

The rest of the paper is organized in the following way: Section 2 presents
related work in group recommendation; Section 3 contains a description of the
system used in this study; Section 4 describes the experiments we conducted,
outlines main results and presents a discussion of the results; Section 5 will draw
conclusions and present future work.

2   Related Work

PolyLens [7], produces recommendations for groups of users who want to see a
movie. A Collaborative Filtering approach is used to produce recommendations
for each user of the group. The movies with the highest recommended ratings are
considered and a "least misery" strategy is used, i.e., the recommended rating for
a group is the lowest predicted rating for a movie, to ensure that every member
is satisfied.

MusicFX [8] is a system that recommends music to the members of a fitness
center. Since people in the room change continuously, the system gives the users
that are working out in the fitness center the possibility to login. The music to
play is selected considering the preferences of each user in a summation formula.

Flytrap [9] similarly selects music to be played in a public room. A 'virtual
DJ' agent is used by the system to automatically decide the song to play. The
agent analyzes the MP3 files played by a user in her/his computer and considers
the information available about the music (like similar genres, artists, etc.). The
song to play is selected through a voting system, in which an agent represents
each user in the room and rates the candidate tracks.

In-Vehicle Multimedia Recommender [10] is a system that aims at selecting
multimedia items for a group of people traveling together. The system aggregates
the profiles of the passengers and merges them, by using a notion of distance
between the profiles. A content-based system is used to compare multimedia
items and group preferences.

FIT (Family Interactive TV System) [11] is a TV program recommender system. The only input required by the system is a stereotype user representation
(i.e., a class of viewers that would suit the user, like *women, businessmen, students*, etc.), along with the user preferred watching time. When someone starts
watching TV, the system looks at the probability of each family member to
watch TV in that time slot and predicts who there might be watching TV. Programs are recommended through an algorithm that combines such probabilities
and the user preferences.

TV4M [12] recommends TV programs for multiple viewers. The system identifies who is watching TV, by providing a login feature. In order to build a
group profile that satisfies most of its members, all the current viewers profiles are merged, by doing a total distance minimization of the features available

(e.g., genre, actor, etc.). According to the built profile, programs are recommended to the group.

In [13] a group recommender system called *CATS (Collaborative Advisory Travel System)* is presented. Its aim is to help a group of friends plan and arrange ski holidays. To achieve the objective, users are positioned around a device called "DiamondTouch table-top" and the interactions between them (since they physically share the device) help the development of the recommendations.

Pocket RestaurantFinder [14] is a system that suggests restaurants to groups of people who want to dine together. Each user fills a profile with preferences about restaurants, like the price range or the type of cuisine they like (or do not like). Once the group composition is known, the system estimates individual preference for each restaurant and averages those values to build a group preference and produce a list of recommendations.

Travel Decision Forum [15] is a system that helps groups of people plan a vacation. Since the system aims at finding an agreement between the members of a group, asynchronous communication is possible and, through a web interface, a member can view (and also copy) other members preferences. Recommendations are made by using the median of the individual preferences.

In [16], Chen and Pu present *CoFeel*, an interface that allows to express through colors the emotions given by a song chosen by the *GroupFun* music group recommender system. The interface allows users to give a feedback about how much they liked the song and the system considers the preferences expressed through the emotions, in order to generate a playlist for a group.

In [17], Jung develops an approach to identify long tail users, i.e., users who can be considered as experts on a certain attribute. So, the ratings given by the long tail user groups are used, in order to provide a relevant recommendation to the non-expert user groups, which are called short head groups.

Our approach differs from the ones in the literature, since none of the existing group recommendation approaches works with automatically detected groups and is able to adapt to constraints imposed by the context in which a system operates. Moreover, no work ever conducted a study to evaluate the impact of novelty on the accuracy of a group recommender system.

3   Group Recommendation with Automatic Detection of Groups

This section describes the group recommender system used for this study, named *Predict&Cluster*, which automatically detects groups by clustering users.

The tasks performed by the systems are the following:

1. *Predictions of the missing ratings for individual users.* Predictions are built for each user with a User-Based Collaborative Filtering Approach.
2. *Detection of the groups.* Considering both the individual preferences expressed by each user and the predicted ratings, groups of similar users are detected with the k-means clustering algorithm.

3. *Group modeling*. Once groups have been detected, a group model is built for each group, by using the *Additive Utilitarian* modeling strategy.

All the tasks are now be described in detail.

3.1 Predictions of the Missing Ratings for Individual Users

The missing ratings for the items not evaluated by each user are predicted with a classic User-Based Nearest Neighbor Collaborative Filtering algorithm, presented in [18]. This has been proved to be the most accurate way to predict ratings in this scenario [19].

The algorithm predicts a rating p_{ui} for each item i that was not evaluated by a user u, by considering the rating r_{ni} of each similar user n for the item i. A user n similar to u is called a *neighbor* of u. Equation (1) gives the formula used to predict the ratings:

$$p_{ui} = \bar{r}_u + \frac{\sum_{n \subset neighbors(u)} userSim(u,n) \cdot (r_{ni} - \bar{r}_n)}{\sum_{n \subset neighbors(u)} userSim(u,n)} \qquad (1)$$

Values $\bar{r}_u$ and $\bar{r}_n$ indicate the mean of the ratings given by user u and user n. Similarity $userSim()$ between two users is calculated using the Pearson's correlation, which compares the ratings of the items rated by both the target user and the neighbor. Pearson's correlation among a user u and a neighbor n is given in Equation (2) (I_{un} is the set of items rated by both u and n).

$$userSim(u,n) = \frac{\sum_{i \subset I_{un}} (r_{ui} - \bar{r}_u)(r_{ni} - \bar{r}_n)}{\sqrt{\sum_{i \subset I_{un}} (r_{ui} - \bar{r}_u)^2} \sqrt{\sum_{i \subset I_{un}} (r_{ni} - \bar{r}_n)^2}} \qquad (2)$$

The metric ranges between 1.0 and -1.0. Negative values do not increase the prediction accuracy [20], so they are discarded by the task.

3.2 Detection of the Groups

In order to respect the constraint imposed by the context, the set of users has to be partitioned into a number of groups equal to the number of recommendations that can be produced. Since in our application scenario groups do not exist, unsupervised classification (*clustering*) is necessary.

In [21] it was highlighted that the sparsity of the rating matrix strongly affects the performances of a group recommender system and, in particular, of the clustering task. Therefore, a group recommender system that clusters users based on the individual ratings has to include personal predictions in the clustering input; this allows to avoid sparsity, overcome the curse of dimensionality, and increase the system accuracy.

In [22], authors highlight that the k-means clustering algorithm is by far the most used clustering algorithm in recommender systems. Moreover, in previous studies we analyzed [23] and compared [24] a different option to group the users, by using the Louvain community detection algorithm, which produces a hierarchical partitioning of the users; however, results showed that k-means is more accurate in this context.

This task clusters users with the k-means clustering algorithm, based on the individual preferences explicitly expressed by them and on the individual predictions built by the previously presented task. The output produced is a partitioning of the users into groups (clusters), such that users with similar models (i.e., similar ratings for the same items) are in the same group and can receive the same recommendations.

3.3   Group Modeling

In order to create a model that represents the preferences of a group, the *Additive Utilitarian* group modeling strategy [4] is adopted. The strategy sums the individual ratings for each item and produces a list of group ratings (the higher the sum is, the earlier the item appears in the list). The ranked list of items is exactly the same that would be produced when averaging the individual ratings, so this strategy is also called 'Average strategy'. An example of how the strategy works is given in Table 1. The example considers three users (u_1, u_2 and u_3) that rate eight items (i_1, ..., i_8) with a rating from 1 to 10.

In a recent work [25] we showed that this metric allows to obtain the most accurate performances for a group recommender system that detects groups. Indeed, results show that:

- since the considered scenario deals with a limited number of recommendations, the system works with large groups. Therefore, an average, which is a single value that is meant to typify a set of different values, is the best way to put together the ratings in this context;
- for groups created with the k-means clustering algorithm, creating a group model with an average of the individual values for each item is like re-creating the centroid of the cluster, i.e., a super-user that connects every user of the group.

In order to have the same scale of ratings both in the group models and in the individual user models, the produced group models contain the average of the individual predictions, instead of the sum.

Table 1. Example of the *Additive Utilitarian* strategy

	i_1	i_2	i_3	i_4	i_5	i_6	i_7	i_8
u_1	8	10	7	10	9	8	10	6
u_2	7	10	6	9	8	10	9	4
u_3	5	1	8	6	9	10	3	5
Group	20	21	21	25	26	28	22	15

4   Experimental Framework

This section presents the framework built for the experiments.

4.1   Experimental Setup

To conduct the experiments, we adopted the MovieLens-1M dataset.

The number of neighbors used by the first task to predict the ratings is 100 (see [19] for the details of the experiments that allowed to set the value).

The clusterings with k-means were created using a testbed program called KMlocal [26], which contains a variant of the k-means algorithm, called *EZ Hybrid*, that was chosen because it returned a lowest average distortion.

The RMSE values obtained by the system considering increasing percentages of the group for which content had to be novel have been compared, by considering different numbers of groups to detect (that correspond to the number of recommendations that can be produced by the system). The choice to measure the performances for different numbers of groups has been made to show how the accuracy of the systems changes as the constraint changes. In order to evaluate the quality of the predicted ratings for different numbers of groups, in each experiment four different clusterings of the users into 20, 50, 200 and 500 groups were created. Moreover, we compared the results obtained with the previously mentioned four clusterings with the results obtained considering a single group with all the users (i.e., we tested the system in a scenario in which just one set of recommendations can be produced, so predictions for an item are calculated considering the preferences of all the users), and the results obtained by the system that calculates individual predictions for each user (i.e., we simulated the case where there is no constraint, in order to compare the performances of the algorithms when they work with groups).

RMSE was chosen to compare the systems because it is widely used, allows to evaluate results through a single number and emphasizes large errors.

In order to evaluate if two RMSE values returned by two experiments are significantly different, independent-samples two-tailed Student's t-tests have been conducted. In order to make the tests, a 5-fold cross-validation was preformed.

In each experiment, we evaluated the system performances considering different values of a *novelty* parameter, which expresses the minimum percentage of users in a group that did not previously rate an item, in order for it to be recommended. For example, if *novelty* was set to 50% and an item was rated by 60% of the group, the predicted rating for that item would be discarded, since it would be novel just for 40% of the group.

4.2   Dataset and Data Preprocessing

The MovieLens-1M[1] dataset contains 1 million ratings, given by 6040 users for 3900 movies. Our framework uses only the file `ratings.dat`, which contains

---

[1] http://www.grouplens.org/

the user ratings. The file was preprocessed by mapping the feature *UserID* into a new set of IDs between 0 and 6039 to facilitate the computation with data structures. In order to conduct the cross-validation, the dataset was split into five subsets with a random sampling technique (each subset contains 20% of the ratings).

4.3 Metrics

The quality of the predicted ratings was measured through the Root Mean Squared Error (RMSE). The metric compares each rating r_{ui}, expressed by a user u for an item i in the test set, with the rating p_{gi}, predicted for the item i for the group g in which user u is. The formula is shown below:

$$RMSE = \sqrt{\frac{\sum_{i=0}^{n}(r_{ui} - p_{gi})^2}{n}}$$

where n is the number of ratings available in the test set. In order to compare if two RMSE values returned by two experiments are significantly different, independent-samples two-tailed Student's t-tests have been conducted. These tests allow to reject the null hypothesis that two values are statistically the same. So, a two-tailed test will evaluate if an RMSE value is significantly greater or significantly smaller than another RMSE value. Since each experiment was conducted five times, the means M_i and M_j of the RMSE values obtained by two systems i and j are used to compare the systems and calculate a value t:

$$t = \frac{M_i - M_j}{s_{M_i - M_j}}$$

where

$$s_{M_i - M_j} = \sqrt{\frac{s_1^2}{n_1} + \frac{s_2^2}{n_2}}$$

s^2 is the variance of the two samples, n_1 and n_2 indicate the number of values considered to build M_1 and M_2 (in our case both are equal to 5, since experiments were repeated five times). In order to determine the $t - value$ that indicates the result of the test, the degrees of freedom have to be determined:

$$d.f. = \frac{(s_1^2/n_1 + s_2^2/n_2)^2}{(s_1^2/n_1)^2/(n_1 - 1) + (s_2^2/n_2)^2/(n_2 - 1)}$$

Given t and $d.f.$, the $t - value$ (i.e., the results of the test), can be obtained in a standard table of significance as: $t(d.f.) = t - value$. The $t - value$ derives the probability p that there is no difference between the two means. Along with the result of a t-test, the standard deviation SD of the mean is presented.

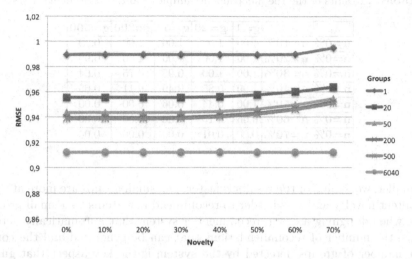

Fig. 1. Performances for different values of novelty

4.4 Experimental Results

Figure 1 shows the RMSE for different values of the *novelty* parameter and each set of groups.

Since the results between the different settings are very close, we performed an independent-samples two-tailed Student's t-test for each point in the figure, in order to compare if the difference in the results between no *novelty* (*novelty* = 0%) and a specific *novelty* value was significant. Because of the amount of tests conducted, we summarized the results in Table 2.

Each cell of the table contains the result of a test, which is the probability p that the difference in the RMSE values obtained with two different *novelty* values for a partitioning is due to chance. For example, $p(n = 0\% \; n = 10\%, g = 1) = 1.00$, means that for one group ($g = 1$), the difference in the RMSE values obtained with *novelty* = 0% and *novelty* = 10% is not significant (in fact the probability p that the two values are the same is 1).

4.5 Discussion

Results show that up to 40% there is no significant worsening of the performances ($p \geq 0.5$). These really good results mean that if we avoid recommending all the items that have already been evaluated by at least 40% of the users in a group, the system is still able to produce accurate recommendations to the group. In other words, our system is still able to keep the same accuracy, while increasing user satisfaction, by filtering the recommended content.

Performances start worsening when *novelty* > 40% and for a high number of groups ($g > 20$ in Table 2). So, when the number of groups is high and groups

Table 2. Results of the the independent-samples two-tailed Student's t-tests

	g=1	g=20	g=50	g=200	g=500
n=0% n=10%	1.00	1.00	1.00	1.00	0.99
n=0% n=20%	1.00	1.00	1.00	1.00	0.99
n=0% n=30%	1.00	1.00	0.95	0.75	0.64
n=0% n=40%	1.00	0.85	0.49	0.11	0.05
n=0% n=50%	1.00	0.44	0.06	0.00	0.00
n=0% n=60%	0.66	0.10	0.00	0.00	0.00
n=0% n=70%	0.00	0.01	0.00	0.00	0.00

get smaller, worsening of the results is faster, i.e., small groups are more affected by content novelty and it is harder to recommend new items to a small group.

So, when designing a group recommender system that automatically detects groups, the number of recommendations that can be generated and the consequent number of groups detected by the system is the key aspect that guides the amount of content novelty that can be provided to users. More specifically, novel content can be provided to at least 40% the group, no matter how many groups are detected by the system. If the system deals with a large amount of groups, the system is not able to recommend novel content without affecting the results.

5   Conclusions and Future Work

In this paper we presented a study related to the recommendation of novel content in a group recommender system that detects groups, in order to respect a constraint on the number of recommendations that can be produced.

We conducted an analysis in order to study how the performances of the system are affected when a constraint on the percentage of a group for which content has to be novel is introduced.

Experimental results, validated through statistical tests, show that our system is able to recommend novel content to nearly half of a group without affecting its accuracy. This means that user satisfaction can be increased for a good part of a group with no cost in terms of performances.

The fact that the same percentage of the *novelty* parameter leads to significantly different performances according to the structure of the group, opens a new research scenario, related to finding the properties of a group that characterize the performances of the system. This analysis is left as future work.

References

1. Ricci, F., Rokach, L., Shapira, B.: Introduction to recommender systems handbook. In: Recommender Systems Handbook, pp. 1–35. Springer, Berlin (2011)

2. Boratto, L., Carta, S.: State-of-the-art in group recommendation and new approaches for automatic identification of groups. In: Soro, A., Vargiu, E., Armano, G., Paddeu, G. (eds.) Information Retrieval and Mining in Distributed Environments. SCI, vol. 324, pp. 1–20. Springer, Heidelberg (2011)

3. Jameson, A., Smyth, B.: Recommendation to groups. In: Brusilovsky, P., Kobsa, A., Nejdl, W. (eds.) Adaptive Web 2007. LNCS, vol. 4321, pp. 596–627. Springer, Heidelberg (2007)

4. Masthoff, J.: Group recommender systems: Combining individual models. In: Recommender Systems Handbook, pp. 677–702. Springer (2011)

5. Ziegler, C.N., McNee, S.M., Konstan, J.A., Lausen, G.: Improving recommendation lists through topic diversification. In: Ellis, A., Hagino, T. (eds.) Proceedings of the 14th international conference on World Wide Web, WWW 2005, Chiba, Japan, May 10-14, pp. 22–32. ACM (2005)

6. Lops, P., de Gemmis, M., Semeraro, G.: Content-based recommender systems: State of the art and trends. In: Ricci, F., Rokach, L., Shapira, B., Kantor, P.B. (eds.) Recommender Systems Handbook, pp. 73–105. Springer (2011)

7. O'Connor, M., Cosley, D., Konstan, J.A., Riedl, J.: Polylens: A recommender system for groups of user. In: Proceedings of the Seventh European Conference on Computer Supported Cooperative Work, pp. 199–218. Kluwer (2001)

8. McCarthy, J.F., Anagnost, T.D.: Musicfx: An arbiter of group preferences for computer supported collaborative workouts. In: Poltrock, S.E., Grudin, J. (eds.) Proceedings of the ACM 1998 Conference on Computer Supported Cooperative Work, CSCW 1998, Seattle, WA, USA, November 14-18, pp. 363–372. ACM (1998)

9. Crossen, A., Budzik, J., Hammond, K.J.: Flytrap: intelligent group music recommendation. In: Proceedings of the 7th International Conference on Intelligent User Interfaces, IUI 2002, pp. 184–185. ACM, New York (2002)

10. Zhiwen, Y., Xingshe, Z., Daqing, Z.: An adaptive in-vehicle multimedia recommender for group users. In: Proceedings of the 61st Semiannual Vehicular Technology Conference, vol. 5, pp. 2800–2804 (2005)

11. Goren-Bar, D., Glinansky, O.: Fit-recommend ing tv programs to family members. Computers & Graphics 28(2), 149–156 (2004)

12. Yu, Z., Zhou, X., Hao, Y., Gu, J.: Tv program recommendation for multiple viewers based on user profile merging. User Modeling and User-Adapted Interaction 16(1), 63–82 (2006)

13. McCarthy, K., Salamó, M., Coyle, L., McGinty, L., Smyth, B., Nixon, P.: Cats: A synchronous approach to collaborative group recommendation. In: Sutcliffe, G., Goebel, R. (eds.) Proceedings of the Nineteenth International Florida Artificial Intelligence Research Society Conference, Melbourne Beach, Florida, USA, May 11-13, pp. 86–91. AAAI Press (2006)

14. McCarthy, J.: Pocket RestaurantFinder: A situated recommender system for groups. In: Workshop on Mobile Ad-Hoc Communication at the 2002 ACM Conference on Human Factors in Computer Systems (2002)

15. Jameson, A.: More than the sum of its members: challenges for group recommender systems. In: Proceedings of the Working Conference on Advanced Visual Interfaces, pp. 48–54. ACM Press (2004)

16. Chen, Y., Pu, P.: Cofeel. Using emotions to enhance social interaction in group recommender systems. In: 2013 Workshop on Tools and Technology for Emotion-Awareness in Computer Mediated Collaboration and Learning, Alpine Rendez-Vous (ARV) (2013)

17. Jung, J.J.: Attribute selection-based recommendation framework for short-head user group: An empirical study by movielens and imdb. Expert Systems with Applications 39(4), 4049–4054 (2012)
18. Schafer, J.B., Frankowski, D., Herlocker, J., Sen, S.: Collaborative filtering recommender systems. In: Brusilovsky, P., Kobsa, A., Nejdl, W. (eds.) Adaptive Web 2007. LNCS, vol. 4321, pp. 291–324. Springer, Heidelberg (2007)
19. Boratto, L., Carta, S.: Exploring the ratings prediction task in a group recommender system that automatically detects groups. In: The Third International Conference on Advances in Information Mining and Management, IMMM 2013, pp. 36–43 (2013)
20. Herlocker, J.L., Konstan, J.A., Borchers, A., Riedl, J.: An algorithmic framework for performing collaborative filtering. In: Proceedings of the 22nd Annual International ACM SIGIR Conference on Research and Development in Information Retrieval, SIGIR 1999, pp. 230–237. ACM, New York (1999)
21. Boratto, L., Carta, S.: Using collaborative filtering to overcome the curse of dimensionality when clustering users in a group recommender system. In: Proceedings of 16th International Conference on Enterprise Information Systems (ICEIS), pp. 564–572 (2014)
22. Amatriain, X., Jaimes, A., Oliver, N., Pujol, J.M.: Data mining methods for recommender systems. In: Ricci, F., Rokach, L., Shapira, B., Kantor, P.B. (eds.) Recommender Systems Handbook, pp. 39–71. Springer, Boston (2011)
23. Boratto, L., Carta, S., Chessa, A., Agelli, M., Clemente, M.L.: Group recommendation with automatic identification of users communities. In: Proceedings of the 2009 IEEE/WIC/ACM International Joint Conference on Web Intelligence and Intelligent Agent Technology, WI-IAT 2009, vol. 03, pp. 547–550. IEEE Computer Society, Washington, DC (2009)
24. Boratto, L., Carta, S., Satta, M.: Groups identification and individual recommendations in group recommendation algorithms. In: Picault, J., Kostadinov, D., Castells, P., Jaimes, A. (eds.) Practical Use of Recommender Systems, Algorithms and Technologies. CEUR Workshop Proceedings, vol. 676 (November 2010)
25. Boratto, L., Carta, S.: Modeling the preferences of a group of users detected by clustering: A group recommendation case-study. In: Proceedings of the 4th International Conference on Web Intelligence, Mining and Semantics (WIMS 2014), pp. 16:1–16:7. ACM, New York (2014)
26. Kanungo, T., Mount, D.M., Netanyahu, N.S., Piatko, C.D., Silverman, R., Wu, A.Y.: An efficient k-means clustering algorithm: Analysis and implementation. IEEE Trans. Pattern Anal. Mach. Intell. 24(7), 881–892 (2002)

Optimizing Queue-Based Semi-Stream Joins with Indexed Master Data

M. Asif Naeem[1], Gerald Weber[2],
Christof Lutteroth[2], and Gillian Dobbie[2]

[1] School of Computer and Mathematical Sciences,
Auckland University of Technology, Auckland, New Zealand
[2] Department of Computer Science,
The University of Auckland, Auckland, New Zealand
mnaeem@aut.ac.nz, {gerald,Lutteroth,gill}@cs.auckland.ac.nz

Abstract. In Data Stream Management Systems (DSMS) semi-stream processing has become a popular area of research due to the high demand of applications for up-to-date information (e.g. in real-time data warehousing). A common operation in stream processing is joining an incoming stream with disk-based master data, also known as semi-stream join. This join typically works under the constraint of limited main memory, which is generally not large enough to hold the whole disk-based master data. Many semi-stream joins use a queue of stream tuples to amortize the disk access to the master data, and use an index to allow directed access to master data, avoiding the loading of unnecessary master data. In such a situation the question arises which master data partitions should be accessed, as any stream tuple from the queue could serve as a lookup element for accessing the master data index. Existing algorithms use simple safe and correct strategies, but are not optimal in the sense that they maximize the join service rate. In this paper we analyze strategies for selecting an appropriate lookup element, particularly for skewed stream data. We show that a good selection strategy can improve the performance of a semi-stream join significantly, both for synthetic and real data sets with known skewed distributions.

Keywords: Real-time Data Warehousing, Stream processing, Join, Performance measurement.

1 Introduction

Real-time data warehousing plays a prominent role in supporting overall business strategy. By extending data warehouses from static data repositories to active data repositories, business organizations can better inform their users and make effective timely decisions. In real-time data warehousing the changes occurring at source level are reflected in data warehouses without any delay. Extraction, Transformation, and Loading (ETL) tools are used to access and manipulate transactional data and then load them into the data warehouse. An important phase in the ETL process is a transformation where the source level changes are

L. Bellatreche and M.K. Mohania (Eds.): DaWaK 2014, LNCS 8646, pp. 171–182, 2014.

mapped into the data warehouse format. Common examples of transformations are unit conversion, removal of duplicate tuples, information enrichment, filtering of unnecessary data, sorting of tuples, and translation of source data keys.

A particular type of stream-based joins called semi-stream joins are required to implement the above transformation examples. In this particular type of stream-based join, a join is performed between a single stream and a slowly changing table. In the application of real-time data warehousing [4,9,10], the slowly changing table is typically a master data table while incoming real-time sales data may form the stream.

Most stream-based join algorithms [9,10,7,3,2,6,5] use the concept of staging in order to amortize the expensive disk access cost over fast stream data. The concept of staging means the algorithm loads stream data into memory in chunks, while these chunks are differentiated by their loading timestamps. To implement the concept of staging these algorithms normally use a data structure called a queue. The main role of the queue is to keep track of these stages with respect to their loading timestamps. Some of these algorithms [3,2,6,5] use elements of the queue to look up and load relevant disk-based master data through an index. In this paper we call these queue elements *lookup elements*.

Most of the above algorithms choose the oldest value of the queue as lookup element. The process of choosing the oldest tuple in the queue is at first glance intuitive: this tuple must be joined eventually to avoid starvation, and since it was not processed before it is now due. Hence choosing the oldest tuple in the queue ensures correctness. However, as we will show in this paper, this strategy is not optimal with regard to join service rate. The optimal lookup element is the element for which the most useful partition of master data is loaded into memory, i.e. the partition which will join the most stream tuples currently in the queue.

This paper addresses the challenge of finding near-optimal strategies for selecting the lookup element. We have identified two *eviction aims* that influence the choice of the stream tuple for master data lookup: 1) to maximize the expected number of stream tuples that are matched and hence removed from the queue, and 2) to limit the cost (or lost opportunity) that a tuple incurs while sitting in the queue. An important property of stream tuple behavior in our context is the average frequency of matches to a whole master data partition, i.e. the average number of tuples that is matched when the partition is loaded from disk.

2   Related Work

In this section, we present an overview of the previous work that has been done in the area of semi-stream joins, focusing on those that are closely related to our problem domain.

A seminal algorithm MESHJOIN [9,10] has been designed especially for joining a continuous stream with disk-based master data, like in the scenario of active data warehouses. The MESHJOIN algorithm is a hash join, where the stream serves as the build input and the disk-based relation serves as the probe input. A characteristic of MESHJOIN is that it performs a staggered execution of the hash table build in order to load in stream tuples more steadily. To implement this staggered execution the algorithm uses a queue. The algorithm makes no assumptions about data distribution or the organization of the master data, hence there is no master data index. The algorithm always removes stream tuples from the end of the queue, as they have been matched with all master data partitions.

R-MESHJOIN (reduced Mesh Join) [7] clarifies the dependencies among the components of MESHJOIN. As a result the performance is improved slightly. However, R-MESHJOIN implements the same strategy as the MESHJOIN algorithm for accessing the disk-based master data, using no index.

Partitioned Join [3] improved MESHJOIN by using a two-level hash table, attempting to join stream tuples as soon as they arrive, and using a partition-based wait buffer for other stream tuples. The number of partitions in the wait buffer is equal to the number of partitions in the disk-based master data. The algorithm uses these partitions as an index, for looking up the master data. If a partition in a wait buffer grows larger than a preset threshold, the algorithm loads the relevant partition from the master data into memory. The algorithm allows starvation of stream tuples as tuples can stay in a wait buffer indefinitely if the buffer's size threshold is not reached.

Semi-Streaming Index Join (SSIJ) [2] was developed recently to join stream data with disk-based data. In general, the algorithm is divided into three phases: the pending phase, the online phase and the join phase. In the pending phase, the stream tuples are collected in an input buffer until either the buffer is larger than a predefined threshold or the stream ends. In the online phase, stream tuples from the input buffer are looked up in cached disk blocks. If the required disk tuple exists in the cache, the join is executed. Otherwise, the algorithm flushes the stream tuple into a stream buffer. When the stream buffer is full, the join phase starts where master data partitions are loaded from disk using an index and joined until the stream buffer is empty. This means that as partitions are loaded and joined, the join becomes more and more inefficient: partitions that are joined later can potentially join only with fewer tuples because the stream buffer is not refilled between partition loads. By keeping the stream buffer full and selecting lookup elements carefully the performance could be improved.

One of our algorithms, HYBRIDJOIN [5], addresses the issue of accessing disk-based master data efficiently. Similar to SSIJ, an index based strategy to access the disk-based master data is used, but every master data partition load is amortized by joining over a full stream tuple queue. HYBRIDJOIN uses the last queue element as lookup element, which means that unlike Partitioned Join it prevents starvation. However, as will be explained in this paper, the choice of the last queue element as lookup element is suboptimal.

CACHEJOIN [6] is an extension of HYBRIDJOIN, which adds an additional cache module to cope with Zipfian stream distributions. This is similar to Partitioned Join and SSIJ, but a tuple-level cache is used instead of a page-level cache to use the cache memory more efficiently. CACHEJOIN is able to adapt its cache to changing stream characteristics, but similar to HYBRIDJOIN, it uses the last queue element as a lookup element for tuples that were not joined with the cache.

Recently, we presented an improved version of CACHEJOIN called SSCJ [8], which optimizes the manipulation of master data tuples in the cache module. While CACHEJOIN uses a random approach to overwrite tuples in the cache when it is full, SSCJ overwrites the least frequent tuples. To the best of our knowledge, the CACHEJOIN/SSCJ class of semi-stream join algorithms is currently the fastest when considering skewed data. SSCJ and CACHEJOIN use the same suboptimal strategy to access the queue, and this can be improved with the approach presented here. Due to space limitations we test this approach using HYBRIDJOIN and CACHEJOIN only.

3   Problem Definition

This section defines the problem we are addressing, using the existing HYBRID-JOIN algorithm as an example for clarification. HYBRIDJOIN has a simple architecture, using a queue to load disk-based master data into memory, and is therefore particularly suitable for illustration.

Fig. 1 presents an overview of HYBRIDJOIN. The master data on disk contains three partitions, $p_1, \ldots, p_3$. Partitions are loaded into memory through the *disk buffer*, which can hold one partition. In the original algorithm the stream tuples are stored in a hash table and the queue stores pointers to these stream tuples. To make our discussion clearer, we will refer to the queue as if the stream tuples are directly contained therein, as our focus is to highlight the behavior of the

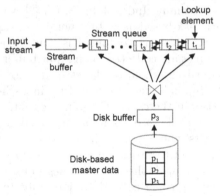

Fig. 1. Overview of HYBRIDJOIN

queue. The queue is implemented as a doubly-linked list, allowing the random deletion of stream tuples.

In each iteration the algorithm loads a chunk of stream tuples into the queue and a partition of master data into the disk buffer. To decide which partition will be loaded into the buffer, HYBRIDJOIN retrieves the oldest (i.e. last) element of the queue and uses it as a lookup element in the index for the master data table. Once the relevant partition is loaded into the disk buffer, the algorithm performs a join between the master data tuples and stream tuples. During the join operation the algorithm removes the matched stream tuples, which lie at random positions, from the queue. With parameter skew some tuples occur more frequently in the stream data. Consequently, there can be many matches of stream tuples in the queue against one master data tuple.

For a master data partition, we define the *stream probability* as the probability that a random stream tuple matches a master data record on that partition. If the stream probability of a partition is at least $1/h_S$, with h_S the size of the queue in number of tuples, we call this a *common partition*. Among the common partitions there are typically some *high-probability partitions*.

The purpose of queueing stream tuples is to amortize a master data partition access by processing several stream tuples with it. After a master data access, all the h_S stream tuples in the queue are processed and potentially joined. We define the *load probability* of a master data partition as the probability that a random stream tuple is *used* to load that partition, i.e. that a lookup element matches the master data partition. The load probability of a partition is smaller than its stream probability because after a partition access all the stream tuples matching it are removed from the queue,

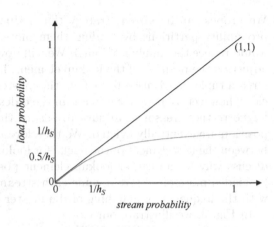

Fig. 2. The load probability, shown as a function of the stream probability

reducing the occurrence of lookup elements that match the partition. Due to the amortization process over a queue of h_S elements, the load probability is in fact always smaller than $1/h_S$. After one lookup for a partition p, the earliest stream tuple that can lead to a lookup of p has to be a fresh stream tuple from the same page that enters the queue after the last lookup. Since HYBRIDJOIN uses the oldest tuple of the queue as lookup element, before a partition is loaded again after a join, new tuples matching that partition need to move all the way from the beginning to the end of the queue. These stream tuples must therefore be more than h_S stream tuples after the last lookup.

Hence there is a *saturation effect*: If we look at partitions in order of increasing stream probability, the load probability first is equal to the average frequency, but can never go beyond $1/h_S$. We furthermore observe that for all common partitions the load frequency will be close to, or larger than $1/(2h_S)$, since after one loookup of a partition, we expect the partition to be again among the next h_S stream tuples entering the queue. Hence we see that, interestingly, the high-probability partitions are loaded not much more often than the common partitions. A qualitative illustration of how the load probability depends on the stream probability is shown in Fig. 2.

It is important to realize that this behavior is not optimal. Although at first glance it seems to optimally amortize disk accesses, this is not the case. For high-probability partitions, a large number of matching stream tuples accumulate in the queue before the partition is loaded again. This takes up space in the queue, therefore these tuples should be evicted earlier. Hence we have to reduce the saturation effect for high-probability partitions, i.e. we have to ensure that high-probability partitions are loaded more frequently.

4   Proposed Solution

We propose an improved strategy that reduces the saturation effect for high-probability partitions by loading them more often, but not too often (which would reduce the number of joins). We will now argue that we can ensure this by adjusting the position of the lookup element, i.e. the position in the queue of the stream tuple that is used to look up the master data partition to load next. It is clear, however, that we have to retain the oldest queue tuple as a lookup element for correctness reasons: we have to ensure that tuples that have come to this position are eventually evicted. We therefore propose a strategy that alternates between the last element of the queue for lookup and an earlier element. We now discuss why for an earlier lookup element position, high-probability pages will be loaded more often. We assume each stream tuple in the queue is annotated with the frequency of matching of the master data partition.

In Fig. 3, we illustrate our expectation for the HYBRIDJOIN strategy at an arbitrary point in time; this is purely an illustration and is not intended to be quantitatively correct. The crucial point is that for each of the high-frequency master data partitions, the time of last lookup (expressed in stream tuples) was on average half a queue length ago. Therefore, the tuples of each of these partitions will only have advanced a fraction of the queue, on average half of the queue. This is a random process, so at the start of the queue, many frequent pages

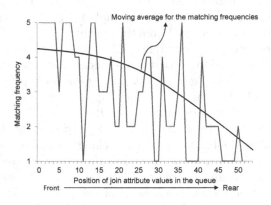

Fig. 3. Queue analysis

will have started to reappear, and their prevalence diminishes towards the end of the queue. The figure shows a made-up example where the queue contains join attribute values of 50 stream tuples and the corresponding matching frequencies lie between 1 and 5. We also have drawn a moving average for the corresponding matching frequencies. Again, the shape of the moving average is only indicating the tendency to drop towards the end of the queue; we do not consider its exact shape here. The figure also illustrates that some of the tuples with corresponding high matching frequencies will come through to the end of the queue. However, HYBRIDJOIN accumulates the least frequent stream tuples towards the end of the queue, where the lookup element is situated. The join attributes towards the

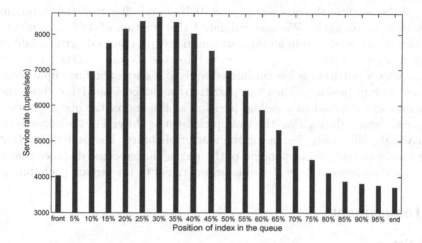

Fig. 4. Performance analysis of HYBRIDJOIN by accessing the queue at different positions

front of the queue have higher matching frequencies as compared to the older stream tuples in the queue. Hence if any algorithm, such as HYBRIDJOIN or CACHEJOIN, chooses only the oldest stream position at the end of the queue as a lookup position, it will under-represent the high-probability pages, as we have predicted in our considerations about the saturation effect.

With the introduction of a second lookup position earlier in the queue we can therefore expect to hit high-probability pages more frequently. The second lookup position will lead to a behavior different from the one in Fig. 3. The front of the queue, i.e. the newest tuples, is not be the best position for the new lookup element since it would over-represent high-probability pages; there would be a high probability that the most frequent page gets loaded before new tuples for that page had an opportunity to accumulate in the queue. As explained previously, the algorithm will alternate between the two lookup positions, the new position and the oldest element in the queue, as some tuples will remain in the queue for a long time otherwise.

We used an empirical approach to find an optimal position for the new lookup element and present the effect of the new lookup position on the performance of HYBRIDJOIN. Again, HYBRIDJOIN is chosen because it is the simplest algorithm this new approach can be applied to. We conducted an experiment in which we measured the performance of the algorithm while varying the new lookup position. We started with a lookup position at the front of the queue and then moved it in 5% steps towards the end of the queue.

The results of our experiment are shown in Fig. 4. From the figure it can be observed that a lookup element at the front of the queue is not optimal, as we predicted, and as we move the lookup position away from the front the performance starts improving. This behavior continues up to a certain fraction of the queue length (about 30% of the queue size in our experiment, i.e. the index of the entry in the queue is approx. $0.3h_s$) and after that the throughput starts to decrease again. We have validated this fraction of 30% with different parameters and used it in in all the performance experiments described later on.

Given a certain index $c \cdot h_s$, $0 < c < 1$ for the lookup position, one could assume that a partition is loaded immediately if a corresponding tuple reaches the new lookup position. Then the partition would be loaded $1/c$ times more frequently as compared to a lookup position at the end of the queue. However, tuples can "sneak through" as the lookup element at the end of the queue is used alternatingly, effectively decreasing the load probability. The probability that a tuple sneaks through is independent of the page. The expected distance that the tuple can then move (this is $c' - c$) is proportional to the stream probability.

5   Experiments

5.1   Setup

We performed our experiments on a *Pentium-i5* with 8GB main memory and 500GB hard drive as secondary storage running Java. We compared the original HYBRIDJOIN [5] and CACHEJOIN [6] algorithms with optimized versions using the proposed approach, using the cost models presented with the original algorithms to distribute the memory among their components. The master data R was stored on disk using a MySQL database with an index.

The two algorithms retrieve master data using a lookup element from a stream tuple queue. Choosing these two algorithms allows us to compare the effects of near optimal lookup elements for algorithms with and without a cache component. HYBRIDJOIN does not have a cache component so all stream tuples are processed through the queue, while CACHEJOIN has an additional cache module that processes the most frequent stream tuples. For this reason, CACHEJOIN is considered an efficient algorithm in comparison to other semi-stream-join algorithms [3,2,9,10,7]. We analyzed the service rates of all algorithms using synthetic, TPC-H and real-life datasets.

Synthetic Datasets. The stream datasets we used is based on a Zipfian distribution, which can be found in a wide range of applications [1]. We tested the service rate of the algorithms by varying the skew value from 0 (fully uniform) to 1 (highly skewed). Details of the synthetic datasets are specified in Table 1.

TPC-H Datasets. We also analyzed the service rates using the TPC-H datasets, which is a well-known decision support benchmark. We created the datasets using a scale factor of 100. More precisely, we used the table `Customer` as master data and the table `Order` as the stream data. In table `Order` there is one foreign key attribute `custkey`, which is a primary key in the `Customer` table, so the two tables can be joined. Our `Customer` table contained 20 million tuples, with each tuple having a size of 223 bytes. The `Order` table contained the same number of tuples, with each tuple having a size of 138 bytes. The plausible scenario for such a join is to add customer details corresponding to an order before loading the order into the warehouse.

Table 1. Data specification of the synthetic datasets (similar to those used for the original HYBRIDJOIN and CACHEJOIN algorithms)

Parameter	Value
Total allocated memory M	50MB to 250MB
Size of disk-based relation R	0.5 *million* to 8 *million tuples*
Size of each disk tuple	120 *bytes*
Size of each stream tuple	20 *bytes*
Size of each node in the queue	12 *bytes*
Data set	Based on Zipf's law (exponent varies from 0 to 1)

Real-Life Datasets. We also compared the service rates of the algorithms using a real-life datasets[1]. These datasets contains cloud information stored in a summarized weather report format. The master data table is constructed by combining meteorological data corresponding to the months April and August, and the stream data by combining data files from December. The master data table contains 20 million tuples and the stream data table contains 6 million tuples. The size of each tuple in both the master data table and the stream data table is 128 bytes. Both tables are joined using a common attribute, longitude (LON). The domain of the join attribute is the interval [0,36000].

Measurement Strategy. The performance or service rate of a join is measured by calculating the number of tuples processed in a unit second. In each experiment, the algorithms first completed a warmup phase before starting the actual measurements. These kinds of algorithms normally need a warmup phase to tune their components with respect to the available memory resources, so that each component can deliver a maximum service rate. The calculation of the confidence intervals is based on 2000 to 3000 measurements for one setting. During the execution of the algorithm, no other application was running in parallel. The stream arrival rate throughout a run was constant.

5.2   Performance Evaluation

We identified three parameters for which we want to understand the behavior of the algorithms. The three parameters are: the total memory available M, the size of the master data table R, and the value of the parameter skew e in the stream data. For the sake of brevity, we restrict the discussion for each parameter to a one-dimensional variation, i.e. we vary one parameter at a time.

Performance Comparisons for Different Memory Budgets. In our first experiment we tested the performance of all algorithms using different memory budgets while keeping the size of R fixed (*2 million tuples*). We increased the available memory linearly from 50MB to 250MB. Fig. 5(a) presents the comparisons of both approaches with and without implementing the optimal lookup element strategy. From the figure the performance improvement in the

---

[1] These datasets are available at: http://cdiac.ornl.gov/ftp/ndp026b/

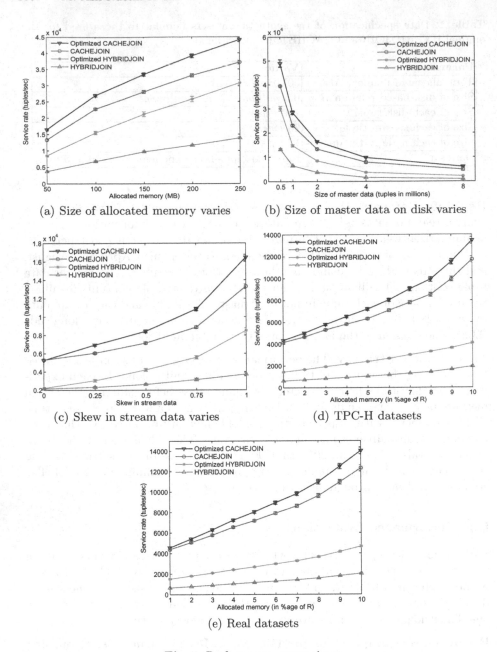

(a) Size of allocated memory varies

(b) Size of master data on disk varies

(c) Skew in stream data varies

(d) TPC-H datasets

(e) Real datasets

Fig. 5. Performance comparisons

case of the both algorithms is clear. More concretely, in the case of Optimized HYBRIDJOIN the algorithm performed 3 times better than the original HY-BRIDJOIN. Although the improvement is comparatively smaller in the case of Optimal CACHEJOIN, it is still considerable. The reason the improvement is smaller is the cache component that processes the most frequent part of the stream data.

Performance Comparisons for Varying the Size of R. In our second experiment we tested the performance by varying the size of the disk-based relation R. We chose values for R from a simple geometric progression. The performance results are shown in Fig. 5(b). From the figure it can be seen that Optimized HYBRIDJOIN performed more than twice as well as the original HYBRID-JOIN. Also in the case of Optimized CACHEJOIN the performance improved considerably.

Performance Comparisons for Different Values of Skew. In this experiment we compared the service rates of both algorithms with and without the optimal lookup element strategy while varying the skew in the stream data. To vary the skew, we varied the Zipfian exponent from 0 to 1. At 0 the input stream S has no skew, while at 1 the stream was strongly skewed. The size of R was fixed at 2 million tuples and the available memory was set to 50MB. The results presented in Fig. 5(c) show that both optimized algorithms (especially Optimized HYBRIDJOIN) performed significantly better than the existing ones, even for only moderately skewed data. This improvement became more pronounced for increasing skew values. At a skew of 1, Optimized HYBRIDJOIN performs approximately 3 times better than the original HYBRIDJOIN. In the case of Optimized CACHEJOIN the improvement was comparatively smaller but is still noticeable. Our strategy does not add any overhead to the processing cost, therefore in the case of fully uniform data, when the skew is equal to 0, the performance is not worse than that of the original algorithms. We do not present data for skew values larger than 1, which models short tails.

TPC-H Datasets. In this experiment we measured the service rates produced by the algorithms at different memory settings. We allocated the size of primary memory as a percentage of the size of R. The results are shown in Fig. 5(d). From the figure it can be noted that the optimized versions of the algorithms performed better than the original algorithms. Especially in the case of Optimized HYBRIDJOIN this improvement is remarkable.

Real-Life Datasets. We also tested our approaches using real data. The details of the datasets were presented in the beginning of this section. In this experiment we also measured the service rate produced by the algorithms at different memory settings, similar to the one using the TPC-H datasets. The results are shown in Fig. 5(e). From the figure, the performance of the optimized algorithms is again significantly better than that of the original algorithms, supporting our argument.

6   Conclusions

Most semi-stream join algorithms amortize disk accesses to master data over a queue of stream tuples in memory. Several of those algorithms use an index to look up master data partitions for particular elements in that queue. We identified the choice of the lookup element, i.e. the queue stream tuple used as a key in such an index, as an important and underutilized issue for such algorithms. For example, HYBIRDJOIN and CACHEJOIN always choose lookup elements from the end of their queues. Because of that they under-represent high-probability partitions of disk-based master data and do not fully exploit the characteristics of skew in stream data, resulting in a suboptimal performance.

As a solution, we have proposed a new approach in this paper for choosing an element for index-based master data lookup from a stream tuple queue, based on the position of the lookup element in the queue. The approach alternates between the last queue element to avoid starvation, and an intermediate queue element, balancing the rate in which high-probability partitions are loaded. We provided a theory for the improved behavior and validated it with experiments using HYBIRDJOIN and CACHEJOIN, showing that the optimized algorithms perform significantly better than the original ones.

References

1. Anderson, C.: The Long Tail: Why the Future of Business Is Selling Less of More. Hyperion (2006)
2. Bornea, M.A., Deligiannakis, A., Kotidis, Y., Vassalos, V.: Semi-streamed index join for near-real time execution of ETL transformations. In: ICDE 2011: IEEE 27th International Conference on Data Engineering, pp. 159–170. IEEE Computer Society (2011)
3. Chakraborty, A., Singh, A.: A partition-based approach to support streaming updates over persistent data in an active datawarehouse. In: IPDPS 2009: IEEE International Symposium on Parallel & Distributed Processing, pp. 1–11. IEEE Computer Society (2009)
4. Karakasidis, A., Vassiliadis, P., Pitoura, E.: ETL queues for active data warehousing. In: IQIS 2005: 2nd International Workshop on Information Quality in Information Systems, pp. 28–39. ACM (2005)
5. Naeem, M.A., Dobbie, G., Weber, G.: HYBRIDJOIN for near-real-time data warehousing. International Journal of Data Warehousing and Mining (IJDWM) 7(4) (2011)
6. Naeem, M.A., Dobbie, G., Weber, G.: A lightweight stream-based join with limited resource consumption. In: Cuzzocrea, A., Dayal, U. (eds.) DaWaK 2012. LNCS, vol. 7448, pp. 431–442. Springer, Heidelberg (2012)
7. Naeem, M.A., Dobbie, G., Weber, G., Alam, S.: R-MESHJOIN for near-real-time data warehousing. In: DOLAP 2010: ACM 13th International Workshop on Data Warehousing and OLAP. ACM (2010)
8. Naeem, M.A., Weber, G., Dobbie, G., Lutteroth, C.: SSCJ: A semi-stream cache join using a front-stage cache module. In: Bellatreche, L., Mohania, M.K. (eds.) DaWaK 2013. LNCS, vol. 8057, pp. 236–247. Springer, Heidelberg (2013)
9. Polyzotis, N., Skiadopoulos, S., Vassiliadis, P., Simitsis, A., Frantzell, N.E.: Supporting streaming updates in an active data warehouse. In: ICDE 2007: 23rd International Conference on Data Engineering, pp. 476–485 (2007)
10. Polyzotis, N., Skiadopoulos, S., Vassiliadis, P., Simitsis, A., Frantzell, N.: Meshing streaming updates with persistent data in an active data warehouse. IEEE Trans. on Knowl. and Data Eng. 20(7), 976–991 (2008)

Parallel Co-clustering with Augmented Matrices Algorithm with Map-Reduce

Meng-Lun Wu and Chia-Hui Chang

National Central University,
Dept. of Computer Science and Information Enginnering,
Taoyuan, Taiwan

Abstract. Co-clustering with augmented matrices (CCAM) [11] is a two-way clustering algorithm that considers dyadic data (e.g., two types of objects) and other correlation data (e.g., objects and their attributes) simultaneously. CCAM was developed to outperform other state-of-the-art algorithms in certain real-world recommendation tasks [12]. However, incorporating multiple correlation data involves a heavy scalability demand. In this paper, we show how the parallel co-clustering with augmented matrices (PCCAM) algorithm can be designed on the Map-Reduce framework. The experimental work shows that the input format, the number of blocks, and the number of reducers can greatly affect the overall performance.

Keywords: Co-clustering, recommender system, Map-Reduce, Hadoop.

1 Introduction

Many e-commerce and social network websites retain records of the pairwise relationships of users and information items to recommend products that are likely of interest to users. For instance, Netflix has achieved 80% purchasing behavior based on movie recommendations by analysing users' clicked movies and content-based information. Google provides good search results based on learning behavior from the user's query log and clicked web pages.

For large number of users and items, the associations between these two types of entities are usually quite sparse. As co-clustering provides simultaneous clustering of both users and items, it has been used in many recommendation systems as a technique to address the sparsity issue. However, many real-world applications require the analysing of multi-relational data with associated covariates or side-information to deal with the cold-start problem, which occurs when a new user or item enters the system. Therefore, two co-clustering algorithms SCOAL [3] and CCAM [11] are designed to handle both dyadic data and content-based information.

However, dealing simultaneously with dyadic data and content-based information presents a challenge for scalability as the computation cost is comparably high when the size of each matrix increases. This scalability issue also concerned by industry vendors since they have to process several terabytes data

L. Bellatreche and M.K. Mohania (Eds.): DaWaK 2014, LNCS 8646, pp. 183–194, 2014.
© Springer International Publishing Switzerland 2014

daily. Therefore, researches for parallel approaches are proposed to improve the efficiency of existing co-clustering algorithms. For example, Deodhar et al. [4] parallelized the simultaneous co-clustering and learning (SCOAL) algorithm under the Map-Reduce platform to handle the big data caused by dyadic data and content-based information. Matrix factorization algorithm [13] has also been considered on multi-core or distributed systems by solving several sub-problems in parallel.

This paper presents our study on the design of a parallel co-clustering algorithm with augmented matrices (PCCAM) to consider both content-based information and the click data for collaborative filtering. We implement PCCAM on Map-Reduce platform and show that the input format as well as parameters can greatly affect the performance.

The rest of the paper is organized as follows. Section 2 compares the existing parallel algorithms for co-clustering. The CCAM algorithm is given in section 3. We propose the PCCAM algorithm in section 4. In section 5, we discuss the scalability issue by comparing run-time and performance tuning. Section 6 presents our conclusion.

2   Related Work

For large amounts of data, there is a compelling need to develop a scalable implementation of such co-clustering algorithms. Divide-and-conquer is a fundamental concept for tackling large-data problems. This idea can be implemented in parallel or distributed by different workers (e.g., multiple cores in a multi-core processor, multiple processors in a machine, or many machines in a cluster). Different distributed computing platforms such as Grid computing, Message Passing Interface (MPI) [8], and Map-Reduce[1] have been proposed to solve large co-clustering problems. For instance, Papadimitriou et al. [9] presented a case study of distributed co-clustering using Map-Reduce. Ramanathan et al. [10] parallelized the information theoretic co-clustering algorithm [5] using a cloud middle-ware. Yu et al. [13] proposed a coordinate descent based method for matrix factorization in both multi-core and distributed MPI systems.

The basic strategy of these parallel co-clustering algorithms is matrix partition which finds the local optimum of each sub-matrix via divide-and-conquer. For example, Banerjee et al. [1] proposed the partitioned Bregman co-clustering formulation from a matrix approximation view point. George et al. [7] discussed a special case of the weighted Bregman co-clustering algorithm, which minimizes the error between the rating matrix and the approximated matrix for a collaborative filtering problem. Bisson et al. [2] combined multiple similarity matrices produced by each sub-matrix using a multi-view learning algorithm (MVSIM).

However, those algorithms focus primarily on distributing the computation of co-clustering on a single dyadic matrix. To deal with additional side-information, two co-clustering techniques have been proposed. For example, the SCOAL algorithm simultaneously groups dyadic data through minimizing the mean squared

---

[1] http://hadoop.apache.org/docs/r1.2.1/mapred_ tutorial.html

error between the original matrix and the approximated user-item rating matrix, where the rating is predicted by a model trained on each co-cluster (sub-matrix) under the consideration of the user and item content-based information. This algorithm has also been parallelized to Map-Reduce (called PSCOAL algorithm) for reducing the computation of row/column assignment and model updating [4]. Nevertheless, a critical point of these partition-based approaches is that the model updating procedure for each co-cluster (sub-matrix) do not parallelize at all.

The co-clustering with augmented matrices (CCAM) algorithm [11] is another approach to handle varied input data in a hybrid framework. The concept of the CCAM algorithm is to minimize the mutual information loss of a linear combination of rating data and content-based information, where the mutual information can measure the dependency between two objects among those matrices. Although the variety issue is addressed by the CCAM algorithm, the scalability issue still exists when facing large data sets. In this paper, we parallelized the CCAM algorithm to achieve this goal. Because some specialized notation will be used in this paper, the detail of CCAM algorithm is given below.

3   Co-clustering with Augmented Matrices Algorithm

Given three input data, the user-item rating matrix, item feature matrix and user profile matrix, three matrices are normalized respectively into $f(A, U)$, $g(A, S)$, and $h(U, L)$ for the optimization task. The goal of CCAM is to find a co-clustering $(\hat{A}, \hat{U})$ that minimizes the weighted mutual information loss among variables A (item) and U (user), A and S (item feature), as well as U and L (user profile). The objective function is given by

$$q(\hat{A}, \hat{U}) = [I(A, U) - I(\hat{A}, \hat{U})] + \lambda \cdot [I(A, S) - I(\hat{A}, S)] + \varphi \cdot [I(U, L) - I(\hat{U}, L)] \\ = D(f(A, U) \| \hat{f}(A, U)) + \lambda \cdot D(g(A, S) \| \hat{g}(A, S)) + \varphi \cdot D(h(U, L) \| \hat{h}(U, L)) \tag{1}$$

where $\hat{f}(A, U)$, $\hat{g}(A, S)$, and $\hat{h}(U, L)$ are the approximated joint probabilities based on co-cluster $(\hat{A}, \hat{U})$ defined by

$$\hat{f}(a, u) = f(\hat{a}, \hat{u}) f(a \mid \hat{a}) f(u \mid \hat{u}), \ where \ f(\hat{a}, \hat{u}) = \sum_{a: C_A(a) = \hat{a}} \sum_{u: C_U(u) = \hat{u}} f(a, u) \tag{2}$$

$$\hat{g}(a, s) = g(\hat{a}, s) g(a \mid \hat{a}), \ where \ g(\hat{a}, s) = \sum_{a: C_A(a) = \hat{a}} g(a, s) \tag{3}$$

$$\hat{h}(u, l) = h(\hat{u}, l) h(u \mid \hat{u}), \ where \ h(\hat{u}, l) = \sum_{u: C_U(u) = \hat{u}} h(u, l) \tag{4}$$

The optimization task can be solved using the following update rules:

$$C_A^{(t+1)}(a) = \underset{\hat{a}}{argmin} \{ f(a) D(f(U|a) \| \hat{f}^{(t)}(U|\hat{a})) + \lambda \cdot g(a) D(g(S|a) \| \hat{g}^{(t)}(S|\hat{a})) \} \tag{5}$$

$$C_U^{(t+2)}(u) = \underset{\hat{u}}{argmin} \{ f(u) D(f(A|u) \| \hat{f}^{(t+1)}(A|\hat{u})) + \varphi \cdot h(u) D(h(L|u) \| \hat{h}^{(t+1)}(L|\hat{u_i})) \} \tag{6}$$

where $\hat{f}^{(t)}(U|\hat{a})$ and $\hat{g}^{(t)}(S|\hat{a})$ are row centroids and row feature centroids at iteration t, which can be acquired from Eq.(2) and Eq.(3) by lift $f(a)$ and $g(a)$ outside:

$$\hat{f}(u|\hat{a}) = f(u \mid \hat{u}) f(\hat{u} \mid \hat{a}) = \frac{f(u)}{f(\hat{u})} \frac{f(\hat{a}, \hat{u})}{f(\hat{a})} \tag{7}$$

$$\hat{g}(s|\hat{a}) = \frac{g(\hat{a}, s)}{g(\hat{a})} \qquad (8)$$

Similarly, $\hat{f}^{(t+1)}(A|\hat{u})$ and $\hat{h}^{(t+1)}(L|\hat{u})$ are column centroids and column feature centroids at iteration $t+1$, which can be derived from Eq.(2) and Eq.(4) by lift $f(u)$ and $h(u)$ outside:

$$\hat{f}(a|\hat{u}) = f(a \mid \hat{a})f(\hat{a} \mid \hat{u}) = \frac{f(a)}{f(\hat{a})} \frac{f(\hat{a}, \hat{u})}{f(\hat{u})} \qquad (9)$$

$$\hat{h}(l|\hat{u}) = \frac{h(\hat{u}, l)}{h(\hat{u})} \qquad (10)$$

4 Parallel Co-clustering with Augmented Matrices (PCCAM) Algorithm

In this section, we present the parallel version of the co-clustering with augmented matrices algorithm using Map-Reduce. Figure 1 illustrates the flowchart of the CCAM algorithm. Step 1 initializes the row and column clusters randomly and computes the row and column cluster centroids based on the initial clusters. Step 2 assign a new row cluster to each row instance based on Eq. (5). Step 3 compute new column cluster centroids according to the new row clusters and last-round column clusters based on Eq.(9) and Eq.(8). Step 4 assigns a new column cluster to each column instance based on Eq. (6). Step 5 computes new row cluster centroids based on the last round row clusters and new column clusters based on Eq.(7) and Eq.(10). The new objective value is calculated, and the algorithm resumes at step 2 until convergence.

The computation bottleneck of CCAM is the calculation of the distance between each row (column) and row (column) cluster centroids. Therefore, we will focus on this task for parallelization. According to Hadoop HDFS policy, each Map-Reduce job can process multiple input data, but each instance with the

Table 1. Two input formats: Format 1 (Top) and Format 2 (Bottom)

(a) Format 1: Two data matrix

| $f(A, U)$ | | | | | | | | | $g(A, S)$ | | | | |
identifier	index	u_1	u_2	u_3	u_4	u_5	u_6	C_A	identifier	index	s_1	s_2	C_A
a	1	0.05	0.05	0	0	0.05	0.05	$\hat{a}_1$	af	1	0.2	0	$\hat{a}_1$
a	2	0.05	0.05	0	0.05	0	0.05	$\hat{a}_1$	af	2	0.2	0.2	$\hat{a}_1$
a	3	0.15	0.15	0	0	0	0.05	$\hat{a}_2$	af	3	0.15	0.15	$\hat{a}_2$
a	4	0	0	0.15	0.05	0.05	0.05	$\hat{a}_2$	af	4	0	0.1	$\hat{a}_2$

(b) Format 2: Concatenated data matrix

| $f(A, U)$ | | | | | | | | $g(A, S)$ | | |
identifier	index	u_1	u_2	u_3	u_4	u_5	u_6	s_1	s_2	C_A
a	1	0.05	0.05	0	0	0.05	0.05	0.2	0	$\hat{a}_1$
a	2	0.05	0.05	0	0.05	0	0.05	0.2	0.2	$\hat{a}_1$
a	3	0.15	0.15	0	0	0	0.05	0.15	0.15	$\hat{a}_2$
a	4	0	0	0.15	0.05	0.05	0.05	0	0.1	$\hat{a}_2$

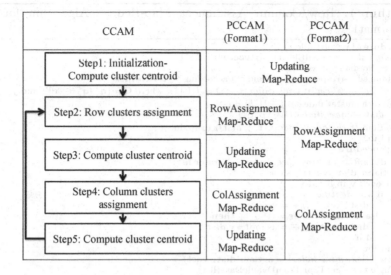

CCAM	PCCAM (Format1)	PCCAM (Format2)
Step1: Initialization- Compute cluster centroid	Updating Map-Reduce	
Step2: Row clusters assignment	RowAssignment Map-Reduce	RowAssignment Map-Reduce
Step3: Compute cluster centroid	Updating Map-Reduce	
Step4: Column clusters assignment	ColAssignment Map-Reduce	ColAssignment Map-Reduce
Step5: Compute cluster centroid	Updating Map-Reduce	

Fig. 1. Overview of CCAM algorithm and PCCAM algorithm

same instance index from multiple input (e.g. $f(A, U)$ and $g(A, S)$) can't be processed in the same Mapper. The reason is that the key of each instance is the line offsets within the input file, which can't be designated as instance index. However, we require both $f(A, U)$ and $g(A, S)$ at the same time for computing the distance for row assignment. Therefore, if three input data matrices (called format 1) are to be used, an additional Map-Reduce procedure is required for step 2 and 3, respectively. On the other hand, if we combine $f(A, U)$ and $g(A, S)$ in a big matrix (called format 2), step 2 and 3 can be processed in one Map-Reduce procedure. Table 1 gives the example of two input formats, in which each instance is denoted as tuple **t** with different kinds of attributes.

4.1   The PCCAM Algorithm for Input Format 1

The spirit of CCAM is to iterate between the row cluster assignment and column cluster assignment until convergence. For the first format, the methodology is to have two Map-Reduce jobs for row (column) assignment and updating separately. Because each set of input data will be delivered to different mappers separately, the new cluster will be acquired in the reducer of the first Map-Reduce job for row assignment stage when both the click and content matrices are processed (by different mappers). For example, step 2 of Figure 1 requires two distances for computing a new row cluster, where one is from each row to row cluster centroid ($\hat{f}(U|\hat{a})$), and the other is from each row feature to row feature centroid ($\hat{g}(S|\hat{a})$). When two mappers finish the calculation, the reducer can assign the new row cluster. Once all new row clusters are acquired, the second Map-Reduce job can start for cluster centroid updating. The procedures for column assignment and updating in step 4 and 5 also require two Map-Reduce jobs.

Algorithm 1. Row/Column Assignment Map-Reduce Algorithm for PC-CAM(format1)

 1: **procedure** MAPPER(tuple **t**)
 2:     $marginal \leftarrow$ SumAllElements(**t**.vector)
 3:     **p** $\leftarrow$ Normalize(**t**.vector)
 4:     Obtained corresponding centroids **Q** according to **t**.$identifier$.
 5:               ▷ e.g. if **t**.$identifier = a$, **Q** $= \hat{f}(U|\hat{a})$, where **Q**$= \{\mathbf{q}_j^T | \mathbf{q}_j^T \in \Re^{|\mathbf{P}|}$ and $j \in [1, k]\}$
 6:     **for** each cluster index j, $j \in [1, k]$ **do**
 7:         $dist \leftarrow marginal * D_{KL}(\mathbf{p}\|\mathbf{q}_j)$
 8:         Emit (**t**.$instance\_index$, (**t**, j, $dist$)) to Reducer
 9:     **end for**
10: **end procedure**

 1: **procedure** REDUCER(Instance_ index i, Iterator list)
 2:     Initialize **dist_ vect**$(1: k) = 0$
 3:     **for** each **v** in list **do**
 4:         **if** **v**.**t**.$identifier = a / u$ **then**
 5:             **dist_ vect**[**v**.j]$+ = $ **v**.$dist$
 6:         **else if** **v**.**t**.$identifier = af / uf$ **then**
 7:             **dist_ vect**[**v**.j]$+ = weight * $ **v**.$dist$
 8:         **end if**
 9:     **end for**
10:     [$min\_ dist$, $min\_ index$] $=$ argmin(**dist_ vect**)
11:     Tuple **t**$_1 \leftarrow$ getTupleFromDyadicData(list)
12:     Feature tuple **t**$_2 \leftarrow$ getTupleFromFeatureData(list)
13:     **t**$_1$.$cluster\_index = min\_index$
14:     **t**$_2$.$cluster\_index = min\_index$
15:     Emit (i, **t**$_1$) to Dyadic+**t**$_1$.$identifier$ file
16:     Emit (i, **t**$_2$) to Feature+**t**$_2$.$identifier$ file
17:     Emit (i, $min\_ dist$) to Objective Value file
18: **end procedure**

The parallel version of step 2 in CCAM algorithm is implemented by Algorithm 1, and step 3 can be achieved by Algorithm 2. Algorithm 1 conducts the row (column) cluster assignment of Figure 1, where Mappers compute the distance of the received tuple **t** to each centroid of the k clusters (For simplicity, we set k as the notation of the number of row cluster and column cluster). The output of Mappers consist of the instance index as Key, and the distance to each cluster $j \in [1, k]$ as Value. Although $f(A, U)$ and $g(A, S)$ are sent to different Mappers, Reducers can aggregate these two distances via the same instance index and assign the new row cluster of a given instance index. In addition to the assignment of the new cluster for each tuple, Reducers of Algorithm 1 also emit the minimal distance to calculate the objective value.

Algorithm 2 is executed in the initialization step and updating step of Figure 1 for cluster centroid updating. In this Map-Reduce job, Mappers simply emit the instances using the cluster index and identifier as Key and tuple **t** as Value, while Reducers compute the co-cluster joint probability $f(\hat{A}, \hat{U})$ based on the new row or column clusters and the marginal probability $f(A)$ or $f(U)$. Those probabilities will be used to calculate the centroids $\hat{f}(U|\hat{a})$ (Eq. (7)) or $\hat{f}(A|\hat{u})$ (Eq.(9)) in the local computer. Due to the limitation of data format, PCCAM algorithm (format1) takes twice times for completing row (column) assignment.

Algorithm 2. Updating Map-Reduce Algorithm for PCCAM(format1)

```
1: procedure MAPPER(tuple t)
2:     Emit ((t.cluster_ index + t.identifier), t) to Reducer
3: end procedure
1: procedure REDUCER(Key key, Iterator tuple_ list)
2:     if key.identifier = a / u then
3:         Initialize array(1:k) = 0
4:         Obtain opposite cluster_ index_ vect C according to key.identifier.
5:                                    ▷ if key.identifier = a, C = C_U; vice versa.
6:         for each tuple t in tuple_ list do
7:             marginal ← SumAllElements(t.vector)
8:             Emit (t.instance_ index, marginal) to Marginal+key.identifier file
9:             array+=Par_ Sum(t.vector, C)
10:        end for
11:        Emit (key.cluster_ index, array) to Joint_ probability+key.identifier file
12:    else if key.identifier = af / uf then
13:        Initialize array(1:|vector|) = 0
14:        for each tuple t in tuple_ list do
15:            array+= t.vector
16:        end for
17:        array ← Normalize(array)                    ▷ based on Eq. (8) and Eq. (10)
18:        Emit (key.cluster_ index, array) to Feature_ centroid+key.identifier file
19:    end if
20: end procedure
```

The complexity for PCCAM algorithm (format1) is $O(T(2mk+2nl))$, where T is the number of iterations, and m is the number of row, k is the number of row cluster, n is the number of column, l is the number of column cluster.

4.2 The PCCAM Algorithm for Input Format 2

For the second format where the clicking data and side-information are concatenated, we can combine the row (column) assignment and updating steps in one Map-Reduce job. The Mappers for this Map-Reduce job compute the distance from each received tuple to the centroids of k clusters and finds the smallest one, to which the instance is assigned. The Mappers then output the new cluster index and identifier as Key and tuple t as Value to Reducers, while the minimum distance will be emitted to the objective file to compute the objective value in the local computer. The Reducers update the marginal probability $f(A)$ and $f(U)$ and joint probability $f(\hat{A}, \hat{U})$ for calculating $\hat{f}(U|\hat{a})$ (Eq.(7)) or $\hat{f}(A|\hat{u})$ (Eq.(9)) in the local computer, and the feature centroids can be directly computed by Eq. (8) and Eq. (10) in Reducers.

In contrast to Algorithm 1, Algorithm 3 only requires one Mapper to complete one Map-Reduce job of Algorithm 1 and one Reducer to complete one Map-Reduce job of Algorithm 2. The complexity for PCCAM algorithm (format2) is $O(T(mk+nl))$.

Algorithm 3. Row/Column Assignment and Updating Map-Reduce Algorithm for PCCAM(format2)

1: **procedure** MAPPER(tuple **t**)
2:     Obtained centroids **Q** and feature centroids **QF** according to **t**.*identifier*.
3:         ▷ e.g. if **t**.*identifier* = a, **QF** = $\hat{g}(S|\hat{a})$, where **QF**= $\{\mathbf{qf}_j^T|\mathbf{qf}_j^T \in \Re^{|\mathbf{P2}|}, j \in [1,k]\}$
4:     $m_1 \leftarrow$ SumAllElements(**t.instance**)
5:     $\mathbf{p}_1 \leftarrow$ Normalize(**t.instance**)
6:     $m_2 \leftarrow$ SumAllElements(**t.feature_ instance**)
7:     $\mathbf{p}_2 \leftarrow$ Normalize(**t.feature_ instance**)
8:     $[min\_dist, j] = argmin_j\{m_1 \cdot D(\mathbf{p}_1\|\mathbf{q}_j) + weight \cdot m_2 \cdot D(\mathbf{p}_2\|\mathbf{qf}_j)\}$

9:   ▷ a general form of Eq.(5) and Eq. (6)
10:     $t.cluster\_ index = j$
11:     Emit ((**t**.*cluster_ index* + **t**.*identifier*), **t**) to Reducer.
12:     Emit (**t**.*instance_ index*, *min_ dist*) to Objective Value file
13: **end procedure**

1: **procedure** REDUCER(Key *key*, Iterator tuple_ list)
2:     Initialize **array**(1:k) = 0
3:     Initialize **feature_ centroids**(1:|**feature_ instance**|) = 0
4:     Obtain opposite cluster_ index_ vect **C** according to *key.identifier*.
5:                                 ▷ if *key.identifier* = a, **C** = $\mathbf{C}_U$; vice versa.
6:     **for each** tuple **t** in tuple_ list **do**
7:         *marginal* $\leftarrow$ SumAllElements(**t.instance**)
8:         Emit (**t**.*instance_ index*, *marginal*) to Marginal+*key.identifier* file
9:         **array**+=Par_ Sum(**t.instance**, **C**)
10:         **feature_ centroids**+= t.**feature_ instance**
11:     **end for**
12:     Emit (*key.cluster_ index*, **array**) to Joint_ probability+*key.identifier* file
13:     **feature_ centroids** $\leftarrow$ Normalize(**feature_ centroids**)   ▷ based on Eq. (8) and Eq. (10)
14:     Emit (*key.cluster_ index*, **feature_ centroids**) to Feature_ centroid+*key.identifier* file
15: **end procedure**

---

5   Experiments

In this experiment, we evaluate the efficiency using Movie-lens data sets which also provide the user feature and movie genres for comparison with PSCOAL [4]. The data comes with different size including ML-100K, ML-1M, and ML-10M [2].

For comparison, we set the same cluster size k as PSCOAL for ML-100K (k=4) and ML-1M (k=5) and k=10 for ML-10M in the following experiment. The row weight λ and column weight φ are set at 0.5 for all three data sets through all experiments. The parameters (k, λ, φ) can be determined by the good effectiveness of validation set, which was discussed in CCAM paper [11][12].

5.1   Evaluation with Different Input Formats

The first experiment evaluates the performance with different data formats. We compare the run-time of two PCCAM algorithms with CCAM (single thread) on the ML-100K data set. The following experiment is conducted in a single cluster Hadoop platform using a computer with an Intel Core2 Quad 2.4GHz CPU and 8GB RAM.

---

[2] http://grouplens.org/datasets/movielens/

Table 2 lists the execution time of each stage for CCAM and two PCCAM algorithms. Format 1 of the PCCAM algorithm takes longer time than CCAM for ML-100K, while format 2 of the PCCAM algorithm has displayed better performance than the CCAM algorithm. The run-time of format 2 is better than format 1, because format 1 requires two Map-Reduce jobs (Algorithm 1+ Algorithm 2) to complete the Row (Column) clustering, which is less efficient than format 2 (requiring only one Map-Reduce job (Algorithm 3)). We can say the overhead of format 1 on ML-100K is higher than the benefit it can obtain from parallelization.

Table 2. Comparing run-time under different input formats. (# block size = default value (9223372036854775807 bytes); # reducer = 1)

Algorithms	CCAM	PCCAM(format1)	PCCAM(format2)
Initialization	0:12	0:30	0:30
Row clustering (avg.)	5:13	7:52	0:36
Column clustering (avg.)	4:51	7:45	0:39
Time for 3 iterations	30:31	47:28	5:30

5.2   Parameters Tuning

In this section, we will identify the parameters affecting the performance. We use 6 computers with Intel Xeon Quad-Core 3.2GHz processors and 16 GB memory for the comparison.

Because the Hadoop platform divides the data into smaller blocks to be sent to each Mapper before starting a Map-Reduce job, we need to know how the block size will affect the performance. We evaluated the run-time on three data sets under different sizes of block. Figure 2 shows that the block size has little affect for ML-100K because the data are overall quite small but has more impact on ML-1M data set where a block size of 1000K bytes offers efficient performance. Thus, the block size also depends on the data size. For example, ML-10M only can be executed when the block size = 100,000K bytes, and blocks that are too small will cause system crash due to the increased number of map tasks.

The numbers of mappers or reducers also affect the efficiency. Because the number of mappers depends on the block size, when the block size is small, the number of mappers is large, and vice versa. Therefore, we do not give the experiments for testing run-time against the number of mappers, as it could be inferred from Figure 2. Instead, we show the number of reducers against the run-time as shown in Figure 3. As we can see, ML-100K, ML-1M and ML-10M have the best performance when the number of reducers is equivalent to the number of desired cluster size. We find that the optimal reducer size is equal to the number of clusters. Thus, the number of reducers could be set to the number of clusters to obtain a better performance.

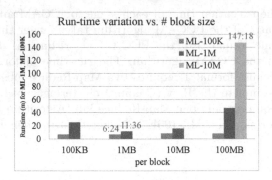

Fig. 2. Run-time performance with different block size

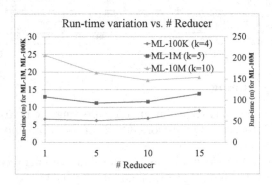

Fig. 3. Run-time performance with various number of Reducers

5.3   Performance Comparison with PSCOAL

We compare the performance with the PSCOAL algorithm shown in [4]. The number of mappers and reducers are set to 10, which is the same as in PSCOAL. Figure 4 shows that PCCAM achieved much better performance than PSCOAL algorithm on the ML-100K and ML-1M data sets. We observe that run-time is decreased as the number of slaves increased because more slaves achieved better parallelization.

Figure 5 gives the scalability among the different slaves in terms of speed-up and run-time. The speed-up value is calculated based on the run-time on n slaves divided by the run-time on 1 slave. As shown in Figure 5, the scalability performs well when number of slaves is greater than 2, while the run-time becomes stable when the number of map tasks is parallelized to a sufficient number of slaves.

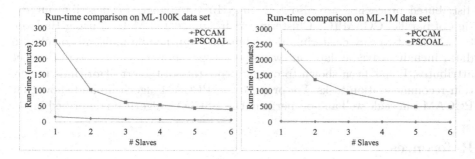

Fig. 4. Run-time comparison between PCCAM and PSCOAL on (a) ML-100K (b) ML-1M

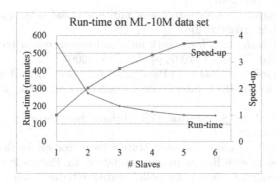

Fig. 5. Run-time and Speed-up on ml-10M

Table 3. Prediction accuracy

ML-100K					ML-1M				
Algorithm	k	λ	φ	MSE	Algorithm	k	λ	φ	MSE
PCCAM	10	0.04	0.01	0.916	PCCAM	10	0.02	0.01	0.833
PSCOAL	4	N/A	N/A	0.915	PSCOAL	5	N/A	N/A	0.856

Table 3 compares the PCCAM algorithm with the PSCOAL algorithm under the mean squared error (MSE)[3] using 10 fold cross validation, where the parameters of both algorithms are used the optimal settings provided by [12] and [4]. PCCAM algorithm performs comparable MSE as PSCOAL algorithm for small data set (ML-100K) and outperforms PSCOAL algorithm for large data set (ML-1M).

6   Conclusion

The uses of information technology in the modern era have produced huge volumes of data. Therefore, scalable and easy-to-use data mining algorithms for

---

[3] http://en.wikipedia.org/wiki/Mean_ squared_ error

distributed processing are also emerging. In this paper, we describe our design of parallel CCAM (Co-Clustering with Augmented Matrix) using Hadoop. CCAM is a two-way clustering algorithm that can handle multiple correlation data, including dyadic data on two object types and the correlation of object attributes. It has been shown to outperform state-of-the-art algorithms in certain recommendation tasks. We propose the PCCAM algorithm and show that PCCAM can scale well and outperform the PSCOAL algorithm described in [4].

References

1. Banerjee, A., Dhillon, I., Ghosh, J., Merugu, S., Modha, D.: A generalized maximum entropy approach to Bregman co-clustering and matrix approximation. JMLR 8, 1919–1986 (2007)
2. Bisson, G., Grimal, C.: Co-clustering of multi-view datasets: a parallelizable approach. In: ICDM 2012 (2012)
3. Deodhar, M., Ghosh, J.: A framework for simultaneous co-clustering and learning from complex data. In: KDD 2007, pp. 250–259 (2007)
4. Deodhar, M., Jones, C., Ghosh, J.: Parallel simultaneous co-clustering and learning with Map-Reduce. In: IEEE International Conference on Granular Computing, pp. 149–154 (2010)
5. Dhillon, I.S., Mallela, S., Modha, D.S.: Information theoretic co-clustering. In: KDD 2003, pp. 89–98 (2003)
6. Forbes, P., Zhu, M.: Content-boosted Matrix Factorization for Recommender Systems: Experiments with Recipe Recommendation. In: RecSys 2011 (2011)
7. George, T., Merugu, S.: A scalable collaborative filtering framework based on co-clustering. In: ICDM 2005 (2005)
8. Pacheco, P.S.: Parallel Programming with MPI (1997) ISBN 1-55860-339-5
9. Papadimitriou, S., Sun, J.: Disco: distributed co-clustering with Map-Reduce. In: ICDM 2008 (2008)
10. Ramanathan, V., Ma, W., Ravi, V.T., Liu, T., Agrawal, G.: Parallelizing an Information Theoretic Co-clustering Algorithm Using a Cloud Middleware. In: 2010 IEEE International Conference on Data Mining Workshops (ICDMW), pp. 186–193 (2010)
11. Wu, M.-L., Chang, C.-H., Liu, R.-Z.: Co-clustering with augmented data matrix. In: Cuzzocrea, A., Dayal, U. (eds.) DaWaK 2011. LNCS, vol. 6862, pp. 289–300. Springer, Heidelberg (2011)
12. Wu, M.L., Chang, C.-H., Liu, R.-Z.: Integrating content-based filtering with collaborative filtering using co-clustering with augmented matrices. Expert Systems with Applications 41(6), 2754–2761 (2014)
13. Yu, H.F., Hsieh, C.J., Si, S., Dhillon, I.: Scalable coordinate descent approaches to parallel matrix factorization for recommender systems. In: ICDM 2012, pp. 765–774 (2012)

Processing OLAP Queries over an Encrypted Data Warehouse Stored in the Cloud

Claudivan Cruz Lopes[1], Valéria Cesário Times[1], Stan Matwin[2],
Ricardo Rodrigues Ciferri[3], and Cristina Dutra de Aguiar Ciferri[4]

[1] Informatics Center, Federal University of Pernambuco, Recife, Brazil
[2] Institute for Big Data Analytics, Dalhousie University, Halifax, Canada
[3] Computer Science Department, Federal University of São Carlos, São Carlos, Brazil
[4] Computer Science Department, University of São Paulo, São Carlos, Brazil
{ccl2,vct}@cin.ufpe.br, stan@cs.dal.ca,
ricado@dc.ufscar.br, cdac@icmc.usp.br

Abstract. Several studies deal with mechanisms for processing transactional queries over encrypted data. However, little attention has been devoted to determine how a data warehouse (DW) hosted in a cloud should be encrypted to enable analytical queries processing. In this article, we present a novel method for encrypting a DW and show performance results of this DW implementation. Moreover, an OLAP system based on the proposed encryption method was developed and performance tests were conducted to validate our system in terms of query processing performance. Results showed that the overhead caused by the proposed encryption method decreased when the proposed system was scaled out and compared to a non-encrypted dataset (46.62% with one node and 9.47% with 16 nodes). Also, the computation of aggregates and data groupings over encrypted data in the server produced performance gains (from 84.67% to 93.95%) when compared to their executions in the client, after decryption.

1 Introduction

One of the services provided by cloud computing is often referred to as *Database as a Service (DAS)*, where data management is outsourced to a cloud provider. This allows customers to create, maintain and query their data in the cloud using their internet connection. Because data are stored in the DAS provider, there are potential risks of sensitive data, such as financial information or medical records, being stored in an untrusted host [16]. For security reasons, sensitive data may be encrypted before being sent to the cloud. However, the execution of queries over these data requires decrypting, which often causes high processing costs and can compromise data privacy if this task is performed in an unsafe data provider. Thus, the execution of queries directly over encrypted data is able to significantly improve query performance, while maintaining data privacy [15].

There are several studies in the literature dealing with mechanisms for query processing over encrypted data [1–7, 9, 13]. Also, in many database applications, data are often aggregated, integrated and stored in a data warehouse (DW), in order to be

L. Bellatreche and M.K. Mohania (Eds.): DaWaK 2014, LNCS 8646, pp. 195–207, 2014.
© Springer International Publishing Switzerland 2014

queried in a suitable manner and to help users in increasing the productivity of their decision-making processes. However, little attention has been devoted to the investigation of *how the dimensional data of a DW hosted in a cloud should be encrypted for allowing the processing of analytical queries.* A method for encrypting and querying such a DW is the focus of this article.

In this article, we investigate the development of a method for encrypting and querying a DW hosted in a cloud. To achieve this objective, we introduce in this article the following contributions:

— We describe performance tests that investigate how dimensional data should be encrypted.
— We propose a novel method for encrypting and querying a DW hosted in a cloud, which generates indistinguishable encrypted data (i.e. encrypted values different from each other) and allows the execution of joins between large fact tables and dimension tables, data aggregations, selection constraints, data groupings and sorting operations over the encrypted dimensional data stored in the cloud.
— We introduce an OLAP system based on the proposed encryption method.
— We validate the proposed OLAP system in terms of query processing.

This article is organized as follows. Section 2 surveys related work. Section 3 presents the results gathered from performance tests that investigate how dimensional data should be encrypted. Section 4 proposes a novel encryption method for DW. Section 5 details the architecture of the proposed OLAP system, which is validated experimentally in Section 6. Section 7 concludes the article.

2   Related Work

Several encryption techniques have been proposed to perform computations over encrypted data, enabling query processing directly over encrypted databases. *Symmetric Encryption* [1, 7, 10] and *Asymmetric Encryption* [1] are used to encrypt attributes that are compared by equality operations; *Homomorphic Encryption* [8–9] is applied to attributes that are used in the computation of aggregation functions, such as *sum* and *average*; *Order Preserving Encryption (OPE)* [7, 12–13] and *Bucketization* [2–3] ensure that their encrypted values maintain the same order as their corresponding original values, and are applied to attributes that are used in the computation of *max* and *min* aggregation functions, or attributes that are compared using relational operators such as $=, >, <, \geq, \leq, \neq$. Also, *Multivalued OPE (MV-OPE)* [3–4] is an OPE encryption that produces a probabilistic encryption schema, where unique values from an original dataset are encrypted to distinct encrypted values with a high probability. In this article, these distinct encrypted values are referred to as *indistinguishable encrypted values.*

Based on the aforementioned encryption techniques, data encryption schemas and encryption systems have been proposed. In [4], an MV-OPE encryption schema is proposed but the processing of grouping operations, used in many database applications, has to be done in the client, after decryption. *CryptDB* [7] is a system that enables the processing of SQL queries over encrypted data. However, range constraints and sorting operations are executed over values encrypted by an OPE encryption, which leaks the order of encrypted data and reveals the distribution of the original values since unique original values are encrypted to the same encrypted value [12].

In [6], a data encryption schema is proposed where each database column is encrypted using homomorphic encryption, an indexing mechanism based on MV-OPE and a secure hash function. This approach does not enable the execution of data groupings and sorting operations in the same query, because these computations must be specified over different columns of database tables, while most DBMS require that these operations are specified in a query over the same columns.

In [5], each record is encrypted as a unique value by using a symmetric encryption, and the values involved in selection constraints are given as input to an encryption function to produce indistinguishable encrypted values. Nevertheless, this data encryption schema does not allow the execution of grouping operations over encrypted values because each record in the database is encrypted as a unique value.

In summary, to the best of our knowledge, all the existing methods suffer from particular limitations that constrain their use in practical DW implementations. We propose a novel method that does not suffer from any of these limitations and scales with respect to the number of processors used in its parallelization.

3   Performance Tests

To compute analytical queries over an encrypted DW, there is a need for: joining large fact tables and dimension tables; performing aggregations that are usually based on the sum aggregate function; computing selection constraints (i.e. range/equality constraints); and executing data groupings and sorting operations. To achieve this, we carried out performance tests that investigated the following hypotheses:

— Hypothesis 1. *Primary keys of dimension tables and fact tables, and foreign keys of fact tables should be left unencrypted.*
— Hypothesis 2. *Sum aggregation functions should be calculated directly over the encrypted data stored in the server.*
— Hypothesis 3. *Encrypted DWs can be used in the processing of selection constraints, data groupings and sorting operations.*

For executing the performance tests, we considered a client-server system architecture. The client is responsible for encrypting the data before sending them to the server, mapping the user analytical queries to queries based on the encrypted DW, decrypting the data returned by the execution of analytical queries over the encrypted DW in the server and computing query operations that were not executed in the server. The server provides all DBMS functionalities by accessing the encrypted data. We executed performance tests based on the Star Schema Benchmark (SSB) [11], which is the standard benchmark for the performance evaluation of DW modeled according to the star schema. We executed all SSB queries over synthetic datasets created with scale factor 1, which produced about 6M records in the fact table. We defined the test configurations of Table 1, and for each of them, each SSB query was executed five times, and the averages of the elapsed time (collected in seconds) were calculated. Experiments were conducted on a laptop with 2.8 GHz Core i7-2640M processor, 6 GB RAM, 5400 RPM SATA 1 TB HD, Debian 7.0 64 bits, PostgreSQL 9.2 and JRE7, which played the role of client and server. Network costs were not computed and the encryption algorithms used to encrypt the datasets were implemented in Java.

Table 1. Test Configurations Used in the Experiments

Hypothesis	Test Configuration	Description
1	KEY-ENC	Primary/foreign keys were encrypted by Blowfish [10].
	KEY-NONENC	Primary/foreign keys were kept unencrypted.
2	MEASURE-SYM	Measures were encrypted by Blowfish.
	MEASURE-HOM	Measures were encrypted by the homomorphic encryption proposed in [8].
3	ALL-MVOPE	Descriptive attributes/measures were encrypted by MV-OPE encryption proposed in [4].
	ALL-OPE	Descriptive attributes/measures were encrypted by OPE encryption defined in [13].
	SYM	Descriptive attributes/measures were encrypted by Blowfish.

In investigating Hypothesis 1, we found that the test configuration KEY-ENC (i.e. primary and foreign keys encrypted) presented drawbacks w.r.t. KEY-NONENC (i.e. keys unencrypted). First, the encryption of primary and foreign keys resulted in an increasing of up 500% in their sizes, degrading the performance of the join operation in 33.26% w.r.t. the use of unencrypted keys, as shown in Figure 1(A). Second, when the SSB queries were executed, primary and foreign keys were used only in the computation of joins, which revealed associations between the data items of the tables associated with the joins being processed, independent of keeping the key attributes encrypted or not. In addition, we can assume that primary and foreign keys of dimensional data schemas are often surrogate keys as is highly recommended by the literature [14]. Therefore, keeping these keys unencrypted in a DW provides better performance than encrypting them, and does not affect data confidentiality because they are often composed by artificial values that do not display any semantic information.

To investigate Hypothesis 2, we collected the elapsed time for computing the sum aggregation function over encrypted measures. Results showed that the encryption of measures by using homomorphic encryption (MEASURE-HOM) produced performance gains of 20.54% when compared to measures encrypted through a symmetric encryption (MEASURE-SYM), as depicted in Figure 1(B). This gain occurred because with homomorphic encryption, the sum aggregation function was calculated directly over the encrypted measure values stored in the server, whereas using the symmetric encryption, the sum had to be executed in the client after decryption since the symmetric encryption does not enable the computation of the sum operation over encrypted data. Therefore, results indicated that the computation of aggregations in the server is more efficient than their calculations in the client after decryption.

To investigate Hypothesis 3, we collected the elapsed time for performing selection constraints, data groupings and sorting operations over encrypted descriptive attributes and encrypted measures, since these operations are usually specified over these attributes in analytical queries. Results showed that ALL-OPE had performance gains of 38.91% and 8.69% w.r.t. SYM and ALL-MVOPE, respectively, as shown in Figure 1(C). When compared to SYM, the ALL-OPE's gains occurred because the encrypted values generated by an OPE encryption (such as ALL-OPE) preserved the same order as their respective original values, enabling the execution of the aforementioned operations directly over the encrypted data in the server. On the other hand, by using a symmetric encryption (SYM), only equality constraints were executed over encrypted data in the server, whereas range constraints, data grouping and sorting operations were performed in the client, after decryption. The ALL-OPE's performance gains w.r.t. ALL-MVOPE were modest because encrypted values produced by an MV-OPE encryption (such as ALL-OPE) also preserves the order of their respective original values, and allows the computation of selection constraints and sorting operations over the encrypted data in the server. However, data groupings were executed in the client after decryption, because the encrypted values produced by MV-OPE encryption are distinct from each other.

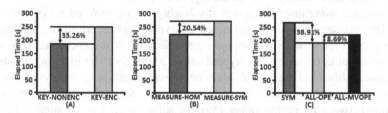

Fig. 1. Query Processing Performance for each Test Configuration

Because a DW is a high-redundant dimensional database, we have that encrypting a DW based on fixed encrypted values (as done by OPE) means a DW with a high degree of redundant encrypted values and a vulnerability that could be explored by statistical attacks [3–4]. However, such vulnerability can be minimized through the use of an encryption technique that produces indistinguishable encrypted values (as performed by MV-OPE). We conclude that MV-OPE is best suited than OPE for encrypting descriptive attributes and measures, since MV-OPE caused a small overhead w.r.t. OPE in the processing of selection constraints, data groupings and sorting operations, and because MV-OPE is more appropriate than OPE to prevent statistical attacks over encrypted data.

4   A Method for Encrypting a DW

Based on the collected results of Section 3, we propose a novel method for encrypting a DW, which is detailed by the Algorithm *Encrypt*. It receives the original value to be encrypted and its respective attribute type as input, and generates as output the respective encrypted value(s). The algorithm tests the attribute type to define the kind of encryption that will be applied over the original value. If the attribute type is a primary key or a foreign key (Line 1), then the original value itself is returned (Line 2). If the attribute type is a measure (Line 3), then two encrypted values are returned: *EncrMeasure1* and *EncrMeasure2* (Line 7). *EncrMeasure1* is obtained by applying a homomorphic encryption over the original value (Line 4), while the *EncrMeasure2* is computed by a MV-OPE encryption that produces an ordered set of integer values for the original value, and an integer value (i.e. encrypted value) is chosen randomly from this set (Lines 5 and 6). If the attribute type is a descriptive attribute (Line 8), then a similar MV-OPE encryption process as outlined before is executed, i.e., an ordered set of integers is generated for the original value (Line 9), and an encrypted value (i.e. *EncrDescriptiveAtt*) is chosen from this set and returned by the algorithm (Lines 10 and 11). To execute selection constraints, data groupings and sorting operations over *EncrMeasure2* and *EncrDescriptiveAtt*, the MV-OPE encryption generated by the Algorithm *Encrypt* satisfies the following properties:

1. *The generation of an ordered set of integer values must be an order preserving process.* This ensures that the generated encrypted values preserve the order of their respective original values, enabling selection constraints and sorting operations be executed directly over the encrypted values.
2. *An encrypted value must be chosen randomly from an ordered set of integer values.* This guarantees that the generated encrypted values are different from each other with a high probability, improving the security of the encrypted DW against statistical attacks because all encrypted values are likely to be indistinguishable.
3. *Different multivalued encrypted values belonging to the same ordered set of integers can be re-encrypted to a unique value.* This is for eliminating the indistinguishability between encrypted values from the same ordered set of integers (as stated in Property 2), ensuring that data groupings are performed over the encrypted values without decrypting them.

```
Algorithm Encrypt
Input: Original Value, Attribute Type
Output: Encrypted Value(s)
1:   if Attribute Type is in {Primary Key, Foreign Key}
2:      return Original Value
3:   if Attribute Type is Measure
4:      EncrMeasure1 = Homomorphic(Original Value)
5:      Set = GenerateOrderedSetOfInt(Original Value)
6:      EncrMeasure2 = ChooseIntFromSet(Set)
7:      return {EncrMeasure1, EncrMeasure2}
8:   if Attribute Type is Descriptive Attribute
9:      Set = GenerateOrderedSetOfInt(Original Value)
10:     EncrDescriptiveAtt = ChooseIntFromSet(Set)
11:     return EncrDescriptiveAtt
```

5   Querying an Encrypted DW Hosted in the Cloud

Our proposed method detailed in Section 4 is used by an OLAP system outlined in this section. The proposed OLAP system allows that analytical queries be processed directly over the encrypted DW without post-processing in the client, and that the analytical query processing makes use of the scalability provided by the cloud. Figure 2 shows the architecture of the proposed OLAP system, which has four main components: *DaaS*, *Scalable Layer*, *Secure Host* and *Client Layer*. The *DaaS* and the *Scalable Layer* are components deployed in the cloud (*server*), and therefore, are considered vulnerable elements, whereas the *Secure Host* and the *Client Layer* are deployed in the user environment (*client*) and are seen as secure elements.

The *DaaS* maintains an encrypted DW whose data are partitioned horizontally among several database management systems *DW-1*, *DW-2*, ..., *DW-n*, so that each *DW-i* has an embedded query processor to compute analytical queries. Also, the *DaaS* contains a *DW Master* which is a repository that contains the addresses of each *DW-i*. The address of a *DW-i* is used to open a connection with the *DW-i* in order to execute analytical queries.

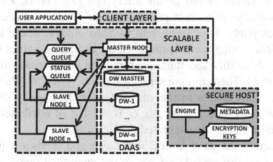

Fig. 2. The Proposed OLAP System Architecture

Client Layer is responsible for receiving user's queries from *User Application* and for sending back the responses of these queries to the *User Application*. The user's analytical queries and their respective answers are built based on the logical data schema of the original DW, ensuring a transparent encryption for *User Application*.

To process these analytical queries over an encrypted DW, *Client Layer* interacts with *Secure Host*, which rewrites and transforms them into queries based on the logical data schema of the encrypted DW. *Secure Host* is also responsible for encrypting query parameters and for decrypting the analytical queries' results received from *Scalable Layer*. For this, *Secure Host* keeps metadata about the logical data schemas of the original DW and of its corresponding encrypted DW, and holds the encryption keys used in the encryption and decryption processes.

Scalable Layer is aimed at providing a scalable mechanism to perform analytical queries over an encrypted DW stored in the *DaaS*. It is composed by a *Master node*, *Query Queue* and *Status Queue*, and several *Slave nodes*. *Master node* is responsible for distributing analytical queries between *Slave nodes*, and for merging the partial results obtained from *Slave nodes* into a merged final result set to be sent to *Client Layer*. *Slave nodes* are responsible for submitting analytical queries to be processed in a specific *DW-i* and for post-processing the retrieved results in order to compute the data groupings. Finally, *Query Queue* is used by *Master node* for sending messages to *Slave nodes* in order to request the execution of analytical queries over the encrypted DW stored in the *DaaS*, while *Status Queue* is used by *Slave nodes* to notify the *Master node* about the ending of an analytical query processing, so that the *Master node* can retrieve the partial results from each *Slave node*.

5.1   Computing OLAP Queries over Indistinguishable Encrypted Data

In detail, *User Application* submits an analytical query to *Client Layer*, which forwards the query to the *Secure Host's Engine* in order to be mapped into an analytical query to be executed over the encrypted DW. To achieve this, the *Engine* searches the metadata for mapping attributes and tables specified in the given analytical query into the corresponding attributes and tables of the encrypted DW. Also, it removes the *group by clause* since this operation cannot be executed directly over the indistinguishable encrypted values stored in the encrypted DW. Next, *Engine* obtains the necessary encryption keys to encrypt each parameter's value defined in the analytical query and generates public keys to be sent to *Slave nodes*. These keys are used by *Slave nodes* for re-encrypting the indistinguishable encrypted values in order to eliminate their indistinguishability and thus, enabling the computation of data groupings (as stated in Section 4). Then, *Engine* sends a set $K = \{QRY, G, PuK\}$ back to *Client Layer*, where *QRY* corresponds to the mapped analytical query (based on the encrypted DW and without data groupings), *G* is the list of attributes specified in the *group by clause*, and *PuK* are the generated public keys.

Master node receives the set *K* from *Client Layer* and sends a request to *DW Master* to obtain the address of each *DW-i* where the analytical query will be executed. The *Master node* then issues a set of messages of the type $\{ID, K, ADDRi\}$ to *Query Queue*, where *ID* is the analytical query identifier, *K* is the set *K* and *ADDRi* is the address of a specific *DW-i*. Further, the *Slave nodes* read the messages queued in the *Query Queue* and send the analytical query *QRY* to the respective *DW-i* to be processed. Each message read from *Query Queue* is dequeued to avoid reprocessing. Further, *Slave nodes* obtain the result set from the *DW-i* and compute the data groupings by executing the Algorithm *ExecuteGrouping*. This algorithm takes the result set, the list of attributes *G* specified in the *group by clause*, and the public keys *PuK* as input, and executes an iterative process (Lines 1 to 8) to generate as output a result set *G'* with the query results grouped by *G* (as specified in the user's analytical query).

For each attribute specified in the *group by clause* (Line 2), the algorithm obtains its respective public key from *PuK* (Line 3) and its respective indistinguishable encrypted value from the record read (Line 4), and uses both to produce a unique encrypted value (Line 5), which is stored in a record of identifiers (Line 6). A unique encrypted value is obtained by mapping the indistinguishable encrypted value selected by a query into a sorted identifier. This identifier is generated for the ordered set of integers to which the indistinguishable encrypted value belongs. Therefore, there is a 1:1 association between ordered sets of integers and their identifiers, and this association eliminates the indistinguishability since indistinguishable encrypted values belonging to the same ordered set of integers will be mapped to the same identifier in a query. Also, a same ordered set of integers is associated with distinct identifiers in different query executions with a high probability because the identifier of an ordered set is randomly generated in different query executions. Further, the original values remain unknown to the server because the generated identifiers do not reveal any information about the original values except their ordering.

Next, the list of measure values are collected from the record read (Line 7) and used together with the identifiers obtained for the attributes in G in order to compose the new result set G' whose sum of measure values is grouped by G (Line 8). A result set G' is represented by a hash table HT composed by two columns denoted by k and v. For each row (k, v) in HT, k is the concatenation of each identifier (obtained for the attributes specified in G), and v is a list of measure values. A new row (k', v') is added to HT when the value k' is not found in HT; otherwise, each value of v' is added to the corresponding value of v stored in an existing row (k, v) of HT, so that $k = k'$. When all records in the result set are processed, the new result set G' is returned (Line 9).

```
Algorithm ExecuteGrouping
Input: Result Set, G, PuK
Output: Result Set With Data Groupings
1:  for each Record in Result Set
2:    for each Attribute in G
3:      PKey = GetPublicKeyFromPuK(PuK, Attribute)
4:      Indist = GetIndistFromRecord(Record, Attribute)
5:      Id = GenerateOrderedSetOfIntID(Indist, Key)
6:      Save(Id, RecordOfIds)
7:    Measures = GetMeasuresFromRecord(Record)
8:    G' = AddGrouping(G', Measures, RecordOfIds)
9:  return G'
```

In the sequence, each *Slave node* notifies the *Master node* by sending a message to *Status Queue*. A message is composed by *{ID, ADDRsh}*, where *ID* is the analytical query identifier and *ADDRsh* is the address of the *Slave node*. Then, *Master node* reads messages queued in *Status Queue* (each message read is dequeued to avoid reprocessing), and obtains the result set from each *Slave node* identified by *ADDRsh*. *Master node* merges all result sets into a merged final result set by executing the Algorithm *MergeResults* detailed as follows. It receives a result set G' produced by a *Slave node* and a merged result set *MRS* as input, and produces a new *MRS* merged with G'. For each row (k', v') in G' (Line 1), it tries to find a row (k, v) of *MRS* so that $k = k'$ (Line 2). If a row is found (Line 3), then each measure value of v' in (k', v') is added to the corresponding measure value of v of the found row (Line 4). Otherwise, the row (k', v') is inserted into *MRS* (Line 6). When all rows of G' are processed, the merged result set *MRS* is returned (Line 7).

After merging all result sets, the *Master node* sends the final result set to *Client Layer*, which forwards it to *Secure Host's Engine* for decryption. *Engine* obtains the necessary encryption keys, decrypts the final result set, and sends it back to *Client Layer*. Finally, *Client Layer* delivers the final results to *User Application*.

```
Algorithm MergeResults
Input: Result Set G', Merged Result Set MRS
Output: Merged Result Set MRS
1:  for each (k', v') in G'
2:      Row = FindRowInMergedResultSet(k', MRS)
3:      if Row is found
4:          AddMeasureValues(v', Row)
5:      else
6:          InsertRowInMergedResultSet((k', v'), MRS)
7:  return MRS
```

6   Performance Evaluation

We defined three test configurations: *All Processing in the Server (ALL-SRV)*, which is the test configuration for the OLAP system proposed in Section 5, where all DW's attributes were encrypted according to the encryption method of Section 4, and the homomorphic encryption defined in [8]; *Non Encrypted Baseline (NONENC-BLN)*, where a DW with unencrypted values was considered; and *Post-Processing in the Client (POST-CLI)*, where primary and foreign keys were kept unencrypted, descriptive attributes were encrypted by the OPE encryption defined in [13], measures were encrypted by using Blowfish [10], and the *Secure Host* was responsible for decrypting the final results and computing the sum aggregation functions and data groupings.

Using the benchmark SSB [11], we created datasets with scale factor 1 for each test configuration. The *Client Layer* and *Secure Host* were deployed on a laptop with 2.10 GHz Pentium T4300 processor, 4 GB RAM, 5400 RPM SATA 500 GB HD, Debian 6.0 32bits and JRE7. The *Scalable Layer* was deployed on the Windows Azure using computers with shared CPUs, 768 MB RAM, 20 GB HD and Windows Server 2008 R2. The *DaaS* was deployed on the Windows SQL Azure. The bandwidth network used between the *Client Layer* and the *Master node* was 6Mbps, while the bandwidth network used between the components of the *Scalable Layer* itself and between the components of the *Scalable Layer* and the *DaaS* was 5Mbps. Network costs between the *Client Layer* and the *Secure Host* were not computed. The encryption algorithms were implemented in Java.

Experiments were performed using 1, 2, 4, 8 and 16 *Slave nodes*, so that the SBB datasets were equally partitioned among 1, 2, 4, 8 and 16 *DW-i*. For each test configuration, each SSB query was executed five times, and the averages of the following metrics were collected in seconds: *Query*: average time spent by each *Slave node* to process an analytical query; *Retrieve*: time spent by the *Master node* to collect the query results obtained from each *Slave node* and merge them; *Total Server*: server time (i.e. sum of *Query* and *Retrieve*); *Download*: time spent by the *Client Layer* to retrieve the final results of an analytical query from the *Master node*; *Post Processing*: time spent by the *Secure Host* to decrypt the analytical query results; and finally, *Total Client*: client time (i.e. sum of *Download* and *Post Processing*).

We investigated the impact of *scaling out* on query processing performance by distributing the execution of analytical queries across *Slave nodes* located in the *Scalable Layer* of the proposed system architecture of Figure 2. Using *ALL-SRV* (i.c. a DW encrypted by the proposed method) and *NONENC-BLN* (i.e. an unencrypted DW), we collected the elapsed time for computing all SSB queries. Results showed that the overhead of *ALL-SRV* w.r.t. *NONENC-BLN* ranged from 9.47% to 46.62%, and decreased with the increase in the number of *Slave nodes* (Figure 3(A)). This decrease occurred because with the increase in the number of *Slave nodes*, the workload of each *Slave node* was minimized shortening the average time spent by each *Slave node* to complete the processing of all SBB queries (i.e. *Query*).

Figure 3(B) depicts the values of *Query* collected for *ALL-SRV* and *NONENC-BLN*. For this metric, the overhead of *ALL-SRV* w.r.t. *NONENC-BLN* ranged from 50.11% (with 1 *Slave node*) to 17.72% (with 16 *Slave nodes*). Results showed that when the proposed system was *scaled out*, the use of an encrypted DW did not impair the processing performance of analytical queries, because the overhead caused by the proposed encryption method was reduced when compared to an unencrypted DW.

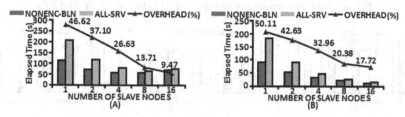

Fig. 3. Query Processing Performance of *ALL-SRV* w.r.t. *NONENC-BLN*

We also compared the elapsed time for computing all SSB queries using *ALL-SRV* (i.e. all query operations were executed in the server) with the elapsed time produced by *POST-CLI* (i.e. the client was responsible for computing sum aggregation functions and data groupings). Figure 4(A) shows that *ALL-SRV* obtained performance gains w.r.t. *POST-CLI* that varied from 84.67% to 93.95%, by ranging the number of *Slave nodes* from 1 to 16. These gains occurred because *POST-CLI* does not compute aggregates and data groupings over the server's encrypted data, and consequently, a large volume of data were transferred from *Slave nodes* to the *Master node*, and from the *Master node* to the *Client Layer*. This increased the workload of the *Master node* that had to merge a greater number of records and the workload of the *Secure Host* that had to decrypt a larger number of records and to compute aggregates and data groupings over the decrypted data records. Figure 4(B) shows that *ALL-SRV* produced performance gains w.r.t. *POST-CLI* of at least: 97.60% in the processing performed by the *Master node* (i.e. *Retrieve*); 97.50% in the time spent for transferring data records from the *Master node* to the *Client Layer* (i.e. *Download*); and 96.20% in the post-processing executed by the *Secure Host* (i.e. *Post Processing*).

We also gathered the elapsed time on query processing distributed between the client and the server. Figure 5(A) shows that *ALL-SRV* produced performance gains w.r.t. *POST-CLI*, which: ranged from 63.21% to 82.13% in the server; and were greater than 97.38% in the client, indicating that the calculation of aggregates and data groupings in the client (after decryption) impaired the overall system's performance on the processing of analytical queries. As shown in Figure 5(B), *POST-CLI* spent from 62.77% up to 77.20% of its runtime in the client, illustrating that the post-processing in the client executed by *POST-CLI* was time consuming. Figure 5(C) shows that *ALL-SRV* spent from 67.33% up to 89.33% of its runtime in the server, because *ALL-SRV* enabled the execution of analytical queries directly over the encrypted data in the server, while the *ALL-SRV*'s remaining time was the time required to transfer the queries' encrypted results to the client, and the time for decrypting these results in the client.

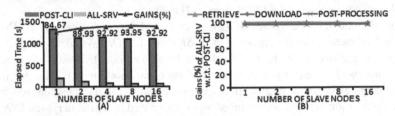

Fig. 4. Query Processing Performance of *ALL-SRV* w.r.t. *POST-CLI*

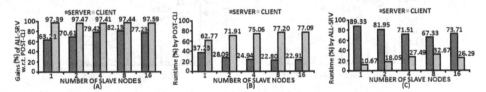

Fig. 5. Distribution of Queries Processing between the Client and the Server

7   Conclusion and Future Work

We proposed a new method for encrypting a DW and enabling the processing of OLAP queries over an encrypted DW that was validated experimentally. Results showed that: primary and foreign keys must be kept unencrypted; the computation of aggregates using measures encrypted by homomorphic encryption favored the analytical query processing performance; and the use of an MV-OPE encryption enables the computation of selection constraints, data groupings and sorting operations over descriptive attributes and measures. Also, we proposed an OLAP system based on the novel encryption method. This system performs analytical queries over encrypted DWs partitioned among several nodes in the DAS. Results showed that the overhead caused by our encryption ranged from to 9.47% (with 16 *Slave nodes*) to 46.62% (with 1 *Slave node*); and that the computation of aggregates and data groupings over encrypted data in the server produced performance gains that ranged from 84.67% to 93.95% when compared to their executions in the client, after decryption. As future work, we intend to evaluate the performance of the proposed OLAP system with different data scalability, and to investigate if data groupings should be computed by the component *DaaS* of the proposed system using indistinguishable encrypted values. A study of the proposed encryption method in terms of security guarantees is another indication of future work.

Acknowledgments. This work has been funded by the Conselho Nacional de Desenvolvimento Científico e Tecnológico (CNPq) under the grants 246688/2012-2 and 246263/2012-1, by the São Paulo Research Foundation (FAPESP) under the grant 2011/23904-7, by the Natural Sciences and Engineering Research Council of Canada, and by the Canadian Bureau for International Education.

References

1. Kadhen, H., Amagasa, T., Kitagawa, H.: A Novel Framework for Database Security Based on Mixed Cryptography. In: Proc. ICIW, Venice, Italy, pp. 163–170 (2009)
2. Hore, B., Mehrotra, S., Canim, M., Kantarcioglu, M.: Secure Multidimensional Range Queries over Outsourced Data. The VLDB Journal 21(3), 333–358 (2012)
3. Kadhen, H., Amagasa, T., Kitagawa, H.: Optimization Techniques for Range Queries in the Multivalued-Partial Order Preserving Encryption Scheme. Knowledge Discovery, Knowledge Engineering and Knowledge Management 272, 338–353 (2013)
4. Kadhen, H., Amagasa, T., Kitagawa, H.: MV-OPES: Multivalued-Order Preserving Encryption Scheme: A Novel Scheme for Encrypting Integer Value to Many Different Values. IEICE Trans. Inf. & Syst. 93-D(9), 2520–2533 (2010)
5. Chen, K., Kavuluru, R., Guo, S.: RASP: Efficient Multidimensional Range Query on Attack-resilient Encrypted Databases. In: Proc. CODASPY, New York, USA, pp. 249–260 (2011)
6. Liu, D., Wang, S.: Programmable Order-Preserving Secure Index for Encrypted Database Query. In: Proc. CLOUD, Washington, USA, pp. 502–509 (2012)
7. Popa, R.A., Redfield, C.M.S., Zeldovich, N., Balakrishnan, H.: CryptDB: Processing Queries on an Encrypted Database. Commun. ACM 55(9), 103–111 (2012)
8. Castelluccia, C., Chan, A.C.F., Mykletun, E., Tsudik, G.: Efficient and Provably Secure Aggregation of Encrypted Data in WSN. ACM Trans. Sen. Netw. 5(3), 1–36 (2009)
9. Liu, D.: Securing Outsourced Databases in the Cloud. In: Security, Privacy and Trust in Cloud Systems, pp. 259–282. Springer, Heidelberg (2014)
10. Schneier, B.: Description of a New Variable-Length Key, 64-bit Block Cipher (Blowfish). In: Fast Software Encryption, Cambridge Security Workshop, London, UK, pp. 191–204 (1993)
11. O'Neil, P., O'Neil, E., Chen, X.: The Star Schema Benchmark. Online Publication of Database Generation Program (2007),
    http://www.cs.umb.edu/~poneil/StarSchemaB.pdf
12. Wang, P., Ravishankar, C.V.: Secure and Efficient Range Queries on Outsourced Databases using Rp-trees. In: Proc. ICDE, Brisbane, Australia, pp. 314–325 (2013)
13. Agrawal, R., Kiernan, J., Srikant, R., Xu, Y.: Order Preserving Encryption for Numeric Data. In: Proc. SIGMOD, Paris, France, pp. 563–574 (2004)
14. Kimball, R., Ross, M.: The Data Warehouse Toolkit: The Definitive Guide to Dimensional Modeling, 3rd edn. John Wiley & Sons (2013)
15. Suciu, D.: SQL on an Encrypted Database: Technical Perspective. Commun. ACM 55(9), 102–102 (2012)
16. Adabi, J.D.: Data Management in the Cloud: Limitations and Opportunities. IEEE Data Eng. Bull. 32(1), 3–12 (2009)

Reducing Multidimensional Data

Faten Atigui, Franck Ravat, Jiefu Song, and Gilles Zurfluh

IRIT - Université Toulouse I Capitole, 2 Rue du Doyen Gabriel Marty
F-31042 Toulouse Cedex 09
{atigui,ravat,song,zurfluh}@irit.fr

Abstract. Our aim is to elaborate a multidimensional database reduction process which will specify aggregated schema applicable over a period of time as well as retains useful data for decision support. Firstly, we describe a multidimensional database schema composed of a set of states. Each state is defined as a star schema composed of one fact and its related dimensions. Each reduced state is defined through reduction operators. Secondly, we describe our experiments and discuss their results. Evaluating our solution implies executing different requests in various contexts: unreduced single fact table, unreduced relational star schema, reduced star schema or reduced snowflake schema. We show that queries are more efficiently calculated within a reduced star schema.

Keywords: multidimensional design, data reduction, experimental assessment.

1   Introduction

Nowadays, decision support systems are based on Multidimensional Data Warehouse (MDW). A MDW schema is based on facts (analysis subjects) and dimensions (analysis axis). By definition, in a MDW, data is stored permanently and new data is periodically added. Hence a DW stores a huge volume of data in which the decision maker may well get lost during his analyses. On the other hand, the pertinence of MDW data decreases with age: while detailed information is generally considered important for recent data [11], it may be of lesser interest for older data. As data value decreases with time, we implement selective deletion at low levels of granularity according to the users' needs. This reduction is achieved mainly through progressive data aggregation: older data is synthesized.

Our objective is to provide a multidimensional analysis environment adapted to decision makers' needs, allowing them to remove the temporal granularity levels which are of little use for analysis.

This paper is composed of the following sections: Section 2 describes a state of the art of data reduction. Section 3 defines our model of multidimensional data based on reductions. Section 4 provides an evaluation of our solution in various implementation environments.

L. Bellatreche and M.K. Mohania (Eds.): DaWaK 2014, LNCS 8646, pp. 208–220, 2014.
© Springer International Publishing Switzerland 2014

2   Related Work

Reducing data allows us both to decrease the quantity of irrelevant data in decision making and to increase future analysis quality [12]. In the context of decision support, data reduction is a technique originally used in the field of data mining [9], [12].

In the DW context, [2] were the first to propose solutions for data deletion. More precisely, they study data expiration in materialized views so that they are not affected and can be maintained after updates.

In the multidimensional area, [11] presents a technique for progressive data aggregation of a fact. This study intends to specify data aggregation criteria of a fact due to higher levels of dimensions. The authors also propose techniques to query reduced multidimensional objects. As mentioned in [6], this work is highly theoretical but it fails to provide us a concrete example of implementation strategy. In [6], a gradual data aggregation solution based on conception, implementation and evaluation is proposed. This solution is based on a table containing different temporal granularities: second, minute, hour, month and year.

This previous work only focuses on the fact table. [5] and [6] use a temporal table for gradual data reduction. Our goal is more ambitious as it aims to study data reduction of the complete multidimensional schema. This reduction depends only on the users' needs. We intend to provide a coherent analysis environment and thus facilitate the decision maker's task by limiting the analysis to semantically coherent data.

3   Our Model

3.1   Case Study

This case study shows a multidimensional schema progression that fulfills the decision maker's needs. During the last four years, sales analysis is carried out with reference to lowest levels of granularity: product, customer and sale date. In the previous period, from 2010 to 2000, analyses are summarized according to product ranges, dates and customer cities because no analysis referring to customers and product codes is required. Before 2000, only annual sales by product ranges make sense.

The following 3 figures represent the evolution of a conceptual multidimensional schema. Each schema represents a state; it is based on a subject of analysis (fact) related to different dimensions. Each fact is composed of one or more indicators. For example, in Figure 1, the fact named "*Sale*" is composed of two indicators: Quantity and Amount. A dimension models an analysis axis; it reflects information according to which subjects of analysis are to be dealt with. For example, in figure 1, the "Sales" fact is connected to 3 dimensions: Products, Customers and Time. Dimension attributes (also called parameters or levels) are organized according to one or more hierarchies. Hierarchies represent a particular vision (perspective) of a dimension. Each schema is based on the graphic notation introduced in [3].

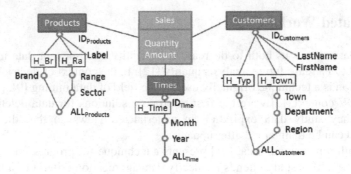

Fig. 1. Current state of the MDW

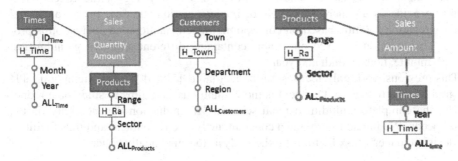

Fig. 2. First reduced state of the MDW.   **Fig. 3.** Second reduced state of the MDW

3.2 Concepts

A MDW is thus modeled as a set of states. The current state corresponds to the present state of the MDW. Past states correspond to a succession of reduced states over time. Each state consists of a star schema composed of a fact and its dimensions.

Definition. A MDW is defined by $S = (n^S ; E ; Map)$ where:

- $n^S \in N$ is the name of the MDW;
- $E = \{E_1 ; \dots ; E_n\}$ is a set of states composing the MDW;
- Map: $E \rightarrow E \mid Map(E_k) = E_{k+1}$ is a derivation function defining the state named E_{k+1} obtained by the reduction of E_k.

Let us define F and D such as $F = \{F_1, \dots, F_n\}$ is a finite set of facts, $n \geq 1$ and $D = \{D_1, \dots, D_m\}$ is a finite set of dimensions, $m \geq 2$.

Definition. A state is a star schema defined for a temporal period such as $E_i = (F_i ; D_i ; T_i)$ where

- $F_i \in F$ is a fact representing a subject of analysis ;
- $D_i = \{D_{TIMES} ; D_1 ; \dots ; D_m\} \subseteq D$ is a set of dimensions associated to the fact with necessarily a temporal dimension denoted D_{TIMES} ;
- $T_i = [t_{start} ; t_{end}[$ is a temporal interval defined on the D_{TIMES} dimension and associated to the state E_i.

To define T_j, we adopt a linear and discrete time model approaching time in granular way through time observation units [13]. A temporal grain is an integer relative to a time unit; we adopt the standard time units manipulated through functions: Year, Quarter, Month, Day... For example, Year (1990) defines the instant "1990" for the year time unit. An instant is a temporal grain. We note T_{now} the current instant which is characterized by its dynamic nature, ie. T_{now} changes constantly depending on the passage of time. A time interval is defined by a couple of instants noted "t_{start}" and "t_{end}". These instants can be fixed (temporal grains) or dynamic (defined with the instant "T_{now}").

Example. The following figure represents the 3 states of our case study. It illustrates the principle of states derived by the reduction. This MDW is defined as follows: $\mathcal{E} = \{E_1 ; E_2 ; E_3\}$ with Map = { (E_1, E_2) ; (E_2, E_3) } where

- $E_1 = (F_{SALES} ;\{D_{PRODUCTS} ; D_{TIMES} ; D_{CUSTOMERS}\} ; [Year(T_{now})-4 ; Year(T_{now})[)$;

- $E_2 = (F_{SALES} ;\{D_{PRODUCTS} ; D_{TIMES} ; D_{CUSTOMERS}\} ; [Year(T_{now})-14 ; Year(T_{now})-4[)$;

- $E_3 = (F_{SALES} ; \{D_{PRODUCTS} ; D_{TIMES}\} ; [Year(1990); Year(T_{now})-14[).$

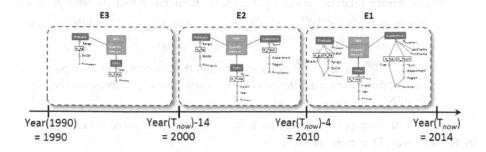

Fig. 4. Reduction principle of multidimensional schemas

The state denoted E1 and called current state, is associated to the validity interval $[Year(t_{now})-4; Year(T_{now})[$ corresponding to $[2010; 2014[$. The instances of this state correspond to sales between 2010 and 2014 only, according to the D_{TIMES} dimension. In the same way, the state named E2 stores data related to sales between 2000 and 2010, whereas the state denoted E3 stores data related to sales prior to 2000.

In Figure 4, 1990 is a fixed instant representing the date when the database was created. In this figure, we can also find time-variant intervals (moving over time) defined by the following instants: $Year(T_{now})-14$, $Year(T_{now})-4$ and $Year(T_{now})$. So, next year, $Year(T_{now}) = 2015$, $Year(T_{now})-4 = 2011$ and $Year(T_{now})-14 = 2001$. At each change of year, the states denoted E1, E2 and E3 will be instantly updated.

Definition. A *fact*, denoted F_i, $\forall i \in [1..n]$, is defined by (n^{Fi}, M^{Fi}) where

- $n^{Fi} \in \mathcal{N}$ is the fact name;
- $M^{Fi} = \{m_1,..., m_{pi}\}$ is a set of *measures* or indicators.

Definition. A *dimension*, denoted D_i, $\forall i \in [1..m]$, is defined by (n^{Di}, A^{Di}, H^{Di}), where

- $n^{Di} \in \mathcal{N}$ is the dimension name;

- $A^{Di} = \{ a_1^{D_i}, ..., a_{r_i}^{D_i} \}$ is the set of the *attributes of the dimension*;

- $H^{Di} = \{ H_1^{D_i}, ..., H_{h_i}^{D_i} \}$ is a set of *hierarchies*.

Hierarchies organize the attributes of a dimension, from the finest graduation (root parameter, ID_{Di}) to the most general graduation (extremity parameter, All_{Di}). Thus, a hierarchy defines the valid navigation paths on an analysis axis.

Definition. A *hierarchy*, denoted H_j (abusive notation of $\boldsymbol{H_j^{D_i}}$, $\forall i \in [1..m]$, $\forall j \in [1..h_i]$) is defined by $(n^{Hj}, P^{Hj}, \prec^{Hj}, Weak^{Hj})$, where:

- $n^{Hj} \in \mathcal{N}$ is the hierarchy name;

- $P^{Hj} = \{ p_1^{H_j}, ..., p_{q_j}^{H_j} \}$ is a set of attributes called *parameters*, $P^{Hj} \subseteq A^{Di}$;

- $\prec^{Hj} = \{ (p^{Hj}_x, p^{Hj}_y) \mid p^{Hj}_x \in P^{Hj} \wedge p^{Hj}_y \in P^{Hj} \}$ is an antisymmetric and transitive binary relation between parameters. Remember that the antisymmetry means that $(p^{Hj}_{k1} \prec^{Hj} p^{Hj}_{k2}) \wedge (p^{Hj}_{k2} \prec^{Hj} p^{Hj}_{k1}) \Rightarrow p^{Hj}_{k1} = p^{Hj}_{k2}$ while the transitivity means that $(p^{Hj}_{k1} \prec^{Hj} p^{Hj}_{k2}) \wedge (p^{Hj}_{k2} \prec^{Hj} p^{Hj}_{k3}) \Rightarrow p^{Hj}_{k1} \prec^{Hj} p^{Hj}_{k3}$.

- $Weak^{Hj} : P^{Hj} \rightarrow 2^{A^{D_i} \backslash P^{H_j}}$ is an application that associates to each parameter a set of dimension attributes, called *weak attributes* (2^N represents the power set of N).

In the rest of the paper we denote each fact F_i that is an abusive notation of $F_{n^{F_i}}$. In the same way, D_i corresponds to $D_{n^{D_i}}$.

Example. The E_3 state of the previous figure is composed of one fact and two dimensions and it is valid from 1990 to 2000. The fact table named SALES contains the measure Amount. The dimension PRODUCTS contains the hierarchy H_Ra on which the parameters are organized according to their granularity level: from the lowest level Range to the highest level $ALL_{PRODUCTS}$. The other dimension is named D_{TIMES}, it is graduated by the attributes Year and ALL_{TIMES} on the hierarchy H_Time.

The abstract representation is as follows:

$E_3 = (F_{SALES} ; \{ D_{PRODUCTS} ; D_{TIMES} \} ; [t_{1990} ; t_{2000}[)$ where:
- $F_{SALES} = (SALES; \{ Amount \})$;

- $D_{PRODUCTS} = (PRODUCTS; \{Range, Sector, ALL_{PRODUCTS} \}; \{H\_Ra\})$;
- $D_{TIMES} = (TIMES; \{ Year, ALL_{TIMES} \}; \{H\_Time\})$.

The H_Ra hierarchy is defined by $(n^{H\_Ra}, P^{H\_Ra}, \prec^{H\_Ra}, Weak^{H\_Ra})$ where:
- $n^{H\_Ra} = H\_Ra$;
- $P^{H\_Ra} = \{ Range, Sector, ALL_{PRODUCTS} \}$;
- $\prec^{H\_Ra} = \{(Range, Sector); (Sector, ALL_{PRODUCTS})\}$;

$Weak^{H\_Ra} = \varnothing$.

3.3   Reduction Operators

Deriving the reduced schema denoted E_{k+1} from a schema denoted E_k is performed through the composition of derivation operators. We define the set of these operators as $O = \{RollUp^{reduce}; Drop^{reduce}; Slice^{reduce}\}$ as the minimum core of elementary operators to define the derivation.

- The $RollUp^{reduce}$ operator provides a new state in which the specified dimension is reduced by removing all the attributes under the parameter that is specified in the operator. If the specified parameter is an extremity parameter like $ALL_{Drollup}$, the dimension is completely removed in the reduced state.
- The $Drop^{reduce}$ operator provides a new state in which the fact is reduced by the deletion of the specified measure.
- The $Slice^{reduce}$ operator provides a reduced state in which the instances of the specified dimension denoted D_{Slice} is reduced. The dimension instances that satisfy the predicate denoted $pred_{slice}$ are kept in the new state.

Table 1. Reduction operators on schemata

Operators	
$RollUp^{reduce}(E_k ; D_{rollup} ; p_{rollup} ; T_{k+1}) = E_{k+1}$	
Inputs	$E_k = (F_k ; \mathcal{D}_k ; T_k)$: initial state;
	$D_{rollup} \in \mathcal{D}_k$: dimension dedicated to a reduction;
	$p_{rollup} \in A^{Drollup}$: reduction parameter of the D_{rollup} dimension.
Output	$E_{k+1} = (F_{k+1} ; \mathcal{D}_{k+1} ; T_{k+1})$ reduced state such as
	- $F_{k+1} = F_k$;
	- $\mathcal{D}_{k+1} = \mathcal{D}_k \setminus \{ D_{rollup} \} \cup \{ D_{new} \}^{(*)}$ with $D_{new} = (n^{Dnew} ; A^{Dnew} ; H^{Dnew})$
	$\quad$ - $n^{Dnew} = n^{Dold}$
	$\quad$ - $A^{Dnew} = \{ a_x \in A^{Drollup} \mid a_x = p_{rollup} \vee \forall H_j \in H^{Drollup}, p_{rollup} \prec^{Hj} a_x \}$
	$\quad$ - $H^{Dnew} = \{ H_x \in H^{Drollup} \mid n^{Hx} = n^{Hj} \wedge P^{Hx} = \{ p_y \in P^{Hj} \mid p_y = p_{rollup} \vee$
	$\quad p_{rollup} \prec^{Hj} p_y \} \wedge \prec^{Hx} := \{ (p^{Hj}_{x1}, p^{Hj}_{x2}) \in \prec^{Hj} \mid p^{Hj}_{x1} = p_{rollup} \vee p_{rollup}$
	$\quad \prec^{Hj} p^{Hj}_{x1} \} \wedge Weak^{Hx} := \{ (p_{x1}, A^{Hx}_{x1}) \in Weak^{Hj} \mid p_y \in P^{Hj} \}.$
$Drop^{reduce}(E_k ; m_{drop} ; T_{k+1}) = E_{k+1}$	
Inputs	$E_k = (F_k ; \mathcal{D}_k ; T_k)$: initial state ;
	$m_{drop} \in M_k$ is a measure of F_k.
Output	$E_{k+1} = (F_{k+1} ; \mathcal{D}_{k+1} ; T_{k+1})$ reduced state such as
	- $F_{k+1} = (n^{Fk}, M^{Fk} \setminus \{ m_{drop} \})$;
	- $\mathcal{D}_{k+1} = \mathcal{D}_k$.
$Slice^{reduce}(E_k ; D_{slice} ; pred_{slice} ; T_{k+1}) = E_{k+1}$	
Inputs	$E_k = (F_k ; \mathcal{D}_k ; T_k)$: initial state ;
	$D_{slice} \in \mathcal{D}_k$: dimension dedicated to a reduction;
	$pred_{slice}$: selection predicate on a domain denoted $dom(D_{slice})$ of D_{slice}.
Output	$E_{k+1} = (F_{k+1} ; \mathcal{D}_{k+1} ; T_{k+1})$ reduced state such as
	- $F_{k+1} = F_k$;
	- $\mathcal{D}_{k+1} = \mathcal{D}_k$ with $dom(D_{slice}) = \{ v_i \in dom(D_{slice}) \mid pred_{slice}(v_i) = TRUE \}.$

$^{(*)}$ If $A^{Dnew} = \{ALL_{Drollup}\}$ then $\mathcal{D}_{k+1} = \mathcal{D}_k \setminus \{ D_{rollup} \}$

Example. In the previous example, we defined two reduced states. Each of them is defined by a derivation function. These functions are defined bellow. The first Map function, composed of two RollUp^{reduce} operators, permits to define the "E2" state. The second Map function, composed of two RollUp^{reduce} operators and one Drop^{reduce} operator, permits to define the "E3" state.

- RollUp^{reduce}(RollUp^{reduce}(E$_1$;   D$_{PRODUCTS}$;   P$_{RANGE}$;   [Year(T$_{now}$)-14 ; Year(T$_{now}$)-4[) ; D$_{CUSTOMERS}$; P$_{TOWN}$; [Year(T$_{now}$)-14 ; Year(T$_{now}$)-4[) = E$_2$;

- RollUp^{reduce}(RollUp^{reduce}(Drop^{reduce}(E$_2$; Quantity ; [Year(1990) ; Year(T$_{now}$)-14[) ; D$_{CUSTOMERS}$; ALL$_{CUSTOMERS}$; [Year(1990) ; Year(T$_{now}$)-14[) ; D$_{TIMES}$; P$_{YEAR}$; [Year(1990) ; Year(T$_{now}$)-14[) = E$_3$.

4   Experimental Assessment

4.1   Data Collection

In order to make experimental assessments, we implement two types of R-OLAP databases with the Oracle DBMS and each type has two different implementations. The first type of MDW corresponds to databases without reduction. Its first implementation is called *Global Star,* consists in an unreduced R-OLAP implementation based on 4 tables (*Products, Customers, Times* and *Sales*). The second implementation is called *Global Table* in which we merge the three analysis axis (dimensions *Products, Customers* and *Times*) with the fact table (*Sales*); consequently this implementation is composed of a single fact table that encompasses all.

The population of the analysis axis was done as follows: (a) the dimension *Times* contains all dates from 01/01/1990 to 31/12/2013, (b) the two other dimensions contain random data defined by generation of synthetic data. Allocation of random data was made so that father attribute of a hierarchy does not have the same number of sons while respecting the integrity constraints of strict hierarchies (any son attribute of a hierarchy has a single father attribute).

We have defined various versions of non-reduced databases by ranging the tuple numbers of the dimensions *Customers* and *Products* from 10 to 40 tuples.

- |*Customers*| = 10, 20, 30, 40 tuples
- |*Products*| = 10, 20, 30, 40 tuples
- |*Times*| = 8401 tuples (from 01/01/1990 to 31/12/2013)
- |*Sales*| = |*Customers*| x |*Products*| x |*Times*| = **840 100** to **13 441 600** tuples.

Even though the dimensions *Customers, Products* and *Times* are integrated in the fact table of *Global Table*, the implementation details of MDW *Global table* are the same as MDW *Global Star*. The following table describes different values associated to the attributes of dimension containing variable data.

The second type of MDW corresponds to reduced databases. This type consists of three states according to the case study presented in this article (see Figure 4). We have defined two implementations of reduced databases:

- a denormalized implementation (*R-OLAP star schema* defined in fig. 5 (a)),
- a normalized implementation (*snowflake schema* defined in fig. 5 (b)).

The operations permitting to get the different states of MDB were implemented with the help of triggers in Oracle DBMS.

Table 2. Implementation details of the dimensions in *Global Star* and *Global Table*.

|Customers| x |Products|	Contents of the dimension Customers	Contents of the dimension Products
10 x 10	2 *Towns*, 2 *Departments* , 1 *Region*, 2 *Types*	2 *Ranges*, 2 *Sectors*, 2 *Brands*
20 x 20	4 *Towns*, 3 *Departments*, 2 *Regions*, 4 *Types*	4 *Ranges*, 3 *Sectors*, 4 *Brands*
30 x 30	6 *Towns*, 4 *Departments*, 2 *Regions*, 6 *Types*	6 *Ranges*, 4 *Sectors*, 6 *Brands*
40 x 40	8 *Towns*, 5 *Departments*, 3 *Regions*, 8 *Types*	8 *Ranges*, 5 *Sectors*, 8 *Brands*

—— E1 ——

SALES(Quantity, Amount, <u>IDProducts#</u>, <u>IDTime#</u>, <u>IDCustomers#</u>)
CUSTOMERS(<u>IDCustomers</u>, Lastname, Firstname, Town,
          Department, Region, Type)
PRODUCTS(<u>IDProducts</u>, Range, Sector, Brand)
TIMES(<u>IDTime</u>, Month, Year)

—— E1 ——

SALES(Quantity, Amount, <u>Range#</u>, <u>IDTime#</u>, <u>Town#</u>)
CUSTOMERS(<u>IDCustomers</u>, Lastname, Firstname, Town#, Type#)
TOWN(<u>Town</u>, Department#)
DEPARTMENT(<u>Department</u>, Region#)
REGION(<u>Region</u>)
TYPE(<u>Type</u>)
PRODUCTS(<u>IDProducts</u>, Range#, Brand#)
RANGE(<u>Range</u>, Sector#)
SECTOR(<u>Sector</u>)
BRAND(<u>Brand</u>)
TIMES(<u>IDTime</u>, Month#)
MONTH(<u>Month</u>, Year#)
YEAR(<u>Year</u>)

—— E2 ——

SALES(Quantity, Amount, <u>Range#</u>, <u>IDTime#</u>, <u>Town#</u>)
CUSTOMERS(<u>Town</u>, Department, Region)
PRODUCTS(<u>Range</u>, Sector)
TIMES(<u>IDTime</u>, Month, Year)

—— E2 ——

SALES(Quantity, Amount, <u>Range#</u>, <u>IDTime#</u>, <u>Town#</u>)
CUSTOMERS(<u>Town</u>, Department#)
DEPARTMENT(<u>Department</u>, Region#)
REGION(<u>Region</u>)
PRODUCTS(<u>Range</u>, Sector#)
SECTOR(<u>Sector</u>)
TIMES(<u>IDTime</u>, Month#)
MONTH(<u>Month</u>, Year#)
YEAR(<u>Year</u>)

—— E3 ——

SALES(Amount, <u>Range#</u>, <u>Year#</u>)
PRODUCTS(<u>Range</u>, Sector)
TIMES(<u>Year</u>)

—— E3 ——

SALES(Amount, <u>Range#</u>, <u>Year#</u>)
PRODUCTS(<u>Range</u>, Sector#)
SECTOR(<u>Sector</u>)
YEAR(<u>Year</u>)

(a) R-OLAP star schema   (b) R-OLAP snowflake schema

Fig. 5. R-OLAP schemata of reduced MDB

4.2   Protocol

The experimental assessment compares the execution time and the cardinalities of queries executed in two unreduced R-OLAP implementations with two types of reduced R-OLAP implementations of the same multidimensional database. This experimental assessment takes into account three criteria:

- Database volumetry: As mentioned above, we will apply queries of 4 versions for the different types of databases.
- Query types: (a) Queries containing only joins and no selection criteria on non-temporal dimensions (querying all the data of reduced database states), (b) Queries containing conditions restrictions on the data (querying certain data in certain states)
- Scope of queries: (a) queries related to one or more dimension tables, (b) queries manipulating 1, 2 or 3 states.

4.3 Results and Discussion

4.3.1 Queries without Restriction Predicates on Non-temporal Dimensions

The first experimental assessment compares the theoretical execution time of queries (explain plan of the Oracle DBMS) by varying the size of the MDW in accordance with the protocol previously described. We have defined 14 queries manipulating different tables and different states. Each query is implemented in SQL.

Table 3. Queries without restriction predicates on non-temporal dimensions

Queries	States	Dimensions
Q_1: Amount of sales for the last three years	E_1	1 D: Time
Q_2: Amount and quantity of sales in 2008	E_2	1 D: Time
Q_3: Amount of annual sales before 2000	E_3	1 D: Time
Q_4: Amount of sales by cities from 2010 to 2012	E_1	2 D: Time, Customers
Q_5: Amount of monthly sales by departments from 2000 to 2005	E_2	2 D: Time, Customers
Q_6: Amount of annual sales by sector before 2000	E_3	2 D: Time, Products
Q_7: Amount of sales by cities, sectors and months in 2012	E_1	3 D: Time, Products, Customers
Q_8: Amount of annual sales by sectors and departments from 2000 to 2005	E_2	3 D: Time, Products, Customers
Q_9: Amount of monthly sales since 2000	E_1; E_2	1 D: Time
Q_{10}: Amount of annual sales per cities from 2002 to 2012	E_1; E_2	2 D: Time, Customers
Q_{11}: Amount of sales per year and range from 1990 to 2009	E_2; E_3	2 D: Time, Products
Q_{12}: Amount of sales by cities and sectors from 2002 to 2012	E_1; E_2	3 D: Time, Products, Customers
Q_{13}: Amount of annual sales	E_1; E_2; E_3	1 D: Time
Q_{14}: Amount of annual sales per ranges	E_1; E_2; E_3	2 D: Time, Products

Remark. It is impossible to define a query manipulating 3 states and 3 dimensions because the state denoted E3 is only composed of 2 dimensions.

Whatever the volumetry of database, the query execution time in a non reduced environment (the column with stripe and the gray column in the figure below) is more important than in a reduced environment (the white and black columns in the figure below). The lowest execution times are performed on the database called *Reduced star*: database content is reduced and the table number is limited. In each version, the average earnings are 89.43% for the size 10 X 10 to 90.26% in size 40 X 40. Whatever the database volumetry, the execution time gain is significant: about 90%. As mentioned in figure 7, the highest average execution time for unreduced MDW ranges from 3432.4 (*global table* database 10 x 10) to 55221 (*global table* database 40 x40) and this average execution time has increased by 1509%. The lowest average execution time is found within the Reduced Star databases: it ranges from 115.6 to 1720, and it has increased by 1388%, lower than 120% compared to the unreduced DW. Thus, the more the datawarehouse volumetry increases, the more the execution time gain on a reduced DW is important.

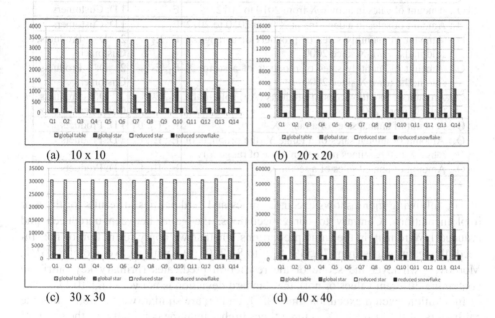

(a)   10 x 10

(b)   20 x 20

(c)   30 x 30

(d)   40 x 40

Fig. 6. Execution time in 4 versions

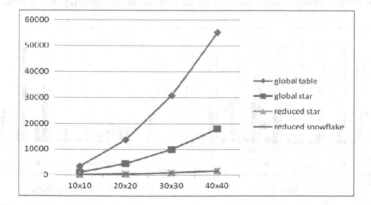

Fig. 7. Average execution time for the different versions of MDW

4.4   Queries with Restriction Predicates on Non-temporal Dimensions

This second experimental assessment aims to analyze the impact of restriction criteria in queries on the execution time. This experimental assessment is only done on the database based on the size 40 x 40. The queries of this second experimental assessment are defined in the following table.

Table 4. Queries with restriction predicates on non-temporal dimensions

Queries	States	Dimensions
Q_1 : Amount of sales of a customer X from 2010 to 2012	E_1	1 D: Customers
Q_2: Amount of sales in a town X from 2010 to 2012	E_1	1 D: Customers
Q_3: Amount of sales in a department X from 2010 to 2012	E_1	1 D: Customers
Q_4: Amount of monthly sales of products of a range X sold in the town X since 2000	E_1, E_2	2 D: Products, Customers
Q_5: Amount of monthly sales of products of a sector X sold in the town X since 2000	E_1, E_2	2 D: Products, Customers
Q_6: Amount of monthly sales of products (All) sold in the town X since 2000	E_1, E_2	2 D: Products, Customers
Q_7: Amount of annual sales of a range X	$E_1; E_2; E_3$	1 D: Products
Q_8: Amount of annual sales of a range Y (the products of range X is three times more than those of range Y)	$E_1; E_2; E_3$	1 D: Products
Q_9: Amount of annual sales of a sector X	$E_1; E_2; E_3$	1 D: Products

The following figure shows execution times and cardinalities of results for the four implementations. Contrary to our expectations, the gains between unreduced and reduced DW remain in the same proportions whether we apply restriction or not. Indeed, this gain ranges from 77.54% (Q_5) to 88.27% (Q_1) with an average of 84.61%. Moreover, whatever the scope of the restriction predicates (primary key, attribute containing different values or not), the standard deviation is not very high (0.1).

In addition, even if execution times of Q_7 and Q_8 are similar, we can notice that the cardinality of the result of Q_7 is three times higher than the cardinality of the result of Q8. This is because the DBMS must review all the tuples of the tables to get the query result. So, the execution time gain is independent of the cardinality of the result.

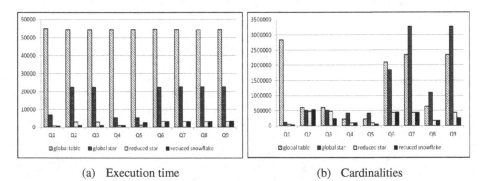

(a)   Execution time   (b)   Cardinalities

Fig. 8. Execution times and cardinalities for 9 queries containing restriction predicates

5   Conclusion

This paper resides within the field of MDW. Our objective is to specify aggregated schema over time in order to retain only the data useful for decision support according to the needs of users. Firstly, we define a conceptual model which allows us to specify MDW schemata composed of a set of states varying over time. Each state consists of a star schema and is defined with a mapping function, itself defined with reduction operators based on an extension of classical OLAP operators adapted to the reduction context. Secondly, we defined experimental assessments. Evaluating our solution consists in executing different queries in various environments: ROLAP schema without reduction, single fact table schema without reduction as well as star and snowflake schemata with reductions. We use multidimensional databases with different sizes; the fact table size ranges from 840,100 to 13,441,600 tuples. Whatever the database volumetry, the execution time gain between unreduced and reduced databases is significant: about 90%. Moreover, the more the datawarehouse volumetry increases, the more the execution time gain is important. These gains remain in the same proportions when we apply restriction predicates or not on the queries. Finally, the execution time gain is independent of the cardinality of the result.

In the future, we intend to extend our conceptual proposal in order to integrate other operators in the definition of the reduction function. We also intend to extend our experiments by combining our own work on reductions with that concerning indexes in a multidimensional context [6]. At last we wish to apply the principles of reduction to a reel data sample of analytic domain such as banking or insurance etc.

References

1. Boly, A., Hébrail, G., Goutier, S.: Forgetting Data Intelligently in Data Warehouses. In: Proceedings of the IEEE International Conference on Research, Innovation and Vision for the Future, pp. 220–227. IEEE Press, New York (2007)
2. Garcia-Molina, H., Labio, W., Yang, J.: Expiring data in a warehouse. In: Gupta, A., Shmueli, O., Widom, J. (eds.) 24th International Conference on Very Large Data Bases (VLDB), pp. 500–511. Morgan Kaufmann (1998)
3. Golfarelli, M., Maio, D., Rizzi, S.: Conceptual design of data warehouses from E/R schemes. In: Proceedings of the 31st Hawaii Int. Conf. on System Sciences (1998)
4. Golfarelli, M., Rizzi, S.: A survey on temporal data warehousing. International Journal of Data Warehouse and Mining 5(1), 1–17 (2009)
5. Iftikhar, N., Pedersen, T.B.: Using a Time Granularity Table for Gradual Granular Data Aggregation. In: Catania, B., Ivanović, M., Thalheim, B. (eds.) ADBIS 2010. LNCS, vol. 6295, pp. 219–233. Springer, Heidelberg (2010)
6. Iftikhar, N., Pedersen, T.B.: A rule-based tool for gradual granular data aggregation. In: Song, Cuzzocrea, Davis (eds.) Proceedings of DOLAP 2011, pp. 1–8. ACM (2011)
7. Kimball, R.: The Data Warehouse Toolkit: Practical Techniques for Building Dimensional Data Warehouses. John Wiley & Sons, USA (1996)
8. Last, M.,, Maimon, O.: Automated dimensionality reduction of data warehouses. In: Jeusfeld, M.A., Shu, H., Staudt, M., Vossen, G. (eds.) Proceedings of DMDW 2010. CEUR Workshop Proceedings, vol. 28, p. 7. CEUR-WS.org (2000)

9. Okun, O., Priisalu, H.: Unsupervised data reduction. Signal Processing 87(9), 2260–2267 (2007)
10. Ravat, F., Teste, O., Tournier, R., Zurfluh, G.: Algebraic and graphic languages for OLAP manipulations. International Journal of Data Warehouse and Mining (4), 17–46 (2008)
11. Skyt, J., Jensen, C.S., Pedersen, T.B.: Specification-based data reduction in dimensional data warehouses. Information System 33(1), 36–63 (2008)
12. Udo, I.J., Afolabi, B.: Hybrid Data Reduction Technique for Classification of Transaction Data. Journal of Computer Science and Engineering 6(2), 12–16 (2011)
13. Wang, X., Bettini, C., Brodsky, A., Jajodia, S.: Logical Design for Temporal Databases with Multiple Granularities. ACM Trans. Database Syst. 22(2), 115–170 (1997)

Towards an OLAP Environment
for Column-Oriented Data Warehouses

Khaled Dehdouh, Fadila Bentayeb, Omar Boussaid, and Nadia Kabachi

Laboratoire ERIC/ Universit de Lyon 2
5 avenue Pierre Mendes-France 69676 Bron Cedex, France
{khaled.dehdouh,fadila.bentayeb,
omar.boussaid,nadia.kabachi}@univ-lyon2.fr

Abstract. Column-oriented database systems offer decision-makers the most appropriate model for data warehouse storage. However, in the absence of on-line analytical operators, the only, very costly, way of constructing OLAP [1] cubes involves using the *UNION* operator for group by queries in order to obtain all the *Group By* required to compute the OLAP cube. To solve this problem, in this article we propose a new aggregation operator, called C-CUBE (*Columnar-CUBE*), which allows data cubes to be computed using column-oriented data warehouses. We implemented the C-CUBE operator within the column-oriented DBMS, *MonetDB* and conducted experiments on the benchmark SSBM (Star Schema Benchmark). Thus we have shown that C-CUBE has OLAP cubes computation times reduced by up to 60% compared with the SQL Server CUBE operator in a 1TB warehouse.

Keywords: OLAP cube, Data warehouses, Columnar databases.

1 Introduction

A column-oriented database system stores table data column by column. This storage technique allows values belonging to a given column to be shared in the same disk space which improves the column access time enormously. The column-store is characterised by a highly effective join operation called an invisible join [1] and seems to be a very useful solution for business intelligence (BI) and, more particular, data warehouses [2]. These are used for on-line analysis which can aid decision-making [3]. Using OLAP operators, the user can extract data cubes corresponding to analytical contexts. The column-store approach is obviously suitable for the structure of the multidimensional data and aggregation performance [4]. However, the basic features of column-oriented database systems are limited. Indeed, column-oriented DBMS do not have OLAP cube computation operators. Nevertheless, the OLAP cube can be computed using the naive method, using a combination of group-by queries. For D dimensions,

---

[1] On-Line Analytical Processing.

L. Bellatreche and M.K. Mohania (Eds.): DaWaK 2014, LNCS 8646, pp. 221–232, 2014.

the execution of 2^D sub-queries to perform the various aggregations, which considerably increases the number of times the database is accessed. Consequently, the naive method reduces DBMS performance, particularly when scaling-up.

In this paper we propose a C-CUBE aggregation operator for column-oriented DBMSs. This operator allows OLAP cubes to be computed using column-oriented data warehouses. C-CUBE extends the invisible join, used in column-oriented database systems, to take into account all dimension combinations. Moreover, unlike the naive method which uses the database to perform the various aggregations, C-CUBE uses a view (result of an extraction query) based on attributes (dimensions and measures) needed to compute the OLAP cube. This strategy means there is no need to return to the warehouse data in order to perform aggregations. We have confirmed this technique on a benchmark SSBM [5], stored in a column-oriented DBMS, MonetDB, and we then conducted experiments which showed that C-CUBE significantly optimises OLAP cube computation times compared with the naive method, the oracle and SQL Server CUBE operator.

The rest of this paper is organised as follows. Section 2 gives the state of the art on column-oriented database systems. Section 3 presents the various concepts connected with our work. Section 4 contains a detailed description of the OLAP cube computation process and an introduction of the C-CUBE operator. In section 5 we implemented our C-CUBE operator and conducted experiments. Finally, we conclude this paper and present some perspectives on our work in section 6.

2   Related Work

Column stores go back to the 1970s with the use of transposed files and the technique of vertical partitioning [6]. It is in the 1980s that the advantage of decomposed storage models was addressed in the literature. These models are 'DSM' and 'NSM', predecessors of column-oriented storage and row-oriented storage respectively [7]. It is only in the 2000s that column stores emerged as an alternative to row-oriented storage adopted by relational DBMSs in particular for storing multidimensional databases [1]. Work carried out in this area has allowed column-oriented database systems to benefit from undeniably improved performance in terms of the join operation, known as the invisible join. Let us recall that the invisible join is a join operation which uses value positions and hash tables to extract data that satisfy the predicates of the query [1]. Moreover, using the later materialisation technique [8] and the data compression technique which operate efficiently on values of the same type, the space required to store data is significantly reduced [9]. Of the column-oriented DBMSs we can cite C-Store[2], MonetDB[3] and Vertica[4]. However, although everyone agrees that the column-store is well suited to multidimensional data and as such to data

---

[2] http://db.csail.mit.edu/projects/cstore/
[3] www.monetdb.org
[4] http://www.vertica.org/

cube computation, column-oriented DBMSs unfortunately do not have OLAP operators.

3   OLAP Cube Computation in Column-Oriented Data Warehouses

In this section we define the OLAP cube and the naive method to compute it.

3.1   OLAP Cube

A data cube is a collection of aggregated and consolidated data used to summarise information and explain the pertinence of an observation. It allows the fact under observation to be represented from several perspectives (dimensions) known as multidimensional [10]. The data cube is explored using many operations which allow it to be handled [11]. Cube computation produces aggregations that are beyond the limits of the *Group by*. For example, in the case of calculation of the sum, it computes in a multidimensional way and returns sub-totals and totals for all possible combinations. This involves performance of all aggregations according to all levels of hierarchies of all dimensions. For a cube with three dimensions A, B and C, the performed aggregations relate to the following combinations: (A, B, C), (A, B, ALL), (A, ALL, C), (ALL, B, C), (A, ALL, ALL), (ALL, B, ALL), (ALL, ALL, C), (ALL, ALL, ALL).

3.2   The Naive Method for OLAP Cube Computation

The naive method for computation of the data cube is proposed by Gray [10] and involves using the UNION operator to group a collection of aggregation queries executed separately using *Group By*. However, this method has the drawback of requiring multiple access to the database in order to perform the various aggregations. Indeed, for a number of dimensions, D, there will be 2^D group-by queries to execute. This further reduces the performance of the database management system. The above clearly shows that the naive method for OLAP cube computation is not suitable for large databases as it is an approach characterised by exponential complexity in terms of the number of dimensions. To solve this problem, given that column-oriented DBMSs do not have OLAP operators, we propose a new aggregation operator, C-CUBE to compute OLAP cubes using column-oriented data warehouses.

4   C-CUBE: CUBE Operator for Column-Oriented Databases

In this C-CUBE section we present our OLAP operator for the cube computation in column-oriented data warehouses. The technique used by the C-CUBE operator we propose is to extract from the warehouse the data which satisfy all the predicates of the query. These data are grouped according to columns which represent dimensions. The result is therefore a relation R made up of columns representing dimensions and column(s) representing the measure(s) to be aggregated. The R relation is an intermediate result based on which all aggregations of the OLAP cube will be performed. At this stage, the intermediate result already allows the total aggregation and the aggregation according to all columns representing dimensions to be produced.

In order to obtain other aggregations that make up the cube, each dimension of relation R is hashed with the values it contains to obtain a list of positions. The values of these lists are binary. They may correspond to 1 or 0, with 1 indicating if the hash value exists at this position and 0 if not. The intersection of lists of the positions of the various columns (dimensions) with a *logical AND function* provides a set of lists of positions which represent the positions of dimension values to be combined with the measure column to be aggregated including at different levels of granularity. The grouping of all of these results is the cube. In order to set out the phases of execution for this technique using an example, in the following section we present a data warehouse benchmark, *SSBM*.

4.1 Star Schema Benchmark

In this paper, we use the Start Schema Benchmark (SSBM) [5] to evaluate the performances of the C-CUBE operator. *SSBM* is a data warehousing benchmark derived from TPC-H[5] model. Unlike TPC-H, SSBM uses the star schema to evaluate the performances of the data warehouse and is easier to implement.

SSBM is a data warehouse which manages line orders according to dimensions, PART, SUPPLIER, CUSTOMER and DATE. It consists of a single fact table called LINEORDER made up of seventeen column to give informations about order, with a composite primary key consisting of the Orderkey and Linenumber attributes, and foreign keys references to the from the dimension tables. This model comes with a simple *Group by* queries. It should be noted that our objective is not to evaluate SSBM as such but to use it to evaluate the performance of our C-CUBE operator in the column-oriented DBMS. As such we will refrain from using the queries that comes with SSBM and instead will create new queries to compute the OLAP cubes.

Queries: In order to evaluate the performances of the C-CUBE operators, which we propose so as to compute the OLAP cube using a column-oriented warehouse, we use four OLAP cube computation queries. These queries increase the number of dimensions involved in cube computations.

Query 1: This is a query allowing a two dimensional OLAP cube to be computed with restrictions on the columns *Nation* of the CUSTOMER dimension, *Nation* of the SUPPLIER dimension and *Year* of the DATE dimension. This query computes the revenue (*sum ((Revenue)*) at different levels of granularity according

---

[5] http:www.tpc.org/tpch

to the attributes, *City* of the CUSTOMER dimension and *City* of the SUPPLIER dimension.

Query 2: This is a query which allows us to compute the three dimensional OLAP cube with the same restrictions as for the first query. It computes the revenue (*sum (Revenue)*) at different levels of granularity according to the attributes *City* of the CUSTOMER dimension, *City* of the SUPPLIER dimension and *Year* of the DATE dimension.

Query 3: It is a query allowing a four dimensional OLAP cube to be computed with restrictions on the columns *Nation* of the CUSTOMER dimension, *Nation* of the SUPPLIER dimension, *Brand1* of the PART dimension and *Year* of the DATE dimension. It computes the revenue (*sum (Revenue)*) at different levels of granularity according to the attributes *City* of the CUSTOMER dimension, *City* of the SUPPLIER dimension, and *Year* of the DATE dimension AND *Brand1* of the PART dimension.

Query 4: This is a query which allows us to compute the five dimensional OLAP cube with the same restrictions as for the previous query (query 3). It computes the revenue (*sum (Revenue)*) at different levels of granularity according to the attributes *City* of the CUSTOMER dimension, *City* of the SUPPLIER dimension, and *Year* and *Month* of the DATE dimension AND *Brand1* of the PART dimension.

4.2   C-CUBE Operator Execution Phases

The C-CUBE computes the OLAP cube in four phases. In order to provide more detail on these phases, we illustrate our explanation with an example of *query 2*, cited in section 4.1. The example computes the sales revenue from products delivered by french suppliers and for which orders were placed by german customers in the years 1996 and 1997.

First Phase: It involves using the data warehouse to extract data which satisfy all the predicates (filters). This phase allows an intermediate result to be obtained to build all parts of the cube. To carry out this phase, query predicates are applied separately to the respective dimensions to obtain lists of the primary keys of dimensions that satisfy the predicates. Given that the keys are foreign keys at the level of the fact table, they are used to extract related position lists at the level of the fact table. Associating these position lists with a *logical AND function* produces a single list of P positions. This represents the position of values in the fact table which satisfy all query predicates at once. The extraction of values from the fact table according to P produces intermediate result R in the form of a view which can be used to compute the OLAP cube.

According to our example, this means selecting the primary keys for french towns (*Nation = France*) of the SUPPLIER dimension, those of german towns of the CUSTOMER dimension (*Nation = Germany*) and those of years 1996 and 1997 of the DATE dimension (*Year* in (1996, 1997)). This operation produces three key lists.

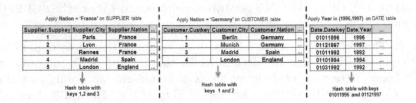

Fig. 1. Extraction primary keys of dimensions that satisfy the predicates

In order to produce position lists corresponding to these keys in the fact table, the dimensions *Suppkey*, *Custkey* and *Orderdate* of the LINEORDER fact table are hashed in the three respective key lists. This operation produces three position lists. Once associated, these produce a position list P which represents the positions of tuples which satisfy the query predicates. This operation produces an intermediate result R. At this stage of execution, the aggregations of the revenue sums of (*ALL, ALL, ALL*) and (*Suppkey, Custkey, Orderdate*) are computed as described in figure 2:

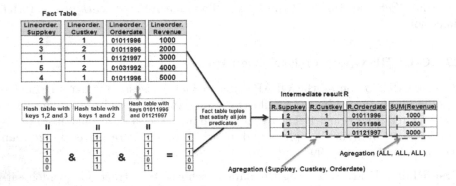

Fig. 2. Intermediate result R produces with values in the fact table which satisfy all query predicates at once

Second Phase: In this phase, each intermediate result R column representing a dimension is hashed with the values which make it up to produce lists of the positions of these values. Indeed, the *Suppkey* dimension of the intermediate result is hashed at values (2, 3 and 1) producing three position lists respectively $P_{Suppkey(2)}$, $P_{Suppkey(3)}$ and $P_{Suppkey(1)}$. *Custkey* is hashed at values (1 and 2) producing $P_{Custkey(1)}$ and $P_{Custkey(2)}$ and finally, *Orderdate* is hashed at (01011996 and 01121997) producing $P_{Orderdate(1996)}$ and $P_{Orderdate(1997)}$. At this stage of execution, these position lists allow values of the column representing the measurement to be aggregated for each dimension separately. Indeed, they produce the aggregations of the following revenue sums: (Suppkey, ALL, ALL), (ALL, Custkey, ALL) and (ALL, ALL, Orderdate).

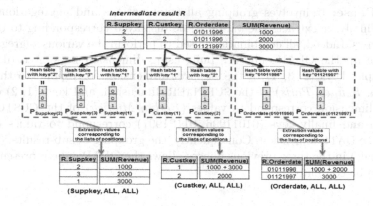

Fig. 3. Aggregations for each dimension separately

Third Phase: At this phase, the list of dimension positions at R are associated through the *logical AND operator*. This operation identifies the values of dimensions to be combined and the values of the measurement to be aggregated which correspond to the different combinations. In this case, in order to identify the various possible combinations between the *Suppkey* and *Custkey* dimensions, the list of positions $P_{Suppkey(2)}$, $P_{Suppkey(3)}$ and $P_{Suppkey(1)}$ are associated with $P_{Custkey(1)}$ and $P_{Custkey(2)}$ using the *logical AND operator*. The results are the three lists $P_{Suppkey(2)-Custkey(1)}$, $P_{Suppkey(3)-Custkey(2)}$ and $P_{Suppkey(1)-Custkey(1)}$. These lists allow the values of the *Suppkey* and *Custkey* dimensions and the measurement of Revenue to be aggregated. These operations produce the aggregations of the following revenue sums: (Suppkey, Custkey, ALL), (Suppkey, ALL, Orderdate) and (ALL, Custkey, Orderdate).

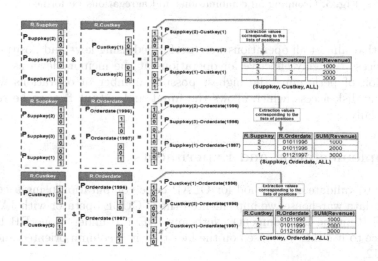

Fig. 4. Aggregations for dimensions combinations

Fourth Phase: It involves grouping all combinations and aggregations performed. Finally, it extracts the values to be displayed corresponding to the dimension keys. Indeed, all combinations used to perform the various aggregations that make up the OLAP cube are built using dimension keys. According to our example, keys (2, 3, 1) of the *Suppkey* dimension correspond to the values (*Lyon, Rennes, Paris*) of the SUPPLIER dimension, and keys (1, 2) of the *Custkey* dimension correspond to values (*Berlin, Munich*) of the CUSTOMER dimension and finally keys (01011996, 01121997) correspond to values (1996, 1997) of the DATE dimension. Consequently, the grouping of sub-results, (totals or sub-totals) in the three previous phases allows the OLAP cube represented in figure 5 to be computed.

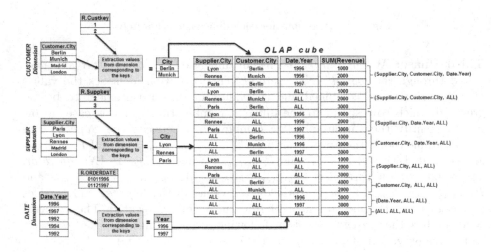

Fig. 5. Grouping all combinations and agregations performed

Note that almost all operations were performed using keys and position lists. This process is highly effective for operations at the memory level as it offers the option of performing the highest possible number of operations without requesting disk access, which allows the input and output flow to be reduced substantially.

5   Implementation and Experiments

In order to validate our method for OLAP cube computation using a column-oriented data warehouse, we implemented the C-CUBE operator with JAVA 1.7 (java SE Runtime Environment), Netbeans IDE 7.4, under MonetDB DBMS, the choice to use this was based on the fact that it is column-oriented and open source [12].

5.1   Environments Setup

The test environment includes an intel-Core TMi3-3220 CPU@3.30 GHZ machine with 4 GB of RAM. This machine uses the Microsoft Windows 7 64bits operating system. In order to evaluate our C-CUBE operator proposal, we setup two others environments namely *ORACLE 11g* [14] and *MS SQL Server 2008 R2* [15] to compare our proposal with the CUBE operator provided by these environments. Moreover, to conduct our experiments, we implemented the warehouse benchmark SSBM, described in section 4.1, under the three environments. In terms of data, we used generate database (dbgen) [16] to populate SSBM data warehouse with data samples until 1 TB.

5.2   Experiments

In this part of the paper, we evaluate the performances of the C-CUBE operator in terms of OLAP cube computation times. To do this, We conducted two experiments. The first assessed the OLAP cube computation time against a gradual increase in the number of dimensions. The second assessed OLAP cube computation times whilst varying the size of the warehouse.

(a) Cube Computation with variations in the number of dimensions: The objective of this experiment is to evaluate C-CUBE operator when faced with variations in the number of dimensions. We used a data sample of the 60 million tuples. Then, we compared the performance of C-CUBE with the naive method of the column-oriented approach and the CUBE operator of the *ORACLE* and *SQL Server* DBMS. To do this, we executed four queries presented in section 4.1. These queries compute the OLAP cubes with a gradually increasing number of dimensions. The results we obtained are presented in figure 6.

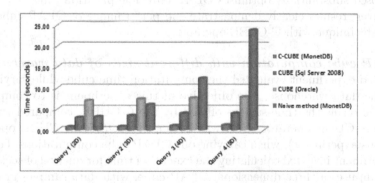

Fig. 6. OLAP cube computations faced with variations in dimensions number

We observed that the OLAP cube computation time when using the naive method varied by an average of 16.2 to 75 seconds. These computation times increase according to the number of dimensions. Indeed, the time required to compute a two dimensional cube is 16.2 seconds, and the time required to compute a three dimensional cube is 30.7 seconds, practically double. Beyond these three dimensions, this method records poor results compared with other methods. This result is due to the fact that for two dimensions, the system executes (2^2) aggregation queries, often with repeated joins and grouped results. However, with three dimensions, it executes eight (2^3) aggregation queries, which is practically double. Thus, an increased number of dimensions involves significant disk access in order to extract warehouse data which leads to the memory having to also cope with intermediate results. Consequently, the memory becomes saturated and the disk is needed to manage intermediate results. As such, it is clear that this process results in a significant increase in the cost of input/output which in turn results in increased execution time and decreased system performance.

On the other hand, we observe a slight variation in execution times for requests producing the OLAP cube, with C-CUBE and CUBE operators whatever the number of dimensions involved. However, C-CUBE performs better than CUBE with the CUBE operator recording times of between 7 and 7, 9 seconds from *ORACLE* and 3 and 4 from *SQL Server*, whereas the C-CUBE operator records times of between 0, 7 and 1, 2 seconds.

The advantage of C-CUBE lies in the use of position lists when computing OLAP cubes. These lists use little in the way of memory and are perfectly suitable for memory-level operations without requesting repeated disk access. Indeed, the technical results of increased number of dimensions when computing the data cube is the creation and handling of bit vectors which represent lists of value positions. This process does not have a heavy impact on memory. Moreover, combinations are produced with lists (bit vectors made up of 1 or 0) and not values and these are only extracted after the position list related to a combination has been built. In terms of the results, we observed that the C-CUBE operator we proposed substantially optimises OLAP cube computation times.

Moreover, results clearly demonstrate the performance of OLAP cube computation technique with C-CUBE operator.

(b)OLAP cube computation with different sizes of data warehouses:

The second experiment evaluated the computation time cube while varying the size of the data warehouse. The objective of this experiment is to compute in percentage terms the time-savings offered by the C-CUBE, compared with the *SQL Server* CUBE operator (wich was better than *ORACLE* CUBE operator, in previous experiment), when carrying out OLAP cube computations. For this, The experiment involved calculating the execution time for query 2 of section 4.1 which computes a three dimensional OLAP cube, with data samples gradually increasing from 100 GB to 1 TB. The results are presented in figure 7.

The figure (a) shows that the curve representing the OLAP cube computation time with the C-CUBE operator, produces better results regardless of the size of the data warehouse. The figure (b) represents the trend in terms of the percentage of OLAP computation times between the C-CUBE and CUBE operators with different sizes of data warehouses. Indeed, the difference between the

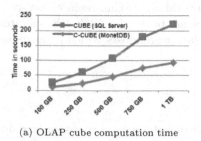

(a) OLAP cube computation time

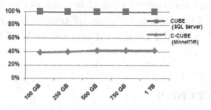

(b) Trend in terms of the percentage of OLAP computation times

Fig. 7. OLAP cube computation with different sizes of data warehouses

OLAP cube computation times and those of the two operators varies in general by 60% for a warehouse size of between 100 GB and 1 TB. This difference in favour of the C-CUBE operator clearly demonstrates its advantage.

The C-CUBE operator's performance is due to the fact that the system uses a view (result of an extraction query) based on attributes (dimensions and measurements) needed to compute the OLAP cube. This strategy allows it to perform different combinations and aggregations to be performed at the level of the memory in order to avoid returning to the warehouse data. Indeed, it is made possible by the use of the value position instead of the value itself. The positions of values represented by lists of vectors do not take up a lot of memory space. This offers the option of performing several operations at the memory level and consequently considerably reduces the input and output flow.

Ultimately, the experiments we conducted clearly demonstrated that the C-CUBE operator we proposed for computing the OLAP cube in column-oriented database systems is efficient when dimension numbers are increased and maintains its performance with different sizes of data warehouses.

6   Conclusion

The major contribution of this work is to extend the use of column-oriented database systems to the data warehouses. In this context, in this paper we have proposed an OLAP cube computation operator, called C-CUBE. The benefit of this operator is that it is based on the principle of the invisible join which is used here to perform the aggregations of the OLAP cube very effectively. Indeed, like the technique used in the invisible join, C-CUBE handles lists of value positions and hash tables containing keys of the different dimensions specified in the query. This approach substantially reduces input and output flows. The implementation of this column-oriented DBMS, MonetDB and the experiments we conducted on the relational data warehouses, SSBM clearly showed the performance of our operator compared with that of the ORACLE and SQL Server CUBE operator.

More broadly, the C-CUBE operator can be applied across the column-oriented DBMS and may even be rolled out to non-relational column-oriented DBMSs.

It is in this context that this work opens up several interesting research perspectives. One potential area of research involves adapting the C-CUBE operator to OLAP cube computations using NoSQL[6] data warehouses which manage big data [13].

References

1. Abadi, D.J., Madden, S.R., Hachem, N.: Column-stores vs. row-stores: how different are they really? In: International Conference on Management of Data, pp. 967–980 (2008)
2. Matei, G.: Column-Oriented Databases, an Alternative for Analytical Environment. Database Systems Journal, 3–16 (2010)
3. Inmon, W.H.: Building the Data Warehouse. Information Sciences, Inc., Wellesley (1992)
4. Stonebraker, V., Abadi, D.J., Batkin, A., Chen, X., Cherniack, M., Ferreira, M., Lau, E., Lin, V., Madden, S., O'Neil, E., O'Neil, P., Rasin, A., Tran, N., Zdonik, S.: C-store: a column-oriented DBMS. In: Proceedings of the 31st International Conference on Very Large Data Bases, pp. 553–564 (2005)
5. O'Neil, P., O'Neil, B., Chen, X.: The Star Schema Benchmark, SSB (2009), http://www.cs.umb.edu/~poneil/StarSchemaB.PDF
6. Batory, D.S.: On searching transposed files, Association for Computing Machinery, pp. 531–544 (1979)
7. Copeland, G.P., Khoshafian, S.N.: A decomposition storage model. Special Interest Group on Management of Data Record, 268–279 (1985)
8. Abadi, D.J., Myers, D.S., Dewitt, D.J., Madden, S.: Materialization strategies in a column oriented. In: International Conference on Data Engineering, pp. 466–475 (2007)
9. Abadi, D.J., Madden, S., Ferreira, M.: Integrating compression and execution in column-oriented database systems. In: Special Interest Group on Management of Data Conference, pp. 671–682 (2006)
10. Gray, J., Chaudhuri, S., Bosworth, A., Layman, A., Reichart, D., Venkatrao, M., Pellow, F., Pirahesh, H.: Data cube: A relational aggregation operator generalizing group-by, cross-tab, and sub total. Journal of Data Mining and Knowledge Discovery 29–53 (1997)
11. Rafanelli, M.: Operators for Multidimensional Aggregate Data, Multidimensional Databases: Problems and Solutions, pp. 116–165. IGI Publishing Group (2003)
12. Idreos, V., Groffen, F., Nes, V., Manegold, S., Mullender, K., Sjoerd, K., Kersten, V.: MonetDB: Two Decades of Research in Column-oriented Database Architectures. Journal IEEE Data Engineering Bull., 40–45 (2012)
13. Jerzy, D.: Business Intelligence and NoSQL Databases. Information Systems in Management, 25–37 (2012)
14. Oracle 11g, http://www.oracle.com
15. SQL Server 2008 R2, http://www.microsoft.com
16. DBGEN: Data generator, https://github.com/electrum/ssb-dbgen

---

[6] Not Only Sql.

Interval OLAP: Analyzing Interval Data

Christian Koncilia[1], Tadeusz Morzy[2,A], Robert Wrembel[2,*], and Johann Eder[1]

[1] Klagenfurt University, Institute of Informatics Systems, Klagenfurt, Austria
[2] Poznan University of Technology, Institute of Computing Science, Poznań, Poland
{koncilia,eder}@isys.uni-klu.ac.at,
{Tadeusz.Morzy,Robert.Wrembel}@cs.put.poznan.pl

Abstract. The ability to analyze data organized as sequences of events or intervals became important by nowadays applications since such data became ubiquitous. In this paper we propose a formal model and briefly discuss a prototypical implementation for processing interval data in an OLAP style. The fundamental constructs of the formal model include: events, intervals, sequences of intervals, dimensions, dimension hierarchies, a dimension members, and an iCube. The model supports: (1) defining multiple sets of intervals over sequential data, (2) defining measures computed from both, events and intervals, and (3) analyzing the measures in the context set up by dimensions.

1 Introduction

It is observed that current applications in use generate huge sets of data. Some of the data have the character of events that last an instant, whereas some of them last for a given time period - an interval. Events typically have a strict order, thus possess a sequential nature. Sequential data can be categorized either as *time point-based* or *interval-based* [13].

Some examples of systems that generate this kind of data include: workflow systems, Web logs, RFID, public transport, and sensor networks. In workflow systems objects arrive to ordered tasks at certain points in time and they are processed there during a certain period of time. By analyzing workflow log data one is able to discover bottlenecks and idle time. In Web log analysis, especially for e-commerce, one may be interested in knowing the navigation path leading to a product purchase. RFID technology is becoming widely used in supply chain management (e.g., just-in-time delivery). Here, by analyzing sequences of events generated by the RFID devices one is able to optimize product transportation routes. In advanced public transportation infrastructures, cf. [12] passenger tracking records are automatically generated by various devices. These records can be used for analyzing the most frequently used routes and, thus, for discovering route bottlenecks, station bottlenecks, and rush hours in various districts. In intelligent installations (e.g., ambient living, jet engines, refineries), numerous

---

* Work supported from the Polish National Science Center (NCN), grant No. 2011/01/B/ST6/05169.

L. Bellatreche and M.K. Mohania (Eds.): DaWaK 2014, LNCS 8646, pp. 233–244, 2014.

sensors supply their data. Based on the chronologically analyzed data, one can discover patterns of behavior or predict device breaks.

There is a substantial demand for models and tools for analyzing sequential data. Most of the existing OLAP techniques, although very advanced ones, allow to analyze mostly set oriented data without exploiting the existing order among the data. For this reason, it was necessary to create new models and techniques that would be able to store and analyze sequential data efficiently.

Paper Contribution In this paper, we contribute a formal and implementation model, called *I-OLAP*, for an OLAP system that enables the user to define and analyze intervals stemming from sequential data. In particular, we present an extension of the S-OLAP concept [3] to achieve the following features: (1) enable the user to easily define multiple *sets of intervals* over sequential data, (2) define *measures* computed from both, events and intervals, (3) analyze these measures easily along multiple *dimensions*.

Analyzing sequences and intervals could also be done with standard SQL queries. However, we will show that this leads to huge query statements which therefore are nearly unreadable and most notably not maintainable. Therefore, we prototypically implemented a query language which enables the user to analyze sequential data and interval based data.

Paper Organization: This paper is organized as follows. In section 2 we will discuss related work. Section 3 presents our running example and define the set of example queries. Section 4 presents the I-OLAP data model. Section 5 shows how to query interval data, based on our data model. In section 6 we will briefly discuss the implementation of our approach. Section 7 summarizes the paper, outlines open issues and research directions for the future.

2   Related Work

The model which will be presented in this paper is - to the best of our knowledge - the first OLAP model focusing on how to analyze interval data. However, our model is building on different approaches which focus on how to analyze sequential data. These approaches will be briefly discussed in this section.

[3] propose a formal model for time point-based sequential data with the definitions of a fact, measure, dimension, and a dimension hierarchy. Thus, the model allows to analyze sequential data in an OLAP style. However, neither a query language nor a prototype system was built on the model.

In the *S-OLAP* approach [12] propose the set of operators for a query language for the purpose of analyzing patterns, whereas [4,5] concentrate on an algorithm for supporting ranking pattern-based aggregate queries and on a graphical user interface. The drawback of this approach is that it is based on relational data model and storage for sequential data.

Stream Cube [9] and *E-Cube* [11,10] implement OLAP on data streams. Their main focus is on providing tools for OLAP analysis within a given time window of constantly arriving streams of data.

[6,7] address interval-based sequential data, generated by RFID devices. In [6] the authors focus on reducing the size of such data. They propose techniques

for constructing RFID cuboids and computing higher level cuboids from lower level ones. Based on this foundation, [7] propose a language for analyzing paths with aggregate measures, generated by RFID devices.

[17,16] focus on mining sequential patterns on interval-based data applying a class of Apriori and PrefixSpan algorithms.

From the commercial systems only *Oracle* [15] and *Teradata Aster* [2] support SQL-like analysis of sequential data in their OLAP engines but they focus on pattern recognition in time-point-bases sequential data.

To the best of our knowledge, the aforementioned contributions do not support the analysis of interval data. With this respect, there is an evident need for developing a model and a query language capable of discovering and analyzing such data in an OLAP style.

3   Running Example

As a running example, we will use sample data acquired by sensors installed in an intelligent building. Let us assume that: (1) the sensors report the status of lights and temperatures in some rooms, and (2) our data warehouse stores events that report changes, i.e. if a light sensor reports a sequence of events $< \{room1, t1, on\}, \{room1, t2, on\}, \{room1, t3, on\}, \{room1, t4, off\} >$, the second and third event will not be stored. Table 1 depicts the data received from the light and heating sensors in two rooms (room_id 100 and 101). Obviously, the light sensors return boolean values (on, off) whereas the heating sensors report the temperature as float values once per hour. Heating sensor H_1 reports a failure at 2013.03.20 14:08:13. This problem has been fixed 3 hours later.

Table 1. Example light and temperature sensor data

room_id	sensor_id	time	value
100	L_1	2013.03.20 10:08:12	on
100	H_1	2013.03.20 10:08:13	19.2
100	H_1	2013.03.20 11:08:13	20.0
100	H_1	2013.03.20 12:08:13	21.2
100	L_1	2013.03.20 12:24:12	off
100	H_1	2013.03.20 13:08:13	18.0
100	L_1	2013.03.20 13:09:12	on
100	H_1	2013.03.20 14:08:13	failure
100	H_1	2013.03.20 17:08:13	21.2
100	L_1	2013.03.20 17:38:12	off
101	H_2	2013.03.20 9:18:13	19.0
101	L_2	2013.03.20 9:19:12	on
101	H_2	2013.03.20 10:18:13	21.5
101	H_2	2013.03.20 11:08:13	21.6
101	L_2	2013.03.20 19:40:12	off

Typical examples of OLAP queries on this kind of interval data could include: (1) find the floor (sum of all rooms on a floor) where light was on for the longest time per day, (2) find all rooms where light was off and heating was on, i.e. the temperature increased, (3) report the average heating costs per room and day, (4) report the five rooms with the longest/shortest period of time light was on, (5) report the largest number of state changes per day per a light sensor.

4 I-OLAP Data Model

In this section we propose a metamodel and a formal model of interval OLAP (I-OLAP). The elements of the *I-OLAP* metamodel are shown in Figure 1. It consists of *events and its attributes, dimensions, hierarchies, and dimension members, intervals, sequences of intervals,* and *iCubes.*

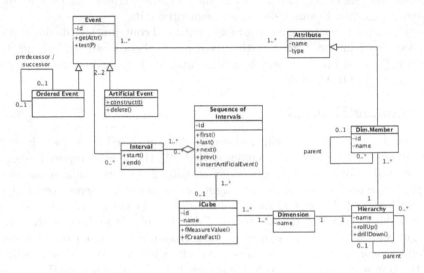

Fig. 1. The I-OLAP metamodel

4.1 Events and Attributes

Event $e_j = (a_{1j}, a_{2j}, \ldots, a_{nj})$, where: a_{ij} is the value of attribute A_i of the j-th elementary event and $a_{ij} \in Dom(A_i)$. $\mathbb{A} = \{A_1, A_2, \ldots, A_n\}$ is the set of attributes of the elementary event, and $Dom(A_i)$ is the domain of the ith attribute (including atomic values plus null). The set of all events $\mathbb{E} = \{e_1, e_2, \ldots, e_m\}$.

Intuitively, we can say that an event is simply a tuple in the original transactional dataset. In our running example the first record could be mapped to event e_1 with assigned values *room_id = 100, sensor_id = L_1, time = 2013.03.20 10:08:12,* and *value = on.* An example set of events is given in Table 2.

In the model we distinguish two specializations of the event, namely: artificial, and consecutive. The artificial event exists temporarily to answer a given query, cf. Section 5. Consecutive events are used to represent intervals, cf. Section 4.2.

4.2 Intervals

Intuitively, intervals correspond to the 'gap' between any two consecutive events. Figures 2(a) and 2(b) depict intervals for 'Heating' and 'Light'. The intervals are defined over attribute *value,* i.e. the current temperature and light status.

Table 2. Events in our running example

Event ID	Event data
e_1	100, L1, 2013.03.20 10:08:12, on
e_2	100, H1, 2013.03.20 10:08:13, 19.2
e_3	100, H1, 2013.03.20 11:08:13, 20.0
e_4	100, H1, 2013.03.20 12:08:13, 21.2
e_5	100, L1, 2013.03.20 12:24:12, off
e_6	100, H1, 2013.03.20 13:08:13, 18.0
e_7	100, L1, 2013.03.20 13:09:12, on
e_8	100, H1, 2013.03.20 14:08:13, failure
e_9	100, H1, 2013.03.20 17:08:13, 21.2
e_10	100, L1, 2013.03.20 17:38:12, off

Two consecutive events form an interval. Events e_1 and e_2 are consecutive if: (1) both of them belong to intervals that belong to the same sequence of intervals and (2) there exists no other event between both events, i.e. $\nexists e_i \in S : e_1.t \leq e_i.t \leq e_2.t$.

Interval $I = <e_n, e_m>$, where $e_n \in \mathbb{E} \wedge e_m \in \mathbb{E}$, e_n and e_m are consecutive events. The set of all defined intervals is denoted as $\mathbb{I} = \{I_1, I_2, ..., I_n\}$.

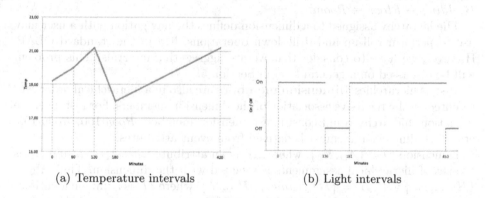

(a) Temperature intervals   (b) Light intervals

Fig. 2. Temperature and Light Intervals

Two basic methods are defined on intervals, namely: (1) `start()` – returns the start event of the interval, (2) `end()` – returns the end event of the interval.

4.3 Sequences of Intervals

Multiple intervals form the sequence of intervals. The order within the sequence is defined by an ordering attribute(s) assigned to all events of all intervals. Hence, within the sequence of intervals all events must have the same ordering attribute(s). Furthermore, an event may be the part of several, different intervals as long as these intervals do not belong to the same sequence of intervals. For example, the user might create three separate sequences of intervals, i.e., about heating, light, and both heating and light.

Sequence of intervals $S = < I_1, I_2, ..., I_n >$, where $I_i \in \mathbb{I}$. The set of all sequences of intervals is denoted as $\mathbb{S} = \{S_1, S_2, ..., S_n\}$.

While defining the **methods on intervals** we were inspired by [3]. The methods include: (1) `first()`, `last()`, `next()`, `prev()` – they allow to iterate over the intervals within a sequence of intervals, (2) `insertArtificialEvent()` – it creates a new, artificial event (cf. Section 5).

4.4   Dimensions, Hierarchies, and Dimension Members

A **dimension** is derived from one attribute assigned to events. Each dimension may have a concept hierarchy associated with it. In order to support galaxy schemas, our model supports 'shared dimensions', i.e. dimensions that may be assigned to multiple cubes. Our running example could for instance have dimensions 'Location', 'Time', and 'Event Type'.

Every dimension consists of one **hierarchy** that represents the root hierarchical element of a cube. This element may consist of multiple sub-elements that, in turn, may consist of multiple sub-elements, thus building a hierarchy. In our running example, the 'Location' dimension could include hierarchy $Building \rightarrow Floor \rightarrow Room$.

The hierarchy assigned to a dimension defines the navigation path a user may use to perform roll-up and drill-down operations, like in the standard OLAP. However, we have to consider that we are aggregating intervals. This problem will be discussed on a general level in Section 5.

Just as hierarchies, **dimension members** are also in a hierarchical order represented by the recursive association of the dimension members. For instance, the 'Location' hierarchy could consist of dimension members $Room100$, $Room102$, etc. Each dimension member is derived from event attributes.

Dimension $D = \{A_D, \mathbb{H}_D\}$, where A_D is an attribute with $A_D \in \mathbb{A}$ and $\mathbb{H}_D$ is the set of hierarchical assignments associated with the dimension. Thus, $\mathbb{H}_D = \{H_1, ..., H_n\}$ with $H_i = \{ID, Name, H.P_{ID}, \mathbb{M}\}$, where ID is a unique identifier, $Name$ is the name of the hierarchy, and $H.P_{ID}$ is the identifier of the parent hierarchy or null if there is no parent.

$\mathbb{M}$ is the set of dimension members assigned to this hierarchy: $\mathbb{M} = \{M_1, ..., M_n\}$, where $M_j = \{ID, Name, M.P_{ID}\}$, where ID is a unique identifier, $Name$ is the name of the dimension member, and $M.P_{ID}$ is the identifier of the parent dimension member or null if there is no parent.

4.5   iCube

iCube is a data cube enabling users to analyze interval-based data. It consists of: (1) entities well known in traditional OLAP, namely dimensions, hierarchies, and dimension members and (2) entities used to analyze interval data, namely sequences of intervals, intervals, events, and attributes.

$iCube = \{\mathbb{S}, \mathbb{D}, \mathbb{F}_{CV}, \mathbb{F}_{FC}\}$ where $\mathbb{S}$ is the set of sequences of intervals, $\mathbb{D}$ is the set of dimensions, $\mathbb{F}_{CV}$ and $\mathbb{F}_{FC}$ are two different sets of functions. Mandatory set $\mathbb{F}_{CV}$, called **compute value functions** includes user defined functions for computing fact values (measures). Optional set $\mathbb{F}_{FC}$, called **fact creating functions** includes user defined functions for creating new measures / facts assigned to an interval.

The compute value functions are used to derive / estimate values from two given consecutive events. For instance, when using the 'Heating' events, the temperatures at $e_1.time$ and $e_2.time$ are defined by $e_1.value$ and $e_2.value$. However, there exists no data for any time point t witch $e_1.time < t < e_2.time$. The user may now define functions to compute values for 'Heating' and 'Light' as shown in Listings 1.1 and 1.2. In these examples, we use simple linear monotonic functions, but any function may be used to compute values.

Listing 1.1. Function computing temperature at a given time point

```
//INPUT: t − timepoint
//OUTPUT: value, in this case the costs
function TempAtT(t) {
  //does t correspond to an event time?
  e = fetchEvent where e.time = t
  if (e != null) return e.value
  //if it does not match any event, find the
  //corresponding interval for t
  i = fetchInterval for Timepoint t
  //if there exists no interval, return null
  if (i == null) return null
  e1 = i.startEvent()
  e2 = i.endEvent()
  //assuming that the temperature rises/falls
  //linear and that ut(t) returns the number
  //of seconds of t
  return (e2.value−e1.value)/
         (ut(e2.time)−ut(e1.time))∗ut(t)}
```

Listing 1.2. Function computing light status at a given time point

```
//INPUT: t the timepoint
//OUTPUT: a value, in this case the costs
function LightStatusAtT(t) {
  //does t correspond to an event time?
  e = fetchEvent where e.time = t
  if (e != null) return e.value
  //if it does not match any event, find the
  //corresponding interval for t
  i = fetchInterval for Timepoint t
  //if there exists no interval, return null
  if (i == null) return null
  e1 = i.startEvent()
  e2 = i.endEvent()
  return e1.value }
```

Now, using functions *LightStatusAtT* and *TempAtT*, we can compute the light status and temperature at any given point in time. For example, fetching the temperature of room 100 for the time point that corresponds to $t = 170$ can be done by calling $TempAtT(170)$.

Listing 1.3. Example function computing energy cost of an interval

```
//INPUT: e1, e2 − two consecutive events
//OUTPUT: value, in this case costs
function costs(e1, e2) {
  //assuming that light sensors are boolean and return only ON or OFF
  if (e1.value == 'on')
  //assuming that the costs for each minute are 0.02 cents
    return (e2.time − e1.time)∗0.02
  else return 0 }
```

The fact creating functions are used to create facts that do not stem from events, but from sequences or intervals. For instance, in our running example we could assign a user defined operation that for 'Light' computes the costs by multiplying minutes between a 'Light on' and a 'Light off' event with a given cost factor (cf. Listing 1.3). Obviously, this fact cannot be derived from a single event but from sequences or intervals.

The two following methods on iCube are available to create new fact creating functions and compute value functions, namely: (1) `fMeasureValue()` – creates new function $f \in \mathbb{F}_{CV}$, (2) `fCreateFact()` – creates new function $f \in \mathbb{F}_{FC}$.

5   Querying I-OLAP Data

In this section, we will discuss how to answer queries on the I-OLAP model. We would like to emphasize that, due to a space limit, we will outline how our query language works, rather than its formal description. Basically, answering an I-OLAP query is done in the three steps discussed in this section.

5.1   Step 1: Getting Query Time Frame

The initial step is to get the time frame defined in a query. We assume that the underlying data warehouse has at least one time dimension (which will usually also serve as an ordering attribute for events). The dimension members selected by the user for the time dimension are extracted. This may be the *All* node (the root node) of the time dimension, i.e. all events, or any subset of dimension members belonging to the time dimension, i.e., the subset of events.

For example, for a given query: 'compute the number of minutes the light has been turned on in room 100 between timestamp 2013.01.01 and 2013.01.31', the time frame would be defined by $t_S = 2013.01.01$ and $t_E = 2013.01.31$.

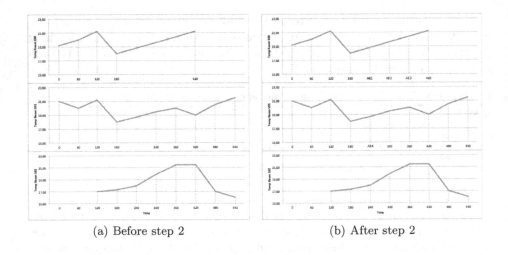

(a) Before step 2   (b) After step 2

Fig. 3. Sequences of intervals before and after applying step 2

5.2   Step 2: Inserting Artificial Events

Artificial events are used to guarantee a uniform distribution of events over all sequences of intervals. For instance, in the sequences of intervals for the temperature depicted in Figure 3(a) there are no such events defined for time point $t = 240$, $t = 300$, and $t = 360$ - for room 101, and for time point $t = 240$ - for room 101. In order to allow queries to aggregate data over multiple sequences of intervals, we extend each interval sequence with artificial events for t_S and t_E returned by step 1.

Extending a sequence of intervals with an artificial event at time point t is done in two steps: (1) inserting new events and (2) adopting all affected intervals. Figure 3(b) depicts the results of inserting artificial events, denoted as AE_i. Artificial events are inserted by the algorithm outlined in Listing 1.4.

We would like to emphasize that this only happens on a conceptual level. Each meaningful implementation of the model would first select all interval sequences affected by the query and enrich by artificial events only these interval sequences.

Next, adopting affected intervals takes the set of sequences of intervals as an input and creates a uniform event distribution over all sequences of intervals, cf. the pseudocode in Listing 1.2. As a result, the following condition is fulfilled: if there exists an event e with $e.t = T$ in any sequence than there also exist events e' in all other sequences of intervals with $e'.t = T$.

Listing 1.4. Creating an artificial event

```
//INPUT: t − time point
// I − sequence of intervals
//OUTPUT: new sequence of intervals
function createArtificialEvent(t, I) {
    e_prev = event e with max(e.t) ≤ t;
    if (e_prev.t = t) return I
    //create new event
    e_new = e_prev
        e_new.t = t
    foreach (f ∈ I.F_CV)
        e_new.value = f(t)
    end foreach
    //insert new event into sequence
    //of intervals
    I_new = Insert(e_new, I)
    return I_new }
```

Listing 1.5. Uniformly distributing an artificial event among sequences of intervals

```
//INPUT: I set of sequences of intervals
//OUTPUT: new set of sequences of intervals
function createUniformEventDistr(I) {
    foreach (SI ∈ I)
    i = SI.first()
    while (i ≠ NULL)
        foreach (SI' ∈ I)
        j = SI'.first()
        while (j ≠ NULL)
            if (j.start() = i.start()) exists = true
            j = SI'.next()
        end while
        if (false = exists)
            I'=I' ∪ createArtificialEvent(i.start(),SI')
        end foreach
        i = SI.next()
    end while
    end foreach
    return I' }
```

5.3   Step 3: Aggregating Measures

In this section we outline how to aggregate data using aggregate functions. Although we illustrate this step with the average function (AVG), the method is applicable to other aggregation functions such as MIN, MAX, SUM, and COUNT. We will show how to aggregate measures using two scenarios, namely: (1) *time point aggregation*, e.g., "what is the average temperature in all rooms at time point t" and (2) *aggregation along time*, e.g. "what is the average temperature in all rooms between time point t_1 and t_2".

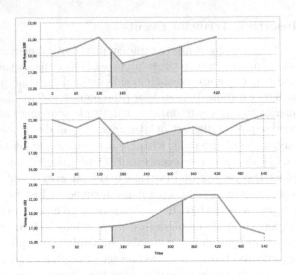

Fig. 4. Aggregation between t_1 and t_2

Table 3. Resulting temperatures for all rooms for $150 \leq t \leq 330$

Room	t=150	t=180	t=240	t=300	t=330	AVG
100	19,2	18,0	–	–	20,0	**19,07**
101	19,1	18,2	18,8	19,2	19,6	**18,98**
102	17,0	17,2	18,0	20,0	20,8	**18,60**
Floor						**18,85**

Answering the first query is straight forward. We simply call the corresponding function defined in $\mathbb{F}_{CV}$ for all facts fulfilling the selection predicate. For instance, in our example we call $TempAtT(t)$ for interval sequences for all rooms.

The second query will be executed as follows. First, we call $TempAtT(t)$ for $t = t_1$ as well as for $t = t_2$. Second, we fetch all events for all sequences of intervals. For each event e, we call $TempAtT(t)$ with $t = e.t$. Figure 4 depicts this technique for $t_1 = 150$ and $t_2 = 330$. The resulting values are given in Table 3.

6   Implementation

We prototypically implemented this approach as a web application using PHP, the *PHP PEG* package[1] (a package used for defining PEGs - parsing expression grammar - and parsing strings into objects) and PostgreSQL.

As we are currently working on a query language enabling the user to analyze sequences in OLAP cubes, called S-SQL (Sequential SQL), we implemented this interval based approach as an extension of S-SQL [1]. Due to space limitations, we cannot give a detailed description of S-SQL. Basically, S-SQL statements enable users to formulate queries in order to analyze sequences using different functions like for instance HEAD(), TAIL() or PATTERN(). The S-SQL prototype consists

---

[1] PHP PEG has been developed by Hamish Friedlander. Available at:
https://github.com/hafriedlander/php-peg

of a parser translating S-SQL into objects and an engine which then creates standard SQL statements out of these objects. As an example the query given in listing 1.6 might be used to fetch all sequences of events that fulfill a given pattern (A,*,B) and where the temperature was below 19 degrees at the start and the end of the day.

Listing 1.6. Sample S-SQL Query

```
SELECT *
FROM t1
WITH PATTERN 'a,*,b' BIND (a,b) TO sensor.heating ON SEQUENCE room
WHERE a.value < 19 AND b.value < 19;
```

This simple query would translate into a SQL query with over 40 lines of code (formatted). Other queries we tested resulted in queries with up to 160 lines of code.

We extended the functionality of S-SQL in order to be able to parse and execute statements using functions defined in $\mathbb{F}_{CV}$ and $\mathbb{F}_{FC}$. In it's current version these functions have to be defined as PL/pgSQL functions. The web service is used to parse the metadata of a given database and apply these functions to the defined intervals. Internally, the three steps as described in section 5 (getting the query time frame, inserting artificial events and aggregating measures) will be applied to the intervals. This enables us to state queries like:

Listing 1.7. Sample I-SQL Query

```
SELECT AVG(value)
FROM t1
WHERE left(sensor_id,1) = 'H' AND
       time >= '2013−03−20 00:00:00' AND time <= '2013−03−20 23:59:59'
```

This query would return the average temperature of all rooms on March, 20th. The implementation would automatically apply the three steps described above to get a correct result.

7 Summary

In this paper we proposed a formal model for processing interval data in an OLAP style and a prototypical implementation. To the best of our knowledge, no such model has been proposed before. The model supports: (1) defining multiple sets of intervals over sequential data, (2) define measures computed from both, events and intervals, and (3) analyze the measures in the context set up by dimensions - to this end we proposed the *iCube*. The formal model was reflected in an implementation model that we also proposed. We shown how to apply the model to querying I-OLAP data. In the next step we will develop physical data structures for supporting I-OLAP queries and evaluate their performance. Future work will focus on analyzing and developing methods to represent interval data by means of functions, similarly as proposed in [8] - for moving objects and in [14] - for interpolating values returned by sensors.

References

1. Retr. March 31, 2014, `http://solap-isys.aau.at`
2. Aster nPath,
   `http://developer.teradata.com/aster/articles/aster-npath-guide`
   (retr. from March 13, 2014)
3. Bębel, B., Morzy, M., Morzy, T., Królikowski, Z., Wrembel, R.: OLAP-like analysis of time point-based sequential data. In: Castano, S., Vassiliadis, P., Lakshmanan, L.V.S., Lee, M.L. (eds.) ER 2012 Workshops 2012. LNCS, vol. 7518, pp. 153–161. Springer, Heidelberg (2012)
4. Chui, C.K., Kao, B., Lo, E., Cheung, D.: S-OLAP: an olap system for analyzing sequence data. In: Proc. of ACM SIGMOD Int. Conf. on Management of Data, pp. 1131–1134. ACM (2010)
5. Chui, C.K., Lo, E., Kao, B., Ho, W.-S.: Supporting ranking pattern-based aggregate queries in sequence data cubes. In: Proc. of ACM Conf. on Information and Knowledge Management (CIKM), pp. 997–1006. ACM (2009)
6. Gonzalez, H., Han, J., Li, X.: FlowCube: constructing RFID flowcubes for multidimensional analysis of commodity flows. In: Proc. of Int. Conf. on Very Large Data Bases (VLDB), pp. 834–845. VLDB Endowment (2006)
7. Gonzalez, H., Han, J., Li, X., Klabjan, D.: Warehousing and analyzing massive RFID data sets. In: Proc. of Int. Conf. on Data Engineering (ICDE) (2006)
8. Güting, R.H., Böhlen, M.H., Erwig, M., Jensen, C.S., Lorentzos, N.A., Schneider, M., Vazirgiannis, M.: A foundation for representing and querying moving objects. ACM Trans. on Database Systems (TODS) 25(1), 1–42 (2000)
9. Han, J., Chen, Y., Dong, G., Pei, J., Wah, B.W., Wang, J., Cai, Y.D.: Stream Cube: An architecture for multi-dimensional analysis of data streams. Distributed and Parallel Databases 18(2), 173–197 (2005)
10. Liu, M., Rundensteiner, E., Greenfield, K., Gupta, C., Wang, S., Ari, I., Mehta, A.: E-Cube: multi-dimensional event sequence analysis using hierarchical pattern query sharing. In: Proc. of ACM SIGMOD Int. Conf. on Management of Data, pp. 889–900. ACM (2011)
11. Liu, M., Rundensteiner, E.A.: Event sequence processing: new models and optimization techniques. In: Proc. of SIGMOD PhD Workshop on Innovative Database Research (IDAR), pp. 7–12 (2010)
12. Lo, E., Kao, B., Ho, W.-S., Lee, S.D., Chui, C.K., Cheung, D.W.: OLAP on sequence data. In: Proc. of ACM SIGMOD Int. Conf. on Management of Data, pp. 649–660 (2008)
13. Mörchen, F.: Unsupervised pattern mining from symbolic temporal data. SIGKDD Explor. Newsl. 9(1), 41–55 (2007)
14. Thiagarajan, A., Madden, S.: Querying continuous functions in a database system. In: Proc. of ACM SIGMOD Int. Conf. on Management of Data, pp. 791–804 (2008)
15. Witkowski, A.: Analyze this! Analytical power in SQL, more than you ever dreamt of. Oracle Open World (2012)
16. Ya-Han, H., Tony Cheng-Kui, H., Hui-Ru, Y., Yen-Liang, C.: On mining multi-time-interval sequential patterns. Data & Knowledge Engineering 68(10), 1112–1127 (2009)
17. Yen-Liang, C., Mei-Ching, C., Ming-Tat, K.: Discovering time-interval sequential patterns in sequence databases. Expert Systems with Applications 25(3), 343–354 (2003)

Improving the Processing of DW Star-Queries under Concurrent Query Workloads

João Pedro Costa[1] and Pedro Furtado[2]

[1] Polytechnic Institute of Coimbra, ISEC, DEIS, Portugal
jcosta@isec.pt
[2] University of Coimbra, Portugal
pnf@dei.uc.pt

Abstract. Currently, Data Warehouse (DW) analyses are extensively being used not only for strategic business decisions by a few, but also for feedback to a wider audience and into daily operational decisions. As a result, there's an increase in the number of aggregation star-queries that are being concurrently submitted. Although such queries require similar processing patterns, they are stressing the database engine ability to deliver timely execution, due to the fact that each query executes independently from the others (query-at-time processing model). Recently, there's an increasing interest in approaches that cooperate to manage large numbers of concurrent aggregation star-queries. We have proposed SPIN in a previous paper [1]. It is a data processing model that shares data and computation in order to handle large concurrent query loads, and its data organization provides almost constant and predictable execution times for all submitted queries. It has a data reader that reads data in circular loop, placing it in a pipeline, before being processed by branches that combine common processing computations. SPIN is IO dependent, i.e. a query is only be answered after a full circular loop, even though tuples and similar predicates have been evaluated in the past. In this paper we propose data processing approach that uses a set of *bitsets*, built on-the-fly, to significantly reduce the query processing time, the tuple evaluation cost and the number of predicates and tuples evaluated, without sacrificing its predictability features. The data read from storage is reduced to the minimum needed by the current query load.

1 Introduction

Common database engines process every query independently without data and processing sharing considerations. In this model (query-at-time) each competes for resources, and thus the execution time increases with the number of queries that are concurrently being executed. While this may not raise performance issues for most operational systems, it is a performance killer when dealing with large Data Warehouses (DW). In this context, large fact and dimension relations are concurrently scanned by each query and tuples are independently filtered, joined and aggregated. This lack of data and processing sharing results in the system inability to provide predictable execution times for scalable data volumes and query workloads.

We have proposed SPIN [1], a data and processing sharing model that delivers predictable execution times to a large set of concurrently running aggregation star queries. SPIN physically stores the star schema as a single de-normalized relation as

L. Bellatreche and M.K. Mohania (Eds.): DaWaK 2014, LNCS 8646, pp. 245–253, 2014.

proposed in [2], [3] to avoid costly join operations. A data reader is continuously spinning, sequentially reading data chunks in a circular loop, placing the data in a base pipeline shared by all running queries. Common filters and computations (with the same operators) from different queries are combined into common pipelines with data switches added to end to share its results. Subsequent data pipelines are then connected as logical branches of this common data pipeline, consuming its output. As a result, pipelines of running queries are split, merged and organized into a workload data processing tree (*WPTree*), with the base pipeline as root. Any query q, Each query starts processing tuples that flow along the pipelines and only stops when all the tuples have being considered for evaluation (after a complete loop). While this improves IO sharing, the query execution is constrained by the time needed to fully read the data. SPIN performance can also be limited by the evaluation costs related to redirecting and filtering tuples as they go along branches, even though most of them already has been evaluated in the past.

The query execution time is constrained by the data reading time (t_{read}) and the query processing time ($t_{process}$), particularly t_{read} since all tuples must be considered for evaluation. While t_{read} is constant and shared among queries, the $t_{process}$ time is influenced by the computation (e.g. aggregation operators) and evaluation of predicates of each query (t_{eval}), and how these can be shared among queries. Since reading and processing is done in parallel, the query execution time is constrained by the largest of these times ($\max(t_{process}, t_{read})$). For wide WPTree (large number of simultaneous queries), the processing time ($t_{process}$) can be larger than the reading time (t_{read}), and thus endangering the objective of execution time predictability.

In this paper we propose a *bitset*-based data processing approach that uses a set of *bitsets* to reduce the overall execution time, by reducing the evaluation time (t_{eval}), the number of evaluated tuples (n_{eval}), and also the time required to read the data (t_{read}).*Bitsets* are built on-the-fly with the results of previous executions, and when built they deliver faster processing times (by using a bit lookup instead of the actual evaluation), and also reduce the number of evaluations and the data read from storage.

2   Related Work

The usage pattern of DWs is changing from the traditional, limited set of simultaneous users and queries, mainly well-known reporting queries, to a more dynamic and concurrent environment, with more simultaneous users and ad-hoc queries. DW query patterns are mainly composed by star aggregation queries, which contain a set of query predicates (filters) and aggregations. The query-at-a-time execution model of traditional RDBMS systems, where each query is executed following its own execution plan, does not provide a scalable environment to handle much larger, concurrent and unpredictable workloads. Analyzing the execution query plan, we observe that the low-level data access methods, such as sequential scan, represent a major weight in the total query execution time. One way to reduce such a burden is to store relations in memory. However, the amount of available memory is limited, insufficient to hold large DW, and is also required for performing join and sort operations. Recently there is increasing interest in approaches that share data and processing among queries.

Cooperative scans [4] enhances performance by improving data sharing between concurrent queries, by performing dynamic scheduling of queries and their data requests taking into account with the current executing actions. While this minimizes the overall IO costs, by mainly using sequential scans instead of a large number of costly random IO operations, and the number of scan operations (since scans are shared between queries), it introduces undesirable delays to query execution and does not deliver predictable query execution times.

QPipe [5] applies on-demand simultaneous pipelining of common intermediate results across queries, avoiding costly materializations and improving performance when compared to tuple-by-tuple evaluation.

CJOIN[6] [7] applies a continuous scan model to the fact table, reading and placing fact tuples in a pipeline, and sharing dimension join tasks among queries, by attaching a bitmap tag to each fact tuple, one bit for each query, and attaching a similar bitmap tab to each dimension tuple referenced by at least one of the running queries. Each fact tuple in the pipeline goes through a set of filters (one for each dimension) to determine if it is referenced by at least one of the running queries. It not, the tuple is discarded. Tuples that reach the en d of the pipeline (tuples not discarded in filters) are them distributed to dedicated query aggregations operators, one for each query. However, its usefulness is limited to small dimensions that can fit entirely in memory and, as recognized in [7], large dimensions may severely impact performance.

SPIN [1] is conceptually related to CJoin, and QPipe in what concerns the continuous scanning of fact data, but it uses a simpler approach with minimum memory requirements and does not have the limitations of such approaches. SPIN uses a de-normalized model, as proposed in [2] as a way to avoid the join costs, at the expense of additional storage costs. Since it has fully data and processing scalability, ONE allows massive parallelization [3], which provides balanced data distribution, scalable performance and predictable query execution times.

3   The *bitset* Branch Processing Approach

The evaluation time (t_{eval}) is constrained by the number of predicates and branches of the current *WPTree*. A submitted query has to process and evaluate each of its predicates, even though similar queries, with common predicates, have been previously processed in the past. As predicates of common queries are associated to common branches, a branch with a given predicate can be repeatedly built or may persist over time while at least one running query uses it. To avoid subsequent evaluation of unchanged data tuples, we propose to maintain the result of the predicate evaluation as tuples flow through data branches. This is particularly relevant for predicates with high evaluations costs. We build a branch *bitset* (bitmap) according to the branch' predicates, where each bit represents the result of the predicate evaluation (true/false) applied to a corresponding tuple index.

As SPIN processes tuples in a circular fashion, when the relation reaches the end, it restarts reading from the beginning. Therefore, future evaluations of a tuple can take advantage of the existence of this *bitset*, since the selection operator that evaluates the predicate can be replaced by a fast lookup operator that look up the corresponding position in the *bitset* to gathers the result. *Bitsets* are small and will be in memory in order to avoid introducing overhead at IO level.

3.1   Creation of Bitsets

A *bitset* is built on-the-fly, as tuples go through the branches and are evaluated, with minimum overhead, since these results are stored in an in-memory data structure. A branch evaluates each tuple and stores the result of the predicate evaluation in the *bitset*, at the tuple index position. This is very important, since it makes it possible in every future evaluation of each row, to decide whether that row should proceed to the next step in the branch, or not, based on the simple lookup of the *bitset*. This avoids most of the evaluation costs. A *bitset* can be built for each value (e.g. 11, 12, 13 ...), sets of values, or ranges of the attribute domain (e.g. [10;13]). To avoid additional overhead, *bitsets* are built according to the selection predicates of the submitted queries. For a new selection operator deployment in the *WPTree*, without a matching *bitset*, a new *bitset* is built with the result of the selection predicate. This bitmap is kept in memory and shared by all branches that can use it, allowing future evaluations of these tuples to be replaced by a fast *bitset* lookup operator.

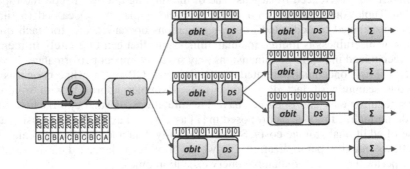

Fig. 1. Branch processing using equivalent BitSets

3.2   Bitset Lookup Operators

Branches can be built, or reorganized, to use *bitsets*. Selection operators (σ) with a matching *bitset* are replaced with a bit-selection operator (σbit), which performs bit lookups to the corresponding index position in the *bitset,* and thus avoiding the evaluation of these tuples. Selection operators without a matching *bitset* can still take advantage of *bitset* evaluation by combining the existing *bitsets* for other values of the attribute domain, by using a bit-selection not operator ($\sigma!bit$) evaluates as true all the index positions in the *bitset* that are 0. $\sigma!bit$ is equivalent to NOT σbit.

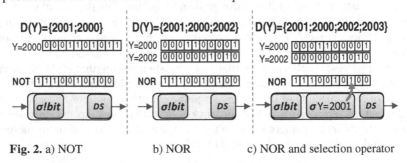

Fig. 2. a) NOT   b) NOR   c) NOR and selection operator

For instance, consider that the domain of the attribute Y (year) is 2000 and 2001, $D(Y) = \{2000, 2001\}$. If there's a *bitset* for 2000, *bitset*(y=2000), the selection operators σ (y=2001) can be replaced with a *σ!bit* (y=2000) operator that applies a bitwise **NOT** to *bitset*(y=2000). The result is equivalent to **NOT**(*σbit* (y=2000)). Fig.2 shows how *bitset* processing can be employed for boost performance even when a perfectly matching *bitset* does not exist. Fig.2-a) depicts the domain complement.

Fig.2-b) depicts the scenario for a wider domain, where a *bitset* exists for each of the values of the attribute domain except for the value Y=2001. In this case, the *σ!bit* operator applies a **NOR**(*bit* (y=2000), *bit* (y=2002)). The domain complement (Fig.2-a)) is a particular case, where the domain has only two distinct values.

Bitset processing can still be used when the number of values, of the attribute domain, without a matching *bitset* is greater than 1 (Fig.2-c). In this case, the selection operator σ is maintained in the branch, but it is preceded by a *σ!bit* that applies a **NOR** to the existing *bitsets*, and thus obtaining a *bitset* with all the index positions that are certainly false, marked as 0. The goal of this *σ!bit* is to avoid the σ operator from evaluating these tuples that are known to be false. The remaining index positions, which can be evaluated as true or false, are market as 1. As the σ operator evaluates these remaining tuples, it the updates and completes the *bitset* with the result of the evaluation (illustrated in the figure with a blue arrow).

3.3   Mixed Branch Processing: Branches with and without *bitsets*

At any given time, SPIN may have branches with *bitset* operators and other branches that have to evaluate the predicates, because there isn't a *bitset* that matches the selection predicate. Branches that use *bitsets* are pushed forward and connected directly to the base data pipeline to maximize the sharing costs, and to reduce the overall number of tuples that have to be evaluated with selection operators (σ). New branches without *σbit* operators are connected as usual to existing pipelines, regardless if they have a *bitset* or not, and the corresponding branch predicate is associated with a *bitset* filled with 1's. This *bitset* can be updated with existing *bitsets* related with the branch' predicate, when exists. Whenever a branch ends building a *bitset*, it is replaced by an equivalent branch with a corresponding *bitset* operator (*σbit*). Since *bitset* processing is faster than the tuple predicate evaluation, branches that contain *bitset* operators are reorganized and pushed forward to the base pipeline.

Over time, predicates more frequently used will have a corresponding *bitset*, and therefore will deliver faster query processing times.

3.4   Merging *bitsets* along the Query Logical Path

When in a logical data path, two or more branches use exclusively *bitsets* to evaluate tuples, then these branches are merged into the later one, composed with a single *bitset* that is a logical AND of all these branches. The resulting branch replaces the merged branches, or is directly connected to the base pipeline. The main goal is to filter as soon as possible the data tuples that are relevant to a query, before reaching the branches that evaluate tuples using selection operators. Fig. 3 depicts the new deployment, where the query logical data paths are built with less data branches, where the last branch of each query path in the previous deployment (Fig. 1) is substituted with a branch that evaluates tuples using single *bitset* that represents the logical evaluation of all the *bitsets* of the logical data path. In the figure, the *bitset* of query 1 (the topmost branch, which has only the first bit set to 1) was built by the selection operator (σ) of the last branch of the logical path, but it could also be built by applying a AND to *bitset(y=2000)* and *bitset(p=b)*. The four queries depicted in the figure are evaluated with dedicated branches, each with a single *σbit* operator. These *bitsets* can be pushed forward to the base pipeline and be used by the data reader to reduce the cost of getting and forwarding the data.

3.5   Pushing forward *bitsets* to the Data Reader

When all the branches connected to the base pipeline use *σbit* operators to filter relevant tuples for each branch, then a *WPbitset* is created by applying a logical OR to all the branches' *bitsets*. The data reader can use this *WPbitset* to control the data to gather from storage and optimize the IO reading cost.

In a mixed environment, with some branches using *σbit* operators and others not, an all 1's *bitset* is considered in the bitwise OR, when exists at least one first level branch that does not use a *σbit* operator to evaluate tuples. For the previous deployment, the result of the merging OR is depicted in Fig. 3.

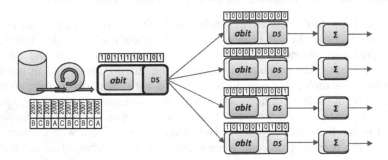

Fig. 3. Data Reader *Bitset* computed as a bitwise OR of the branches bitsets

WPbitset allows the Data Reader to control the data to gather from storage and optimize the IO reading cost, by skipping data chunks that are not relevant for the workload processing tree. For relevant chunks, the data reader uses the *WPbitset* to decide which tuples to place in the pipeline (only tuples pinpointed by the *WPbitset*). Queries, with predicates that are evaluated exclusively with *σbit* operators, can early end its execution and return the result, without having to wait for the full loop to be completed. As soon as the number of processed tuples reaches the *bitset count* (hamming distance), then the query can stop execution since all the tuples relevant for the query (which satisfies the selection predicates) were processed. When that occurs, the query execution time fall below the barrier imposed by the IO cost of reading the full relation (in a circular loop), since with *bitsets* a query can end as soon as all the set positions of the *bitset* as been processed.

4   Evaluation

We extended our SPIN engine (release 1.9.1), which is implemented in Java, and incorporated the *bitset*-based processing approach, presented in this paper, and used the TPC-H benchmark with a scale factor (SF) of 10 to evaluate its performance and scalability capabilities. We used an Intel i5 processor, with 8GB of RAM and 3 SATAIII disks with 2Terabytes each, running a default Linux Server distribution. An additional server submits 1000 random variations of Q5, with different selectivity, generated by a varying number of simultaneous concurrent clients. For this setup, we compare the number of evaluations carry out to tuples as they flow and are switched along the branches in the *WPTree* when using the base SPIN setup (SPIN), the *bitset* operators (*σbit*) (SPIN-WP bitset), and when the Data Reader uses the *WPbitset* (SPIN-DR bitset). Fig. 4 depicts the results for a *WPTree* with 10 branches.

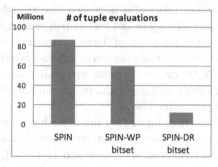

Fig. 4. Number of evaluations

The number of evaluations with SPIN-WP bitset, because *bitsets* are merged along the logical path, is utmost equal to the number of tuples, while the base SPIN requires more evaluations to filter and redirect tuples to appropriated branches. With SPIN-DR bitset, the number of evaluates drops significantly, with some chunks aren't read from storage and uninteresting tuples are not placed into pipelines, and thus reducing the number of evaluations, and improving the average execution time (depicted in Fig.5).

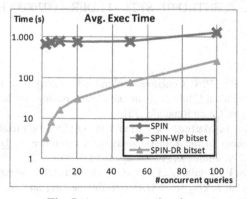

Fig. 5. Average execution time

The results shows that even though *bitset* evaluation is faster than tuple evaluation, the SPIN-WP bitset setup only yields a slightly improvement in the average execution time in comparison with SPIN (in the graph they are almost the same). This is because this query workload results in a *WPTree* with utmost 50 branches, with low selectivity operators and the IO reading costs represents a large percentage of the overall execution cost. Therefore, for more complex, and more processing intensive evaluations, SPIN-WP bitset will deliver better results that the base SPIN. The results also show that when we apply the *WPbitset* to the data reader (SPIN-DR bitset) the execution time is significantly lower since it avoids reading large number of tuples, and consequently reduces the number of evaluations and the evaluation time.

5   Conclusions

We present a *bitset*-based data processing approach that extends the SPIN processing model, a data and processing sharing model that deliver predictable execution times to star-join queries even in the presence of large concurrent workloads. It replaces selection operators, or sets of selection operators, with fast *bitset* selection operators.

Bitset speeds up SPIN by minimizing the processing costs, replacing selection operators with fast bit-selectors, and thus reducing the number of evaluated tuples and consequently the overall processing cost.

Since the number of bitsets is limited by the available memory, we currently are employing run-length encoding compression algorithms, such as the Byte-aligned Bitmap Code (BBC), and Enhanced Word-Aligned Hybrid (EWAH) which require very little effort to compress and decompress and can be used in bitwise operations without decompression. We are evaluating in algorithms that take into account factors such as usage, rebuildability, predicate evaluation costs and hamming distance for managing the *bitsets* that are maintained in memory. We are also working on a column-oriented data organization that can be combined with *bitset* processing for further improvement in IO reading costs and query execution times.

Acknowledgments. This work was partially financed by iCIS – Intelligent Computing in the Internet Services (CENTRO-07- ST24 – FEDER – 002003), Portugal.

References

[1] Costa, J., Furtado, P.: SPIN: Concurrent Workload Scaling over Data Warehouses. In: Proc. of 15th International Conference on Data Warehousing and Knowledge Discovery - DaWaK 2013, Prague, Czech Republic (2013)

[2] Costa, J.P., Cecílio, J., Martins, P., Furtado, P.: ONE: a predictable and scalable DW model. In: Proceedings of the 13th International Conference on Data Warehousing and Knowledge Discovery, Toulouse, France, pp. 1–13 (2011)

[3] Costa, J.P., Martins, P., Cecílio, J., Furtado, P.: A Predictable Storage Model for Scalable Parallel DW. In: 15th International Database Engineering and Applications Symposium (IDEAS 2011), Lisbon, Portugal (2011)

[4] Zukowski, M., Héman, S., Nes, N., Boncz, P.: Cooperative scans: dynamic bandwidth sharing in a DBMS. In: Proceedings of the 33rd International Conference on Very Large Data Bases, Vienna, Austria, pp. 723–734 (2007)

[5] Harizopoulos, S., Shkapenyuk, V., Ailamaki, A.: QPipe: A Simultaneously Pipelined Relational Query Engine. In: Proceedings of the 2005 ACM SIGMOD International Conference on Management of Data, pp. 383–394 (2005)

[6] Candea, G., Polyzotis, N., Vingralek, R.: A scalable, predictable join operator for highly concurrent data warehouses. Proc. VLDB Endow. 2, 277–288 (2009)

[7] Candea, G., Polyzotis, N., Vingralek, R.: Predictable performance and high query concurrency for data analytics. The VLDB Journal 20(2), 227–248 (2011)

Building Smart Cubes for Reliable and Faster Access to Data

Daniel K. Antwi and Herna L. Viktor

School of Electrical Engineering and Computer Science, University of Ottawal,
800 King Edward Rd, Ottawa, Canada
{dantw006,hviktor}@uottawa.ca
http://www.uottawa.ca

Abstract. In data warehousing, selecting a subset of views for materialization has been widely employed as a way to reduce the query evaluation time for real-time OLAP queries. However, materialization of a large number of views may be counterproductive and may exceed storage thresholds, especially when considering very large data warehouses. Thus, an important concern is to find the best set of views to materialize, in order to guarantee acceptable query response times. It further follows that the best set of views may change, as the query histories evolve. To address these issues, we introduce the Smart Cube algorithm that combines vertical partitioning, partial materialization and dynamic computation. In our approach, we partition the search space into fragments and proceed to select the optimal subset of fragments to materialize. We dynamically adapt the set of materialized views that we store, as based on query histories. The experimental evaluation of our Smart Cube algorithm shows that our work compare favorably with the state-of-the-art. The results indicate that our algorithm materializes a smaller number of views than other techniques, while yielding fast query response times.

Keywords: Smart data cubes, dynamic cube construction, real-time OLAP, partial materialization.

1   Introduction

Data Warehouses and Online Analytical Processing (OLAP) are widely used to aid us in the understanding of Big Data, and aim to increasing the productivity in decision-making processes that are supported by business applications. A data warehouse often consists of a number of data marts, modeled through a star schema composed of fact and dimension tables [7]. While *fact tables* store numeric measures of interest, *dimension tables* contain attributes that contextualize these measures. Often, attributes of a dimension may have a relationship with other attributes of the dimension through concept hierarchies. A higher level in a hierarchy contains concise data and several intermediate levels representing increasing degrees of aggregation. All the levels together compose a data cube that allows the analyses of numerical measures from different perspectives.

L. Bellatreche and M.K. Mohania (Eds.): DaWaK 2014, LNCS 8646, pp. 254–265, 2014.
© Springer International Publishing Switzerland 2014

View materialization refers to storing precomputed aggregated data to facilitate multi-dimensional analysis in real-time. It eliminates overheads associated with expensive joins and/or aggregations required by analytical queries. An important issue is the selection of the best set of views to materialize. Thus, given a certain storage cost threshold, there is the need to select the "best" set of views that fits the storage cost as well as provides an acceptable query response time to process OLAP queries. A number of solutions have been proposed to address this problem. These solutions may be categorized into cube computation, cube selection and hybrid approaches. Some of the proposed solutions include partitioning the dimension space and materializing smaller local cubes for each partition fragment [9]. Other techniques include greedily selecting a subset of views based on a given heuristic [6]. Finally others are based on dynamic materialization of views based on user query pattern [8]. However, we are in the era of Big Data. Using approaches based only on partitioning, greedy selection or dynamic computation alone will not suffice. This is because the drawbacks of these techniques will be compounded by the sheer size and dimensionality of the source data.

In this paper, we propose a dynamic, real-time data cube computation approach to address this problem. Our Smart Cube method is based on the combination of *vertical partitioning, partial materialization* and *dynamic view construction*. Our aim is to reduce storage space while guaranteeing fast query response time. To this end, the algorithm partitions the dimension space vertically into disjoint sets and computes localized cube for each fragment. The computation of each local cube is implemented using a partial materialization algorithm. The result is a set of fragmented cubes with non-overlapping attributes. In order to link the localized data cubes together, we dynamically compute and materialize views from attributes of different fragments by continuously monitoring user queries in real time.

The paper is organized as follows. In Section 2, we briefly formalize the problem setting. Section 3 details our Smart Cube algorithm. Section 4 contains our experimental evaluation and Section 5 concludes the paper.

2   Problem Statement

Consider a data warehouse that contains one or more multidimensional data marts. Following Kimball et. al. [8], each data mart may be presented as consisting of a single central fact table and multiple dimensional tables. Suppose the fact table T is defined as a relation where the set of attributes is divided into two parts, namely the set of dimensional key attributes D_i and the set of measures M_i. In general, the set D_i is the key of T. A data cube built from T is obtained by aggregating T and grouping its tuples in all possible ways. Each grouping is a so-called cuboid. Assuming that each c corresponds to a cuboid, we perform all *Group By* c where c is a subset of D_i. Let the data cube computed from T be denoted by C and let the dimensions of the data cube be denoted by D_i. If C is a data cube, and D_i is its dimensions, then $|D_i| = D$. The set of cuboids of

C is denoted by v. Clearly $|v| = 2^D$. However, suppose T is partitioned into a disjoint set $T_1, T_2, \cdots T_n$. Let C_1 be the data cube computed for T_1, C_2 the cube computed for T_2 and so on, up to C_n. The set of cuboids of C_i for $i \in 1 \cdots n$ is denoted by V. Subsequently, $|V| \approx \frac{D}{F}(2^F - 1)$, where F is the number of dimensions for each partition assuming it is uniformly distributed. The size of a cuboid c is expressed by the number of its rows and is denoted by $size(c)$. The size of a set of cuboids S is denoted by $size(S)$. Note that, if $s \preceq w$ then cuboid s can be computed from cuboid w. The fact table T is a distinguished cuboid of v and it is called the base cuboid denoted by c_b. For example, Figure 1 shows a data cube example with four dimensions, namely A, B, C and D, where the full cube contains 2000 million rows.

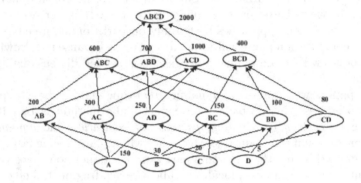

Fig. 1. A data cube example (numbers indicate millions of rows)

In many data warehouses, a single data mart together with its associated cuboids exceed the available main memory storage. In this setting, the aim is to find a set S of cuboids to materialize which satisfies two requirements. Firstly, when materialized, the overall size of the solution is minimized in order to limit the cost associated with I/O operations. Secondly, it is ensured that the evaluation cost of queries does not exceed an application-dependent threshold value. This implies that, for a given available storage space, we need to find a set S such that, the size is less than the available storage space, while maintain high query response. In the next section we introduce our *Smart Cube* algorithm, which addresses these requirements.

3   Smart Cube Algorithm

Our Smart Cube algorithm proceeds as follows. We first partition the multidimensional data mart into a disjoint set of views called fragments. Next, we compute a local cube from each fragment. During this Smart Materialization phase we use partial materialization, where each fragment view is considered as a base table. These views can be used to answer queries against one of the fragments. Further, we also materialize views that involve queries posed against multiple fragments (Smart Views). Our algorithm is dynamic in the sense that we constantly monitor the query history to detect frequent query patterns. We use these patterns to dynamically update the list of the local cubes that are materialized, as the query history evolves.

Algorithm 1. The Smart Cube Construction Algorithm
___
Input: Multidimensional datamart T
        Number of Partitions N,
Output: A set of fragment partitions $\{P_1, \cdots, P_n\}$

1: Partition T into fragments $\{P_1, \cdots, P_n\}$;
2: Scan fact table once and do the following
3: Insert each (tid, measure) into $ID\_measure$ array
4: for each dimension row in each partition
5: Build an Inverted Index entry (rowvalue, tidlist) or a bitmap entry
6: for each fragment Partition P_i;
7: Build local fragment cube by materializing the most beneficial cuboid using Smart_Materialization
8: Propagate bitmap index or inverted indexes to all materialized views using set intersection;
9: for each in-coming query Evaluate it candidacy for materialization using Smart_View algorithm

---

Algorithm 1 illustrates the Smart Cube approach. Our algorithm starts by partitioning T into n fragments. In order to select an optimal set of attributes for each partition we followed the approach of [5]. Given a fragment f of size n, f is made up of dimensional attributes d_i for $i = \{1 \cdots n\}$ and measure attributes m_i for $i = \{1 \cdots n\}$. We select d_i and m_i such that $card(d_i)$ and $sel(f)$ is minimized, where $card(d_i)$ is cardinality of dimension attributes and $sel(f)$ is the number of tuples of the aggregated fragment. The advantage of partitioning is to reduce the storage complexities of the resultant data cubes. For example, the data cube computed from a datamart with fact table T and dimensions A, B, C and D will generate $2^4=16$ cuboids. However, assume that we partition it into two, such that, we have the first fragment data cube with dimensions A and B and second fragment data cube with dimensions C and D. In this case, it will require $2^2 + 2^2 = 8$ cuboids to store the data. After partitioning the multidimensional space, we compute each local cube using our Smart Materialization algorithm.

3.1   Smart Materialization

The Smart Materialization algorithm aims to materialize promising views, as based on a single fragment. In this approach, we are interested in views whose sizes are at most M/pf and not less than r. Here, M is the size of the base cuboid of each fragment, $pf > 1$ is a real number called the performance factor and r is the minimum row threshold. The maximum threshold is used to ensure that we avoid the creation of very large views. The minimum row threshold prevents the materialization of a large number of very small views, which may easily and efficiently be computed from a parent view, without causing any increase in query processing time.

Definition 1 (Performance Factor). *Let S be the set of materialized cuboids and c be a cuboid of C. The performance factor of S with respect to c is defined by $pf(c, S) = \frac{cost(c,S)}{size(c)}$. The average performance factor of S with respect to $C' \subseteq C$ is defined by $\tilde{pf}(C', S) = \frac{\sum_{c \in C'} pf(c,S)}{|C'|}$.*

Definition 2 (Row Threshold). *Let S be the set of materialized cuboids and c be a cuboid of C. If the performance factor of S with respect to c is defined by $pf(c, S) = \frac{cost(c,S)}{size(c)}$. The row threshold r of S is the $size(c')$ such that $pf(c', S) = minF$. Where $minF$ is the minimum acceptable performance Loss.*

Intuitively, the performance factor measures the ratio between the response time for evaluating a query using S, a given materialized sample, in constrast to materializing the entire cube. This implies that the minimal cost to evaluate a query c corresponds to $size(c)$. On the other hand, since performance is related to the number of tuples accessed by c, eliminating all cuboids whose size is below a given threshold implies reducing the number of cuboids to store while maintaining a given minimum performance threshold. Algorithm 2 details our Smart Materialization algorithm.

Algorithm 2. The Smart Materialization Algorithm

Input: Partition P
　　　　Number of Partitions N,
　　　　Parameter pf,
　　　　Row Threshold r,
Output: S Set of materialized cuboids

1: for each P in partitions
2: Initialize the set of materialized views to an empty set ϕ
3: Initialize C_1 to the set of all cuboids under base cuboid thus all children of c_b
4: Initialize L_1 to the set of all cuboids c in C_1 such that $|c| \le M/pf$
5: $S = S \cup L_1$
6: **for** $i = 1; L_i \ne \phi; i + +$ **do**
7: 　$C_{i+1} = \{candidates\ generated\ from\ L_i\}$
8: 　for each $c \in C_{i+1}$do compute_size(c)
9: 　$L_{i+1} = c \in C_{i+1}s.t|c| \le M/pf$
10: 　for each $c \in L_i$ do
11: 　if $c' > r$ s.t $c' \preceq c$ Then
12: 　　$S = S \cup \{c\}$
13: **end for**
14: Return S

To further explain the Smart Materialization algorithm, we again use an example as depicted in Figure 1. In this example, we assume that the full data cube contains 2000 millions rows. The fragment cube ABCD is computed using performance factor $pf = 5$ and row threshold $r = 300$ million (i.e. 15%). The shaded candidates, for example BCD, are materialized because $2000/5=400$ million. AD is also materialized because it meets the two required criteria, namely $250 < 2000/50 = 400$ million. The procedure *compute_size* of the Smart Materialization algorithm is implemented either by calculating the actual size of the cuboid argument or by using estimating size techniques, as discussed in [2]. The cuboid size estimation technique is preferred when we want to reduce computation time. The maximal complexity is far less than 2^D, since the algorithm does not go further down the cube lattice when the row number threshold is reached. Indeed, even when $pf = 1$ and all cuboids have sizes less than M/pf, not all cuboids will be computed unless the minimum row number threshold is 1. Of course, in practice the minimum row number threshold is far greater than 1 and $pf > 1$, therefore the actual complexity is much less than 2^D. The interesting point to note is that the algorithm is run for each partition, therefore reducing the overall complexity to much less than 2^D [2].

After partitioning and Smart Materialization, the size of the overall data cube computed is considerably reduced. However, there are some design issues that needs to be addressed. The major issue is that user queries are mostly unpredictable and in most cases these queries may require combining attributes from the cubes constructed from different partitions. In such a situation, a possible solution is an online computation to answer these queries. However, online computations can be a very expensive process. In order to solve this problem, we introduce the idea of a *Smart View*.

3.2 Smart View Algorithm

As stated above, answering certain OLAP queries may require the joining of views associated with multiple fragments. For example, consider two fragments ABCD and EFGH, and a query that uses say BDF. It may be beneficial to also materialize such virtual views, in order to further minimize query costs. We refer to these views as *Smart Views*, as discussed next.

Our Smart View algorithm constantly monitors incoming queries and materializes a subset of the multi-fragment views subject to space and update time constraints. When a query is posed which cannot be answered by existing cuboids, inverted indexes is used to compute the query online from different fragment cubes. The resultant virtual view may or may not be materialized (stored). The decision is based on two criteria. The first is the number of queries answered using that view must meet a frequency threshold and the second is whether the materialization minimizes the overall query cost. Note that the algorithm is dynamic, since it adapts to new query patterns by creating new views and removing less frequently used views. Since the space for the smart views is limited, there must be a strategy in place when it becomes full. That is, a replacement policy must be applied to make space for new views. In our case, the *Least Recently*

Used (LRU) method is used. The following algorithm summarizes the process of adding views to the smart view set.

Algorithm 3. Smart view algorithm

Input: incoming query q
        frequency threshold w,
Output: B Set of smart views

1: Increment view frequency by 1
2: Search existing View pool to answer query
3: If no cuboid is found to answer q
4: Compose View from fragment cube using set intersection
5: Store View schema
6: compute size of view
7: Qualify view for addition if frequency is at least equal to w and view size is at most equal to M/pf
8: If space is available add view, otherwise call replacement policy
9: Add view to B

Figure 2 shows a sample Smart Cube made up of two fragment cubes and the Smart Views computed from fragment ABCD and EFGH. Each fragment is partially materialized with the views shaded in gray. For example, AGH is a view computed using dimension attribute A from fragment ABCD and dimension attributes GH from fragment EFGH. Notice that, although, a query with attributes AGH can be computed from ACGH, the cost of computing it from AGH is less than the threshold value M/pf. Assume the limit of the view pool storage for the Smart View is reached. Then, any extra view that meets the frequency and cost threshold will have to replace an existing view or views. The choice of the frequency threshold value (w) is determined by the expected view refresh rate. In order to replace an existing view, we check the number of times each view has been used to answer a query and sort them in ascending order. Subsequently, the view with the lowest frequency is removed. The size available is checked to see if there is enough space to add the new view. If not, the next least frequently used view is then removed. This process continues until there is enough space for the new views to be added to the pool.

Recall that the materialized cubes are implemented using inverted indexes. This facilitates easy online computation of fragment cubes using set operations, such as intersections and unions when need be. Inverted indexes are well-known methods in information retrieval. An inverted index for an attribute in a data mart consists of a dictionary of the distinct values in the attribute, with pointers to inverted lists that reference tuples with the given value through tuple identifiers (tids) [3]. In its simplest form it is constructed as follows. For each attribute value in each dimension, we register a list of tuple id (*tid*) or record id (*rid*) associated with it. The inverted index for multiple dimensions can also be computed by using the Intersection and Union set operations. For example, the

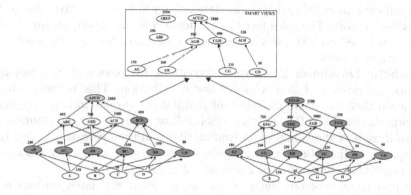

Fig. 2. A sample Smart View

computation of a two dimensional cuboid AB uses set operations to compute the intersection of the tid-lists of dimension A, and the tid-list of dimension B. In our approach, the computation of the two dimensional cuboid from a one dimensional cuboid is done in a bottom-up fashion, using the Apriori algorithm for finding frequent item-sets [1]. That is, our algorithm finds and maintains the frequently co-occurring tid values that satisfy given minimum support and confidence thresholds during the view construction process.

4   Experimental Evaluation

The experiments were conducted on a computer with 2.30 GHZ Intel Core i5 processor, 6GB of RAM, using Microsoft SQL Server 2008 R2 and the Fastbit open source library. We created our inverted indexes using the PFORDELTA compression scheme. The prototype of the cube computation algorithms were implemented in the C# programming language. In our work, we make use of query profiles, which defines a sequence of queries randomly chosen from a set of distinct query types [4]. Query profiles emulate real word query pattern where a set of queries may have frequencies of submission that are greater than, or less than, another set of queries. A query profile is the frequency distribution of query types found in the sequence. For instance, in query profile 10_90, 10% of the query types have 90% of the frequency, i.e. in a sequence composed of 100 queries and for 10 query types, the frequency of one query type is equal to 90, while the total frequency of the other types of queries is 10.

We used a number of databases in our work. Due to space constraints, we only include the results against two in this paper.

TPC-DC Database: TPC-DS is a decision support benchmark that models several generally applicable aspects of a decision support system, including queries and data maintenance [11]. This data mart contains nine (9) dimension table schema with each of their key constraints referentially connected to a fact table schema. The combined dimension attributes in the dimension tables is 80 and the number of measure attributes in the fact table is 10. This implies that a full cube materialization of the data mart will be prohibitive, due to the 2^{80} lattices to store. The sales fact table schema contains 3,000,000 tuples. The workload is based on 700 query submissions randomly chosen from a set of 30 MDX distinct queries.

Synthetic Database: This database contains 10^6 rows and 20 dimensions and was generated to follow a power law distribution. This is based on the observation that the attribute values of real dataset do not follow, in general, a uniform distribution. Rather, they often follow a power law distribution [6]. That is, if we sort the values of a given attribute in the decreasing order, then the frequency of the value of rank i is proportional to $\frac{1}{i^a}$, $a > 0$. a belongs mostly to the range [2,3]. An a value of 2 was set by inspection.

The performance of our Smart Cube algorithm is evaluated on two main criteria based. These criteria are, the overall space consumed by the data cube and the query processing time. We analyze the effect of various parameters on these criteria and also compare our approach to some state-of-the-art approaches based on these criteria, as discussed next.

4.1 Experimental Results

In this section we present the results of our experiments. First, we study the effect of various parameter changes such as performance threshold (f), fragment size, dimensionality and row threshold on the overall performance of our algorithm.

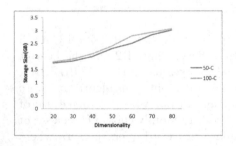

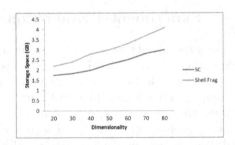

Fig. 3. Storage size of materialized cube with smart views: (50-C) T=2000000, C=50, S=0, F=3. (100-C) T=2000000, C=100, S=2, F=10

Fig. 4. Storage size of Smart Cube with Smart View (SMNV) compared with Shell Fragment (SF): (SMNV) T=2,000,000, C=50, S=0, F=8. (SF) T=2,000,000 C=50, S=0, F=3

Figure 3 and Figure 4 depict how dimensionality affect the space required to compute the data cube. In Figure 3 we analyze the effects of dimensionality on the Smart Cube. In this experiment we used synthetic databases. The first database 50-C, has per attribute cardinality of 50 and skew of 0 and the second database, 100-C, has per attribute cardinality of 100 and skew of 2. The number of tuples in each of the two databases is 2,000,000. The first database is partitioned into 8 fragments whereas the second database is partitioned into 10 fragments. These values were set by inspection. The result of this comparison shows a linear growth in storage size as dimensions grow. As can be seen from Figure 3, the size of the data cube computed for the 100-C database is slightly larger than that of the data cube for 50-C database. This is because, even though the fragment cubes for the 100-C database was larger, it had bigger fragments. Therefore, most of the auto-generated queries did not create new materialized views. Secondly the 100-C database has a cardinality of 100 as compared to the 50-C database which has a cardinality of 50. This implies that the views created by the 100-C database were larger and therefore fewer views were created for the allocated storage space. After adding the Smart View to the Smart Cube, we compared the Smart Cube to the Shell Fragment approach in terms of storage size as dimensionality grow. The Shell Fragment method has shown to reduce the size of data cubes even when dimensionality increases [6]. Figure 4 shows the result of the comparison. The database and parameter settings is the same as in the previous comparison. Note that, the space allocated for smart views affects the size of the overall cube and the computation time of the data cube. If a small space is allocated for the smart views, smaller views are added and replaced quickly since the space is quickly used up resulting in faster computation time but poor query performance. If a larger space is allocated, larger views may be added and replacement may not be that frequent. This may result in slower computation but improved performance and maintenance. Because the size of the smart view is fixed, its addition to the Smart Cube still results in smaller cube size than the Shell Fragment cube.

In order to assess how fast our algorithm is able to process queries, we executed different query profiles against our system and compared it with the state-of-the-art PickBorders and PBS algorithms. PBS is known to be a fast algorithm, and has broadly been used in comparative analysis, providing good results [6][10]. PickBorders on the other hand, has shown to produce better results than PBS while storing a fewer number of cuboids. The database used in this experiment was the TCP-DS database. A number of query profiles were used to evaluate the response time under different scenarios of query submissions. For the TCP-DS database, using $f = 11.39$, as selected by inspection, we conducted a test to compare the query processing speeds of the data cube generated. The query profiles selected include 10_90, 20_80, 50_80, 66_80 and *UNIFORM*. The *UNIFORM* profile means that all query types have the same frequency. Our Smart Cube was configured using row threshold $r = 500$, and recall that we used 30 distinct MDX queries that were generated randomly. The values of f and r were selected by inspection.

Figure 5 shows the results, which indicate that the runtime for executing queries using views created by Smart Cube is comparable to the runtime for executing queries using views selected by PickBorders, when considering all query profiles. This indicates that although Smart Cube uses less materialized views and thus less storage space, PickBorders does not outperform Smart Cube in terms of overall query execution time. This, therefore, means that the selection of larger number of views does not necessarily imply a large time reduction to process queries. Recall that we also compared the Smart Cube technique with

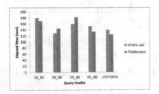
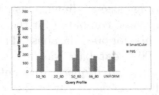
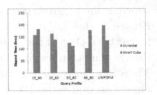
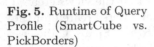

Fig. 5. Runtime of Query Profile (SmartCube vs. PickBorders)

Fig. 6. Runtime of Query Profile (SmartCube vs. PBS)

Fig. 7. Runtime of Query Profile (SmartCube vs. DynaMat)

the PBS algorithm. In Figure 6, the result shows that executing queries using the views that were created using Smart Cube is more efficient than executing queries using views selected by the PBS algorithm. This is because PBS naively selects the smallest unselected view for materialization, without taking into consideration the cost of materializing the view. The result further corroborated our initial assumption that the selection of large number of views does not necessarily imply a large time reduction to process queries.

In Figure 7, we compared our Smart Cube approach with the DynaMat algorithm, against the TPC-DS database. Dynamat is a dynamic view selection algorithm that has shown increased performance over time. The View Pool of DynaMat was simulated as was done in [8]. That is, it utilizes dedicated secondary storage for managing the cache information. In order to make the two approaches comparable, we executed each profile 50 times on the DynaMat to ensure that the Dynamat has adopted as much as possible to the queries. The average execution time was used for the comparison. The result showed that Smart Cube performed favorably when compared with DynaMat. While the Smart View method materializes the most frequent views used to compute queries, DynaMat materializes the results of the most frequent queries. DynaMat performs relatively better in situations where user queries are answered using the view pool with the results of the exact same query already stored in the pool. The query processing time of Smart Cube is relatively better than DynaMat when the view pool cannot be used to compute a user query. The query processing time is similar when user queries are answered using the DynaMat Pool but the exact same query is not stored, and thus has to be computed from existing parent view. The performance of the Smart Cube is better than DynaMat because, even when the smart views are not able to answer queries the partitioned fragment cubes equally performs well. Further, the smart views stores the base view used for answering queries, while the DynaMat Pool stores query results.

5   Conclusion

This paper introduced a novel cube computation algorithm that capitalizes on partitioning and dynamic partial materialization. Our approach facilitates the dynamic construction of materialized views, based on the user query pattern. It adapts to new query patterns by creating new views and removing less frequent ones. Our results show that our Smart Cube algorithm compared favorably to state-of-the-art methods. The approach presented in this paper can take many future directions. One direction for future research is parallelizing the fragment cube computation, thus reducing the Smart Cube construction time.

References

1. Agrawal, R., Srikant, R.: Fast algorithms for mining association rules in large databases. In: Proceedings of the 20th International Conference on Very Large Data Bases, VLDB 1994, San Francisco, CA, USA, pp. 487–499. Morgan Kaufmann Publishers Inc. (1994)
2. Aouiche, K., Lemire, D.: A comparison of five probabilistic view-size estimation techniques in OLAP. In: Proceedings of the ACM Tenth International Workshop on Data Warehousing and OLAP, DOLAP 2007, pp. 17–24. ACM, New York (2007)
3. Bjørklund, T.A., Grimsmo, N., Gehrke, J., Torbjørnsen, Ø.: Inverted indexes vs. bitmap indexes in decision support systems. In: CIKM 2009, pp. 1509–1512. ACM, New York (2009)
4. da Silva Firmino, A., Mateus, R.C., Times, V.C., Cabral, L.F., Siqueira, T.L.L., Ciferri, R.R., de Aguiar Ciferri, C.D.: A novel method for selecting and materializing views based on OLAP signatures and grasp. JIDM 2(3), 479–494 (2011)
5. Golfarelli, M., Maio, D., Rizzi, S.: Applying vertical fragmentation techniques in logical design of multidimensional databases. In: Kambayashi, Y., Mohania, M., Tjoa, A.M. (eds.) DaWaK 2000. LNCS, vol. 1874, pp. 11–23. Springer, Heidelberg (2000)
6. Hanusse, N., Maabout, S., Tofan, R.: A view selection algorithm with performance guarantee. In: Proceedings of the 12th International Conference on Extending Database Technology: Advances in Database Technology, EDBT 2009, pp. 946–957. ACM, New York (2009)
7. Kimball, R., Ross, M.: The Data Warehouse Toolkit: The Complete Guide to Dimensional Modeling, 2nd edn. John Wiley & Sons, Inc., New York (2002)
8. Kotidis, Y., Roussopoulos, N.: Dynamat: a dynamic view management system for data warehouses. SIGMOD Rec. 28(2), 371–382 (1999)
9. Li, X., Han, J., Gonzalez, H.: High-dimensional OLAP: a minimal cubing approach. In: Proceedings of the Thirtieth International Conference on Very Large Databases, VLDB 2004, vol. 30, pp. 528–539. VLDB Endowment (2004)
10. Lijuan, Z., Xuebin, G., Linshuang, W., Qian, S.: Research on materialized view selection algorithm in data warehouse. In: International Forum on Computer Science-Technology and Applications, IFCSTA 2009, vol. 2, pp. 326–329 (2009)
11. T.: Transaction processing performance council (1.1.0) (April 2013), http://www.tpc.org/tpcds/

Optimizing Reaction and Processing Times in Automotive Industry's Quality Management
A Data Mining Approach

Thomas Leitner, Christina Feilmayr, and Wolfram Wöß

FAW, Institute for Application Oriented Knowledge Processing,
Altenberger Straße 69, 4040 Linz, Austria
{thomas.leitner,christina.feilmayr,wolfram.woess}@jku.at
http://www.faw.jku.at/

Abstract. Manufacturing industry has come to recognize the potential of the data it generates as an information source for quality management departments to detect potential problems in the production as early and as accurately as possible. This is essential for reducing warranty costs and ensuring customer satisfaction. One of the greatest challenges in quality management is that the amount of data produced during the development and manufacturing process and in the after sales market grows rapidly. Thus, the need for automated detection of meaningful information arises. This work focuses on enhancing quality management by applying data mining approaches and introduces: (i) a meta model for data integration; (ii) a novel company internal analysis method which uses statistics and data mining to process the data in its entirety to find interesting, concealed information; and (iii) the application Q-AURA (*quality - abnormality and cause analysis*), an implementation of the concepts for an industrial partner in the automotive industry.

Keywords: Data Mining, Quality Management, Apriori Algorithm, Automotive Industry.

1 Introduction

Numerous methods and concepts exist for establishing and structuring quality management in companies (e.g., Six Sigma [10]), but typical systems are not capable of integrating and analyzing information of the entire business process. Such an integrated data set is necessary to extract new, unknown meaningful information. Specific approaches, such as data mining methods, enable analyzing of large data sets in order to identify concealed relationships.

Q-AURA[1] (*quality - abnormality and cause analysis*) is being developed in cooperation with *BMW Motoren GmbH*, located in Steyr, Austria, which is part of the *BMW Group*, a major player in the premium automotive industry. *BMW Motoren GmbH* as a component producer builds most of the benzine and diesel

---

[1] Qualität - Auffälligkeiten und Ursachenanalyse (German).

L. Bellatreche and M.K. Mohania (Eds.): DaWaK 2014, LNCS 8646, pp. 266–273, 2014.
© Springer International Publishing Switzerland 2014

engines for BMW automobiles. The competence center of the diesel engine development is also located there. In recent years, innovation cycles became shorter, resulting in reduced development and test time periods. Consequently, the number of engines manufactured within a year has increased: while 700,000 engines were built in 2009, the production volume reached 1.2 million units in 2011 [3].

With the higher manufacturing demands, also those of the quality management rise. As a result of a higher production rate, potentially more faulty engines could be produced, especially if the time between the fault causation in the engine development process and the productive correction is not decreased. The main goal of Q-AURA is to accelerate the problem solving process, which comprises diagnostics, reaction, and processing time. Q-AURA reduces both reaction and processing times by performing the following tasks: (i) detecting relevant engine faults (i.e., those that occurred more frequently in recent weeks), (ii) detailed fault analysis to find interesting attribute distributions in car and engine features, and, most importantly, (iii) identifying the critical technical engine modifications that may have caused the faults.

This paper is organized as follows: Section 2 discusses the central problem and associated challenges. Section 3 describes relevant approaches. In Section 4, an explanation of Q-AURA, its underlying concepts, and the evaluation of the quality management experts is provided. Section 5 concludes with future research directions and improvements to Q-AURA.

2   Problem Statement

This work makes three contributions, of which each addresses a particular problem. The first one is *overcoming the complex and heterogenous structure of the underlying information systems (IS)*. Many systems at different process steps store important data. Each individual (database) system has a different scope, which results in individual independent data models that, in turn, hinder a global (integrated) view of the entire data. Consequently, the demand for data integration is very high. For this purpose, we developed a meta model which structures the information and supports integration. We focus on an engine life cycle, which contains (i) the development phase, containing bills of materials (BOM) and technical modifications to the engine, (ii) the engine production phase followed by the car manufacturing phase, which store additional information about the engine and the car (e.g., time stamps), and (iii) the after-sales phase, which contains information about engine and car faults gathered by the dealers' workshops.

The second task is the development of a *data analysis process*. Its main purpose is to identify the technical modifications that most likely provoke a specific technical vehicle fault. Additionally, faults that progress negatively are automatically detected and analyzed in detail using statistical and data mining methods. All time-based calculations are done automatically based on the engine production time, which is also a new feature accomplished by Q-AURA. The information about the exact engine production time enables (i) to calculate relevant technical modifications that might be the cause of an analyzed fault and (ii) consequently, it enables to determine the exact engine production volume that is affected by a particular fault. The determination of technical modifications that may be relevant for a fault (based on their effective date in the assembly process) was done manually e.g., by contacting experts of the development department and using different systems for car and fault information. These manual tasks are now replaced by the proposed tool, Q-AURA, which needs a fraction of the processing time. The trends of entire faults are calculated weekly, the critical ones are automatically processed, and the results are presented to the quality management expert. Q-AURA also provides a user-driven analysis in order to start additional executions.

The final task is *implementation and verification of the invented analysis concept* to demonstrate the improvement in the quality management process(es). Q-AURA was implemented for, and is now successfully used by, the BMW quality management.

3   Related Work

Statistics and data mining techniques are a field of competence, which has been gaining more popularity in recent years. Work related to the subject of this paper can be considered under two headings: (i) state-of-the-art data mining concepts that are applied in the proposed approach (Q-AURA) and (ii) approaches that have a similar objective or apply data mining concepts similar to those in the proposed approach.

3.1   State-of-the-Art Data Mining Algorithms

Data mining applications are divided into four classes according to the type of learning: *classification, clustering, association rule mining,* and *numeric prediction* [11]. Q-AURA uses methods from the association rule mining and numeric prediction.

The purpose of *association rule mining* is to find relevant associations between features. A typical application area is market basket analysis: the goal is to find rules between sets of attributes with some defined minimum confidence, for example, that 90 % of the people who purchase bread and butter also buy diet coke. In this case bread and butter are called the *antecedent* and diet coke is the *consequent*. The relative amount is the *confidence factor*. An association rule is in accordance with well-known IF-THEN rules and should be as specific as possible while satisfying the specified minimum support [1]. Q-AURA uses the association rule mining algorithm *Apriori* [2] which is a fast discovery approach with the special characteristic that the possible item sets (set of attributes) are determined in a one-step way to improve the performance [1] [2].

Numeric prediction is used when the predicted outcome is not a discrete class but a numeric quantity. Consider, for example, two sets of values of which each value corresponds to a value in the other set. In a diagram, each pair of values is represented by a point. Depending on the application it may be attested that these points are not randomly distributed, but range more or less around a smoothed curve called *curve of regression*. If the graph resembles a straight line, the term *line of regression* is used [12]. Q-AURA uses linear regression, a numeric prediction concept, for identifying fault trends.

3.2 Related Approaches

Buddhakulsomsiri et al. [4] introduced an approach whose objective is similar to that of Q-AURA. They used association rules on warranty data in the automotive industry with the goal to identify significant relationships between attribute values and specific problem-related *labor codes*. They use various attributes of automobiles and warranty costs. The algorithm yields rules that have a variable set of attributes in the IF clause and a specific *labor code* in the THEN clause.

The difference between the Q-AURA approach and the method introduced by Buddhakulsomsiri et al. is that the main goal of Q-AURA is to find relevant technical modifications, which increases the task's complexity because of the lower ratio between attributes and fault code instances; identification of attribute characteristics that describe a specific fault is an additional benefit. Warranty cost information is ignored in the Q-AURA analysis since it does not help with finding the fault's cause.

There are several approaches that address the manufacturing industry in general, among them Harding et. al. [6], which categorized the projects by application area. To the best knowledge of the authors, there exists no approach that focuses on product improvement by analyzing faults of cars in the after sales market and finding potential faulty modifications. Q-AURA addresses this problem by using data mining concepts as described in the following section.

4 Supporting Quality Management with Data Mining Methods

This chapter describes the extent to which data mining methods can be applied to enhance quality management. Q-AURA is a system that monitors faults gathered from warranty information systems, calculates relevant attribute distributions, and identifies technical modifications that may be responsible for increasing fault rates in a specific time period. The process addressed by Q-AURA encompasses phases from early development of an engine to the after sales market. During the development phase, information about technical modifications and their position in different BOMs is documented. Data Sources of the engine production phase are queried to identify produced engines (engine serial numbers), their production time stamps, and the manufacturing BOM, which establishes the connection between an engine and its technical modifications. The following phase is the automobile production, where the connection between the engine and the car as well as attributes of the car itself (e.g., vehicle order country, vehicle type) are retrieved. The after sales information comprises warranty claims, fault codes, and other attributes that describe a particular fault. For Q-AURA, we developed a meta model to structure and harmonize this distributed and heterogeneous information.

4.1   Q-AURA Approach

The Q-AURA process is divided into six steps, illustrated in Figure 1. The first step reduces reaction time, while steps 2 to 6 reduce processing time. The most recent six weeks of warranty claims (for cars that were produced in the past three years) are used to detect current problems, which are classified by engine type, fuel type, and car brand (cf. Figure 1-1). *Regression Analysis* [12] is used to calculate the characteristics of the six-week fault pattern. We tested three different approaches to regression analysis: first, we used various convex functions to generate regression curves. Then we tested a three-point smoothing function. However, this method eliminates the first and the last value in the dataset, which are both important in Q-AURA because a sufficiently large and recent time slot must be considered to derive reliable conclusions about the trend of the fault patterns. Finally, we tested regression analysis using straight lines, which achieved the best outcome (Equation (1)):

$$y = k * x + d, \tag{1}$$

where y and x are the coordinates of a data point on the regression line, and k is the gradient. d is equal to the value of y at the point $x = 0$. We calculated the *gradient* k, the *mean value* $\bar{y}$ and the *coefficient of determination* R^2 and use these for evaluation. The gradient indicates how much the number of faults increases from one week to the next. The mean value shows the average number of faults per week, and the coefficient of determination describes the steadiness of the curve. Equation (1) shows that the regression line depends on only one variable. For this special case the coefficient of determination is equal to the square of *Pearson's Correlation Coefficient* r_{xy}^2 (Equation (2)) [8]:

$$R^2 = r_{xy}^2 = \frac{s_{xy}^2}{s_x^2 s_y^2}. \tag{2}$$

Thresholds for measures are individually defined, evaluated, and adjusted in cooperation with the quality management experts according to their domain knowledge. A fault is considered significant and analyzed further if all of these thresholds are exceeded (in this case: $k \geq 1.8$, $\bar{y} \geq 10$, $R^2 \geq 0.2$). In the second step (cf. Figure 1-2) the production week histogram of the engines with a particular fault is generated. It plots the two-year failure characteristics of up to three-year-old cars and is normalized by the total number of produced engines belonging to the same class (vehicles with the same car brand, fuel type, and engine type). Finally, the result is smoothed by a 5-point smoothing function to eliminate the outliers. Significant increases must be identified to determine the starting point of periods in which more than the average number of faulty engines were produced. In the next step (cf. Figure 1-3) significant decreases are determined, resulting in a set of time periods that are bounded by a significant increase and the subsequent decrease. Each period is analyzed in detail by initially generating a BOM distribution and normalizing it in order to identify the critical BOMs (cf. Figure 1-4). Then, the technical modifications are retrieved from the BOMs and limited by those that were set operational in a time period before and after the corresponding significant increase (cf. Figure 1-5). Finally, the subset of modifications that most likely caused the fault is computed. Two methods were implemented for determining candidate technical modifications: the first determines technical modifications which are included in most of the critical BOMs, and the second uses the *Apriori* [2] algorithm for the same task.

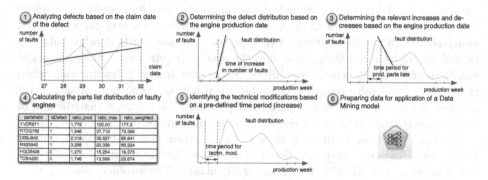

Fig. 1. Q-AURA process in detail

A more detailed description of the input and output is given in Figure 2. The first column in the input table identifies an engine (engine serial number). The next column *isDefect* determines whether the record is a faulty one or not followed by the relevant technical modifications (mod) of the corresponding BOMs. The dataset is processed by the Apriori algorithm and results in an output table, which consists of different rules that satisfy a given threshold for the calculated quality metrics *support* and *confidence*. A rule consists of at least one technical modification (mod). The modifications of a particular rule are aggregated by a conjunction operator and are those which take the Boolean value *true*. The conclusion (right-hand side of rule; currently analyzed fault code) gets true, if and only if, the premise (left-hand side of rule; technical modifications) is also true.

4.2 Application Scenarios

Q-AURA can be applied in two scenarios, which determine, on the one hand, how the user interacts with the application and, on the other, the complexity of the provided Q-AURA functionalities. The first one periodically calculates new results and the second one is a user-driven approach.

Periodical analysis is executed weekly, and therefore user interaction is not necessary. The process steps are analogous to those depicted in Figure 1, but

Input							Rules	Output			

engineNr	isDefect	GHST321	DKSCT65	AKSDT23	Z2374A0	...
73829418	1	x	x			...
10294832	0			x	x	...
23402103	1		x		x	...
30493829	1	x	x		x	...
17549219	0	x		x		...
29384753	0		x		x	...
...	...	...	...	...	...	...

rule$_1$: mod$_x$(DKSCT65) ∩ mod$_x$(GHST321) ⇒ FaultCode$_1$
rule$_2$: mod$_x$(DKSCT65) ⇒ FaultCode$_1$
rule$_3$: mod$_x$(Z2374A0) ⇒ FaultCode$_1$
rule$_x$: mod$_x$(DKSCT65) ∩ mod$_x$(Z2374A0) ⇒ FaultCode$_1$
...

mod = {mod$_1$ ... mod$_n$}

rules	support	confidence	description	...
rule$_1$	65,8	45,23	test1	...
rule$_2$	94,3	67,23	test2	...
rule$_3$	23,42	54,23	test3	...
rule$_4$	10,32	23,12	test4	...
rule$_5$	8,32	12,32	test5	...
rule$_6$	27,23	34,21	test6	...
...	...	...	...	...

Fig. 2. Input table and output table of the Apriori approach

additionally different attribute distributions are calculated to derive which cars are mainly affected by the fault (e.g., vehicle order country). For users who have some prior knowledge what the corresponding problem could be – before the system identifies them as critical – Q-AURA provides a user-parameterized analysis. We distinguish between two types of analysis: the first relies (like the periodical analysis) on fault codes, and the second requires a set of engine serial numbers or a set of vehicle identifiers. The latter one is advantageous when richer contextual information is available since it takes time until a significant number of customers observe a particular problem. Therefore, a distribution of fault codes is calculated to get deeper insights into the problem history.

Q-AURA has been evaluated, since the application is used by BMW quality management experts every day. Due to the success beside Steyr the application is deployed at the Munich and Hams Hall plants. The problem solving time was recorded before Q-AURA was applied by the quality management experts in Steyr. After Q-AURA has been used for one year it was measured again. The result was that the problem solving time concerning engines which were produced in Steyr has been reduced approximately by 2%.

5   Future Research Directions

Validation of the context matching of technical modifications with those of analyzed faults is completely subject to user interpretation. We have thus started to implement two improvements to Q-AURA. The first aims for earlier fault detection and will therefore reduce reaction time, while the second improves the accuracy of the analysis and will reduce processing time.

The first approach uses warranty and fault information available in different databases and at different stages of approval. Combining fault trends at various stages will increase the accuracy compared to using only one source, but the data must be evaluated/qualified using different data quality metrics [7]. Failure curves of the data sets are calculated and future values are predicted, which are weighted by the data quality of the corresponding data source. Significant faults are calculated by combining the predictions. Afterwards, the predictions are compared with new values of the following production week and weightings are adjusted.

The current approach identifies relevant technical modifications by means of statistics and heuristics. The second optimization incorporates also context information to improve accuracy. First, relevant data sources with additional information about technical modifications are identified and quality metrics are calculated. Afterwards, technical modifications are classified on the basis of their meaning. We propose two different models: (i) a *knowledge-based model*, based on string matching [5] and a semantic network, and (ii) a *data mining model* enriched with technical modification attributes based on classification [9]. Afterwards the *assignment* between faults and modification categories is performed. Two different approaches are proposed: (i) a *supervised* machine learning approach integrating user feedback, and (ii) an *unsupervised* machine learning approach, using string comparisons [5] and a semantic network. In the final step, the Q-AURA result tables with the modifications are filtered by context.

References

1. Agrawal, R., Mannila, H., Srikant, R., Toivonen, H., Verkamo, A.I.: Fast discovery of association rules. In: Fayyad, U., et al. (eds.) Advances in Knowledge Discovery and Data Mining, pp. 307–328. MIT Press (1996)
2. Agrawal, R., Imieliński, T., Swami, A.: Mining association rules between sets of items in large databases. In: Proceedings of the 1993 ACM SIGMOD International Conference on Management of Data, SIGMOD 1993, pp. 207–216. ACM, New York (1993)
3. BMW Motoren GmbH: Benchmark information of production plant Steyr (2012), http://www.bmw-werk-steyr.at/ information aquisition: December 10, 2012. website
4. Buddhakulsomsiri, J., Siradeghyan, Y., Zakarian, A., Li, X.: Association rule-generation algorithm for mining automotive warranty data. International Journal of Production Research 44(14), 2749–2770 (2006)
5. Hall, P.A.V., Dowling, G.R.: Approximate String Matching. ACM Comput. Surv. 12(4), 381–402 (1980)
6. Harding, J.A., Shahbaz, M., Srinivas, S., Kusiak, A.: Data Mining in Manufacturing: A Review. Journal of Manufacturing Science and Engineering 128, 969–976 (2006)
7. Heinrich, B., Kaiser, M., Klier, M.: How to measure Data Quality? A Metric Based Approach. In: Proceedings of the 28th International Conference on Information Systems (ICIS), Association for Information Systems, Montreal (December 2007)
8. Mittlböck, M., Schemper, M.: Explained Variation for Logistic Regression. Statistics in Medicine 15(19), 1987–1997 (1996)
9. Read, J., Pfahringer, B., Holmes, G., Frank, E.: Classifier chains for multi-label classification. In: Buntine, W., Grobelnik, M., Mladenić, D., Shawe-Taylor, J. (eds.) ECML PKDD 2009, Part II. LNCS, vol. 5782, pp. 254–269. Springer, Heidelberg (2009)
10. Smith, B.: Lean and Six Sigma – A One-Two Punch. Quality Progress 36(4), 37–41 (2003)
11. Witten, I.H., Frank, E., Hall, M.A.: Data Mining: Practical Machine Learning Tools and Techniques, 3rd edn. Morgan Kaufmann (2011)
12. Yule, G.U.: On the Theory of Correlation. Journal of the Royal Statistical Society 60(4), 812–854 (1897)

Using Closed n-set Patterns for Spatio-Temporal Classification

S. Samulevičius[1], Y. Pitarch[2], and T.B. Pedersen[1]

[1] Department of Computer Science, Aalborg University, Denmark
{sauliuss,tbp}@cs.aau.dk
[2] Université of Toulouse, CNRS, IRIT UMR5505, F-31071, France
pitarch@irit.fr

Abstract. Today, huge volumes of sensor data are collected from many different sources. One of the most crucial data mining tasks considering this data is the ability to predict and classify data to anticipate trends or failures and take adequate steps. While the initial data might be of limited interest itself, the use of additional information, *e.g.*, latent attributes, spatio-temporal details, etc., can add significant values and interestingness. In this paper we present a classification approach, called Closed n-set Spatio-Temporal Classification (CnSC), which is based on the use of latent attributes, pattern mining, and classification model construction. As the amount of generated patterns is huge, we employ a scalable NoSQL-based graph database for efficient storage and retrieval. By considering hierarchies in the latent attributes, we define pattern and context similarity scores. The classification model for a specific context is constructed by aggregating the most similar patterns. Presented approach CnSC is evaluated with a real dataset and shows competitive results compared with other prediction strategies.

Keywords: Pattern mining, time series, classification, prediction, context, latent attributes, hierarchy.

1 Introduction

Huge volumes of sensor data are collected from many different sources. As a result, enormous amounts of time series data are collected constantly. Raw data without post-processing and interpretations has little value, therefore multiple algorithms and strategies have been developed to deal with knowledge and behavior discovery in the data, *i.e.*, classification, pattern mining, etc. The discovered knowledge allows predicting the behavior of the system being monitored, *i.e.*, identify potential risks or trends, and perform optimization according to them. When the data distribution is stationary and the values are discrete, this is thus a classification problem. Latent attributes, *i.e.*, attribute values inferred from the raw data, has proven to be a good strategy for enhancing data mining performance. Latent attributes can represent different types of information such as temporal, *e.g.*, day and hour, spatial, *e.g.*, POI, spatio-temporal, *e.g.*, events,

L. Bellatreche and M.K. Mohania (Eds.): DaWaK 2014, LNCS 8646, pp. 274–287, 2014.

etc. Such attributes form a context for the raw data. Similar raw data records will (typically) have similar contexts.

In this paper we present a spatio-temporal classification approach, called Closed n-set Spatio-Temporal Classification (CnSC) which is based on closed n-set pattern mining in hierarchical contextual data and subsequent aggregation of the selected mined patterns. The proposed approach extends traditional classification approaches in several directions. Firstly, we utilize hierarchies in the latent attributes, *i.e.*, if none of the mined patterns match a specific context, we instead use the most similar (according to the hierarchical level). Secondly, pattern mining operations are the most time-consuming part and require special treatment. We thus propose persistent pattern storage using a NoSQL graph database (DB) and an effective scheme for mapping the mined patters to graphs. These solutions support efficient pattern storage and retrieval. CnSC is experimentally evaluated using the real-world sensor dataset from a mobile broadband network. The evaluations show that properly configured CnSC, outperforms the existing solutions in terms of classification accuracy.

The remainder of the paper is organized as follows. In Section 2 we present the relevant related work. Background and problem definitions are stated in Section 3. In Section 4 we describe CnSC. The graph database solutions and mappings are described in Section 5. We experimentally evaluate CnSC in Section 6 and conclude with our contributions and future work in Section 7.

2   Related Work

Different approaches considering patterns have been analyzed in a number of papers the recent years. The two main application areas where patters have been used are prediction and classification. Spatio-temporal pattern-based **prediction** often analyzes trajectory patterns of the moving user. User profiles defined using historical locations enable estimate future, *e.g.*, places that potentially will be visited in the city [10], opponent actions in games [5], or mobility patterns for personal communication systems [15]. In this paper we analyze the prediction potential using mined patterns.

A survey of recent pattern-based **classification** papers is presented in [1], where classification strategies are compared considering a) efficient pattern storage and operation methods, *e.g.*, use of trees or graphs; b) pattern selection and model construction for classification, *e.g.*, iterative pattern mining for the optimal model construction. Most of the pattern-based papers address a common problem, *i.e.*, huge amount of mined patterns, which requires efficient pattern operations. Rule-based classifiers C4.5 [11] and CMAR [8] store patterns in decision trees for higher efficiency. In CnSC we consider graph database [6], which allows efficiently manipulate mined patterns using SQL like queries. *Top-K* most similar pattern use for the pattern-based classification is a common strategy [1, 14]. Other strategies, such as, similarity between patterns [2] or emerging patterns [4], are used for the optimal pattern selection. In this paper we define similarity metric for pattern and context which incorporates hierarchical attribute structure.

The selected patterns further has to be aggregated into a single classification model [8]. To achieve higher classification results, pattern selection, feature extraction, and model construction can be done iteratively [1], *i.e.*, pattern mining is continued until the classification model meets required threshold. In this paper we run non-iterative process and optimize classification results by considering hierarchical structures in the data. Most often classification of the unknown values returns a default class [9] value. In CnSC this problem is solved considering hierarchical structures.

3   Background and Problem Definition

3.1   Data Format

Let S be a set of time series such that each S_i in S is on the form $S_i = (id_i, seq_i)$ where id_i is the time series identifier and $seq_i = \langle s_{i,1}, \ldots, s_{i,N} \rangle$ is the sequence of records produced by id_i[1]. Each $s_{i,j} = (tid_{i,j}, v_{i,j})$ conveys information about the timestamp, $tid_{i,j}$, and the value, $v_{i,j}$. The time series identifiers are denoted by id and the set of timestamps are denoted by tid. Seasonality can often be observed in the time series. However, since we aim at exploiting intra-seasonal patterns to model time series behaviors, this seasonality aspect is disregarded by splitting time series according to an application-dependent temporal interval T, e.g., one day, one week. This results in a bigger set of shorter time series, denoted by S^T, such that each element is on the form $S_i^k = (id_i^k, seq_i^k, sid_i^k)$ where sid_i^k is the section identifier of S_i^k and indicates which part of the original time series, S_i, is represented by it, i.e., $sid_i^k = k$.

Considering latent attributes in the mining process has often proven to be a good strategy to enhance the result quality [12]. Latent attributes can be *temporal*, e.g., the sensor value captured during the night, *spatial*, e.g., is the sensor near to some restaurants, or *spatio-temporal*, e.g., is there any traffic jam next to a sensor. The set of latent attributes, denoted by $\mathcal{A} = \{A_1, \ldots, A_M\}$, can be derived from the time series identifier and the timestamp of a record using a function denoted Map, i.e., $Map : id, tid \rightarrow \mathcal{A}$. Latent attributes can be hierarchical[2]. The hierarchy associated with the attribute A_i is denoted by $H(A_i) = A_{i,0}, \ldots, A_{i,ALL_i}$ where $A_{i,0}$ is the finest level and A_{i,ALL_i} is the coarsest level and represents all the values of A_i. $Dom(A_i)$ is the definition domain of A_i and $Dom(A_{i,j})$ is the definition domain of A_i at level $A_{i,j}$. An instance $a \in Dom(A_1) \times \ldots \times Dom(A_M)$ is called a *context* and it is a low level context if $a \in Dom(A_{1,0}) \times \ldots \times Dom(A_{M,0})$. Some notations are now introduced: $Up(a_i, A_{i,j})$ returns the unique generalization (if it exists) of a_i at level $A_{i,j}$; $Down(a_i, A_{i,j})$ provides the set of specializations (if it exists) of a_i at level $A_{i,j}$[3]; $NbLeaves(a_i)$ returns the number of specializations of attribute value a_i at the

---

[1] For ease of reading and whiteout loss of generality, we assume that each time series has the same length N.

[2] No restriction is made on the type of hierarchies.

[3] These two functions can be straightforwardly extended to contexts.

finest level $A_{i,0}$; and, given an attribute value a_i, $Lv(a_i)$ returns its hierarchical level. Finally, given a context, denoted by a, the function $LowLevelContext(a)$ returns the set of low level contexts which are a specialization of a.

A user-defined discretization function $Disc$ is introduced to map time series values into a set of L classes, denoted by $C = \{c_1, \ldots, c_L\}$. Finally, to provide the information about how often a context, denoted by a, is associated to a class value, c, we introduce the function $Support$ that is defined as $Support(a, c) = \{sid_i^j | \exists(tid_{i,k}^j, v_{i,k}^j)\ s.t.\ Map(T_i^j, tid_{i,k}^j) = a\ and\ Disc(v_{i,k}^j) = c\}$. The input dataset structure is now formally defined.

Definition 1 (Input dataset structure). *Let S be a time series dataset, T be a temporal interval, A be a set of latent attributes, and C be a set of classes derived from the time series value discretization. The input dataset, denoted by $\mathcal{D} = \{d_1, \ldots\}$, is such that each d_i is on the form $d_i = (a_i, c_i, Supp_i)$ where a_i is a context, $c_i \in C$ is a class value, and $Supp_i = Support(a_i, c_i)$.*

Example 1. *Assuming $T = 1$, time series S_1 and S_2 are split in 2 shorter time series, i.e., S_1 into S_1^1 and S_1^2, and S_2 into S_2^1 and S_2^2, see Fig.1(a). The set of latent attributes $A = \{A, B\}$ with associated hierarchies is shown in Fig.1(b) and Fig.1(c). Generalization of attribute a_{11} at level A_1 is $Up(a_{11}, A_1) = a_1$ and specialization of attribute a_1 at level A_0 is $Down(a_1, A_0) = \{a_{11}, a_{12}\}$.*

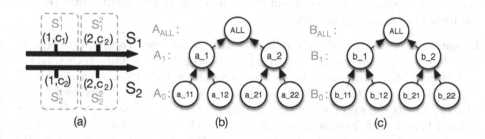

Fig. 1. (a) Time series splitting strategy with $T = 1$; (b) Hierarchy associated with the latent attribute A; (c) Hierarchy associated with the latent attribute B

3.2 $n - ary$ Closed Sets

Our classification approach relies on n-ary closed sets which are described here. Let $\mathcal{A}^c = A_1^c, \ldots, A_n^c$ be a set of n discrete-valued attributes whose domains are respectively $Dom(A_1^c), \ldots, Dom(A_n^c)$ and R be a n-ary relation on these attributes, i.e., $R \subseteq Dom(A_1^c) \times \ldots \times Dom(A_n^c)$. One can straightforwardly represent this relation as a (hyper-)cube where the measure m associated with the cell $(a_1, \ldots, a_n)$, such that $a_i \in A_i^c$ with $1 \leq i \leq n$, equals 1 when $R(a_1, \ldots, a_n)$ holds and 0 otherwise. Intuitively, a closed n-set can thus be seen as *a maximal subcube of 1's*. More formally, a n-set $H = \langle X_1, \ldots, X_n \rangle$ such that $X_i \subseteq Dom(A_i^c)$

is a closed n-set iff (a) all elements of each set X_i are in relation with all the other elements of the other sets in R, and (b) the X_i sets cannot be enlarged without violating (a). In [3], the algorithm DATA-PEELER has been proposed to efficiently solve the pattern mining problem. In this paper, this approach will be used to mine n-ary closed sets.

We consider $\mathcal{A}^c$ as the union of latent attributes and the class attribute. It thus enables the discovery of patterns that can be seen as a maximal spatio-temporal context having the same class value. Since DATA-PEELER does not consider hierarchical attributes, only low-level latent attribute values are considered during the mining phase. Moreover, even if n-ary closed patterns do not convey any information about the support of a pattern, it is possible to incorporate it very straightforwardly by considering the *Support* function as an attribute of $\mathcal{A}$.

Definition 2 (Pattern). *A pattern is in the form* $P = (A_1^P, \ldots A_n^P, C^P, Supp^P)$ *such that* $A_i^P \subseteq Dom(A_{i,0})$, $C^P \subseteq C$ *and* $Supp^P = \bigcap Support(a, c)$ *with* $a \in (A_1^P \times \ldots \times A_n^P)$ *and* $c \in C^P$. *The set of patterns is denoted by* $\mathcal{P}$.

Finally, DATA-PEELER allows to set some thresholds to prune the search space. Among them, the *minSetSize* parameter is a vector whose length is the number of attributes, and specifies the minimum size of each attribute set that compose a pattern. In this study, this parameter is used to fulfill two goals. First, mined patterns must include at least one class value, i.e., $|C^P| > 0$ for all $P \in \mathcal{P}$. Second, as shown in our experiment study analysis, it can significantly reduce the number of extracted patterns.

Example 2. *Let us assume the following Map functions:* $Map(S_1, 1) = (\{a_{11}, a_{21}\}, \{b_{12}, b_{21}\})$, $Map(S_1, 2) = (\{a_{21}\}, \{b_{22}\})$, $Map(S_2, 1) = (\{a_{12}\}, \{b_{11}, b_{12}\})$, *and* $Map(S_2, 2) = (\{a_{11}, a_{12}\}, \{b_{11}\})$. *Input for* DATA-PEELER *is shown in Fig. 2 (left). For instance, the first four cells in the left table are associated with* $Map(S_1, 1)$. *Its class is* c_1 *and after splitting T=1,* $(1, c_1)$ *belongs to* S_1^1. *Running* DATA-PEELER *with parameter* $minSetSize = (1\ 1\ 1\ 1)$ *mines patterns with minimum one value of each attribute, see in Fig. 2 (right). For instance,* P_1 *means that the context* (a_{12}, b_{11}) *is associated with the class value* c_2 *in the two time intervals, i.e., 1 and 2.*

S_1	S_2
$\{a_{11}\}, \{b_{12}\}, \{c_1\}, \{1\}$	$\{a_{12}\}, \{b_{11}\}, \{c_2\}, \{1\}$
$\{a_{11}\}, \{b_{21}\}, \{c_1\}, \{1\}$	$\{a_{12}\}, \{b_{12}\}, \{c_2\}, \{1\}$
$\{a_{21}\}, \{b_{12}\}, \{c_1\}, \{1\}$	$\{a_{11}\}, \{b_{11}\}, \{c_2\}, \{2\}$
$\{a_{21}\}, \{b_{21}\}, \{c_1\}, \{1\}$	$\{a_{12}\}, \{b_{11}\}, \{c_2\}, \{2\}$
$\{a_{21}\}, \{b_{22}\}, \{c_2\}, \{2\}$	

$$P_1 = (\{a_{12}\}, \{b_{11}\}, \{c_2\}, \{1, 2\})$$
$$P_2 = (\{a_{11}, a_{12}\}, \{b_{11}\}, \{c_2\}, \{2\})$$
$$P_3 = (\{a_{21}\}, \{b_{22}\}, \{c_2\}, \{2\})$$
$$P_4 = (\{a_{11}, a_{21}\}, \{b_{12}, b_{21}\}, \{c_1\}, \{1\})$$
$$P_5 = (\{a_{12}, \{b_{11}, b_{12}\}, \{c_2\}, \{1\})$$

Fig. 2. (left) The dataset $\mathcal{D}$ which serves as input for DATA-PEELER ; (right) n-ary closed set extracted from $\mathcal{D}$.

3.3 Problem Definition

Given a set of closed n-sets, $\mathcal{P}$, extracted from $\mathcal{D}$ with $minSetSize = v$ and an evaluation dataset, $\mathcal{D}_E$, with unknown class values, our objective is two-fold:

1. Accuracy. Given a context, denoted by $a \in \mathcal{D}_E$, performing an accurate classification requires (1) identifying an appropriate subset of $\mathcal{P}$ and (2) combining these patterns to associate the correct class to the context a.
2. Efficiency. Pattern mining algorithms often generate a huge amount of patterns. Dealing with up to millions of patterns require efficient storage and fast pattern retrieval techniques.

4 Pattern-Based Classification

We first provide a global overview of CnSC and then detail its two main steps: (1) finding patterns that are similar to the context we would like to predict and (2) combining these patterns to perform the prediction.

4.1 Overview

For each context, denoted by a, the first step is to identify similar patterns. The similarity between a pattern P and a context a is formally defined in Subsection 4.2. In a few words, similarity between P and a will be total if there exists a perfect matching between the context a and the contextual elements of P. Otherwise, the similarity will be high if P involves elements that are *hierarchically close* to the members of context a, *i.e.*, few generalizations are needed to find the nearest common ancestors of each a_i in a. Searching for these similar patterns might be particularly tricky when dealing with a huge amount of patterns. We propose an algorithm to efficiently retrieve these patterns. The similar patterns are aggregated to support the final decision according the following principle. Patterns are grouped depending on their class value(s). For each class, a score is calculated based on both the support and similarity of associated patterns. The class with the highest score is then elected.

4.2 Getting the Most Similar Patterns

Our similarity measure relies on the hierarchical nature of latent attributes and makes use of the nearest common ancestor definition.

Definition 3 (Nearest common ancestor). *Let a_i and a_i' be two elements of $Dom(A_{i,0})$. The nearest common ancestor of a_i and a_i', denoted by $nca(a_i, a_i')$, is minimum attribute at level $A_{i,j}$ which generalizes a_i and a_i':*

$$nca(a_i, a_i') = Up(a_i, A_{i,j}) = Up(a_i', A_{i,j}) \, with \, j = \operatorname*{arg\,min}_{x \in [0, ALL_i]} (x)$$

Definition 4 (Similarity measure). *Let* $P = (A_1^P, \ldots A_n^P, C^P, Supp^P)$ *be a pattern and* $a' = (a_1', \ldots, a_n')$ *be a low-level context. The similarity between* P *and* a', *denoted by* $Sim(a', P)$, *is defined as:*

$$Sim(a', P) = \frac{1}{\prod_{i=0}^{n} \min_{a_i \in A_i^P} (|Down(nca(a_i', a_i), A_{i,0})|)}$$

Example 3. *Consider the hierarchies in Fig.1 and pattern P_4 in Fig.2, we estimate the similarity between the context $a' = (a_{22}, b_{12})$ and P_4. $nca(a_{22}, a_{11}) = A_{ALL}$, i.e, $|Down(A_{ALL}, A_0)| = 4$, $nca(a_{22}, a_{21}) = a_2$, i.e, $|Down(a_2, A_0)| = 2$, and $nca(b_{12}, b_{12}) = b_{12}$, i.e, $|Down(b_{12}, B_0)| = 1$, thus $Sim(a', P_4) = \frac{1}{2*1} = 0.5$.*

Searching for these most similar patterns can be problematic when dealing with huge amount of patterns. Interestingly, by analyzing our similarity function behavior, an efficient algorithm can be designed to retrieve the patterns. The most similar patterns are obviously patterns having the same context as the one we are interested in. To avoid cases where no pattern perfectly matches the context, i.e., no classification could be performed, patterns being similar enough are also considered. To this aim, a user-defined numerical threshold, denoted by $minSim$, is introduced to retain all the patterns having similarity greater or equal than $minSim$. There are as many choices as the number of latent attributes. Since we aim at finding the most similar patterns, it is necessary to minimize the denominator of the similarity function. Thus, the generalization must be performed on the attribute value whose generalization has the smallest number of leafs. This one-by-one generalization process is repeated iteratively until no more similar pattern is found. The set of the extracted most similar patterns is denoted by $\mathcal{P}(a, minSim)$. Algorithm 1 formalizes this process.

Algorithm 1: Calculate $\mathcal{P}(a, minSim)$

Data: $a = (a_1, \ldots, a_n)$ a context, $minSim$, and $\mathcal{P}$ a set of n-ary closed sets
$\mathcal{P}(a, minSim) \leftarrow \{P \mid context(P) = a\}$;
$currSim \leftarrow 1$;
$a' \leftarrow a$;
while *true* **do**
    $next \leftarrow \arg\min(nbLeaf(Up(a_i', Lv(a_i') + 1)))$;
    $a^2 \leftarrow (a_1^2, \ldots, a_n^2)$ *s.t.* $a_i^2 = Up(a_i', Lv(a_i') + 1)$ *if* $i = next$ *and* $a_i^2 = a_i'$ *otherwise*;
    foreach $a'' \in LowLevelContext(a^2)$ **do**
        $currSim \leftarrow sim(a, P)$ *with* $context(P) = a''$;
        if $currSim < minSim$ **then**
            return $\mathcal{P}(a, minSim)$;
        else
            $\mathcal{P}(a, minSim) \leftarrow \{P \mid context(P) = a''\}$;
    $a' \leftarrow a^2$;

4.3   Combining Patterns

Once similar patterns have been found, the classification can be performed. For each class, $c \in C$, a score is calculated based on the patterns associated with c. This score is based on both the support of the pattern and its similarity with the context to classify. Algorithm 2 formalizes this process.

Algorithm 2: Classification

Data: $a = (a_1, \ldots, a_n)$ a context and $\mathcal{P}(a, minSim)$ similar patterns

foreach $c \in C$ **do**
  | $Score(c) \leftarrow 0$;
  foreach $P = (A_1^P, \ldots, A_n^P, C^P, Supp^P) \in \mathcal{P}(a, minSim)$ **do**
    | **foreach** $c \in C^P$ **do**
      | | $Score(c) \leftarrow Score(c) + (Sim(a, P) \times |Supp^P|)$;
return $\underset{c \in C}{\arg\max}(Score(c))$

Example 4. *Classification process of the context $a = (a_{22}, b_{12})$ is now illustrated considering the set of patterns shown in Fig. 2 and $minSim = 0.2$. Only P_4 is associated with the class value c_1. The score associated with this class value is thus 0.5×1, i.e., the product between the similarity between the context and the pattern and the size of the pattern support set size. For class value c_2, its score is $\frac{1}{4} \times 1 = 0.25$ (coming from P_5) since the others patterns have a similarity with a being lower than $minSim$, e.g, $Sim(a, P_1) = \frac{1}{8} < minSim$. The class value c_1 will be the one predicted here.*

5   Efficient and Compact Pattern Storage

5.1   Requirements

A well-known result in pattern mining, as well as in DATA-PEELER , is the generation of a huge amount of patterns, *i.e.*, millions or billions of n-ary closed sets. Such output is definitely non-human-readable and, even worse, it can significantly reduce its potential for being used in more complex analysis. In this study, two requirements should be met to make CnSC tractable:

- Persistent storage. As stated in the experiment results, DATA-PEELER is the most time consuming step within the whole process. For this reason, it is run only once. To avoid pattern recompilation in case of breakdown, it is preferable not to store patterns in main memory but rather persistently.
- Fast point-wise pattern query. As detailed in the previous section, the most critical operation of our method is searching for the most similar patterns of a given context. To do so, Algorithm 1 extensively queries the pattern set using point-wise queries in order to test if a pattern exists. Therefore, the data structure should be optimized for such queries.

5.2   Graph Definition

Our method is based on a graph structure for two main reasons. First, hierarchical and membership relationships can be naturally expressed in a graph-based fashion. Second, the rise of NoSQL databases, and especially graph databases, enables a persistent storage of graphs. Such graph databases are particularly of interest in our scenario since we can benefit from indexing techniques, and powerful and efficient SQL-like data manipulation languages. We now detail our model choices.

Vertices are of one single type and represent values of the latent attributes. The set of vertices is denoted V. Given $v \in V$, the function $label(v)$ returns the attribute value represented by v. Two types of relationship co-exist in the graph structure. First, an arc from v_1 to v_2 exists if v_2 is a direct generalization of v_1. Second, there exists an edge between two data nodes if (1) they do not belong to the same attribute and (2) if there exists at least one pattern where these two nodes co-occur. Some properties are attached to these edges: the class, a list of pattern identifier(s) and a list of pattern support(s). Note that this modeling enables the existence of L edges between two vertices where L is the number of classes. The pattern modeling is less intuitive and will be discussed in Section 5.4. We now formally define these relationship categories.

Definition 5 (The *IS A* relationship). *Let v and v' be two vertices. There exists a directed so-called IS A relationship between v and v', denoted by $e = (v, v')$, if Up(label(v), lv(label(v))+1) = label(v'). The set of IS A relationships is denoted by E_I.*

Definition 6 (The *CO-OCCURING* relationship). *Let v (resp. v') be a vertex with $label(v) \in Dom(A_i)$ (resp. $label(v') \in Dom(A_j)$) and $\mathcal{P}$ be a set of n-ary closed patterns. An undirected so-called CO-OCCURING relationship, denoted by $e = (v, v')$, exists between v and v' if a pattern $P = (A_1^P, \ldots, A_n^P, C^P, Supp^P)$ can be found in $\mathcal{P}$ such that $A_i \neq A_j$. Some properties are attached with such relationships: Class represents one class value associated with P, i.e., $Class \in C^P$; tPattern is a list of pattern identifiers associated with the class value Class where v and v' co-occur; and tSupp is a list that contains the support of patterns that are stored in tPattern. The set of CO-OCCURING relationships is denoted by E_C.*

Definition 7 (Graph structure). *Let $\mathcal{A}$ be a set of attributes and $\mathcal{P}$ be a set of n-ary closed sets. The graph indexing both attribute values and patterns is defined as $G = (V, E)$ with $E = E_I \cup E_C$.*

Example 5. *Figure 3 provides a graphical representation of an example graph structure considering two dimensions and a given set of n-ary closed sets.*

5.3   The Context Matching Operation

The most frequent query that is run to perform classification is to identify patterns that perfectly match a given context $a = (a_1, \ldots, a_n)$. Algorithm 3 details

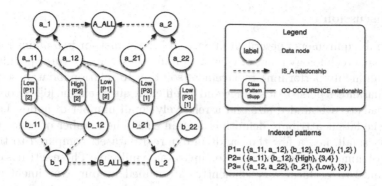

Fig. 3. Graph structure indexing two hierarchical latent attributes and three patterns

the procedure. For ease of reading, it is assumed that we are looking for patterns with one specific class i.e., $Class = c$. The algorithm can, however, be straightforwardly extended to an arbitrary number of classes by repeating this algorithm as many times as the number of class values. Intuitively, the returned set of patterns is the result of the intersection between the edges connecting every pairs of context attribute values. If there exists a_i and a_j in a such that $v = (node(a_i), node(a_j))$ is not in V_C, it implies that no pattern match the context a and the algorithm can stop by returning the empty set.

Algorithm 3: Find patterns

Data: $a = (a_1, \ldots, a_n)$ a context, c a class value, and G graph structure

$patterns \leftarrow \emptyset$;

$i \leftarrow 1$;

$First \leftarrow true$;

while $i < n$ **do**

  $j \leftarrow i + 1$;

  while $j \leq n$ **do**

    if $\exists e_C = (v, v') | v = node(a_i),\ v' = node(a_j)\ and\ class(e_C) = c$ **then**

      if $First$ **then**

        $First \leftarrow false$;

        $patterns \leftarrow tPattern(e_C)$;

      else

        $patterns \leftarrow patterns \cap tPattern(e_C)$;

        if $|patterns| = 0$ **then**

          return $\emptyset$

    else

      return $\emptyset$

    $j \leftarrow j + 1$;

  $i \leftarrow i + 1$;

return $patterns$

5.4 Discussion

The initial requirements described in Section 5.1 are met for two mains reasons. First, from a technology point of view, graph databases are mature enough to be used as an highly performing replacement solution to relational databases. This choice guarantees persistent storage and an efficient query engine [6]. Second, our model leads to a bounded size and a relatively small number of nodes, i.e., the sum of the latent attribute definition domain size. The number of edges is also relatively small, i.e., a few thousands in our real dataset, compared to the the number of mined n-ary closed sets, i.e., up to some millions. These settings enable Algorithm 3 to perform very efficiently as outlined by our experiment results presented in the next section. Finally, we discuss the *non-adoption* of another more intuitive graph model. In such an alternative model, patterns would be also represented by vertices and a relationship would exist between a data node and a pattern node if the attribute value represented by the data node belongs to the patterns represented by the pattern node. Despite this alternative model is intuitive, it would significantly increase the time needed to classify a single context. Indeed, in this scenario, the number of vertices in the graph database would be huge. Even if modern servers can deal with such amount of data, this point would be very critical regarding the query phase. Indeed, graph databases are not designed to optimally handle so-called super nodes, i.e., nodes having tens of thousands of relationships. This model has been tried out and most of data nodes were indeed super-nodes, i.e., the density of the graph was closed to 1 (complete graph) leading to very slow (up to some minutes) point-wise query response time. Since the time performance of the approach mainly depends on this query response time, the model has been abandoned.

6   Experimental Evaluation

This section is dedicated to CnSC evaluation. We first detail the adopted protocol and then present and discuss our result.

6.1   Protocol

Dataset. We consider a real dataset from Mobile Broadband Network (MBN). MBN is composed of 600 network nodes (time series producers) which monitor traffic for 5 consecutive weeks at an hourly basis. Traffic load level which represents mobile user activity, allows discretizing nodes into *low* and *high* according to traffic level at node. Technical MBN design requires dense node network for *high* traffic loads, while during *low* loads some of the MBN nodes potentially could be turned off for a temporal period. MBN traffic classification allows estimating network load at individual network elements and performing network optimization. Three latent attributes and their hierarchical structures are considered: (1) the cell identifier that can be geographically aggregated into sites; (2) the day of the week that is aggregated in either week days or week-end days;

and (3) the hour that is aggregated in periods, i.e., the night, the morning, the afternoon and the evening.

Evaluation Metrics. From an effectiveness point of view, CnSC will be evaluated using the standard classification measures, the F-measure. From an efficiency point of view, we will study running time and particularly the model building time, i.e., pattern mining and pattern insertion in the graph database, and the average evaluation time per context to classify.

Competitors. CnSC is compared with a previous work, STEP [13], and the Weka implementation of the Naive Bayes classifier [7].

Evaluated Parameters. We evaluate $minSim$ and $minSetSize$ parameter impact on CnSC. $minSim$ is evaluated in the range from 0.1 to 1 with step of 0.1 (0.8 serves as the default value). $minSetSize$ combinations, *i.e.*, $v_{Strong} = (3\ 5\ 3\ 4\ 1)$, $v_{Normal} = (2\ 3\ 2\ 3\ 1)$, and $v_{Soft} = (2\ 2\ 2\ 2\ 1)$ (v_{Normal} serves as the default value) investigate how the number of mined patters affects results.

Implementation and Validation Process. A 10-fold cross validation strategy has been used. Results presented in this section are thus averaged. The approach has been implemented in Java 1.7. DATA-PEELER software is implemented in C++[4]. The graph database used in this study is Neo4J (community release, version 2.0.0)[5] as it has been showed in [6] that it offers very good performances.

6.2 Results

$minSetSize$ **Impact.** We first evaluate the impact of this parameter on the model construction time (Fig. 4 (left)). It can be decomposed in two main steps: pattern mining and pattern insertion in the database. The less stringent this constraint, the more important the total time. Most of the time is spent to mine patterns. DATA-PEELER extracts around 40M patterns for v_{Soft}, around 10M for v_{Normal}, and only 1K for v_{Strong}. Due to a high number of extracted patterns with v_{Soft}, the proposed solution failed maintaining the graph structure. For this reason, setting v_{Soft} is not evaluated in the remaining experiments. The number of patterns indexed in the graph structure has a big impact on the classification time as stated in Fig. 4 (center). The less patterns, the lower the time to retrieve similar patterns. Moreover, this result shows the great performances of both our algorithms to find similar patterns within the graph and the Neo4j graph databases. Indeed, in the worst case, only 6ms are needed to retrieve patterns with similarity greater than 0.8 within the 10M patterns. From an effectiveness point of view, Fig. 4 (right) shows that the more patterns, the higher the F-measure. While F-measure is quite low (0.65) with v_{Strong}, it is significantly better than the two competitors when $minSetSize = v_{Normal}$ (0.85 against 0.81 for STEP and 0.78 for the Naive Bayes classifier).

---

[4] It can be downloaded at
    http://homepages.dcc.ufmg.br/~lcerf/fr/prototypes.html.
[5] http://www.neo4j.org

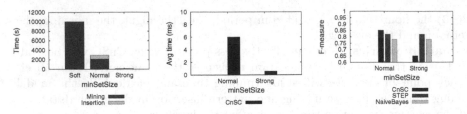

Fig. 4. *minSetSize* impact: decomposition of the running time during the model construction (left); average time to classify a single context (center); F-measure (right)

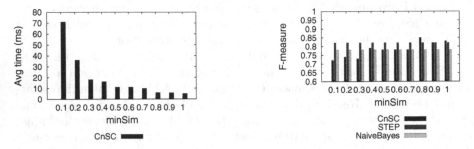

Fig. 5. *minSim* impact: average time to classify a single context(right); F-measure(left)

minSim **Impact.** The average time required to classify a context is shown in Fig. 5 (left). Good performances of the average classification time are confirmed. Moreover, as it can be expected, when similarity goes lower, more patterns need to be considered, leading to higher, though acceptable, average classification time. From an effectiveness point of view, two main conclusions can be drawn from Fig. 5 (right). First, when high *minSim* values are used, too few patterns are considered. This proves the usefulness of taking hierarchies into account to enlarge the context to also include similar ones. Conversely, too low *minSim* values leads to consider non relevant patterns in the classification process and reduces CnSC accuracy. From Fig. 5 (right) it can be seen that optimal value is *minSim* = 0.8. Second, overall performances of CnSC are comparable for most settings and are significantly better when using the optimal value of *minSim*.

7   Conclusion

This paper addressed the problems of classifying time series data and presented CnSC, a novel pattern-based approach. Indeed, closed n-ary sets are extracted from latent attributes to serve as maximal context to describe the class value. Hierarchical aspect of latent attributes is considered to incorporate similar contexts within the classification process. Use of the graph structure and NoSQL graph database allows a very efficient pattern management and classification time. With a good parameter setup, *i.e.*, *minSim* = 0.8 and *minSetSize* = v_{Normal}, CnSC offers a significantly better accuracy than two state-of-the-art approaches.

Several directions can be taken to extend this work. Among them, we can mention automatic estimation of the best parameter values and concept drift detection in the time series distribution to trigger the classification model reconstruction.

References

1. Bringmann, B., Nijssen, S., Zimmermann, A.: Pattern-based classification: A unifying perspective. CoRR, abs/1111.6191 (2011)
2. Bringmann, B., Zimmermann, A.: One in a million: picking the right patterns. Knowl. Inf. Syst. 18(1), 61–81 (2009)
3. Cerf, L., Besson, J., Robardet, C., Boulicaut, J.-F.: Closed patterns meet n-ary relations. ACM Trans. Knowl. Discov. Data 3(1), 1–3 (2009)
4. Dong, G., Li, J.: Efficient mining of emerging patterns: Discovering trends and differences. In: Proceedings of the Fifth ACM SIGKDD International Conference on Knowledge Discovery and Data Mining, pp. 43–52 (1999)
5. González, A.B., Ramírez Uresti, J.A.: Strategy patterns prediction model (SPPM). In: Batyrshin, I., Sidorov, G. (eds.) MICAI 2011, Part I. LNCS, vol. 7094, pp. 101–112. Springer, Heidelberg (2011)
6. Holzschuher, F., Peinl, R.: Performance of graph query languages: Comparison of cypher, gremlin and native access in neo4j. In: Proceedings of the Joint EDBT/ICDT 2013 Workshops, pp. 195–204 (2013)
7. John, G.H., Langley, P.: Estimating continuous distributions in bayesian classifiers. In: Eleventh Conference on Uncertainty in Artificial Intelligence, pp. 338–345 (1995)
8. Li, W., Han, J., Pei, J.: Cmar: accurate and efficient classification based on multiple class-association rules. In: Proceedings of the 2001 IEEE International Conference on Data Mining, pp. 369–376 (2001)
9. Liu, B., Hsu, W., Ma, Y.: Integrating classification and association rule mining. In: Proceedings of the Fourth International Conference on Knowledge Discovery and Data Mining, pp. 80–86 (1998)
10. Monreale, A., Pinelli, F., Trasarti, R., Giannotti, F.: Wherenext: A location predictor on trajectory pattern mining. In: Proceedings of the 15th ACM SIGKDD International Conference on Knowledge Discovery and Data Mining, pp. 637–646 (2009)
11. Quinlan, J.R.: C4. 5: programs for machine learning, vol. 1. Morgan Kaufmann (1993)
12. Rao, D., Yarowsky, D., Shreevats, A., Gupta, M.: Classifying latent user attributes in twitter. In: Proceedings of the 2nd International Workshop on Search and Mining User-generated Contents, pp. 37–44 (2010)
13. Samulevicius, S., Pitarch, Y., Pedersen, T.B., Sørensen, T.B.: Spatio-temporal ensemble prediction on mobile broadband network data. In: 2013 IEEE 77th Vehicular Technology Conference, pp. 1–5 (2013)
14. Wang, J., Karypis, G.: Harmony: Efficiently mining the best rules for classification. In: Proceedings of the Fifth SIAM International Conference on Data Mining, pp. 205–216
15. Yavas, G., Katsaros, D., Ulusoy, Ö, Manolopoulos, Y.. A data mining approach for location prediction in mobile environments. Data Knowl. Eng. 54(2), 121–146 (2005)

Short Text Classification Using Semantic Random Forest

Ameni Bouaziz[1], Christel Dartigues-Pallez[1], Célia da Costa Pereira[1],
Frédéric Precioso[1], and Patrick Lloret[2]

[1] Université Nice Sophia Antipolis, Laboratoire I3S (CNRS UMR-7271), France
bouaziz@i3s.unice.fr, precioso@polytech.unice.fr,
{christel.dartigues-pallez,celia.pereira}@unice.fr
[2] Semantic Group Company, Paris, France
plloret@succeed-together.eu

Abstract. Using traditional Random Forests in short text classification revealed a performance degradation compared to using them for standard texts. Shortness, sparseness and lack of contextual information in short texts are the reasons of this degradation. Existing solutions to overcome these issues are mainly based on data enrichment. However, data enrichment can also introduce noise. We propose a new approach that combines data enrichment with the introduction of semantics in Random Forests. Each short text is enriched with data semantically similar to its words. These data come from an external source of knowledge distributed into topics thanks to the Latent Dirichlet Allocation model. Learning process in Random Forests is adapted to consider semantic relations between words while building the trees. Tests performed on search-snippets using the new method showed significant improvements in the classification. The accuracy has increased by 34% compared to traditional Random Forests and by 20% compared to MaxEnt.

Keywords: Short text classification, Random Forest, Latent Dirichlet Allocation, Semantics.

1 Introduction

In the past few years, an expansion of interactive websites usage has been seen allowing users to contribute to their content. The users contributions are generally short texts (comments in social networks, feedback in e-commerce websites, etc.). This has led to huge amounts of stored short texts with an increasing need to classify them and to extract the knowledge they contain.
Traditional text classification process usually implies 3 steps:

- *Pre-processing among which feature extraction from texts*: a set of words are carefully chosen from the texts in order to be used in the document representation. The bag-of-words model is the most common process for this purpose. In such a model, the occurrence (or frequency) of each word in the document is used for representing the documents.

L. Bellatreche and M.K. Mohania (Eds.): DaWaK 2014, LNCS 8646, pp. 288–299, 2014.

- *Learning*: the set of document representations obtained after the pre-processing phase is used for training a classifier (or classification model).
- *Classification*: in this last step the classification model is used on new texts to determine which class they belong to.

As pointed out by [1] and others, the performance of using traditional text classification methods, (like random forest for example), for short text classification is limited, because of the word sparseness, the lack of context information and informal sentence expressiveness. A common method to overcome these problems is to enrich the original texts with additional information. However, data enrichment can also produce noise and it is then important to take the real benefit of these new data in the classification process into consideration before using them.

Some authors [1,2,3,4] proposed new methods which are essentially based on the enrichment of short texts , on "intelligent" reduction of the number of irrelevant document features and/or on the combination of them. However, none of those methods proposed to combine a semantic enrichment both for each word in the text and for the text as a whole entity. Furthermore, to the best of our knowledge there is no work in the literature exploring the semantic behind the short texts through out the construction of the trees in a random forest.

We propose a new Random Forest method impacting on the first two previously mentioned classification steps as follows:

- At the pre-processing step, we use the Latent Dirichlet Allocation (LDA)[11] model to derive topics from existing texts. The short texts are then enriched, in a two-level process, in order to increase the number of features (words). More precisely, first, the text is enriched with the topics that are more similar to each word in the text. Besides, in order to also take into consideration both the overall context information and the informal expressiveness of the short text, we further enrich the text with the words from the four topics which are more relevant with respect to the whole text.
- At the learning step, the construction of the trees is based on a random selection of the features obtained in the previous step. The information gain concerning these features determines the feature to be chosen at each node. We adapt the learning step in order to also exploit the semantic relations between the features.

The above two-level enrichment provides a first improvement in the quality of the classification model. Exploiting semantics when constructing the trees provides a further improvement. These are the two original contributions of our work.

The rest of the paper is organized as follows Section 2 presents related works, Section 3 introduces our classification method. Experimental results are presented in Section 4. Finally, Section 5 concludes the paper and presents some potential ideas for future work.

2　Related Works

Random Forest introduced by Breiman in [5] are one of the most known and effective machine learning algorithms in data mining and particularly in text classification. They base their classification on a majority vote of decision tree classifiers. Many implementations were proposed in the literature depending on the nature of data to classify.

Forest Random Combination [5] is an extension of the traditional implementation that creates new features by combining linearly existing ones, this helps in increasing the size of the features space in case of small datasets. Geurts et al. proposed in [6] the Extremely Randomized Trees where they modify the feature selection process by making it totally random instead of based on a partitioning criterion. Geurts et al. increase diversity in trees but make them bigger and more complex. To solve classification problems for unbalanced datasets, Chen et al. in [7] proposed two solutions: Balanced Random Forest, based on sampling to balance data, and Weighted Random Forest which assigns weights to features according to their representativeness. This weighting is taken into account at feature selection and at majority votes.

All these algorithms gave good performances when applied on data they were designed to classify. However, when it comes to short text classification, they are less efficient, indeed short texts have many characteristics that make their classification a challenge. The most obvious characteristic is the shortness of the texts [8] because they are composed of few words up to few sentences at maximum (tweets, for instance, do not exceed 140 characters). As a consequence those texts do not provide enough data to measure similarities between them. For example, there is no way to know that two short texts, one composed of synonyms of the other, share the same context without considering semantic relations between the words which compound them.

Short texts are characterized by their data sparseness [8]. As input of learning algorithms a text may be represented by a vector containing the weights of its words. Because of that sparseness, vectors representing short texts are nearly empty. This lack of information implies a poor coverage of the classes representing the dataset, thus complicating the correct classification of new data.

Most of the works on short text classification propose to use semantics to overcome these challenges by either reduction or enrichment of features. Both of them use an external source of knowledge to provide a semantic network that builds relations between features. The semantic network can be an ontology (WordNet or Wikipedia) [9,10] or it can use topic model techniques (LDA, Latent Semantic Indexing [12], ...) on large scale data collections to semantically group them into topics.

Yang et al. [1] come with a reduction approach based on topic model, they combine semantic and lexical aspects of short text words. With this approach the feature space dimension is no longer equal to the whole number of words but is reduced to the number of topics. A mapping based on semantic and lexical features allows to transform the vector representing a text in the word space to a smaller representation in the topic space. This method achieved efficient classification, however it seems to be only efficient on equally distributed data over the different classes.

In [8] Phan et al. introduce a new method of short texts enrichment, they apply LDA to generate topics, then they integrate these topics in the short texts. This method combined with the Maximum Entropy (MaxEnt) learning algorithm [13] gave a high classification accuracy even when tested on small size datasets. However as pointed out by Chen et al. in [3] the Phan's approach [8] may have limits owing to the fact that generated topics are considered of a single level. To tackle this problem, Chen et al. propose an algorithm that selects the best subset of all generated topics. That subset contains topics which help more discriminating short text classes while being separated enough from each other. They therefore ensure that short text enrichment is done only based on relevant topics and that no noise is introduced by adding less important topics.

All works above improve short text classification but we noticed that they focused exclusively on the texts pre-processing phase, no study has been carried out to adapt the various learning algorithms to use semantics during the classification phase, in particular with random forests. In some works interested in Random Forest prediction tasks like [14] and [15], some attempts have been made to benefit from semantic relations between features during the learning process.

Caragea et al. in [14] organize features in a hierarchical ontology and combine it with four learning algorithms to predict additional friendship links in a social network. Best results were obtained with Random Forests. In [15] Chen and Zhang propose a new Random Forest implementation named module Guided Random Forests in order to predict biological characteristics. They create a correlated network of features and group them into modules which are used to guide the feature selection in all the nodes of the trees.

Although these works are not really linked to text classification , their results on prediction tasks encourage to implement a new Random Forest method that takes benefit from semantic links between words to improve short text classification.

3   Description of the Proposed Method

In this section we will describe in detail: (i) how the short texts are enriched, (ii) how the document's features are selected, and (iii) how the semantic of the words are considered in the Semantic Random Forests.

3.1   Short Text Enrichment

To overcome the short texts classification issues mentioned above, like in [8], we propose to enrich short texts with words semantically related to their words. Thus we transform the vector representing a short text in a larger one containing more information.

The proposed enrichment method relies on the semantic relations between the words in the short texts, we use an external semantic network as a source for these relations. The network must satisfy two conditions, first it has to be specific to our dataset and second it has to cover all our data domains. To build this semantic network we apply the Latent Dirichlet Allocation (LDA)[11] on an external source of data. LDA is a topic model method that allows to discover underlying structure of a big set of data. It is a generative probabilistic method that uses Dirichlet distribution to identify hidden topics of a dataset. Then, based on similarity calculation it associates each word to one or many of the generated topics. Each word has a weight representing its importance in a topic. Thanks to LDA we obtain a semantic network where nodes are topics and words, and edges are links between topics and word. The edge's weight is the corresponding word weight in the topic. The obtained network is used in enriching short texts following this algorithm:

Enrichment Algorithm

```
program EnrichTexts (Output: set of  enriched texts)
   Input: set of short texts, semantic network
   for shortText in short texts:
      // words enrichment Procedure
      topicsForAllTextWords = []
      for each word in shortText:
         topicWeight = [] //map of topics and the
         //weight of the current word in them
         for each topic in topics:
            if word in topic:
               topicWeight[topic] = wordWeight
            else:
               topicWeight[topic] = 0
         bestTopicForWord = max(topicWeight)
         topicsForAllTextWords += bestTopicForWord

      // whole short text enrichment procedure
      similarities = []
      bestTopics=[]
      for topic in topics:
         similarity = computeSimilarity(shortText,topic)
         similarities.add(similarity)
       bestTopics = find_N_BestTopics(similarities)

      // Adding found topics to current short text
      for topic in topicsForAllTextWords:
         shortText.add(topic)
      for topic in bestTopics:
         shortText.add(topic)
end.
```

To enrich a short text, our algorithm considers the short text into 2 different ways:

- a text as a set of words taken separately: for each word, the algorithm looks for the nearest topic to the word, which means the topic in which this word has the biggest weight. Then we add to this word in the short text all the words of the found topic. This process transforms short texts in a set of general contexts. It allows to build a generic model that is able to classify any text related to the domains of the initial short texts,
- a text as one entity: the goal here is to give more importance to the general meaning of the text. The algorithm computes the similarities between all the topics and the text. The similarity is defined as the number of common words between a text and a topic. Then the algorithm chooses the N nearest topics, which means the N topics with the highest similarities. Finally, the algorithm adds all the words of the chosen topics to the text.

To illustrate the enrichment algorithm, let us imagine we want to enrich the following short text (obtained after lexical pre-processing):

national football teams world

The enrichment will rely on a set of four topics defined as follows (the numbers are the weights of the words in the topics):

- Topic 1: sport (0.6), football (0.2), teams (0.15), goal (0.05).
- Topic 2: music (0.4), instrument (0.3), songs (0.2), piano (0.1).
- Topic 3: volleyball (0.5), teams (0.3), win (0.12), basketball (0.08).
- Topic 4: web (0.45), site (0.3), programs (0.2), computer (0.05)

Words *national* and *world* do not belong to any topic, they are not enriched, however the word *football* is enriched with topic 1 as this is the only topic it belongs to. The word *teams* belongs to topics 1 and 3, but only topic 3 is used in enrichment as the weight of *teams* in topic 3 is bigger than its weight in topic 1. After enrichment at word level the short text becomes:

national football sport football teams goal **teams** volleyball teams win
basketball **world**

For the whole text enrichment process, if we consider adding the two nearest topics to the texts, topics 1 and 3 are used as they contain the biggest number of common words with the text. At the end, the enriched text is:

national football sport football **teams** goal teams volleyball teams win
basketball **world** sport football teams goal volleyball teams win basketball

3.2   Semantic Random Forest

After having enriched short texts, we focus now on the Random Forest learning process. Random Forest are sets of decision trees whose nodes are built from a set of features obtained at pre-processing step. In our case features are all the words of the enriched texts. We propose a new implementation named Semantic Random Forest that reduces the random feature selection in favor of a semantic driven feature selection. Indeed, in traditional Random Forest, all the features of the corpus are used in building the trees, for each node the algorithm selects randomly K features, then it calculates information gain of each feature (using Gini criterion or Entropy). The feature that provides the best gain is chosen for the node. In Semantic Random Forest this process is slightly modified to become semantically driven: our goal is to have trees whose nodes are semantically linked. We need in a first step to organize the whole set of features in a way where those that are semantically linked are grouped together. We use again LDA to group features into topics. Again, each feature is assigned to a weight representing its importance in a topic. In a second step we modify the tree building process as follows: instead of using all the features to construct a tree, the algorithm makes a first random selection of a small set of features, then for each of these chosen features it looks for the topic where this feature has the maximum weight, then it adds from this topic all features that have weights bigger than a given threshold. Finally the tree is built based on this new set of features by applying the same random feature selection method.

This method allowed us to obtain trees composed of nodes belonging to the same topics, so semantically linked, nevertheless, the initial small feature set selection ensures that we keep the randomness of the forest and avoid then correlation between trees. It also ensures that features not belonging to any topic are not discarded.

Semantic Random Forest

```
program SemanticRandomForest (Output:Semantic Trees)
    Input: set of features extracted from enriched texts
           L, //number of features to be chosen randomly
           K, //total number of features
           threshold //minimum weight of considered words
    // group all features semantically into topics
    Topics = apply LDA on the features set
    For each tree in trees:
        select L features < K
        // build a new set of semantically linked
        //features and store it totalList
        totalList= []
        For each feature in features:
            // look for features from the nearest topic to
            //the current feature
            featuresToAdd = []
            featuresToAdd =  SFS (feature, topics, threshold)
```

```
          totalList += featuresToAdd
       Construct tree using totalList
       //using standard tree building algorithm
end.
```

Semantic Features Selection

```
program SFS (Output:featuresList)
   Input: feature, topics, threshold
   nearestTopic = FindNearestTopic (feature,topics)
   //the topic in which the feature has the biggest weight
   for each feature in nearestTopic:
       if feature.weight >= threshold
           featuresList += feature
end.
```

4 Experiments and Results

4.1 Experimental Dataset and evaluation criterion

To validate our new method we tested it on the "search snippets" dataset. This dataset was collected by Phan [8] and is composed of:

- *short texts corpus*: built from top 20 to 30 responses given by Google search engine for different queries. Each response is composed of an URL, a title and a short description. Each short text is labeled by a class according to the submitted query. The whole corpus contains 8 categories and is divided into training data and test data as shown in table 1.

Table 1. A summary of the distribution of the short texts in the corpus

Categories	Train	Test
Business	1200	300
Computer	1200	300
Culture Art Entertaiment	1880	330
education science	2360	300
engneering	220	150
Health	880	300
Politics Society	1200	300
Sport	1120	300

- *Universal dataset*: this dataset is composed of a set of documents collected from Wikipedia as a response to queries containing some specific keywords. The application of LDA on this document set generated 200 topics with 200 words each. We used these topics as a semantic network for short text enrichment.

To evaluate our algorithm we used the accuracy, defined as follows:

$$Accuracy = \frac{TP + TN}{TP + TN + FP + FN}$$

where:

TP: Number of true positive; TN: Number of true negative
FP: Number of false positive; FN: Number of false negative

4.2   Results and Interpretations

Before applying Semantic Random Forest on search snippets dataset, we tested in a first step the enrichment of texts taken as whole entities combined with traditional Random Forest. The goal is both to evaluate this enrichment's contribution in improving the accuracy and also to find the best number of topics to be used for enriching the short texts. We run tests using a traditional Random Forest implementation provided in the scikit learn library [16], the tests were run first on the search snippets texts without enrichment, then enriched with respectively 1, 2, 3, 4 and 5 topics. All this was repeated using 10 then 20 up to 100 trees. Table 2 and figure 1 summarize the obtained results.

Table 2. Classification accuracy depending on trees number with enrichment of 1 to 5 topics

	10	20	30	40	50	60	70	80	90	100
No Enrichment	0.57	0.592	0.598	0.589	0.589	0.592	0.594	0.595	0.6	0.59
1 topic	0.64	0.643	0.669	0.68	0.679	0.681	0.679	0.684	0.678	0.68
2 topics	0.662	0.666	0.667	0.668	0.674	0.675	0.675	0.675	0.673	0.676
3 topics	0.66	0.667	0.671	0.673	0.675	0.672	0.67	0.672	0.672	0.67
4 topics	0.674	0.696	0.697	0.7	0.704	0.706	0.705	0.707	0.703	0.707
5 topics	0.675	0.693	0.694	0.7	0.702	0.705	0.705	0.705	0.703	0.702

These results show that enriched texts classification is better than short texts classification whatever the number of trees considered. Indeed, the accuracy increased from 0.59 to 0.707 when using 100 trees and an enrichment of 4 topics. As we can see in the graph above, the experience shows also that the best number of topics to add to each short text is 4. Indeed, starting from the fifth topic, texts and topic become semantically far from each other, adding topics is then a noise introduction rather than an enrichment. We evaluate now the added value of the full enrichment (text as an entity and as a set of words) and the semantic random forest. We enrich the text by our two enrichment processes: adding the 4 nearest topics to whole text and nearest topic to each word. We apply both traditional and semantic random forest on these enriched texts. These tests are repeated 10 times varying the tree number each time. The obtained results are shown in the table 3 and figure 2.

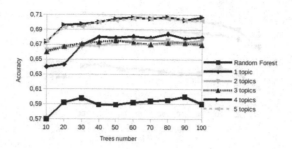

Fig. 1. Variation of short texts classification accuracy depending on trees number for short texts enriched from 1 to 5 topics.

Table 3. Classification accuracy depending on trees number with Random Forest, Random Forest and enrichment and with Semantic Random Forest.

	10	20	30	40	50	60	70	80	90	100
Random Forest	0.57	0.592	0.598	0.589	0.589	0.592	0.594	0.595	0.6	0.59
Enrichment only	0.728	0.745	0.753	0.755	0.758	0.758	0.76	0.761	0.76	0.766
Enrichment + SRF	0.73	0.761	0.771	0.776	0.7763	0.78	0.784	0.784	0.786	0.789

The results in table 3 show the classification improvement obtained thanks to our two enrichment methods. By combining whole text enrichment and word by word enrichment the accuracy increased up to 0.766 for a test with 100 trees, this is higher than the accuracy of 0.707 obtained in our first experiment where only whole text enrichment was done. After this second test, compared to classification without enrichment, we achieved already an accuracy increase from 0.59 to 0.766 which represents a 30% improvement.

Applying the Semantic Random Forest instead of the traditional Random Forest on the same enriched texts, we obtained our best classification results, the accuracy reached the value of 0.789 for 100 trees. SRF allowed an additional improvement of 3% compared to the classification of enriched texts using traditional Random Forests. Our method combining text enrichment and SRF allowed a total increase of 34% compared to traditional short text classification. Results show also that almost the same improvement rates were obtained regardless of the forest trees number which confirms the efficiency of our method.

The classification of the search snippets dataset was also tested on the Maximum Entropy method [13]. The accuracy obtained with MaxEnt was 0.657 which is better than traditional Random Forest but lower than the accuracy obtained by applying our method whose improvement reaches 20%.

In addition to the classification improvement, our method allowed a significant reduction in the trees size. Table 4 shows the evolution of the trees size concerning the 10 trees test. We can notice that the average number of the nodes of Semantic Random Forest trees (2826) is less than the half of those of Random Forest trees(6272).

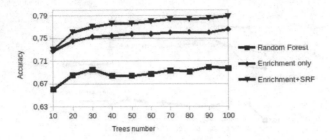

Fig. 2. Variation of short texts classification accuracy depending on trees number for short texts with traditional Random Forest, enriched texts with traditional Random Forest and Enriched texts with enriched Random Forest.

Table 4. Number of tree nodes for Random Forest and Semantic Random Forest.

Tree	1	2	3	4	5	6	7	8	9	10	average
Nodes in RF	6355	6451	6473	6119	5935	6143	6193	6339	6061	6657	6272
Nodes in SRF	2957	2883	2657	2653	2879	2851	2873	2717	3037	2753	2826

5   Conclusion and Future Work

In this work, we proposed a new two-step short text classification approach. The first step is dedicated to text enrichment. The enrichment algorithm relies on Latent Dirichlet Allocation for generating, from an external source of data, topics which are further added to the short text. In our method, topics are considered at word-level, for their similarities with the words in the text, and at text-level for their relevance with the whole text content. The second step is our new Random Forest algorithm that we call Semantic Random Forest (SRF). The SRF learning process is different from the traditional Random Forest in that it considers the semantic relation between features to select the features involved in decision tree construction. We applied our method to the search-snippets dataset and obtained results showing a classification improvement which reaches 34% compared to traditional Random Forest and 20% compared to MaxEnt. We provide a further improvement by exploiting semantics when constructing the trees. We think that by introducing semantic relations at node-level for feature selection could lead to new improvements. Our method is also fully compliant with the Weighted Random Forest approach, thereby allowing extension of our semantic-based approach for unbalanced datasets, with under- and over-represented categories, such as the search-snippets classic dataset. It would be also interesting to apply our method on different datasets. This will allow us to explain how to generally define the parameters of our algorithms and study their complexities.

Acknowledgments. This work has been co-funded by Région Provence Alpes Côte d'Azur (PACA) and Semantic Grouping Company (SGC).

References

1. Yang, L., Li, C., Ding, Q., Li, L.: Combining Lexical and Semantic Features for Short Text Classification. In: 17th International Conference in Knowledge Based and Intelligent Information and Engineering Systems - KES (2013)
2. Amaratunga, D., Cabrera, J., Lee, Y.S.: Enriched Random Forests. Bioinformatics 24(18), 2010–2014 (2008)
3. Chen, M., Jin, X., Shen, D.: Short Text Classification Improved by Learning Multi-Granularity Topics. In: 22nd International Joint Conference on Artificial Intelligence (2011)
4. Song, Y., Wang, H., Wang, Z., Li, H., Chen, W.: Short Text Conceptualization using a Probabilistic Knowledge base. In: 22nd International Joint Conference on Artificial Intelligence, pp. 2330–2336 (2011)
5. Breiman, L.: Random Forests. Machine Learning 45, 5–32 (2001)
6. Guerts, P., Ernst, D., Wehenkel, L.: Extremely randomized trees. Mach. Learn. 63, 3–42 (2006)
7. Chen, C., Liaw, A., Breiman, L.: Using Random Forest to Learn Imbalanced Data (2004)
8. Phan, X.H., Nguyen, L.M., Horiguchi, S.: Learning to Classify Short and Sparse Text & Web with Hidden Topics from Large-scale Data Collections. In: www 2008 Data Mining-Learning, Beijing, China (2008)
9. Hu, X., Zhang, X., Caimei, L., Park, E.K., Zhou, X.: Exploiting Wikipedia as External Knowledge for Document Clustering. In: KDD 2009, Paris, France (2009)
10. Hu, X., Sun, N., Zhang, C., Tat-Seng, C.: Exploiting Internal and External Semantics for the Clustering of Short Texts Using World Knowledge. In: CIKM 2009, Hong Kong, China, pp. 2–6 (2009)
11. Blei, D., Ng, A., Jordan, M.: Latent Dirichlet Allocation. Journal of Machine Learning Research, 993–1022 (2003)
12. Dumais, S.T.: Latent Semantic Indexing. In: TExt REtrieval Conference, pp. 219–230 (1995)
13. Berger, A., Pietra, A., Pietra, J.: A maximum Entropy Approach to Natural Language Processing. Computational Linguistics 22(1), 39–71 (1996)
14. Caragea, D., Bahirwani, V., Aljandal, W., Hsu, W.: Ontology-Based Link Prediction in the LiveJournal Social Network. In: 8th Symposium on Abstraction, Reformulation and Approximation (2009)
15. Chen, Z., Zhang, W.: Integrative Analysis Using Module-Guided Random Forests Reveals Correlated Genetic Factors Related to Mouse Weight. Plos Computational Biology 9, e1002956 (2013)
16. Scikit-Learn Machine Learning in Python, http://scikit-learn.org

3-D MRI Brain Scan Classification Using A Point Series Based Representation

Akadej Udomchaiporn[1], Frans Coenen[1],
Marta García-Fiñana[2], and Vanessa Sluming[3]

[1] Department of Computer Science, University of Liverpool, Liverpool, UK
{akadej,coenen}@liv.ac.uk
[2] Department of Biostatistics, University of Liverpool, Liverpool, UK
m.garciafinana@liv.ac.uk
[3] School of Health Science, University of Liverpool, Liverpool, UK
vanessa.sluming@liv.ac.uk

Abstract. This paper presents a procedure for the classification of 3-D objects in Magnetic Resonance Imaging (MRI) brain scan volumes. More specifically the classification of the left and right ventricles of the brain according to whether they feature epilepsy or not. The main contributions of the paper are two point series representation techniques to support the above: (i) Disc-based and (ii) Spoke-based. The proposed methods were evaluated using Support Vector Machine (SVM) and K-Nearest Neighbour (KNN) classifiers. The first required a feature space representation which was generated using Hough signature extraction. The second required some distance measure; the "warping path" distance generated using Dynamic Time Warping (DTW) curve comparison was used for this purpose. An epilepsy dataset used for evaluation purposes comprised 210 3-D MRI brain scans of which 105 were from "healthy" people and 105 from "epilepsy patients". The results indicated that the proposed process can successfully be used to classify objects within 3-D data volumes.

Keywords: Image mining, Point series based representation, Image classification, 3-D Magnetic Resonance Imaging (MRI).

1 Introduction

Image mining is concerned with the extraction of useful information and knowledge from image data. The representation of the raw image data in a format that allows for the effective and efficient application of data mining techniques is key to the success of applied image mining. The need for appropriate representation is particularly acute with respect to 3-D image data because the size of 3-D image sets, in comparison with 2-D image sets, are significantly larger. An important challenge for 3-D image mining is thus the need for techniques that can cope with the large amount of data that has to be considered.

The domain of image mining can be divided into whole image mining and Region Of Interest (ROI) and whole image mining where ROI image mining is

L. Bellatreche and M.K. Mohania (Eds.): DaWaK 2014, LNCS 8646, pp. 300–307, 2014.

directed at some specific sub-image that exists across an image collection. In this paper ROI image mining is considered, more specifically 3-D ROI image mining, or Volume Of Interest (VOI) image mining. Two VOI image representation techniques are proposed: Disc-based and Spoke-based. Both are point series representation techniques where the 3-D volume of interest is captured in terms of a series of points referenced using a 2-D coordinate system so that the point series can be plotted as a curve. Using a K-Nearest Neighbour (KNN) technique point series derived from unlabelled images can be compared with point series from images where the label is known and consequently a label can be identified for the new image. With respect to the work described in this paper the "warping path" distance, generated using Dynamic Time warping curve comparison [1], is proposed as an appropriate KNN distance measure. Alternatively the curves can be processed further so that a feature vector representation is generated compatible with standard classifier generation techniques. In this paper it is suggested that Hough signature extraction [5] be used for this purpose. Both approaches are explored in this paper.

To act as a focus for the work 3-D Magnetic Resonance Imaging (MRI) brain scan data was used. The 3-D MRI scans consisted of a sequence of 2-D "slices" in three planes: Sagittal (SAG), Coronal (COR), and Transverse (TRA). The VOIs in this case were the lateral (left and right) ventricles. The ventricles are fluid-filled open spaces at the centre of the brain; there are four ventricles in a human brain, but in this paper only the lateral ventricles are considered. Examples of a number of 3-D MRI brain scan slices, one in each plane, are shown in Figure 1 where the lateral ventricles are the dark areas at the centre of the brain.

The rest of the paper is organised as follows. Section 2 describes the point series generation process. This is followed in Section 3 with a description of the suggested classification processes to be adopted. The experimental set-up and the evaluation of the proposed techniques is then presented in Section 4. Finally, the paper is summarised in Section 5.

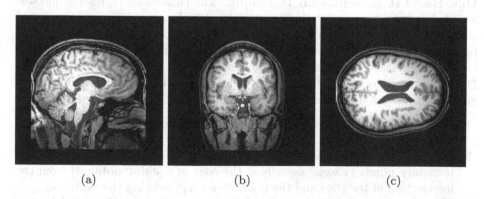

(a)   (b)   (c)

Fig. 1. Example of a 3-D brain MRI scan. (a) Sagittal (SAG) plane; (b) Coronal (COR) plane; (c) Transverse (TRA) plane

2   Point Series Model Generation

The proposed point series model generation process is illustrated in Figure 2. From the figure it can be seen that a segmentation process is first applied to the raw MRI brain scan data so as to identify the objects of interest (the left and right lateral ventricles). The authors' thresholding technique, described in [8], was used for this purpose. Next point series curves were generated to represent the identified ventricles. This process is the central contribution of this paper and is described in detail in Sub-section 2.1. The next step is the extraction of signatures from the generated curves. This was done using Hough signature extraction [5] which is described further in Sub-section 3.

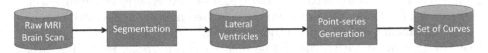

Fig. 2. Point series generation process (rectangular shapes indicate processes)

2.1   Point Series Generation

The two techniques whereby a given object can be defined in terms of a point series are presented in this section: (i) Disc-based and (ii) Spoke-based. The input for both is a binary-valued point cloud where the voxels are labelled as being either black (belonging to the object of interests) or white (not part of the object of interest). The output in both cases was a set of point series curves describing the object of interest. In total six curves were generated for each MRI image, three describing the left ventricle and three the right ventricle. For each ventricle (left and right) one curve was generated with respect to each plane: (i) Sagittal (xy), (ii) Coronal (yz) and (iii) Transverse (xz). The distinction between the two techniques is how the curves are generated.

Disc-Based Representation Technique. The Disc-based representation was founded on the idea of collecting point series data by considering a sequence of slices slice by slice (in some dimension) and collecting point information from the boundary where each slice and the object of interest intersect. The intersection is usually described by a circular shape hence the technique is referred to as the Disc based technique. The technique is illustrated in Figure 3. The point series generation process using the Disc-based technique is as follows:

1. Find the geometric centroid of the ventricle under consideration (left or right).
2. Define a slice and calculate the distance from the identified centroid to the boundary points (voxels) describing the edge of the disc obtained from the intersection of the slice and the point cloud representing the ventricle.
3. Move the slice one pixel along the selected axis and repeat (2) until the entire object has been defined.
4. Use the collected distances to define a curve (point series) with distance along the Y-axis and the sequential point number along the X-axis.

Spoke-Based Representation Technique. The Spoke-based representation techniques is illustrated in Figure 4. The technique involves measuring the distance from the geometric centroid of the object of interest to points on the boundary, in a given plane. The effect is that of a sequence of spokes of different length radiating from the centroid hence the name of the technique. A point series curve is again generated which can be plotted as a curve with distance (length of spoke) plotted along the Y-axis and the sequential point number on the X-axis. The generation process is as follows:

1. Finding the centroid of the ventricle.
2. Generate a spoke, in some pre-define plane, radiating from the centroid to the edge of the object of interest (a ventricle in our case) and measure and record its length.
3. Repeat (2) with a new spoke generated by rotating the previous spoke by an angle of $\theta°$. Continue in this manner until all $360°$ have been covered.
4. Use the collected distances to define a curve (point series) with spoke-distance along the Y-axis and the sequential point number along the X-axis.

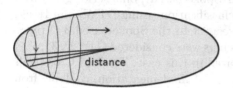   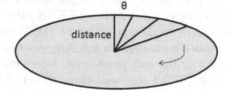

Fig. 3. The Disc-based technique   **Fig. 4.** The Spoke-based technique

3 The Classification Process

Once the set of curves defining the object of interest had been identified these can be used directly to classify unseen data using a KNN style classification approach [2]. Alternatively the curves can be processed further to create a vector space model. The first required some kind of measure to determine the similarity between curves in the curve base and a new unseen curve. To this end Dynamic Time Warping (DTW) was used [1]. The second required some mechanism for generating the desired feature space. A signature based approach, founded on Hough signature extraction [5], is advocated in this paper.

DTW is a well tried and tested technique first described in [1]. DTW operates as follows. Given two curves X with length m and Y with length n, a matrix A is constructed with m rows and n columns. Each element (i, j) within matrix A describes the distance between point i on curve X and the point j on curve Y. The goal is to find the "warping path through this matrix describing the shortest distance from $(0,0)$ to (m, n). To improve the efficiency with which this warping path can be identified the "Sakoe-Chiba" band [7] was used to define a "constraint region".

A signature is a set of feature values that can be used to describe some entity, a curve in our case. The feature values encompassed by a set of signatures thus describe a feature space. The desired signature generation was achieved using Hough signature extraction [5] which in turn is founded on the Hough Transform. The Hough Transform is a parameter-based transformation derived from the Hough Concept. The process commences by transforming the curves into a parameter space (accumulator matrix) A, comprised of m rows and n columns where m is the difference between minimum and maximum collected distance and n is the total number of points on the point series. The process for generating a signature from an accumulator matrix is described in [9]. After this process, the extracted signatures were stored together with an associated class label to which any established classifier generation mechanism could be applied.

4   Experimentation and Evaluation

This section described the experimental set up and the evaluation of the proposed process described above. The experimentation were designed to compare the operation of both the Disc-based and Spoke-based point series generation technique coupled with both KNN direct classification (using DTW) and Hough signature extraction (using SVM). With respect to the Spoke-based 3-D representation technique for different spoke spacings were considered ($\{1°, 2°, 3°, 4°\}$), thus four distinct point series were generated in this case. Ten-fold Cross Validation (TCV) was used through out. The SVM implementation available from the Waikato Environment Knowledge Analysis (WEKA) data mining workbench [4] was adopted. The image set used for evaluation purposes was composed of 210 MRI volumes obtained from the Magnetic Resonance and Image Analysis Research Centre at the University of Liverpool. Each scan consists of 256 slices in three planes (thus 768 slices in all). The resolution of each image slice is 256 x 256 pixels with colour defined using 8-bit gray scale (256 colours). All the experiments were conducted using a 2.9 GHz Intel Core i7 with 8GB RAM on OS X (10.9) operating system.

Table 2 shows classification results obtained using the Hough feature extraction approach, while Table 1 shows the results obtained using the KNN approach

Table 1. Classification results obtained using SVM and Hough

Technique	Accu.	Sens.	Spec.
Disc	59.43	58.49	**60.38**
Spoke ($x = 1°$)	62.20	67.50	57.14
Spoke ($x = 2°$)	**64.63**	**70.00**	59.52
Spoke ($x = 3°$)	61.32	62.26	**60.38**
Spoke ($x = 4°$)	58.49	62.26	57.42
Average	61.21	64.10	58.97

Table 2. Classification results obtained using KNN and DTW

Technique	Accu.	Sens.	Spec.
Disc	62.20	67.50	57.14
Spoke ($x = 1°$)	64.15	66.04	62.26
Spoke ($x = 2°$)	**69.81**	71.70	**67.92**
Spoke ($x = 3°$)	68.87	**75.47**	62.26
Spoke ($x = 4°$)	60.98	67.50	57.14
Average	65.20	69.64	61.34

(with $K = 1$). The metrics used to evaluate calcification performance were accuracy, sensitivity and specificity. From Table 1 it can be seen that with respect to Hough signature extraction the best classification accuracy and sensitivity were obtained using the Spoke-based image representation with a spoke spacing of 2°, while the best classification specificity was obtained using a spacing of 3°. With respect to the KNN approach, from Table 2 it can be seen that the best classification accuracy and specificity were again obtained using the Spoke-based approach with a spacing of 2°, while the best sensitivity was again obtained using the Spoke-based approach and a spacing of 3°. When using the Spoke based approach the classification accuracy, regardless of the technique adopted, tended to peak when using a spacing 2° and then decreased when the spacing increased. It was conjectured that this was because as the representation got coarser details concerning the shape of the ventricles began to be missed, while at lower levels of spacing the amount of detail collected tended to clutter the representation. The relationship between classification accuracy and different spoke spacings is presented in Figure 5. On average the classification results obtained using KNN with DTW were slightly better than those obtained using SVM with Hough signature extraction in terms of all accuracy, sensitivity, and specificity.

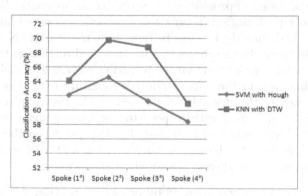

Fig. 5. The Relation between classification accuracy and spoke spacing when using the Spoke-based representation technique

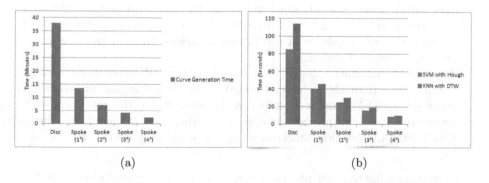

Fig. 6. Performance for: (a) Curve generation; (b) Classification

The performance of the image classification approach based on the proposed point series representation techniques is shown in Figure 6. Figure 6a shows the curve generation time and Figure 6b shows total classification time using Ten-fold Cross Validation (TCV). Note that the time complexity includes the Hough signature extraction process (for SVM) and Dynamic Time Warping process (for KNN). From the figure it can be seen that the Spoke-based representation technique (4°) was the most efficient with respect to both curve generation and classification, while the Disc-based representation was the least efficient. This was because in the case of Spoke-based generation the point series contained fewer points than Disc-based generation. The number of point series for Spoke-based generation was constant for all ventricles (360 point series for 1°, 180 for 2°, 120 for 3° and 90 for 4°) but varies depending to the size of the ventricles in the case of Disc-based representation (over a thousand point series for most ventricles).

5   Conclusions

Two classification motivated approaches to describing 3-D volumes have been described. A disc-based approach and a Spoke-based approach. Both resulted in a point series representation (a curve) describing a volume of interest. These curves can be used directly to classify new data using a KNN classification mechanism. Alternatively, the curves can be processed further so that a feature space representation is generated that is compatible with most standard classifier generation techniques. In the paper a Hough signature extraction mechanism was used to generate the desired feature vectors. In the reported evaluation SVM classification was used with respect to the feature space representation because earlier experiments had indicated that this produced the best results. For evaluation purposes the lateral ventricles within 3-D MRI brain scan data were used as the object of interest. The main findings were:

1. The Spoke-based representation technique, with a spacing of 2° coupled with KNN (and the warping path obtained using DTW as the distance measure) produced the best result in term of classification accuracy.
2. The results obtained using the Spoke-based representation were, on average, better than those obtained using the Disc-based representation. It is therefore concluded that the Spoke-based representation produced a better reflection of the shape of the lateral ventricles than in the case of the Disc-based representation.
3. The results obtained using KNN coupled with DTW tended, on average, to be better than the results obtained using Hough signature extraction coupled with SVM classification. It was conjectured that this was because the signature extraction process introduced a further level of complexity during which some information concerning the shape of the object of interest was lost.

When comparing to other previous works the results reported in [3] were better than those reported here (a best accuracy of 77.27% was reported), however the work in [3] was directed at using the corpus callosum to predict epilepsy which may be a better indicator of epilepsy. In the context of the authors own work the results reported in [8] which used the same dataset but coupled with an oct-tree representation technique the reported effectiveness was marginally better (a best accuracy of 72.34%) than those reported here, however the oct-tree technique was considerably less efficient than those reported in this paper (this is an important factor in the context of "time of consultation diagnosis"). The work reported in [6] also used an oct-tree representation technique and produced slighter better results (a best accuracy of 77.2% but using both the lateral and a "third ventricle"), however this work was directed at predicting Alzheimer's disease and level of education (as opposed to epilepsy); the lateral and third ventricles may be better indicators of Alzheimer's disease and level of education than epilepsy. The approach proposed by Long et al [6]. was also less efficient compared to the work reported here.

For future work the authors intend to focus on further alternative methods of representing 3-D MRI brain scan features so that machine learning techniques can be applied. The intention is also to consider the use of dynamic thresholding techniques, as used in [6], to determine whether this helps improve the effectiveness of the segmentation process.

References

1. Bellman, R., Kalaba, R.: On Adaptive Control Processes. IRE Transactions on Automatic Control 4(2) (1959)
2. Dougherty, G.: Digital Image Processing for Medical Applications. Cambridge Press (2009)
3. Elsayed, A., Hijazi, M.H.A., Coenen, F., García-Fiñana, M., Sluming, V., Zheng, Y.: Time series case based reasoning for image categorisation. In: Ram, A., Wiratunga, N. (eds.) ICCBR 2011. LNCS, vol. 6880, pp. 423–436. Springer, Heidelberg (2011)
4. Hall, M., Frank, E., Holmes, G.: The WEKA Data Mining Software: An Update. ACM SIGKDD Explorations Newsletter 11(1), 10–18 (2009)
5. Hough, P.V.C.: Method and Means for Recognizing Complex Patterns, US Patent (1962)
6. Long, S., Holder, L.B.: Graph-based Shape Shape Analysis for MRI Classification. International Journal of Knowledge Discovery in Bioinformatics 2(2), 19–33 (2011)
7. Sakoe, H., Chiba, S.: Dynamic programming algorithm optimization for spoken word recognition. IEEE Transactions on Acoustics, Speech and Signal Processing 26(1), 43–49 (1978)
8. Udomchaiporn, A., Coenen, F., García-Fiñana, M., Sluming, V.: 3-D MRI Brain Scan Feature Classification Using an Oct-Tree Representation. In: Motoda, H., Wu, Z., Cao, L., Zaiane, O., Yao, M., Wang, W. (eds.) ADMA 2013, Part I. LNCS, vol. 8346, pp. 229–240. Springer, Heidelberg (2013)
9. Vlachos, M., Vagena, Z., Yu, P.S., Athitsos, V.: Rotation Invariant Indexing of Shapes and Line Drawings. In: Proceedings of the 14th ACM International Conference on Information and Knowledge Management, CIKM 2005, pp. 131–138 (2005)

Mining Interesting "Following" Patterns
from Social Networks

Fan Jiang and Carson Kai-Sang Leung[*]

University of Manitoba, Canada
kleung@cs.umanitoba.ca

Abstract. Over the past few years, social network sites (e.g., Facebook, Twitter, Weibo) have become very popular. These sites have been used for sharing knowledge and information among users. Nowadays, it is not unusual for any user to have many friends (e.g., hundreds or even thousands friends) in these social networks. In general, social networks consist of social entities that are linked by some interdependency such as friendship. As social networks keep growing, it is not unusual for a user to find those frequently followed groups of social entities in the networks so that he can follow the same groups. In this paper, we propose (i) a space-efficient bitwise data structure to capture interdependency among social entities and (ii) a time-efficient data mining algorithm that makes the best use of our proposed data structure to discover groups of friends who are frequently followed by social entities in the social networks. Evaluation results show the efficiency of our data structure and mining algorithm.

1   Introduction and Related Works

Social networks are generally made of social entities (e.g., individuals, corporations, collective social units, or organizations) that are linked by some specific types of interdependency (e.g., kinship, friendship, common interest, beliefs, or financial exchange). A social entity is connected to another entity as his next-of-kin, friend, collaborator, co-author, classmate, co-worker, team member, and/or business partner. Social computing aims to computationally facilitate social studies and human-social dynamics in these networks, as well as to design and use information and communication technologies for dealing with social context.

Various social networking websites or services—such as Facebook, Flickr, Google+, LinkedIn, Twitter, and Weibo [11, 14, 16]—are in use nowadays. For instance, Twitter is an online social networking and blogging service that allows users to read the tweets of other users by *following* them. As such, a Twitter user A may be interested in knowing the popular followees. In other words, for any Twitter user A, if many of his friends follow some individual users or groups of users, then A might also be interested in following the same individual users or groups of users.

---

[*] Corresponding author.

L. Bellatreche and M.K. Mohania (Eds.): DaWaK 2014, LNCS 8646, pp. 308–319, 2014.

Similarly, Facebook is another social networking site, in which users can create a personal profile, add other Facebook users as friends, and exchange messages. In addition, Facebook users can also join common-interest user groups and categorize their friends into different customized lists (e.g., classmates, co-workers). The number of (mutual) friends may vary from one Facebook user to another. It is not uncommon for a user p to have hundreds or thousands of friends. Note that, although many of the Facebook users are linked to some other Facebook users via the mutual friendship (i.e., if a user A is a friend of another user B, then B is also a friend of A), there are situations in which such a relationship is no longer mutual. To handle these situations, Facebook added the functionality of "subscribe" in 2011. This functionality was later relabelled as "follow" in 2012. Specifically, a user can subscribe or follow public postings of some other Facebook users without the need of adding them as friends. So, for any Facebook user C, if many of his friends followed some individual users or groups of users, then C might also be interested in following the same individual users or groups of users.

As a third example, in social network like Weibo, relationships between social entities are mostly defined by following (or subscribing) each other. Each user (social entity) can have multiple followers, and follows multiple users at the same time. The follow/subscribe relationship between follower and followee is not the same as the friendship relationship (in which each pair of users usually know each other before they setup the friendship relationship). In contrast, in the follow/subscribe relation-ship, a user A can follow another user B while B may not know A in person (e.g., we may follow the Royal family or our Prime Minister, but the Royal family members or the Prime Minister may not know us). This creates a relationship with direction in a social network. We use A→B to represent the follow/subscribe (i.e., "following") relationship that A is following B.

In recent years, the number of users in Twitter and Weibo has grown rapidly (as on January 1, 2014, over 600 million active users for Twitter, and over 500 million active users for Weibo). This massive number of users creates an even more massive num-ber of "following" relationships. Hence, a natural question to ask is: (1) For this huge amount of data containing these "following" relationships, can data mining techniques be applied? In response, over the past few years, several data mining algorithms and techniques [1, 4, 7] have been proposed. Many of them are applicable to mine social networks (e.g., discovery of special events [2], detection of communities [9, 10, 13], sub-graph mining [15], as well as discovery of popular friends [5], influential friends [8] and strong friends [12]). In this paper, we focus on the data mining task of finding frequent patterns [3, 6]—specifically, *frequently followed patterns*.

Another question to ask is: (2) Do these "following" relationships appear in certain ways/trends? For example, users who follow the twitter of PlayStation are usually also following Xbox360's at the same time. There could be other interesting hidden patterns in these huge amounts of following relationships. Discovering these patterns will greatly benefit both the social network users as well as service providers like Twitter or Weibo. The next logical question is: (3) How can these interesting "follow-ing" patterns be discovered? How to discover these *frequently followed individuals or groups*? Manually going through the entire social networks seem to be impractical due to the huge amount of social data. A more systematic approach is needed. In response to the above questions, we propose a data mining algorithm to discover frequently followed patterns in social networks. Our *key contributions* are: (i) a space-efficient bitwise data structure for capturing interdependency among social entities and (ii) a time-efficient data mining algorithm that makes the best use of our proposed data structure to discover groups of friends who are frequently followed by social entities in the social networks.

The remainder of this paper is organized as follows. The next section introduces the concept of "following" patterns. In Section 3, we present our new space-efficient data structure to capture important information from social networks. Frequent "following" patterns can then be mined using a novel time-efficient algorithm that we proposed in Section 4. Finally, conclusions are given in Section 5.

2   The Notion of "Following" Patterns: "Followed" Groups

In this section, we introduce the notion of "following" patterns (i.e., "followed" groups) in social networks. In social network sites like Twitter or Weibo, social entities (users) are linked by "following" relationships (e.g., a user p follows another user q). Compared with the friendship relationships in Facebook, this "following" relationship is different in the way that, the "following" relationships are directional. For instance, a user A may be following another user B while B is not following A. This property increases the complexity of the problem because of the following two reasons. First, the group of users followed by A (e.g., A→{B, C, ...}) may not be same group of users as those who are following A (e.g., {..., X, Y, Z}→A). Due to asymmetry in the "following" relationships (cf. the symmetric friendship relationships), we cannot use the usual back-tracking methods to determine the reverse relationships (e.g., cannot determine whether or not E→A if we only know A→E). In this case, we need to check both directions to get relationship between pairs of users. Second, in terms of computation, due to asymmetry in the "following" relationships, the computation time and space is doubled. For each pair of users, A→B (A follows B) does not automatically imply B→A. It requires double the amount of memory space to store the relationship between two users (i.e., storing two *directional* "following" relationships A→B and B→A), whereas one only needs to store one *unidirectional* friendship relationship A—B. This also means that, if there is a change in the dataset (e.g., A unfollows B), then we cannot remove the linkage between A and B because B may still be following A (cf. when A "unfriends" B, the friendship linkage between A and B disappears).

As the number of Twitter users is growing explosively nowadays, the number of relationships between Twitter users is also growing. One of the important research problems with regard to this massive amount of data is to discover frequent "following" patterns.

A "following" pattern is a pattern representing the linkages when a significant number of users follow the same combination/group of users. For example, users who follow the twitter of NBA also follow the twitter of Adam Silver (current NBA commissioner). If there are large numbers of users who follow the twitters of both NBA and Adam Silver together, we can define this combination (NBA and Adam Silver) of followees as a frequent "following" pattern (i.e., a frequently followed group).

2.1   Example of "Following" Patterns

Let us consider an example on a social network that is represented as a graph shown in Fig. 1, in which (i) each node represents a user (i.e., a social entity) in the social network and (ii) the follow/subscribe relationships between pairs of users are represented by directed edges between the corresponding nodes. The arrow on an edge represents the "following" direction (e.g., Bob→Carol represents that Bob follows Carol, Alice↔Bob represents that Alice and Bob are following each other). Then, we aim to find *frequent "following" patterns* (or frequently "followed" groups of friends).

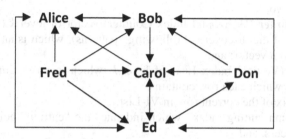

Fig. 1. A sample social network contains six users. Directed edges between users represent the "following" relationship between users. (e.g., Bob→Carol represents that Bob follows Carol, and Alice↔Bob represents that Alice and Bob are following each other).

The above social network can also be represented in an alternative form—a tabular form, as shown in Table 1. To save space, we use the initials of social entities' names in Table 1 and the remainder of this paper (e.g., A for Alice, B of Bob, C for Carol, etc.). An entry of "1" in (row X, column Y) indicates the presence of a link from X to Y. In the social network context, it means X follows Y. An entry of "0" in (row X, column Y) indicates the absence of a link from X to Y. As the "follow" action is not symmetric, X follows Y does not guarantee that Y follows X. For example, Fred follows Alice (as indicated by the presence of "1" in (row Fred, column A), but Alice does not follow Fred (as indicated by the presence of "0" in (row Alice, column F).

Table 1. An alternative bitwise representation of the sample social network in Fig. 1

Follower	Followees					
	A	B	C	D	E	F
Alice	0	1	0	0	1	0
Bob	1	0	1	0	1	0
Carol	1	0	0	0	1	0
Don	0	1	1	0	1	0
Ed	1	1	1	1	0	0
Fred	1	1	1	0	1	0

3   Our FoP-Structure and Its FoP-Miner Algorithm

For discovering frequent "following" patterns in social networks, we propose a new data structure—called **FoP-Structure**—to capture important social information (especially, "**Fo**llowing" **P**atterns or relationships among social entities) in the social network. Recall from the previous section, a social network capturing the "following" relationships can be represented in a bitwise tabular form. Such a table forms the basis of our FoP-Structure, which contains the following three parts:

1. A **SocialTable**, which is the main bitmap based structure, in which each row contains information about one particular user (a follower and its followees) in a social network;
2. An **HI-Counter** (Horizontal Index Counter) vector of level k (where k is the cardinality of the discovered "following" patterns), which is an index counter list stored as a vector; and
3. A **VI-List** (Vertical Index List) of level k, which is a two-dimensional data structure in which each row contains:
    (a) the prefix of the current row in VI-List,
    (b) a "column cutting index" which indicates the "cutting" points during the mining process, and
    (c) a list of "row cutting indices".

3.1   Construction of a FoP-Structure

Given a social network (e.g., the one in either a graph representation as in Fig. 1 or the tabular form as in Table 1), we construct a FoP-Structure by applying the following steps to capture important "following" relationships between pairs of social entities so that frequent "following" patterns can then be discovered from the constructed FoP-Structure:

1. For each node (representing a social entity) in the social network, we create a bitmap row with size N, where N is the number of active users in the social network. Each column (represented by a bit) in the row represents the follow/subscribe relationship between the current user (the follower) and its followee in the social network. In other words, we put a "1" in the ith bit (where $1 \le i \le N$) if the follower follows the ith user in the social network; we put a "0" in the ith bit (where $1 \le i \le N$) if the follower does not follow the ith user in the social network. When we repeat the aforementioned actions in this step, we get multiple bitmap rows to form a *SocialTable*.
2. Then, for each row r of the SocialTable, we create a *top-level VI-List*. Specifically, we record the "row cutting index" (i.e., the column index for the first occurrence of the "1" bit in row r) in the VI-List.
3. Moreover, for each column c of the SocialTable, we also create a *top-level HI-Counter*. Specifically, we count the number of 1s in column c and put the count in the cth position/entry of the HI-Counter.

Note that all above steps can be completed in only one scan of the social network. See the following example for more details.

3.2   Example of the FoP-Structure: HI-Counter and VI-List

Let us continue with the example in Section 2.1. The SocialTable of the FoP-Structure can be constructed—by scanning the social network only *once*—to capture the contents in Fig. 1 (which shows a graph representing the social networks). The resulting SocialTable is a bitwise data structure as shown in Table 1. Based on this SocialTable, we then construct both the HI-Counter and the VI-List of the highest level (i.e., top level) as shown in Figs. 2 and 3, respectively.

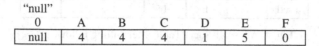

	"null" 0	A	B	C	D	E	F
	null	4	4	4	1	5	0

Fig. 2. HI-Counter for Level 0

Here, the HI-Counter is a vector that stores the count of each column of the SocialTable. For example, the number of "1" in column 2 (for "B") of HI-Counter is 4, which means that four social entities are following user B. Notice that the followees D and F had frequency support values of 1 and 0 (i.e., less than user specified *minsup* of 2) respectively in the level-0 SocialTable, they are not further considered (i.e., all entries in columns D and F can be ignored).

0	"A"	Col 1	Rows 2, 3, 5, 6
1	"B"	Col 2	Rows 1, 4

Fig. 3. VI-List for Level 0

The VI-List, on the other hand, contains several rows of social entity's first appearance. For example, in Fig. 3, the VI-List captures the information: (i) rows 2, 3, 5 and 6 (representing four followers Bob, Carol, Ed and Fred) contain lists of social entities that start with A; (ii) rows 1 and 4 (representing two followers Alice and Don) contain lists of social entities that start with B.

Once the FoP-Structure is constructed, frequent "followed" patterns can be mined from the FoP-Structure. See the following example.

3.3   Example of the FoP-Miner

Let us continue with our example to illustrate the mining process. Assume that the user specified minimum support threshold (*minsup*) is 2, the mining process starts from the first row of the top-level VI-List as shown in Fig. 3. Recall that each row of the VI-List consists of three parts:

(a)   The prefix of the current row in the VI-List (e.g., "A");
(b)   a "column cutting index" (e.g., "Column 1"); and
(c)   an array of "row cutting indices" (e.g., "Rows 2, 3, 5 and 6").

By using the "column cutting index" and "row cutting indices", we can "cut" the SocialTable into a smaller piece (as shown in Fig. 4), to which the mining process for discovering frequent "followed" groups with prefix "A" can be applied. As shown in Fig. 4, original SocialTable is cut into the smaller section in the right hand side, named level-1 SocialTable.

	A	B	C	D	E	F
Alice						
Bob		0	1		1	
Carol		0	0		1	
Don						
Ed		1	1		0	
Fred		1	1		1	

Fig. 4. A level-1 SocialTable after cutting followee A

Then, the next step is to construct a level 1-HI-Counter and level 1-VI-List from the level-1 SocialTable. As shown in Fig. 5, the newly constructed level-1 HI-Counter records all the column counts of the level-1 SocialTable. These counts are frequency support values of all "followed" patterns with prefix "1". In this example, as we proceed to this step, we have frequency support values for "followed" patterns: "AB: 2", "AC: 3", and "AE: 3" (which represent groups {A, B} followed by two social entities, {A, C} followed by three social entities, and {A, E} followed by two social entities).

"A"

0	B	C	D	E	F
null	2	3		3	

Fig. 5. The level-1 HI-Counter for A

After constructing level-1 HI-Counter, we then construct the level-1 VI-List. See Fig. 6, which shows the level-1 VI-List that contains three rows.

0	"AB"	2	5 6
1	"AC"	3	2
2	"AE"	5	3

Fig. 6. The level-1 VI-List for A

Based on this level-1 VI-List, we learn that we can construct a level-2 SocialTable (for AB) and its associated level-2 HI-Counter, and level 12-VI-List as shown in Figs. 7, 8, and 9, respectively.

	A	B	C	D	E	F
Alice						
Bob						
Carol						
Don						
Ed			1		0	
Fred			1		1	

Fig. 7. The level 2-SocialTable after cutting group AB

"AB"

0	C	D	E	F
null	2	1	1	0

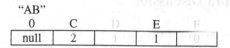

Fig. 8. The level-2 HI-Counter for AB

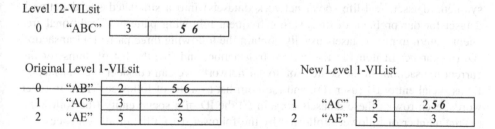

0	"ABC"	3	*5 6*

Fig. 9. The level-2 VI-List for AB

Then, our algorithm returns the result ("ABC: 2"), which represents that group {A, B, C} are followed by two social entities. At this point, we observe that no more frequent result can be mined from the remainder of the level-2 SocialTable for AB. Therefore, the recursive process backtracks to the previous level. Afterwards, the algorithm checks the following two conditions:

(1) Check if the previous VI-List has more rows.
(2) If no, return.
(3) If yes, check if all other rows in previous VI-List have the same "cutting column index" as the current VI-List rows have. If yes, update the new row in previous VI-List. Let us check the example to explain this process in details.

In Fig. 10, after processing the first row of the level-1 VI-List, there are two more rows remaining. Notice that the cutting column index in the level-1 VI-List[1] is the same as the level-2 VI-List[0], therefore we need to update the level-1 VI-List[1] as shown in Fig. 10. After that, the algorithm continues with the new level 1-VI-List. The process is shown in Fig. 10. The result of this step is "1 3 5" with frequency support 1 (i.e., group {A, C, E} are followed by only one social entity). As such a group is only followed by one social entity, which is less than the user-specified *minsup*. Such a group is infrequent (or not too popular).

Level 12-VILsit

0	"ABC"	3	*5 6*

Original Level 1-VILsit

~~0~~	~~"AB"~~	~~2~~	~~5 6~~
1	"AC"	3	2
2	"AE"	5	3

New Level 1-VIList

1	"AC"	3	2 *5 6*
2	"AE"	5	3

Fig. 10. An updated VI-List

The algorithm follows the above process. Consequently, we obtain our mining result, i.e., in the context of social networks, we obtain frequently followed groups.

4   Evaluation and Discussion

For evaluation, we first analyzed the memory space consumption of our FoP-Miner algorithm (and its associated structures) when compared with related works. Recall that the FP-growth algorithm [3] is the tree-based divide-and-conquer approach to mine frequent itemsets from shopper market basket datasets. The algorithm first builds a global FP-tree to capture important contents of the dataset in the tree. The number of tree nodes is theoretically bounded above by the number of occurrences of all items (say, $O(N_{occurrence})$) in the dataset. Practically, due to tree path sharing, the number of tree nodes (say, $O(N_{tree}) < O(N_{occurrence})$) is usually smaller than the upper bound. However, during the mining process, multiple smaller sub-trees need to be constructed. Specifically, for a global FP-tree with depth d, it is not usual for $O(d)$ sub-trees to coexist with the global tree, for a total of $O(d \times N_{tree})$ nodes. In contrast, on the surface, our SocialTable may appear to take up more space (due to the lack of tree path sharing). The SocialTable contains $O(N_{occurrence})$ entries. However, it is important to note that each entry in our SocialTable is just a single bit, instead of an integer for an item ID. In other words, the SocialTable requires $O(N_{occurrence})/8$ bytes of space. Moreover, unlike FP-growth, we do not need to build any sub-SocialTable. In other words, the same SocialTable is used throughout the entire mining process. Thus, our SocialTable requires $O(N_{occurrence})/8$ bytes $\ll O(d \times N_{tree})$ bytes in FP-growth.

In the above analysis, we focused on SocialTable. How about VI-Lists and HI-Counters? For any interesting "followed" patterns of length k, we need to create $O(k)$ VI-Lists and HI-Counters. Note that FP-growth also creates $O(k)$ header tables for the $O(k)$ sub-trees. In other words, the amount of space required by VI-Lists and HI-Counters is similar to that by sub-trees. But, our SocialTable requires much less space than all FP-trees.

Next, we compared the performance of our FoP-Miner algorithm with related works. First, we discuss the datasets that we used for evaluation. In our research, we realized that we can transfer traditional frequent pattern mining datasets (e.g., IBM synthetic dataset, real-life social network datasets) into a simulated social network dataset for our problem of discovering frequent following patterns. Traditional frequent pattern mining datasets usually contain the following three parts: (i) transaction ID, (ii) number of items in the current transaction, and (iii) the list of items for the current transaction. In the context of social network, we can consider the transaction ID as social entity ID (user ID), and the item list as a list of followees' IDs. In other words, each row in these datasets contain (i) the ID of a social entity (the follower), (ii) the number of followees followed by the follower, (iii) a list of all followees followed by the follower. The first dataset we used for evaluation is an IBM synthetic dataset. The second dataset we used is from the Stanford Large Network Dataset Collection (http://snap.stanford.edu/data/). Specifically, the Twitter dataset was used.

All experiments were run in a time-sharing environment in a 1 GHz machine. The reported figures are based on the average of multiple runs. Runtime includes CPU and I/Os; it includes the time for both tree construction and frequent pattern mining steps.

To evaluate the efficiency of our proposed FoP-Miner algorithm and its associated FoP-Structure, we compared our algorithm with the famous frequent pattern mining algorithm FP-growth [3]. Fig. 11 shows the runtime result of our proposed FoP-Miner against FP-growth over different data sizes (in terms of number of social entities). The result shows that FoP-Miner is more time efficient than FP-growth.

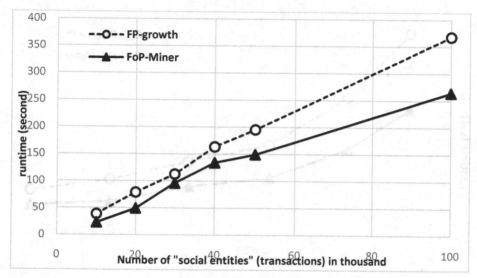

Fig. 11. Runtimes of FP-growth [3] and our proposed FoP-Miner vs. different data size of the IBM synthetic dataset (when using a minimum support threshold set of 1)

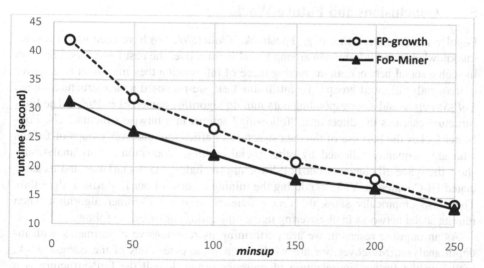

Fig. 12. Runtimes of FP-growth [3] and our proposed FoP-Miner vs. minimum support threshold (*minsup*) values on the IBM synthetic dataset

Figs. 12 and 13 show the runtime result of our proposed FoP-Miner against FP-growth over different minimum support threshold values. As expected, for IBM both synthetic data sets (Fig. 12) and real-life social network data (Fig. 13), our proposed FoP-Miner algorithm is more time-efficient than FP-growth. In other words, FoP-Miner requires less runtime.

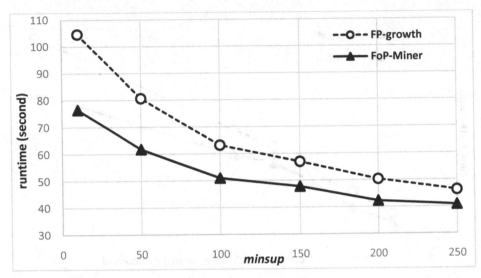

Fig. 13. Runtimes of FP-growth [3] and our proposed FoP-Miner vs. minimum support threshold (*minsup*) values on real-life Stanford large network dataset

5   Conclusions and Future Work

Popular social network sites (e.g., Facebook, Twitter, Weibo) have been used for sharing knowledge and information among social entities over the past few years. For users in such a social network, an interesting piece of information they may want to discover is frequently followed groups. To fulfill this task, we proposed a data structure named *FoP-Structure* and a corresponding data mining algorithm named *FoP-Miner*. The FoP-Structure captures the directional "following" actions in a bitwise structure. The FoP-Miner makes the best use of the data structure to efficiently discover groups of friends who are frequently followed by other social entities. Evaluation results analytically show the space-efficiency of our FoP-Structure (including the SocialTable and its associated HI-Counters & VI-Lists) during the mining process of our FoP-Miner algorithm. They also empirically show the time-efficiency of our FoP-Miner algorithm when mining social networks in discovering interesting groups of friends to follow.

As an ongoing research, we are performing more extensive experiments with in-depth analysis. Moreover, we are working on some extensions of the current work, including the further optimization of memory usage. Recall the FoP-Structure is a bitwise data structure. In the current work, we analyzed the memory usage as $O(N_{occurrence})/8$ bytes, which is smaller than other data structures (e.g., trees). However, as the sizes of social networks are growing explosively, this might not be good enough in the near future. We are working on a further optimization in terms of memory usage of FoP-Structure that will allow us capture more information.

Acknowledgements. This project is partially supported by NSERC (Canada) and University of Manitoba.

References

1. Cuzzocrea, A., Leung, C.K.-S., MacKinnon, R.K.: Mining constrained frequent itemsets from distributed uncertain data. Future Generation Comp. Syst. 37, 117–126 (2014)
2. Dhahri, N., Trabelsi, C., Ben Yahia, S.: RssE-miner: A new approach for efficient events mining from social media RSS feeds. In: Cuzzocrea, A., Dayal, U. (eds.) DaWaK 2012. LNCS, vol. 7448, pp. 253–264. Springer, Heidelberg (2012)
3. Han, J., Pei, J., Yin, Y.: Mining frequent patterns without candidate generation. In: ACM SIGMOD 2000, pp. 1–12 (2000)
4. Jiang, F., Leung, C.K.-S.: Stream mining of frequent patterns from delayed batches of uncertain data. In: Bellatreche, L., Mohania, M.K. (eds.) DaWaK 2013. LNCS, vol. 8057, pp. 209–221. Springer, Heidelberg (2013)
5. Jiang, F., Leung, C.K.-S., Tanbeer, S.K.: Finding popular friends in social networks. In: CGC (SCA) 2012, pp. 501–508 (2012)
6. Leung, C.K.-S., Jiang, F.: Frequent pattern mining from time-fading streams of uncertain data. In: Cuzzocrea, A., Dayal, U. (eds.) DaWaK 2011. LNCS, vol. 6862, pp. 252–264. Springer, Heidelberg (2011)
7. Leung, C.K.-S., Tanbeer, S.K.: Mining popular patterns from transactional databases. In: Cuzzocrea, A., Dayal, U. (eds.) DaWaK 2012. LNCS, vol. 7448, pp. 291–302. Springer, Heidelberg (2012)
8. Leung, C.K.-S., Tanbeer, S.K., Cameron, J.J.: Interactive discovery of influential friends from social networks. Social Netw. Analys. Mining 4(1), art. 154 (2014)
9. Lin, W., Kong, X., Yu, P.S., Wu, Q., Jia, Y., Li, C.: Community detection in incomplete information networks. In: WWW 2012, pp. 341–350 (2012)
10. Ma, L., Huang, H., He, Q., Chiew, K., Wu, J., Che, Y.: GMAC: a seed-insensitive approach to local community detection. In: Bellatreche, L., Mohania, M.K. (eds.) DaWaK 2013. LNCS, vol. 8057, pp. 297–308. Springer, Heidelberg (2013)
11. Schaal, M., O'Donovan, J., Smyth, B.: An analysis of topical proximity in the twitter social graph. In: Aberer, K., Flache, A., Jager, W., Liu, L., Tang, J., Guéret, C. (eds.) SocInfo 2012. LNCS, vol. 7710, pp. 232–245. Springer, Heidelberg (2012)
12. Tanbeer, S.K., Leung, C.K.-S., Cameron, J.J.: Interactive mining of strong friends from social networks and its applications in e-commerce. J. Org. Computing and E. Commerce 24(2-3), 157–173 (2014)
13. Wei, E.H.-C., Koh, Y.S., Dobbie, G.: Finding maximal overlapping communities. In: Bellatreche, L., Mohania, M.K. (eds.) DaWaK 2013. LNCS, vol. 8057, pp. 309–316. Springer, Heidelberg (2013)
14. Yang, X., Ghoting, A., Ruan, Y., Parthasarathy, S.: A framework for summarizing and analyzing Twitter feeds. In: ACM KDD 2012, pp. 370–378 (2012)
15. Yu, W., Coenen, F., Zito, M., El Salhi, S.: Minimal vertex unique labelled subgraph mining. In: Bellatreche, L., Mohania, M.K. (eds.) DaWaK 2013. LNCS, vol. 8057, pp. 317–326. Springer, Heidelberg (2013)
16. Yuan, Q., Cong, G., Ma, Z., Sun, A., Magnenat-Thalmann, N.: Who, where, when and what: discover spatio-temporal topics for twitter users. In: ACM KDD 2013, pp. 605–613 (2013)

A Novel Approach Using Context-Based Measure for Matching Large Scale Ontologies

Warith Eddine Djeddi and Mohamed Tarek Khadir

LabGED, Computer Science Department, University Badji Mokhtar,
Po-Box 12, 23000 Annaba, Algeria
{djeddi,khadir}@labged.net

Abstract. Identifying alignments between ontologies has become a central knowledge engineering activity. In ontology matching the same word placed in different textual contexts assumes completely different meanings. This paper proposes an algorithm for ontologies alignment named XMap ++ (eXtensible Mapping), applied in an ontology mapping context. In XMap++ the measurement of lexical similarity in ontology matching is performed using synset, defined in WordNet. In our approach, the similarity between two entities of different ontologies is evaluated not only by investigating the semantics of the entities names, but also taking into account the context, through which the effective meaning is described. We provide experimental results that measure the accuracy of our algorithm based on our participation with two versions (XMapSig and XMapGen) at the OAEI campaign 2013.

Keywords: Ontology alignment, WordNet's synset, context measures, semantic relationships.

1 Introduction

Ontology matching is one of the most plausible solutions to cope with heterogeneity problems in ontological contents [1]. Ontology matching refers to the process of finding relations or correspondences between similar elements of different ontologies.

The problem of finding the semantic mappings between two given ontologies lies at the heart of numerous information processing applications. Virtually any application that involves multiple ontologies must establish semantic mappings among them using external resources like domain ontology, corpus, thesaurus (e.g., WordNet, Wikipedia), to ensure interoperability. Algorithms that are used in ontology alignment may be very complex and may contain many features and parameters that can affect the performance even of commonly accepted and string metrics, when they are used in new contexts.

The context is the set of information (partly) characterizing the situation of some entity [2]. The notion of context is not universal but relative to some situation, task or application [3][4]. Easy access to this information is essential in enabling the user to verify candidate mappings. In particular, the neighborhood of a term (immediate parent and children in the "is-a" hierarchy) may be especially important. Understanding the

L. Bellatreche and M.K. Mohania (Eds.): DaWaK 2014, LNCS 8646, pp. 320–331, 2014.

correct meaning of words, that have different meanings in different contexts, reduces myriad of semantic problem (i.e., the polysemy and synonymy problem).

In this paper, in order to deal with lexical ambiguity, we introduce the notion of scope belonging to a concept which represents the context where it is placed. WordNet [5] is the semantic networks (thesaurus) exploited in our approach. The similarity between two entities of different ontologies is evaluated not only by investigating the semantics of the entities names, but also taking into account the local context, through which the effective meaning is described. Increasing the radius means enlarging the scope (i.e. this area) and, consequently, the set of neighbour concepts that intervene in the description of the context. Next, we present our two flexible and self-configuring matching tools XMapGen (eXtensible Mapping using Genetic) and XMapSig (eXtensible Mapping using Sigmoid) [6]. Then, we provide the results of our experiments on the OWL ontologies of OAEI campaign 2013 and we compare these results with those of other algorithms that competed in the OAEI challenge.

The rest of this paper is organized as follows. Section 2 reviews some related work in the area of ontology alignment. Section 3 presents in details, the novel approach using context-based measure. Section 4 defines the matching process strategy. Section 5 describes our evaluation methodology and discusses experimental results. Finally, section 6 concludes with an outline of future work.

2   Related Work

Many similarity measures have been adapted for use in matchers in various categories depending on the context of the similarity measurement, such as lexical, structural, or extensional matchers [7]. Lexical database such as WordNet [5] have been used to find synonyms for differing concept string labels. The string-based matchers then work not only on the specific concept labels but also on the corresponding synonyms in the WordNet Lexicon [8].

In [9], the authors introduce two techniques, which exploit the ontological context (the analysis of contexts based on the recurrence of nearby components) of the matched and anchor terms, and the information provided by WordNet, can be used to filter out mappings resulting from the incorrect anchoring of ambiguous terms.

BOwL exploits Lesk algorithm [10] for tagging each word belonging to an entity name with its most likely sense. During the matching stage, semantic and Boolean tags are exploited for obtaining effective mappings: reliable semantic tags are used during the ontology matching stage for identifying homonyms which do not share the same meaning, whereas Boolean tags are exploited for matching composite entity names as if they were Boolean propositions.

S-Match [11] works on lightweight ontologies, namely graph structures where each node label is translated into propositional description logic (DL) formula, which univocally codifies the meaning of the node, and then a propositional satisfiability (SAT) solver is used to check the validity of these formulas. The output of S-Match is a set of semantic correspondences called mappings attached with one of the following semantic relations: disjointness ($\perp$), equivalence ($\equiv$), more specific ($\sqsubseteq$) and less specific ($\sqsupseteq$). S-Match is extendable to host new algorithms and uses a predefined set of background knowledge sources, such as WordNet and UMLS.

More recent systems incorporate background knowledge sources to improve the ontology alignment process [12].

Regarding the proposed techniques devoted to develop highly sophisticated tools for performing ontology matching, we propose an ontology matching system that uses a technique for resolving the ambiguity of concepts names by taking the context of a concept into account. The context of a concept is defined by the set of those concepts that are close to the concepts in terms of shortest path. The large size of existing ontologies and the application of complex match strategies for obtaining high quality mappings makes ontology matching a resource- and time-intensive process, so most of the proposed ontology matching system rely on merely a defined depth (e.g, depth = 2). However the more interesting focus in our technique is to explore more in depth all the composed concepts that are connected directly or indirectly to the central node while taking into account the size of the compared ontologies. Finally to overcome the problem of computationally when aligning medium-sized and large-scale ontologies, we use a particular parallel matching on multiple cores or machines for dealing with the scalability issue.

3   Ontology Matching

Ontology matching tries to establish semantic relations between similar elements in different ontologies to provide interoperability. Ontology matching takes a pair of ontologies as an input and creates the semantic correspondence relationships between these ontologies.

3.1   Exploring Ontological Context of a Concept

This approach aims at discovering linguistic similarities between the involved entities. In general, linguistic similarities are based on morphology and semantics, which are associated to the words that describe the relative entities.

Often the same word placed in different textual contexts assumes completely different meanings. In addition, lexicons are not able to disambiguate situations in which homonyms occur. In order to deal with lexical ambiguity, this approach introduces the notion of "scope" of a concept which represents the context where the concept is placed.

Definition 1. *Let O be ontology and $c \in O$. The scope of c, with radius r, $scope(c,r)$ is a set of all the concepts outgoing from c included in a path of length r, with center c. More formally:*

$$scope(c,r) = \left\{ c' | c' \in O, \; dist(c,c') < r \right\}, \tag{1}$$

where $dist(c,c')$ is the number of edges that are in the path from concept c to concept c'. Let us note that $dist(c,c') = 0$, when $c = c'$.

Definition 2. *Let α and β respectively concepts of the ontology O and O'. Let $Name_\alpha = label(\alpha)$ and $Name_\beta = label(\beta)$ be respectively the linguistic labels associated to the concept α and β. Let $lex(Name_\alpha, Name_\beta) \in [0,1]$ be a lexical similarity associated to* the pair of concept names α and β, with $\alpha \in O$, $\beta \in O'$. The set L is composed of all pairs, defined as follows:

$$L = \left\{ \left(\alpha', \beta' \right) | \forall \alpha \in scope \left(\alpha', r \right), \forall \beta \in scope \left(\beta', r \right) \right\}$$
$$and \left\{ \exists lex \left(Name_\alpha, Name_\beta \right) \neq 0 \right\}. \tag{2}$$

"The need to determine the degree of semantic relatedness between lexically expressed concepts is a problem that pervades much of the computational linguistics" [13]. Recent research on the topic in computational linguistics has emphasized the perspective of semantic relatedness of two words in a lexical resource, or its inverse semantic distance. A natural way to compute the similarity measure of words given a semantic network is to evaluate the distance of nodes corresponding to words being compared to the shortest path from one node to another, the more related the words are. Thus the length of the shortest path in a semantic network is named semantic distance. The first step towards defining a method for measuring the semantic similarity between a pair of concepts (or terms) using the lexicon WordNet is to define how the distance between two WordNet synsets can be measured. The following three similarity measures are based on path lengths between a pair of concepts: *lch* [14], *wup* [15], and *path*. *lch* finds the shortest path between two concepts, and scales that value by the maximum path length found in the is-a hierarchy in which they occur. *wup* finds the depth of the least common subsume (lcs) of the concepts, and then scales that by the sum of the depths of the individual concepts. The depth of a concept is simply its distance to the root node. The measure *path* is a baseline that is equal to the inverse of the shortest path between two concepts. We have adopted the Leacock-Chodorow (*lch*) and Wu and Palmer (*wup*) conceptual distance measures.

Leacock & Chodorow (*lch*) propose a normalized path-length measure which takes into account the depth of the taxonomy in which the concepts are found [14].

$$lch(c_1, c_2) = -log \frac{length(c_1, c_2)}{2D}, \tag{3}$$

where $length(c_1, c_2)$ is the number of nodes along the shortest path between the two nodes (as given by the edge counting method), and D is the maximum depth of the taxonomy.

Wu & Palmer (*wup*) present instead a scaled measure which takes into account the depth of the nodes together with the depth of their *least common subsumer* (*lcs*).

$$wup(c_1, c_2) = \frac{depth(lcs(c_1, c_2))}{depth(c_1) + depth(c_2)}. \tag{4}$$

Definition 3. *Let c_1 and c_2 be two ontology concepts in the ontology O. The distance between two nodes $\delta(c_1, c_2)$ is represented by the minimum number of edges that connect them.*

Definition 4. *Given an ontology O formed by a set of nodes and a root node R. Let c_1 and c_2 be two ontology concepts of which we will calculate the similarity. Then, g is* *the common ancestor of c and c'. The Wu-Palmer similarity is defined by the following expression [15]:*

$$sim(c_1,c_2) = \frac{2 \times \delta(g,R)}{\delta(c_1,R) + \delta(c_2,R)}. \tag{5}$$

3.2 Sense Disambiguation to Improve Matching

Through the given definitions, it is possible to individuate the meaning of a name associated to an ontology concept. Given a word, WordNet provides a list of all the synsets and word senses, related to that word. Firstly the following pseudo-code details the *Algorithm 1* which discriminate the actual sense of a word associated to a given concept.

---

input : Ontology O, concept $\alpha \in O$, radius r and the word w
output: important synset of w which express the exact meaning

1 *build an Array T [|synset(w)|]*;
2 **foreach** t_1 *in synset(w)* **do**
3   *Initialize V [w, similarity(t_1)] = 0*;
4   **for** $i \leftarrow 1$ **to** r **do**
5   *Set M = ∅*;
6   *Set GM = ∅*;
7   **foreach** c *in scope(α, i) − scope(α, i − 1)* **do**
             // Collection of similarity values
8   *Set S = ∅*;
9   **foreach** t_2 *in synset(label(c))* **do**
10   *sim ← similarity&lch(t_1, t_2)*;
11   *S ← S ∪ (sim)*;
12   **end**
13   *max ← maximum in S*;
14   *max ← $\frac{max}{[scope(\alpha, i)]}$*;
15   *M ← M ∪ (max)*;
16   **end**
17   *GM ← maximum in M*;
18   *GM ← $\frac{GM}{i}$*;
19   *V [w, similarity(t_1)] ← GM*;
20   **end**
21 **end**
22 result ← synset of the word w with highest similarity in V [ ];
23 **return** result;

Algorithm 1. Exploring the true sense of $\alpha \in O$ is determined by its place in the hierarchy of its ontology up to a certain depth.

The *Algorithm 1* takes as input an ontology O, a reference concept α in that ontology and a word w. The word w represents the name associated to the concept α (i.e. $w = label(\alpha)$). Secondly, the pseudo-code details also the *Algorithm 1* which takes as input

```
input  : Ontology O, concept α ∈ O
         Ontology O′, concept β ∈ O′
output: similarities number of t₁ compared to t₂
// important synset of α discovered by Algorithm 1
1 t₁ ∈ synset(α);
// important synset of β discovered by Algorithm 1
2 t₂ ∈ synset(β);
3 sim ← similarity&wup(t₁,t₂);
4 return sim;
```

Algorithm 2. Comparing the two synsets associated for a concept $c \in O$ and a concept $c' \in O'$

the output of the *Algorithm 1* (The two selected synset of concept c and c'), and calculate the similarities between the two synset using the measure proposed by Wu and Palmer [15]. Let us note the semantic difference between the concept (or class) in an ontology and the label associated to that concept. The algorithms replies to question like *"I would like to know the effective sense of the word w, placed in the context (or scope) of the concept α"*.

First step (line 1) is to declare a vector structure whose size corresponds to the number of synsets (or senses) associated to the given word w. Goal is to maintain in each cell of the vector a pertinence value that represents how much the word w is semantically related to that sense (or belongs to that synset). *Algorithm 1* selects all the concepts in the scope of α (belonging to the reference ontology O) by varying the radius (lines 4-7), in order to get different set of terms. Then, the *lch* similarity between two terms coming from the concept name of α and the word w (line 10) is computed. For each concept c in the ring shaped area (computed as the difference of the areas between two successive radius, see line 7), the max similarity values between the synset name associated to c and a concept name in synset of α are maintained in M (line 15). At the end of the three loops (lines 5-16) the vector $V[w, similarity(t_1)]$ contains the max similarity values in the variable GM computed for each couple of terms coming from the fixed term $t_1 \in synset(w)$ and all the terms which occur in the scope of α, with respect to the current radius (line 18). This is repeated for each $t_1 \in synset(w)$ in the synset of w. At the end, there is a value (line 19) in each cell of the vector V associated to each term in the synset of w, which the algorithm uses it to judge the important synset with the highest similarity number reflecting the correct meaning of concept α.

Algorithm 2 takes as input the two selected synset; $t_1 \in synset(\alpha)$ fom the ontology O and $t_2 \in synset(\beta)$ fom the ontology O', discovered by *Algorithm 1*, and calculate the similarities between them using the measure proposed by Wu and Palmer [15]. The final semantic similarity value is added to the linguistic matcher or structure matcher in order to enhance the semantic ambiguity during the comparison process of entities names.

4   Proposed Algorithm

4.1   Ontology Matching Process of XMap++

XMap++ is a system for ontology alignment that performs semantic similarity computations among terms of two given OWL ontologies. XMap++ view match as an operator that takes two graph-like structures (e.g., classifications, RDF schemas) and produces a mapping between the nodes of these graphs that correspond semantically to each other. Our semantic matching approach include three matchers. The *String Matcher* based on linguistic matching compares the textual descriptions of the concepts associated with the nodes (labels, names) of each ontology. The *Linguistic matcher* jointly aims at identifying words in the input strings, relaying on WordNet [5] which provide additional information towards unveiling mappings in cases where features such as labels are missing or in cases where names are replaced by random strings. Finally the *Structural matcher* aligns nodes based on their adjacency relationships. The relationships (e.g., subClassOf and is-a) that are frequently used in the ontology serve, at one hand, as the foundation of the structural matching [16]. On the other hand, the structural rules are used to extract the ontological context of each node, up to a certain depth (radius). This process is enriched by applying a transitive inference mechanism, in order to add more semantic information that is not explicit in the asserted ontologies.

Alignment suggestions are then determined by combining and filtering the results generated by one or more matchers. Their values are combined using a differents aggregation strategies (Weighted sum, ANN, GA, MAX, etc.) [16]. The filtering consists of retaining the pairs of terms with a similarity value above a certain threshold as align-

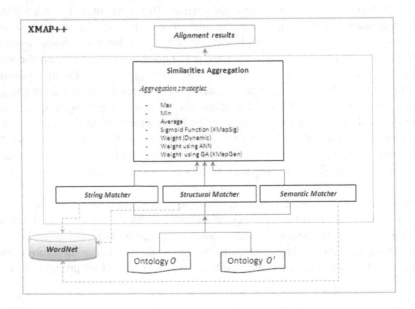

Fig. 1. Sketch of Architecture for XMap++

ment suggestions. The output of the alignment algorithm is a set of alignment relationships between terms from the source ontologies (see Fig. 1).

4.2   XMap++ Using WordNet's Synset

As expressed earlier in section 4.1, the ontology mapping outlines correspondences or matches between concepts coming from two different ontology. In this approach, the concept matching is measured by computing a similarity between concepts at linguistic level. The two algorithms described in the previous section has been exploited to evaluate the concept matching. More formally, given two ontologies O and O' and two concepts c and c', belonging respectively to these two ontologies, there is a match between c and c', if a similarity between them exists, computed as follows:

1. *Algorithm 1* is invocated twice: taking, in turn the concept c as an input and the ontology O, then by taking the concept c' and the ontology O'. Outputs of these two independent executions are two indexes, i for the synset of c and i' for the synset of c'. As said, they identify the important synset with highest similarity number that reflects the correct meaning of the two compared concepts.
2. Once discovered the two synset of involved concepts, the *Algorithm 2* compute the affinity between the concepts by a similarity measure between c and c' using the Wu-Palmer similarity.

In order to obtain all the semantic correspondences among the concepts in the two ontologies, this procedure can be applied for each pair of concepts coming from two ontologies. The final result is an ontology alignment; a similarity value is assigned to each discovered correspondence between concepts in the two ontologies.

5   Evaluation

5.1   Implementation and Setting

A detailed description of the implementaion of XMapGen and XMapSig is out of the scope of this paper, but we summarize here their principles characteristics : XMapGen and XMapSig, a new and lighter implementations of their ancestor XMap++ [16]. XMapGen uses Genetic Algorithm (GA) as a machine learning-based method to ascertain how to combine multiple similarity measures into a single aggregated metric. However XMapSig uses sigmoid function [16] for combining the corresponding weights for different semantic aspects, reflecting their different importance.

5.2   Data Sets and Evaluation Criteria

To evaluate our approach in OAEI 2013[1], we participated in five tracks including *Benchmark*, *Conference*, *Library*, *Anatomy* and *Large Biomedical Ontologies* tracks (see Table .1). We follow the evaluation criteria of OAEI, calculating the precision, recall and f-measure of each test case.

Table 1. Statistics of the OAEI 2013 tracks

Test set	Comparison size	depth (max)
Benchmark (Biblio)	$\cong 9409$ element pairs	8
Anatomy Track	$(2.800 \times 3.300) \cong 9 \cdot 10^6$ element pairs	17
Conference Track	$\cong 11025$ element pairs	7
Library Track	$(25,000 \times 12,000) \cong 300 \cdot 10^6$ element pairs	64
FMA-NCI small fragments	$(3,696 \times 6,488) \cong 10 \cdot 10^6$ element pairs	20
FMA-SNOMED small fragments	$(10,157 \times 13,412) \cong 310 \cdot 10^6$ element pairs	20

Table 2. Depth value variation and its effect on f-measure of XMapSig and XMapGen

	biblio	Anatomy	Conference	Library	FMA/NCI	FMA/SNOMED
Baseline	0.41 (edna)	0.77 (StrEqv)	0.52 (StrEqv)	0.57 (MaPrefEN)	0.81 (Average)	0.55 (Average)
YAM++	0.89	0.90	0.71	0.74	0.91	0.84
MaasMatch	0.69	0.41	0.36	-	0.46	-
Depth = (all the depth)						
XmapSig	0.58	0.75	0.53	0.45	0.60	0.23
XmapGen	0.54	0.75	0.53	0.06	0.61	0.24
Depth = 6						
XmapSig	0.52	0.533	0.37	0.369	0.489	0.154
XmapGen	0.49	0.567	0.38	0.375	0.501	0.173
Depth = 2						
XmapSig	0.46	0.49	0.23	0.24	0.41	0.18
XmapGen	0.43	0.51	0.26	0.28	0.42	0.19
Depth = 0						
XmapSig	0.41	0.33	0.22	0.19	0.14	0.06
XmapGen	0.38	0.345	0.26	0.21	0.19	0.09

Table 3. Execution time (in seconds) of XMapSig and XMapGen vs. OAEI 2013 entrants

	biblio	Anatomy	Conference	Library	FMA/NCI	FMA/SNOMED
YAM++	702	62**	600	731.86	94	100
MaasMatch	173	8532	76	-	12,410	-
Depth = (all the depth)						
XMapSig	612	393	600	2914.167	1.477	11.720
XMapGen	594	403	600	3008.82	1.504	12.127

5.3 Experimental Results

To gauge the effect of our semantic verification process, we have run the experiments both using the full XMapGen and XMapSig implementation (comparing all the depth), as well as using a system varying the level of the depth in the ontological structure. To facilitate our analysis, we assign each concept within its hierarchy a depth value (e.g, depth =0, depth =2 and depth =6). The concepts at the top level of the hierarchy

---

[1] http://OAEI2013.ontologymatching.org/2013/

(depth=0) do not possess any parent concepts. Then, the depth of the concept C is determined as the length of the path from C to the top level concept that is associated with C using the "subClassOf" relation.

The results of these experiments are shown in Table .2. It can be clearly seen that, as expected, the accuracy of our two systems decreases as the value of the depth decreases. It can clearly be seen that the use of all the depth in the hierarchy of a given ontology produces a more accurate result than the use of any less depth value. Whereas for a benchmark track the variation of the depth value has no effect on the f-measure, this may be a matter of just a few classes before the top level class is reached. Additionally, to tackle the large ontology matching problem we improved the runtime of the algorithm using a divide-and-conquer approach that can partition the execution of the matchers into small threads was improved and joins their results after each similarity calculation. A direct comparison between the XMapGen and XMapSig shows that the addition of GA does not has a negative effect on the algorithm but, on the contrary, leads to slightly better results, especially in terms of recall. In most tracks, XMapSig supplies high precision than XMapGen. Whereas using Genetic Algorithm (XMapGen) performs quite high in terms of recall than using sigmoid function (XMapSig)[6][17].

XMapGen and XMapSig produced fairly consistent alignments when matching the five tracks. Some reasons are related to: (a) the absence of domain and range definitions (in fact, of properties in general), as for anatomy, and the presence of multi-lingual labels; (b) fixed threshold used as a filter in the selection module. Different tests require different thresholds; (c) XMapSig does not respect languages, this may lead to false positives; (d) XMapGen and XMapSig exploit only the superclass-subclass relationships (subsumption relationships) that are frequently used in ontologies when the total number of entities is bigger than 1500 entities in each ontology. *We restrict the contextual similarity computing; only the value of the semantic relation between two concepts without taking in consideration the types of cardinality constraints and values between their properties.*

5.4   The Comparison against Other Systems in OAEI Campaign 2013

A detailed comparison against the 23 systems participated in the campaign OAEI 2013 is beyond the scope of this paper, but we selected two systems, YAM++ and Maas-Match, because they use WordNet as background knowledge [17].

We have tabulated the f-measure values for our two systems against YAM++ and MaasMatch in Table .2. As can be seen, the f-measure confirms the good performances of YAM++. In terms of baseline comparison, our two systems perform better than Maas-Match on 5 tracks. Whereas our two systems failed to provide an f-measure higher or roughly equal than the baseline for *Library* and *Large Biomedical* tracks. While Maas-Match performed worse than baseline for the *Anatomy, Conference, Library* and *Large Biomedical* tracks.

The results of the mappings runtime are presented in Table .3: a) *Benchmark track*, it ranges from less than a 3 minutes for MaasMatch and to nearly 11 minutes for both YAM++ and our two systems; b) *Anatomy track*, due to some software and hardware incompatibilities, YAM++ had to be run on a different machine and therefore its runtime (indicated by **) is not fully comparable to the other matchers. Meanwhile, the total time elapsed for the execution of our two systems was approximately 7 minutes against 2 hours and 22 minutes to MassMatch for finishing this track; c) *Conference track*, MaasMatch finished all 21 tests around 1 minute. Whereas, 10 minutes are enough for the three matchers (YAM++, XMapGen and XMapSiG); d) *Library track*, our two systems and YAM++ were able to generate an alignment within 12 hours. MaasMatch doesn't finish in the time frame; e) *Large biomedical ontologies (largebio)*, YAM++ provided the best results on FMA/NCI track 1 and FMA/SNOMED track 1.

Regarding performance we can conclude that YAM++ has a significant improvement from our two systems and MassMatch in terms of both f-measure and runtime, especially for very large scale tasks. This proves the effectiveness and efficiency of the strategy implemented by YAM++ which consists of implementing a disk-based method for storing the temporary information of the input ontology during the indexing process in order to save main memory space. Nevertheless, there is room for continued improvements in our algorithms because our approach lead to an important increase in precision, without having too negative an impact on recall which reflect the main objective of this approach [17].

Generally, according to our results in OAEI 2013, our two systems delivered fair results comparatively to other participants. The aim of this development experience was not to deliver a tool to compete with others in terms of precision and recall. Instead, we aimed at the development of a new and stable version of XMap++ using new and state-of-the-art technologies and alignment methods [17]. In case of scalability (see results of running tests on six dataset in [17]), our two systems are placed above eight out of twenty-three matchers participated in OAEI 2013. Matching larger ontologies still takes significantly longer time when parsing ontologies with Alignment API. We plan to solve this problem using an ontology parser which permits to load multiple ontologies in parallel via threading.

6   Conclusion and Future Works

This paper has presented a novel approach using context-based measure for semantic matching. The proposed ontology matching method based on the context of ontologies solves difficult problems of lexicography, polysemy and synonymy which occur in the phase of ontology mapping. The proposed method based on the variation of the radius value, which induces the variation of the scope area, in order to detect the correct meaning of the context description. The preliminary results were quite good to encourage us to continue seeking better solutions. In future work, how to customize the value of radius for each entity will be studied. That is, this research will be extended to build an algorithm that takes into account the type and the depth of taxonomies related to each entity, in order to automatically tune the radius value parameter. Moreover, our framework is very flexible: many semantic measures may be used in the future (JCN (Jiang and Conrath), Lin, Resnik, Lesk, hst(HIRST and ST-ONGE), etc.).

References

1. Euzenat, J., Shvaiko, P.: Ontology Matching. Springer, Heidelberg (2007)
2. Dey, A., Salber, D., Abowd, G.: A conceptual framework and a toolkit for supporting the rapid prototyping of context-aware applications. Human-Computer Interaction 16, 97–166 (2001)
3. Dourish, P.: Seeking a foundation for context-aware computing. Human-Computer Interaction 16, 2–3 (2001)
4. Chalmers, M.: A Historical View of Context. Computer Supported Cooperative Work 13(3), 223–247 (2004)
5. Miller, G.: WordNet: An electronic Lexical Database. MIT Press (1998)
6. Djeddi, W., Khadir, M.T.: XMapGen and XMapSiG results for OAEI 2013. In: Proceedings of the 8th International Workshop on Ontology Matching co-located with the 12th International Semantic Web Conference (ISWC 2013), pp. 203–210. CEUR-WS.org, Sydney (2013)
7. Sabou, M., Aquin, M., Motta, E.: Exploring the Semantic Web as Background Knowledge for Ontology Matching. J. Data Semantics 11, 156–190 (2008)
8. Lin, F., Sandkuhl, K.: A Survey of Exploiting WordNet in Ontology Matching. In: Bramer, M. (ed.) IFIP AI, pp. 341–350. Springer, Heidelberg (2008)
9. Gracia, J., Lopez, V., d'Aquin, M., Sabou, M., Motta, E., Mena, E.: Solving semantic ambiguity to improve semantic web based ontology matching. In: The 2nd International Workshop on Ontology Matching 2007, Busan, South Korea (November 11, 2007)
10. Mascardi, V., Locoro, A.: BOwL: exploiting Boolean operators and lesk algorithm for linking ontologies. In: Ossowski, S., Lecca, P. (eds.) SAC, pp. 398–400. ACM (2012)
11. Giunchiglia, F., Yatskevich, M., Shvaiko, P.: Semantic Matching: Algorithms and Implementation. J. Data Semantics 9, 1–38 (2007)
12. Zablith, F., d'Aquin, M., Sabou, M., Motta, E.: Investigating the use of background knowledge for assessing the relevance of statements to an ontology in ontology evolution. In: 3rd International Workshop on Ontology Dynamics (IWOD 2009) at ISWC-2009, Washington, DC, USA (2009)
13. Budanitsky, A., Hirst, G.: Semantic Distance in WordNet: An Experimental, Application oriented Evaluation of Five Measures. In: Workshop on WordNet and Other Lexical Resources, in the North American Chapter of the Association for Computational Linguistics (NAACL-2000), Pittsburgh, PA (2001)
14. Leacock, C., Chodorow, M.: Combining local context and WordNet similarity for word sense identification. In: Fellbaum, C. (ed.) WordNet: An Electronic Lexical Database, pp. 265–283. MIT Press (1998)
15. Wu, Z., Palmer, M.: Verb semantics and lexical selection. In: 32nd Annual Meeting of the Association for Computational Linguistics, Las Cruces, New Mexico, pp. 133–138 (1994)
16. Djeddi, W., Khadir, M.T.: Ontology alignment using artificial neural network for large-scale ontologies. the International Journal of Metadata, Semantics and Ontologies (IJMSO) 8(1), 75–92 (2013)
17. Shvaiko, P., Euzenat, J., Srinivas, K., Mao, M., Jiménez-Ruiz, E.: Proceedings of the 8th International Workshop on Ontology Matching co-located with the 12th International Semantic Web Conference (ISWC 2013), CEUR-WS.org, Sydney (2013)

ActMiner: Discovering Location-Specific Activities from Community-Authored Reviews

Sahisnu Mazumder[1], Dhaval Patel[1], and Sameep Mehta[2]

[1] Indian Institute of Technology, Roorkee, Uttarakhand, India
sahisnumazumder@gmail.com, patelfec@iitr.ac.in
[2] IBM Research Lab, Delhi, India
sameepmehta@in.ibm.com

Abstract. Location-specific community authored reviews are useful resource for discovering location-specific activities and developing various location-aware activity recommendation applications. Existing works on activity discovery have mostly utilized body-worn sensors, images or human GPS traces and discovered generalized activities that do not convey any location-specific knowledge. Moreover, many of the discovered activities are irrelevant and redundant and hence, significantly affect the performance of a location-aware activity recommender system. In this paper, we propose a three-phase Discover-Filer-Merge solution, namely ActMiner, to infer the location-specific relevant and non-redundant activities from community-authored reviews. The proposed solution uses Dependency-aware, Category-aware and Sense-aware approaches in three sequential phases to accomplish its objective. Experimental results on two real-world data sets show that the accuracy and correctness of ActMiner are better than the existing approaches.

Keywords: Activity Discovery and Recommendation, Review Mining.

1 Introduction

Movement is an integral part of human daily life. Most of the times, people visit various locations to perform some activity according to their preferences. For example, people visit a restaurant with the purpose of having food, visit a shopping mall to do shopping and so on. And whenever they feel something interesting about those locations, they share those experiences with their friends using location-based social networking (LBSN) platforms like *Yelp*, *Foursquare*, *Brightkite* in terms of location-specific reviews. As of June, 2013, Yelp has been populated by over 42 million reviews for various locations in US [1]. These reviews mainly contain information about users personal experiences about locations in the form of textual description. The location-specific reviews act as a great resource for discovering activities supported by various locations (location-specific activities) and can lead to the successful development of a location-aware activity recommender system. This kind of recommender system can help people to easily figure out the best nearest location for performing a certain activity

L. Bellatreche and M.K. Mohania (Eds.): DaWaK 2014, LNCS 8646, pp. 332–344, 2014.

without wasting time in asking people and making decision from their diverse and biased suggestions.

But, the success of such a recommender system depends on the accuracy of its knowledge base which consists of a set of location-specific activities. We can utilize an approach suggested by Dearman et. al. in [2] to infer the location-specific activities in the form of (*verb, noun/noun phrase*) pairs from community-authored reviews. However, this approach doesn't consider two important issues such as relevancy and non-redundancy of the inferred activities. For example, (*watch, cricket match*) is an *irrelevant activity* for location like *"restaurant"*. Similarly, activities (*have, chicken*) and (*eat, chicken*) are *redundant* to each-other. Presence of irrelevant and redundant activities in knowledge base has an adverse effect on the performance of activity recommendation. For example, separately, redundant activities like (*have, chicken*) and (*eat, chicken*) may not be the most frequent activity performed at a given restaurant. But, if we merge them together as they are redundant, their individual frequencies get added and the resultant activity may become the most frequent activity. So, presence of irrelevant and redundant information can cause incorrect activity ranking. Moreover, irrelevant and redundant information may limit the amount of useful information pushed from the recommender system to the user in broadcast environment [3].

In this paper, we propose a three-phase Discover-Filter-Merge solution, namely ActMiner, to infer the location-specific activities from community-authored reviews. In the first phase, ActMiner uses NLP techniques to discover potential (*verb, noun/noun phrase*) pairs that represent meaningful activities. In the second phase, relevant activities are discovered from the output of the first phase, using ConceptNet [4] and category information of the location. In the last phase, the redundant activities are merged if they are associated with similar sense. The major technical contributions in developing ActMiner are as follows:

- **Problem Formulation**: We formulate the problem of discovering location-specific relevant and non-redundant activities from community-authored reviews. To the best of our knowledge, ActMiner is the first activity discovery technique that addresses the issues of relevancy and non-redundancy of the discovered activities.

- **Novel Techniques**: (1) We propose novel dependency-aware activity extraction technique for meaningful activity discovery. Experimental results show that the proposed method extracts more accurate and meaningful activities compared to those inferred by the existing approach [2]. (2) We introduce novel category-aware relevant activity discovery technique, where we build **Category-aware Concept Hierarchy** (CCH) for each location using ConceptNet and category of the location. CCH contains relevant concepts for a given location and is utilized in the relevant activity discovery process. (3) We develop novel sense-aware approach for minimizing redundancies present in the discovered activities. We use ConceptNet and CCH to discover the activities having same hidden sense and merge them into a single activity.

– **Real-world Experiments**: Experiments are performed on two real data sets to verify the accuracy and correctness of activities discovered by `ActMiner`. We also build a location-aware activity recommender system to evaluate the effectiveness of the proposed solution.

2   Preliminaries and Related Work

Let $L = \{L_1, L_2, L_3, ..., L_m\}$ be the set of m locations and cat_i be the list of categories for location $L_i \in L$. In our database, each location L_i is associated with a set of reviews $Rset_i = \{R_i^1, R_i^2, R_i^3, ..., R_i^n\}$, where R_i^j denotes the jth review of location L_i. The review R_i^j is written in the form of textual description.

Definition 1. Activity. *An activity performed at a particular location is defined as the combination of a verb with a noun or noun phrase. An activity is meaningful if it represents some "doing" sense. It is represented as $A_i^j = (verb, noun/noun\ phrase)$ and stands for the jth activity performed at Location L_i. Here, a noun phrase is a collocation of nouns with which a verb forms association.*

Definition 2. Activity Frequency. *The frequency of an activity A_i^j, denoted as $AF(A_i^j)$, is the number of distinct reviews that has mentioned about A_i^j at location L_i.*

Definition 3. Concept. *The concept associated with an activity A_i^j is defined as the noun or noun phrase part of an activity.*

Problem Statement. Given the categories and review sets for all m locations, i.e. $\{(Cat_i, Rset_i) \mid 1 \leq i \leq m\}$, our proposed solution discovers set of *relevant* and *non-redundant* location-specific activities $Aset_i = \{A_i^j \mid 1 \leq j \leq r\}$ for each location L_i, where A_i^j is an activity performed at location L_i.

Related Work. Majority of the research works done in the area of activity recognition in past decades, have only discovered human physical activities [5],[6] and general activities at a particular location [7], [8], [9] by analysing human body movement, gesture, GPS trajectories etc. And so, these solutions are not effective in location-specific activity discovery purpose. The approach, closest to our proposed solution and concerns about inferring location-specific activities from location-specific reviews, is paper [2] which we have considered as the base paper. The approach uses sentence tokenizer to parse the review text into its individual sentences and then employ part-of-speech tagger to identify verbs and nouns in each sentence. Next, verb-noun pairs are discovered such that noun is located within 5 words following the verb. The discovered verb-noun pairs are represented in their base form and declared as the potential activities supported by the location. Although the procedure is very simple, it is accompanied with three major limitations: **(1)** The approach doesn't discover the complete set of activity. For example, the approach doesn't generate verb-noun pair if the noun occurs before verb. **(2)** The approach has only used the *"distance"* between a verb and a noun for pair generation without considering the *"dependency between words"* which is the key factor for ensuring meaningfulness of the discovered activities. Moreover, **(3)** the approach has not addressed the issues with relevancy, and non-redundancy as highlighted in the introduction section.

3   Solution Overview

Given the categories and review sets for all m locations, `ActMiner` generate activities for each location using Discover-Filter-Merge technique (See Figure 1). The Discover-Filter-Merge technique process review set $Rset_i$ for location L_i in three sequential phase, where activities are discovered in the first phase, irrelevant activities are filtered in the second stage, and redundant activities are merged in the final phase. In the remaining sections, we explain details of each phase using $Rset_i$ for location L_i.

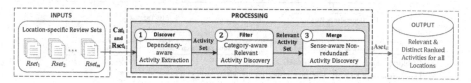

Fig. 1. Working of ActMiner

3.1   Dependency-Aware Activity Extraction

We can process each review $R \in Rset_i$ and discover any *(verb, noun)* or *(verb, noun phrase)* present in the review message. However, this simple approach generates a large number of spurious activities if the relationship between verb and noun is not taken into account. Thus, we develop a novel dependency-aware activity extraction technique that utilizes the typed dependency and proximity information between verb and noun to discover the activity. In particular, we employ a series of NLP operations on each review R in $Rset_i$, as follows:

1. We use OpennlpSentenceDetector [10] to extract individual sentences from review R.
2. Next, we parse each sentence using Stanford Typed dependency parser [11] and detect all activities in terms of *(verb, noun)* pairs that has *dobj, nsubjpass, ccomp* or *prepositional grammatical relations* between verb and noun. Although dependency parser can report around 51 dependencies [12], we have observed that the aforementioned four dependencies help us to capture most the meaningful activities.
3. Dependency parser doesn't generate activity in a form of *(verb, noun phrase)*. To address this issue, we extract noun phrases from each sentence using Stanford POS Tagger [13] and replace noun part of the detected (verb, noun) pair with the corresponding noun phrase part.

We explain above procedure with a help of a sample review R as shown in the Figure 2. First step extracts four sentences from R. In the next step, total five activities are discovered using dependance parsing, where (*trying, food*) activity is discovered from sentence-1 using dependency relation "*dobj(trying, foods)*", (*came, dinner*) is discovered from sentence-2 using "*prep_for(came, dinner)*", and so on. Simultaneously, POS Tagger detects *chicken tikka masala* as a noun phrase from sentence-4. Thus, pair (*enjoyed, masala*) is converted into the pair (*enjoyed, chicken tikka masala*).

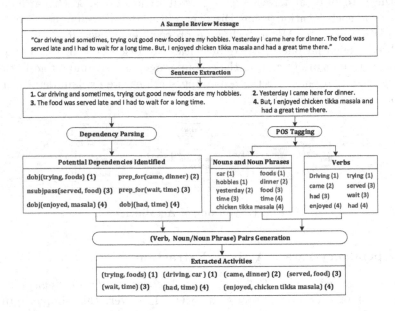

Fig. 2. An Example of Activity Extraction from Review R. The number given in "()" associated with each activity, extracted verb and noun/noun phrase indicates the sentences from which they are extracted.

Sometimes, combination of two nouns, where one noun is derived from a verb in its base form by adding "*ing*" to it, also represent a meaningful activity. In Figure 2, the word "*driving*" in sentence-1 is actually tagged as noun by pos-tagger, but it is derived from the base verb "*drive*". To deal with this case, while POS tagging a sentence, we have used WordNet [14] to check whether a noun that ends with "*ing*" has a base verb form in its synset or not. If there exists a base verb form for that noun, (for example, "*drive*" is the base verb form present in the synset of noun "*driving*"), we treat such noun as verb and generate pair by associating it with the nearest noun or noun phrase in the sentence. Thus, (*driving, car*) has been inferred as an activity from sentence-1.

Once, activities are discovered from each review R of $Rset_i$, all the verbs and nouns in the (*verb, noun/noun phrase*) pairs are converted into their base forms using WordNet. For example, (*ate, food*) is converted into (*eat, food*). Then, *AF* values of each of the discovered activities are calculated and discovered activities are stored in $Aset_i$ in the form $[(verb, noun/noun\ phrase), AF]$.

3.2 Category-Aware Relevant Activity Discovery

The activities discovered in the previous stage are syntactically correct, but may not represent a meaningful activity from the semantic point of view. In particular, there may be activities that cannot be performed at a given location and thus are not relevant. In this paper, we leverage ConceptNet to validate whether the concept associated with an activity conforms to the category of L_i or not. For example, given the activity $(eat,\ food)$ at location L_i, the concept "food" is associated with location L_i, if category of L_i is "restaurant". By exploring ConceptNet, we find that concept "food" is related to the concept "restaurant" by the relation $\{food \overset{At\ Location}{\rightarrow} restaurant\}$. However, in most of the cases, the concept associated with an activity doesn't have any direct relationship with the category of the location. For example, concept "chicken" is not directly associated with the concept "restaurant" in ConceptNet. However, "chicken" is associated with "food" and "food" is intern related to "restaurant". Hence, a simple lookup in ConceptNet doesn't solve the the problem. However, considering the size of the ConceptNet, it is not feasible to explore all possible indirect connections. We present a systematic way of finding out the chain of relations that associates a given concept to the category of a location.

Given a set of activities $Aset_i$ for location L_i, we extract concept by processing each activity in $Aset_i$ and output the *noun or noun phrase* part of an activity as a concept. Let, $Cset_i$ be the set of discovered concepts. Then, we learn a *Category-aware Concept Hierarchy*, denoted as CCH, using $Cset_i$, category list Cat_i and ConceptNet. We use only "IsA", "AtLocation", "DerivedFrom", "UsedFor" and "RelatedTo" relations of ConceptNet to learn the CCH. The first three relations "IsA", "AtLocation" and "DerivedFrom" capture generalization-specialization relationship between two concepts. For example, $\{novel \overset{IsA}{\rightarrow} book\}$ and $\{shopper \overset{DerivedFrom}{\rightarrow} shop\}$. The remaining two relations "UsedFor" and "RelatedTo" are used for linking related concepts. For example, $\{kitchen \overset{UsedFor}{\rightarrow} cook\}$ and $\{cake \overset{RelatedTo}{\rightarrow} birthday\}$.

Definition 4. *Category-Aware Concept Hierarchy (CCH).* *A Category-aware Concept Hierarchy for a given location is a tree-based hierarchy and represented by triplet $< Lv, C, E >$, where $Lv = \{lv_1, lv_2, lv_3, ..., lv_k\}$ is the collection of levels , $C = \{C_1, C_2, C_3, ..., C_k\}$ is a collection of concept sets with C_i be the set of concept at level lv_i, and E is the set of labeled arcs that connect concepts lying in the same level or in successive levels. The structure satisfies following 3 properties:*

- *$C_1 = Cat_i$ and $C_i \subseteq Cset_i$ where $2 \leq i \leq k$.*
- *If any two concepts c and c' at the same level are connected by a labeled arc e_r, then the label of $e_r \in \{$ "RelatedTo", "UsedFor"$\}$.*
- *If any two concepts c and c at level lv_i and $lv_{(i+1)}$ respectively, are connected by a labeled arc e_r, then label of $e_r \in \{$ "IsA", "AtLocation", "DerivedFrom"$\}$.*

CCH Formation. The First level of CCH is initialized with the concepts from Cat_i. Next, we perform two operations *Expand* and *Extend* iteratively to grow the hierarchy in horizontal and vertical dimensions respectively. The *Expand* operation uses "*UsedFor*" and "*RelatedTo*" relations to add concepts from $Cset_i$ into the current level. The *Extend* operation uses "*IsA*", "*AtLocation*" and "*DerivedFrom*" relations to add concepts from $Cset_i$ into the next level. In summary, the *Expand* operation adds concepts that are associated with the concepts lying in the same level and the *Extend* operation adds concepts that are specialized in sense with respect to the concepts lying in just upper level. This iterative procedure stops when the hierarchy cannot grow further.

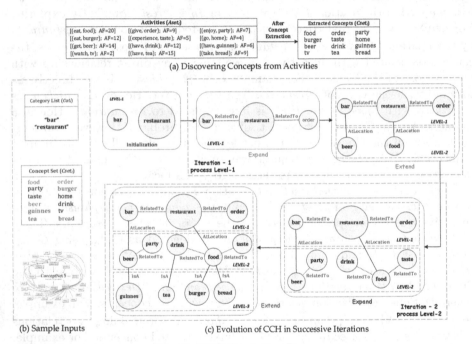

Fig. 3. CCH Formation Process

Example 1. Figure 3(a) provides a set of input activities $Aset_i$ and the set of extracted concepts $Cset_i$ from $Aset_i$. Figure 3(b) shows the sample inputs for CCH formation process which consist of $Cat_i = \{$"*restaurant*", "*bar*"$\}$, $Cset_i$ and the ConceptNet. At first, the CCH is initialized with two concepts "*restaurant*" and "*bar*" at level-1 (See Figure 3(c)). Next, the *expand* operation is performed on level-1 and concept "*order*" from $Cset_i$ is added and linked with "*restaurant*" by relation name "*RelatedTo*" at level-1. Along with that, concept "*bar*" gets associated with concept "*restaurant*" by relation "*RelatedTo*". Next, *extend* operation is performed on the expanded level-1 which adds "*beer*" and "*food*" from $Cset_i$ as a child node of "*restaurant*", links them by "*AtLocation*" relations and forms level-2. At this point *iteration-1* ends and *iteration-2* starts with execution of *expand* operation on level-2 causing expansion of level-2 by adding and linking concepts *"party"*, *"drink"* and *"taste"* with existing concepts in level-2. Next, *extend* operation starts and additional concepts are added and linked with existing concepts as shown in the Figure 3(c) marking the end of *iteration-2* and begining of *iteration-3*. But in *iteration-3*, none of the concepts from $Cset_i$ gets added into the CCH and so, the CCH doesn't grow further and the process terminates.

Once construction of CCH is over, We process each activity in $Aset_i$ and declare it as relevant if the concept of activity is present in the CCH. In summary, given a set of activities, we discover concepts associated with those activities, organize them into the CCH and use this hierarchy to infer relevant activities.

3.3   Sense-Aware Non-redundant Activity Discovery

The activities discovered in the previous stage are relevant but many of them may be redundant with respect the sense of an activity. For example, both the activities (*have, food*) and (*take, food*) are associated with the same concept *"food"* and hidden sense *"eat"* and so, they are redundant to each other. To handle the activity redundancy issue, we discover the activities (having same concept) which are associated with a common hidden sense and merge them into a single activity. This process can be thought as the *sense-based activity clustering*, where each clusters is composed of a set of redundant activities. For example, considering the common hidden sense *"eat"*, activities (*take, food*), (*get, food*) and (*have, food*) form a single activity cluster and is represented by (*have/get/take, food*). Similarly, example of another activity cluster is (*make/prepare, food*) formed based on the common hidden sense *"cook"*. However, activities (*have, chicken*) and (*have, food*) have common hidden sense *"eat"*, but as their concepts are different, they are not redundant and not merged.

Given a set of relevant activities $Aset_i$ for location L_i, we iteratively merge a pair of activities having common hidden sense and same concept. Note that, the concept of an activity is known, but activity sense is unknown. Here, we use *"RelatedTo"*, *"IsA"* and *"UsedFor"* relations of ConceptNet to discover the hidden sense of an activity. For example, the common hidden sense associated with activity (*take, food*) and (*have, food*) is *"eat"* based on the relationship $\{eat \overset{RelatedTo}{\rightarrow} take\ food\}$ and $\{have\ food \overset{UsedFor}{\rightarrow} eat\}$ respectively. So, both activities are merged into single activity (*have/take, food*).

Above procedure merges activities that are mostly associated with generalized concepts such as *"food"*. However, the procedure fails in most of the cases when activities are associated with specialized concepts. For example, redundant activities (*have, burger*) and (*take, burger*) are not merged using the above mentioned procedure. But, if we take the generalized concept of *"burger"* into account, we can merge (*have, burger*) and (*take, burger*) into a single activity. According to the CCH shown in Figure 3(c), the generalized concept of *"burger"* is *"food"*. Since, (*have, food*) and (*take, food*) are already merged based on the common hidden sense *"eat"*, we can also merge (*have, burger*) and (*take, burger*) based on the same hidden sense. This example suggests that we can apply the idea of utilizing generalized concept to merge the specialized activities. In summary, we iteratively merge two redundant activities having common concept c if the activities associated with the *generalized concept of c* are already merged. In this iterative process, we reuse CCH to infer the generalized concept.

At the end of this phase, the discovered relevant and non-redundant activities in $Aset_i$ are ranked based on their AF values and stored in a repository. Note that, merging of redundant activities increases their overall support (i.e., AF value) and helps in obtaining correct activity ranking based on their AF values.

4   Experimental Evaluation

Our experiments evaluate ActMiner in term of its correctness, accuracy of discovered activities, and usefulness in building real world applications using two real-world review data sets, namely yelp and Roorkee. Yelp dataset contains 229,907 reviews from 43,873 users about 11,537 locations from the greater Phoenix, AZ metropolitan area. We choose 7 locations having more than 200 reviews (see Figure 4(a) and prepare a review set for each of the selected locations. Roorkee data set contains 25 locations of Roorkee that are frequently visited by many students and institute staffs from IIT-Roorkee. In total, this local data set contains 686 review messages for 25 different locations as shown in the Figure 4(b).

Loc_ID	Location Name & Address	Categories	No. of Reviews
1	960 W University Dr Tempe, AZ 85281, USA.	Pubs", "Bars", "Nightlife", "Restaurants"	575
2	2611 N Central Ave, Phoenix, AZ 85004, USA.	"Steakhouses", "Restaurants",	278
3	401 E Jefferson St, Phoenix, AZ 85004, USA.	"Arts & Entertainment", "Stadiums & Arenas"	216
4	5701 N Echo Canyon Pkwy, Phoenix, AZ 85073, USA.	"Active Life", "Climbing", "Hiking", "Parks"	210
5	7107 E McDowell Rd, Scottsdale, AZ 85257, USA.	"Food", "Sandwiches", "Breweries", "Pizza", "Restaurants"	232
6	Galvin Bikeway, Phoenix AZ 85008, USA.	"Arts & Entertainment", "Botanical Gardens", "Music Venues", "Nightlife"	260
7	1514 N 7th Ave. 2nd Fl, Phoenix, AZ 85007, USA.	"Bars", "Nightlife", "Lounges"	232

Loc_ID	Location Name	Categories	No. of Reviews
1	Prakash Sweets.	"Dessert Shop", "Snacks"	61
2	Kundan Sweets.	"Dessert Shop", "Snacks"	28
3	Prakash Hotel.	"Hotel", "Restaurant"	55
4	Hotel Royal Palace.	"Hotel", "Restaurant", "bar"	38
5	Dominos Roorkee.	"Pizzarias"	46
6	Sizzlers.	"Restaurant"	13
7	Food Point.	"Restaurant"	19
8	Motel Polaris.	"Hotel", "Restaurant"	28
9	NEEDS.	"Convenience Store"	30
10*	The Pentagon Mall.	"Shopping Mall"	35
11*	Vishal Mega Mart.	"Shopping Mall"	22
12	Woodland Exclusive Store.	"Garment Shop"	18
13*	Reebok Store.	"Shoe Store"	16
14	The Raymond Shop.	"Suits", "Garments shop"	14
15*	Nature Park.	"Park", "Hiking"	09
16*	Solani Park.	"Park", "Hiking"	09
17*	Crystal World.	"Water Park"	16
18	Hobbies Club.	"club", "Recreation"	26
19	NESCAFE@IIT Roorkee.	"Cafe", "Snacks"	41
20	Alpahar Canteen.	"Cafe", "Snacks"	37
21	Mahatma Ghandhi Central Library.	"Library"	35
22	Sports Complex.	"Sports"	19
23	PNB/SBI Bank.	"Bank"	27
24	Computer Centre.	"Cyber cafe", "Computer"	21
25*	Railway Reservation Centre.	"Ticket Reservation"	23

* These locations are not shown in the map due to space limitations

(a)   (b)

Fig. 4. (a) Selected Locations from Yelp Data set. (b) Summary of Roorkee Data set.

Evaluation 1: Validating Correctness of Inferred Activities. We prepare heat maps to compare the popularity of a set of inferred activities among various locations. Figure 5 shows the heat map of activity frequency of 10 selected activities. Considering Yelp data set, *(take, picture)* is the most popular activity at location ID 6 which is a *botanical garden*. Similarly, *(serve, food)* and *(have, lunch)* activities are mostly popular in location ID 1 which is a restaurant cum bar. Considering Roorkee data set, activity *(eat, pizza)* has got the highest popularity at location ID 5 which is a pizzeria and activity (recommend, veg food)

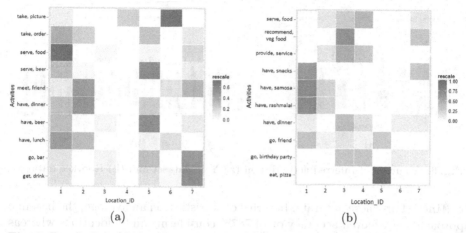

Fig. 5. Popularity of Inferred activities on (a) Yelp dataset and (b) Roorkee dataset

has got the highest popularity in location ID 3 which is the most famous hotel cum restaurant for selling vegetarian foods. In summary, we observe that the discovered information about the location-specific activities conform to the facts of the real-world scenario.

Evaluation 2: Measuring Accuracy of ActMiner. In order to measure the accuracy of discovered activities, we manually obtain the ground truth in a form of a list of activities for each location. Let, GT_i be the set of activities inferred using human perception for a location L_i and the set of activities discovered by an approach is $Aset_i$. Then, the accuracy of activity discovery for location L_i is given as $Accuracy_i = \frac{|ASet_i \cap GT_i|}{|ASet_i|}$.

For comparative study, we obtain the set of activities using baseline and two versions of ActMiner. ActMiner-1 discovers activities using dependency-aware activity extraction technique, whereas ActMiner-2 discovers activities using the idea of dependency-aware and category-aware activity discovery techniques. We have not incorporated the third, i.e., *"Merge"* phase of ActMiner for accuracy evaluation purpose as redundancy minimization does not effect the accuracy. On Yelp data set, we discover top 500 activities for each location using ActMiner-1, ActMiner-2 and Baseline. Figure 6(a) shows that ActMiner-1 outperforms the baseline method in terms of accuracy. More specifically, using Yelp data set, on an average, the baseline approach has achieved accuracy of 68.6% considering all 7 locations whereas dependency-aware activity extraction method, i.e., ActMiner-1 has obtained an average accuracy of 82% showing significant improvement of 13.4%. Moreover, considering top 500 relevant activities discovered using category-aware relevant activity discovery approach, ActMiner-2 has obtained an average accuracy of 85.23% which implies 3.23% average improvement in accuracy to that obtained using ActMiner-1.

On Roorkee data set, we discover all activities for each location using ActMiner-1 and Baseline. Figure 6(b) shows the comparison of accuracies for both the approaches. Again, we observe that the activities discovered by

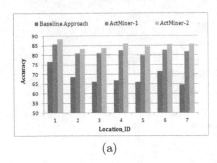
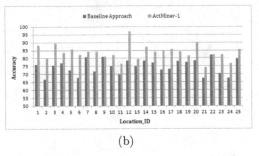

(a)   (b)

Fig. 6. Accuracy of Inferred activities on (a) Yelp dataset and (b) Roorkee dataset

`ActMiner-1` are more accurate. In terms of statistics, on an average, the baseline approach has achieved accuracy of 74.787% considering all 25 locations whereas `ActMiner-1` has obtained an average accuracy of 83.728% showing significant improvement of 8.94% in accuracy compared to the baseline approach. We have not obtained the accuracy of `ActMiner-2` on Roorkee dataset as the reviews in this dataset contain local or Indian concepts that are mostly not available in the ConceptNet and hence, are not suitable for the relevant activity detection purpose.

In summary, `ActMiner` performs more accurately than the baseline approach. The reason behind this is that we have used the dependency relations between words as the metric for activity extraction which ensures the pairing of words to be meaningful. Apart from that, most of the irrelevant activities thrown out after category-aware relevant activity discovery phase, cause improvement in accuracy in `ActMiner-2`.

Evaluation 3: Qualitative Analysis I: Broadcast Service. Now, we investigate the usefulness of sense aware redundant activity discovery phase in terms of advertising popular activities in broadcast environment. We manually obtain the list of redundant activities from the output of `ActMiner-2`. We consider these set of marked activities as the ground truth for redundancy checking purpose. Next, we run the process of sense-aware non-redundant activity discovery on the output of `ActMiner-2`. Figure 7(a) shows the number of redundant activities before and after the sense aware redundancy minimization process. On an average, for all 7 locations, we have observed 51.22% redundancy elimination done by the said process. So, in summary, we can conclude that our sense-aware redundancy elimination approach has successfully eliminated almost half of the redundancies present in the discovered activities. So, this analysis indirectly ensures that using `ActMiner`, we can reduce bandwidth wastage and push more unique information to the user while recommending in broadcast environment.

Evaluation 4: Qualitative Analysis II: Recommendation System. We have developed a location-aware activity recommendation system using the inferred activities of `ActMiner`. Given an activity A and a set of locations $Lset$, the recommender system recommends a location $L \in Lset$ such that $AF(A)$ is highest for location L. We have also developed similar recommendation system

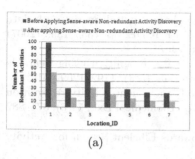
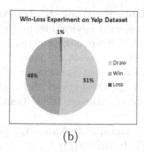
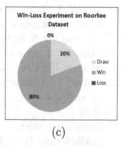

| (a) | (b) | (c) |

Fig. 7. (a) Performance of `ActMiner` in redundancy minimization. Result of Win-Loss Experiment, on (b) Yelp data set and (c) Roorkee data set.

using the baseline approach and evaluated both the recommender systems using *"Win-Loss Experiment"*. In this evaluation, if the location IDs recommended by both the recommender systems are same, we have declared the result as *"Draw"*. Otherwise, the recommender system which recommends the location with higher activity frequency value, wins in the experiment. We have discovered the set of distinct activities for for all locations in each data set and used them as a query input for the evaluation.

In the Figure 7(b), we observe that, the `ActMiner`-based recommender system wins 48% cases and loses in only 1% case while competing with the recommender system formed using baseline approach, whereas in 51% cases, the results are *"Draw"*. Considering Roorkee dataset, `ActMiner`-based recommender system wins 80% cases and makes a draw in 20% cases without any loss as shown in Figure 7(c). From these results, it is quite clear that `ActMiner` outperforms the baseline approach with respect to the performance recommendation and this also proves the efficacy of our proposed solution.

5   Conclusion

In this paper, we propose a Discover-Filter-Merge based technique `ActMiner` to infer the location-specific relevant and non-redundant activities from community-authored reviews. The proposed solution has successfully achieved its objective using novel Dependency-aware, Category-aware and Sense-aware approaches. Experimental analysis shows that `ActMiner` discovers location-specific activities more accurately compared to the baseline approach and proves effectiveness of the solution in providing location-aware activity recommendations.

References

1. Yelp. An Introduction to Yelp: Metrics as of June 30, 2013 (2013),
   http://www.yelp.co.nz/html/pdf/Snapshot_2013_Q2_en.pdf
2. Dearman, D., Truong, K.N.: Identifying the activities supported by locations with community-authored content. ACM UbiComp, 23–32 (2010)

3. Ku, W.S., Zimmermann, R., Wang, H.: Location-based spatial queries with data sharing in wireless broadcast environments. In: IEEE ICDE, pp. 1355–1359 (2007)
4. ConceptNet 5, http://conceptnet5.media.mit.edu/
5. Lara, O., Labrador, M.: A survey on human activity recognition using wearable sensors. IEEE Communications Surveys and Tutorials 15(3), 1192–1209 (2012)
6. Pawar, T., Chaudhuri, S., Duttagupta, S.P.: Body movement activity recognition for ambulatory cardiac monitoring. IEEE Tran. on Biomedical Eng. 54(5), 874–882 (2007)
7. Zheng, K., Shang, S., Yuan, N.J., Yang, Y.: Towards Efficient Search for Activity Trajectories. In: IEEE ICDE (2013)
8. Furletti, B., Cintia, P., Renso, C., Spinsanti, L.: Inferring human activities from GPS tracks. In: ACM SIGKDD International Workshop on Urban Computing, p. 5 (2013)
9. Zheng, V.W., Zheng, Y., Xie, X., Yang, Q.: Collaborative location and activity recommendations with GPS history data. In: ACM WWW, pp. 1029–1038 (2010)
10. Apache OpenNLP Developer Documentation, https://opennlp.apache.org/
11. The Stanford Parser: A statistical parser,
    http://nlp.stanford.edu/software/lex-parser.shtml
12. De Marneffe, M.C., Manning, C.D.: Stanford typed dependencies manual (2008),
    http://nlp.stanford.edu/software/dependenciesmanual.pdf
13. Stanford Log-linear Part-Of-Speech Tagger,
    http://nlp.stanford.edu/downloads/tagger.shtml
14. WordNet, http://wordnet.princeton.edu/

A Scalable Algorithm for Banded Pattern Mining in Multi-dimensional Zero-One Data

Fatimah B. Abdullahi, Frans Coenen, and Russell Martin

The Department of Computer Science, The University of Liverpool, Ashton Street,
Liverpool, L69 3BX, United Kingdom
{f.b.abdullahi,coenen,russell.martin}@liverpool.ac.uk

Abstract. A banded pattern in "zero-one" high dimensional data is one where all the dimensions can be organized in such a way that the "ones" are arranged along the leading diagonal across the dimensions. Rearranging zero-one data so as to feature bandedness allows for the identification of hidden information and enhances the operation of many data mining algorithms that work with zero-one data. In this paper an effective ND banding algorithm, the ND-BPM algorithm, is presented together with a full evaluation of its operation. To illustrate the utility of the banded pattern concept a case study using the GB Cattle movement database is also presented.

Keywords: Banded Patterns, Zero-One data, Pattern Mining.

1 Introduction

Zero-one data occurs in many real world datasets, ranging from bioinformatics [3] to information retrieval [6]. The identification of patterns in zero-one data is an important task within the field of data mining, for example association rule mining [1]. In this paper, we study banded patterns in high dimensional zero-one data. Examples illustrating 2D and 3D bandings are presented in Figures 1 and 2. In practice data can typically not be perfectly banded, but in many cases some form of banding can be achieved. This paper presents a novel N-dimensional Banded Pattern Mining algorithm (ND-BPM) for the identification of banded patterns in zero-one ND data. The operation of the ND-BPM algorithm differs from previous work on banded patterns, such as the MBA [12] and BC [16] algorithms, that allowed for the discovery of banding in only 2D data.

Fig. 1. 2D Banding Example

Fig. 2. 3D Banding Example

L. Bellatreche and M.K. Mohania (Eds.): DaWaK 2014, LNCS 8646, pp. 345–356, 2014.

While the concept of banded matrices has its origins in numerical analysis [17], it has been studied within the data mining community [12,11]. The advantages of banding may be summarized as follows:

1. Banding may be an indication of some interesting phenomena which is otherwise hidden in the data.
2. Working with banded data is seen as preferable from a computational point of view; the computational cost involved in performing certain operations falls significantly for banded matrices leading to significant savings in terms of processing time [10].
3. Related to 2, when a matrix is banded, only the non-zero entries along the diagonal needs to be considered. Thus, when using banded storage schemes the amount of memory required to store the data is directly proportional to the bandwidth. Therefore finding a banding that minimizes the bandwidth is important for reducing storage space and algorithmic speed up [15].

The main issue with the identification of banding in data is the large number of permutations that need to be considered. There has been some research in the context of 2D data focused on minimizing the distance of non-zero entries from the main diagonal of the matrix (bandwidth) by reordering the original matrix [5,10,17]. The current (2D) state-of-the-art algorithm, MBA [13], focuses on identifying banding in binary matrices by flipping zero entries (0s) to one entries (1s) and vice versa, assuming a fixed column permutation.

The rest of this paper is organized as follows. Section 2 discuss related work. A formalism for the banded pattern problem is then presented in Section 3. Section 4 provide an overview of the proposed scoring mechanism and the ND-BPM algorithm is presented in Section 5. Section 6 provides a worked example illustrating the algorithm in the context of 2D. The evaluation of the ND-BPM Algorithm with respect to both 2D and 3D zero-one data is reported in Section 7. Finally, in Section 8 some conclusions are presented.

2   Related Work

From the data analysis perspective banded patterns can occur in many applications, examples can be found in paleontology [4], Network data analysis [7] and linguistics [13]. The property of bandedness with respect to data analysis was first studied by Gemma et al. [12]. They addressed the minimum banding problem by computing how far a 2D data "matrix" is from being banded. The authors in [12] defined the banding problem as: given a binary matrix M, find the minimum number of 0 entries that needs to be modified into 1 entries and the minimum number of 1 entries that needs to be modified into 0 entries so that M becomes fully banded. Gemma et al. fixed the column permutations of the data matrix before executing their algorithm [12]. As noted in the introduction to this paper the current state of the art algorithm is the Minimum Banded Augmentation (MBA) algorithm [13] which uses the principle of assuming "a fixed column permutation" over a given Matrix M. The basic idea is to solve optimally the consecutive one property on the permuted matrix M and then resolve "Sperner conflicts" between each row of the permuted matrix M, by going through all the extra rows and making them consecutive. While it can be argued that the fixed column permutation assumption is not a very realistic assumption with respect to many real world situations, heuristical methods were proposed in [12] to determine a suitable fixed column permutation. Another banding strategy that transposes a matrix is the Barycentric (BC) algorithm that was originally designed for graph drawing and more recently used to reorder binary matrices [16]. The distinction between these previous algorithms and that presented in this paper is that the previous algorithms were all directed at 2D data, while the proposed algorithm operates in 3D. It should also be noted that Bandwidth minimization of binary matrices is known to be NP-Complete [10] as it is related to the reordering of binary matrices [15].

Given the above the MBA and BC algorithms are the two exemplar banding algorithms with which the operation of the proposed ND-BPM algorithm is compared and evaluated as discussed later in this paper (see Section 7).

3   Problem Definition

Let Dim be a set of dimension $\{Dim_1, Dim_2, \ldots, Dim_n\}$. Each dimension comprises a set of k indexes such that $Dim_i = \{a_{i_1}, a_{i_2}, \ldots, a_{i_k}\}$. Thus in 2D space the indexes associated with Dim_1 might be record numbers and the indexes associated with Dim_2 may be attribute value identifiers. In 3D Dim_3 might equate to time and the indexes to discrete time slots, and so on. Note that we will indicate a particular index j belonging to a dimension i using the notation a_{i_j}. Note also that dimensions are not necessarily of equal size. Given a zero-one data set D that corresponds to the data set defined by Dim we can think of this data space in terms of an ND grid with the "ones" indicated by "dots" (ND spheres) and the "zeroes" by empty space. Individual dots can thus be referenced using the ND coordinate system defined by Dim. Such a data space can be "perfectly banded" if there exists a permutation of the indexes such that: (i) $\forall a_{i_j} \in dim_i$ the dots occur consecutively at indexes $\{a, a + 1, a + 2, \ldots\}$ and the "starting index" for dim_i is less than or equal to the starting index for dim_{i+1}.

4   The N Dimensional Banding Mechanism

The discovery of the presence of banding in a zero-one ND space requires the rearrangement of the indexes in each dimension so as to "reveal" a banding (or at least an approximate banding). This is a computationally expensive task especially in the context of ND space. In the case of the ND-BPM algorithm it is proposed that this be achieved using the concept of banding scores. Given a particular dimension Dim_i each index a_{i_j} will have a banding score BS_{i_j} associated with it. These banding scores are then used to rearrange the ordering of the indexes in Dim_i so that the index with the greatest banding score is listed first. Individual banding scores are calculated by considering dimension pairs.

Thus given two dimensions Dim_p and Dim_q we calculate the banding scores for all $a_{p_j} \in Dim_p$ with respect to Dim_q. We use the notation BS_{pq_j} to indicate the banding score of index a_j in Dim_p calculated with respect to Dim_q as follows:

$$BS_{pq_j} = \frac{\sum_{k=1}^{k=|W|}(|Dim_q| - W_k + 1)}{\sum_{k=1}^{k=|W|}(|Dim_q| - k + 1)} \tag{1}$$

where the set W is the set of Dim_q indexes representing "dots" whose coordinate set feature the index x_{p_j} from Dim_p. However, if $n > 1$ we need to do this for all instances of the Dim_p and Dim_q pairings that can exist across the space. Thus the set of dimensions identifiers, I, that excludes the identifiers for Dim_p and Dim_q. Thus:

$$BS_{pq_j} = \frac{\sum_{i=1}^{i=z} BS_{pq_j} \; for \; Dim_i}{z} \tag{2}$$

where $z = \prod_{i=1}^{i=|D|} |Dim_{I_i}|$.

We can also use the banding score concept as a measure of the goodness of a banding configuration. By first calculating the Dimension Banding Score (DBS) for each dimension p with respect to dimension q (DBS_{pq}) as follows:

$$DBS_{pq} = \frac{\sum_{j=1}^{j=|Dim_p|} BS_{pq_j}}{|Dim_p|} \tag{3}$$

The Global Banding Score (GBS) for the entire configuration is then calculated as follows:

$$GBS = \frac{\sum_{p=1}^{p=n-1} \sum_{q=p+1}^{q=n} DBS_{pq}}{\sum_{i=1}^{i=n-1} n - i} \tag{4}$$

5   N Dimensional Banded Pattern Mining (ND-BPM) Algorithm

The ND-BPM algorithm is presented in algorithm 1. The inputs are (line 3): (i) a zero-one data set D and (ii) the set DIM. The output is a rearranged data space that maximizes GBS. The algorithm iteratively loops over the data space. On each iteration the algorithm attempts to rearrange the indexes in the set of dimensions DIM. It does this by considering all possible dimension pairings pq. For each pairing the BS value for each index j in dimension Dim_p is calculated (line 11) and used to rearrange the dimension (line 13). If a change has been effected a change flag is set to $TRUE$ (line 15) and a DBS value calculated (line 17). Once all pairings have been calculated a GBS_{new} value is calculated (line 20). If GBS_{new} is worse than the current GBS value (GBS_{sofar}), or there has been no change, we exit with the current configuration D (line 23). Otherwise we set D to D', and GBS_{sofar} to GBS_{new} and repeat.

Algorithm 1. The ND-BPM Algorithm

```
 1: Input D, DIM
 2: Output Rearranged data space that serves to maximize GBS
 3: change = FALSE
 4: n = |Dim|
 5: GBS_sofar = 0
 6: loop
 7:     D' = D
 8:     for p = 1 to p = n − 1 do
 9:         for q = p + 1 to q = n do
10:             for j = 1 to |DIM_p| do
11:                 Calculate BS_pq_j using Equations (1) and (2) as appropriate
12:             end for
13:             D'' = D' with indexes in Dim_p reordered according to the set BS_pq_j
14:             if D' ≠ D'' then
15:                 change = TRUE
16:                 D' = D''
17:             end if
18:             Calculate DBS_pq using Equation(3)
19:         end for
20:         Calculate GBS_new using Equation(4)
21:     end for
22:     if change = FALSE or GBS_new < GBS_sofar then
23:         exit with current configuration D
24:     else
25:         D = D'
26:         GBS_sofar = GBS_new
27:     end if
28: end loop
```

6 Worked Example

To illustrate the operation of the ND-BPM algorithm a worked example, using the 5×4 data space given in Figure 3, is presented here. We commence by calculating the set of scores BS_{1_j} for dimension 1 (Dim_1) to obtain: $BS_{1_1} = 0.9166$, $BS_{1_2} = 0.5000$, $BS_{1_3} = 0.6666$ and $BS_{1_4} = 0.7777$. Using this set of scores the indexes in Dim_1 are rearranged to produce the configuration shown in Figure 4. The Dim_1 banding score is then $DBS_1 = 0.7277$. Next we calculate the set of scores BS_{2_j} for (Dim_2) to obtain: $BS_{2_1} = 0.8888$, $BS_{2_2} = 1.0000$, $BS_{2_3} = 0.6666$, $BS_{2_4} = 0.6666$ and $BS_{2_5} = 0.5555$. Using this set of scores the indexes in Dim_2 are rearranged to produce the configuration shown in Figure 5. The Dim_2 banding score is then $DBS_2 = 0.8221$ and the global banding score is:

$$GBS = \frac{\sum_{p=1}^{p=n-1} \sum_{q=p+1}^{q=n} DBS_{pq}}{\sum_{i=1}^{i=n-1} n - i} = 0.7749$$

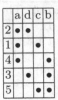

Fig. 3. Input data

Fig. 4. Input data with Dim_1 rearranged

Fig. 5. Input data with Dim_2 rearranged

We repeat the process since changes were made. The set of scores for Dim_1 are now: $BS_{1_1} = 1.0000$, $BS_{1_2} = 0.7777$, $BS_{1_3} = 0.5555$ and $BS_{1_4} = 0.5000$; with this set of scores Dim_1 remains unchanged. However, the Dim_1 banding score DBS_1 is now 0.7944 because of changes to Dim_2 (previously this was 0.7277). The set of scores for Dim_2 are now: $BS_{2_1} = 1.0000$, $BS_{2_2} = 0.8888$, $BS_{2_3} = 0.6666$, $BS_{2_4} = 0.6666$ and $BS_{2_5} = 0.5555$. Again, with this set of scores Dim_2 remains unchanged, thus the configuration shown in Figure 5 remains unchanged. The Dim_2 banding score is now $DBS_2 = 0.8296$ (was 0.8221). The global banding score is now:

$$GBS = \frac{\sum_{p=1}^{p=n-1} \sum_{q=p+1}^{q=n} DBS_{pq}}{\sum_{i=1}^{i=n-1} n - i} = 0.8120$$

On the previous iteration it was 0.7749, however no changes have been made on the second iteration so the algorithm terminates.

7 Evaluation

To evaluate the ND-BPM algorithm its operation was compared with the established MBA and BC algorithms, two exemplar algorithms illustrative of the alternative approaches to identifying banding in zero-one data as described in Section 2. Because MBA and BC were designed to operate using 2D, the evaluation was conducted in these terms. Eight data sets taken from the UCI machine learning data repository [8] were used. The first set of experiments, reported in sub-section 7.1 below, considered the efficiency of the ND-BPM algorithm in comparison with the MBA and BC algorithms. The second set of experiments (Section 7.2) considered the effectiveness of ND-BPM algorithm, again in comparison with the MBA and BC algorithms, with respect to the bandings produced. The third set of experiments, reported in sub-section 7.3 below, considered the effectiveness of banding with respect to a Frequent Itemset Mining (FIM) scenario. To determine the effectiveness of the ND-BPM algorithm with respect to a higher number of dimensions further experiments were conducted using the GB cattle movement database. This is described in Section 7.4.

7.1 Efficiency

To determine the efficiency of the proposed ND-BPM algorithm in the context of 2D and with respect to the MBA and BC algorithms, we recorded the run time required to maximize the banding score GBS in each case. The data sets were normalized and discretized using the LUCS-KDD ARM DN Software[1] to produce the desired zero-one data sets (continuous values were ranged using a maximum of five ranges). Table 1 shows the results obtained. Table 1 presents run-time and the final GBS value obtained in each case. The table also records the number of attributes (after discretization) and the number of records for each data set. From the table it can be observed that there is a clear correlation between the number of records in a dataset and run time as the number of records increases the processing time also increases (this is to be expected). The table also demonstrates that the ND-BPM algorithm requires less processing time than the other two algorithms considered.

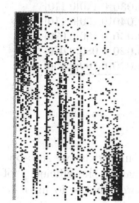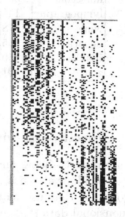

Fig. 6. Banding resulting from ND-BPM algorithm as applied to the Wine dataset ($GBS = 0.7993$)

Fig. 7. Banding resulting from MBA algorithm as applied to the Wine dataset ($GBS = 0.7123$)

Fig. 8. Banding resulting from BC algorithm as applied to the Wine dataset ($GBS = 0.7021$)

7.2 Effectiveness with Respect to Global Banding Score

It was not possible to identify a perfect banding with respect to any of the UCI data sets, this was to be expected. However, in terms of GBS, Table 1 clearly shows that the proposed ND-BPM algorithm outperformed the previously proposed MBA and BC algorithms (best scores highlighted using bold font). Figures 6, 7 and 8 show the bandings obtained using the wine data sets and the ND-BPM, MBA and BC algorithms respectively. Inspection of these Figures indicates that banding can be identified in all cases. However, from inspection of the figures it is suggested that the banding produced using the proposed

---

[1] http://www.csc.liv.ac.uk/$\sim$/frans/KDD/Software/LUCS_KDD_DN_ARM.

ND-BPM algorithm is better. For example considering the banding produced when the MBA algorithm is applied to the wine dataset (Figure 7) the resulting banding includes dots ("1"s) in the top-right and bottom-left corners while the ND-BPM algorithm does not (it features a smaller bandwidth). When the BC algorithm is applied to the wine dataset (Figure 8) the banding is less dense than in the case of the ND-BPM algorithm.

Table 1. Efficiency Experimental Results (best results presented in bold font), GBS = Global Banding Score, RT = Run time (secs.)

Datasets	# Rec s	# Cols	ND-BPM GBS	MBA GBS	BC GBS	ND-BPM RT	MBA RT	BC RT
annealing	898	73	**0.8026**	0.7305	0.7374	**0.150**	0.260	0.840
heart	303	52	**0.8062**	0.7785	0.7224	**0.050**	0.160	0.170
horsecolic	368	85	**0.8152**	0.6992	0.7425	**0.070**	0.200	0.250
lympography	148	59	**0.8365**	0.7439	0.7711	**0.030**	0.140	0.110
wine	178	68	**0.7993**	0.7123	0.7021	**0.040**	0.150	0.110
hepatitis	155	56	**0.8393**	0.7403	0.7545	**0.050**	0.150	0.090
iris	150	19	**0.8404**	0.8205	0.7516	**0.020**	0.080	0.060
zoo	101	42	**0.8634**	0.7806	0.7796	**0.020**	0.100	0.050

7.3   Effectiveness with Respect to FIM

In addition to being an indicator of some pattern that may exist in zero-one data, banding also has application with respect to increasing the efficiency of algorithms that use matrices or tabular information stored in the form of n-dimensional data storage structures. One example is algorithms that use $n \times n$ affinity matrices, such as spectral clustering algorithms [14], to identify communities in networks (where n is the number of network nodes). Another example is Frequent Itemset Mining (FIM) [1,2] where it is necessary to process large binary valued data collections stored in the form of a set of feature vectors (drawn from a vector space model of the data). To test the effectiveness of the bandings produced as a result of the experiments reported in Sub-section 7.1 above, a FIM algorithm was applied to the banded data sets produced using the ND-BPM algorithm (the TFP algorithm [9] was actually used, but any alternative FIM algorithm would equally well have sufficed). The results are presented in Table 2. From the table it can be seen that FIM is always much more efficient when using banded data than when using non banded data if we do not include the time to conduct the banding. If we include the banding time, in 8 out of the 12 cases, it is still more efficient. Similarly, when the FIM algorithm was applied to the banded data sets produced using the MBA and BC algorithms, it was also observed that FIM was more effecient using banded data than when using non banded data without the banding time, with the banding time in 4 (MBA) and 5 (BC) out of the 12 cases, FIM is still more effecient.

Table 2. FIM runtime with and without banding ($\sigma = 2\%$)

Datasets	#Rows	#Cols	Banding Time(s)	FIM time (s) with Banding	Total	FIM time (s) without Banding
adult	48842	97	346.740	**2.274**	349.014	**5.827**
anneal	898	73	0.150	**0.736**	**0.086**	2.889
chessKRvk	28056	58	95.370	**0.082**	95.452	**0.171**
heart	303	52	0.050	**0.294**	**0.344**	0.387
hepatitis	155	56	0.030	**0.055**	**0.085**	22.416
horseColic	368	85	0.070	**0.899**	**0.969**	1.242
letRecog	20000	106	42.420	**3.004**	45.424	**6.763**
lympography	148	59	0.030	**7.997**	**8.022**	12.658
mushroom	8124	90	14.400	**874.104**	**888.504**	1232.740
penDigits	10992	89	21.940	**2.107**	24.047	**2.725**
waveForm	5000	101	3.030	**119.220**	**122.250**	174.864
wine	178	68	0.010	**0.155**	**0.165**	0.169

7.4 Large Scale: Cattle Movement Database

To illustrate the utility of the proposed ND-BPM algorithm, the authors have applied the algorithm to a 3 dimensional data set constructed from the GB Cattle movement data base. The GB cattle movement database records all the movements of cattle registered within or imported into Great Britain. The database is maintained by the UK Department for Environment, Food and Rural Affairs ($DEFRA$). For the analysis reported in this work, data sets for the months of January to December 2003 to 2006, for one county (Lancashire in Great Britain), was used. Each record comprises: (i) Animal Gender, (ii) Animal age, (iii) the cattle breed type, (iv) sender location in terms of easting and northing grid values, (v) the type of the sender location, (vi) receiver location in terms of eastings and northings grid values, (vii) receiver location type and ($viii$) the number of cattle moved. Discretization and Normalization processes were used to convert the input data into the desired zero-one format. As a result the GB dataset comprised 80 items distributed over four dimensions: records, attributes, easting values and northing values. For ease of understanding and so that results can be displayed in a 2D format only three dimensions were considered at any one time (records, attributes and eastings; and records, attributes and northings).

The results obtained are presented in Tables 3 and 4. The tables record the number of attributes (after discretization) representing attribute information, the number of records and the number of slices used to represent the discretized sender eastings and northings. The tables also record the run-times required by the algorithms in order to maximize the global banding score GBS and the final GBS value arrived at in each case.

Figures 9 and 11 shows the sampled data before banding and Figures 10 and 12 shows the sampled data after banding using a subset of the data for the month of January 2003. Inspection of the figures indicates that banding can clearly be identified. More specifically, there are certain movement patterns, that can be identified

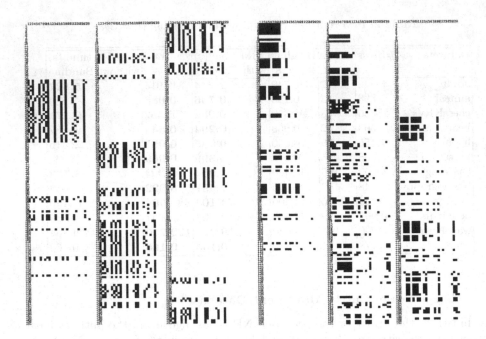

Fig. 9. January Raw data, with Eastings **Fig. 10.** Data set from Figure 9 after as 3rd dimension, before Banding   Banding

Table 3. Experimental Results for GB Summary for Easting locations

Years	Datasets	# Recs	# Attrs	# Slices	GBS	Run time
2003	Jan-Dec	167919	70	10	0.3902	2485.99
2004	Jan-Dec	217566	72	10	0.2823	5475.51
2005	Jan-Dec	157142	72	10	0.3093	2114.09
2006	Jan-Dec	196290	72	10	0.3075	3856.83

Table 4. Experimental Results for GB Summary for Northing locations

Years	Datasets	# Recs	# Attrs	# Slices	GBS	Run time
2003	Jan-Dec	167919	70	10	0.4239	2393.50
2004	Jan-Dec	217566	72	10	0.3101	4786.66
2005	Jan-Dec	157142	72	10	0.3632	1232.09
2006	Jan-Dec	196290	72	10	0.3525	4162.71

from the generated banding. For example, from Figure 10, it can be observed that male cattle breeds are moved more often in the east of the country than in the west. Similarly, from Figure 12, it can be observed that male cattle of (age = 1) are more frequently moved in the north than in the south of the country.

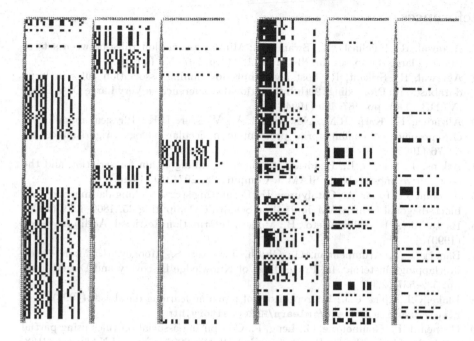

Fig. 11. January Raw data, with Northings as 3rd dimension, before Banding

Fig. 12. Data set from Figure 11 after Banding

8   Conclusions

In this paper the authors have described an approach to identifying bandings in zero-one data using the concept of banding scores. More specifically the ND-BPM algorithm has been presented. This algorithm operates by iteratively rearranging the items associated with individual dimensions according to the concept of banding scores. The operation of the ND-BPM algorithm was compared with the operation of the MBA and BC algoithms in the context of 2D using eight data sets taken from the UCI machine learning repository. In the context of 3D, it was tested using sample data taken from the GB cattle movement database for the months of January to December 2003 to 2006. The reported evaluation established that the proposed approach can reveal banded patterns within zero-one data reliably and with reasonable computational efficiency and able to handle even higher dimensions in reasonable time. The evaluation also confirmed that, at least in the context of FIM, efficiency gains can be realized using the banding concept. For future work the authors intend to extend their research to address situations where we seek to establish banding with respect to a subset of the available dimensions (maintaining the position of indexes in the other dimensions). Whatever the case, the authors have been greatly encouraged by the results produced so far, as presented in this paper.

References

1. Agrawal, R., Imielinski, T., Swami, A.: Mining association rules between sets of items in large databases. In: SIGMOD 1993 pp. 207–216 (1993)
2. Agrawal, R., Srikant, R.: Fast algorithms for mining association rules in large databases. In: Proceedings 20th International Conference on Very Large Data Bases (VLDB 1994), pp. 487–499 (1994)
3. Alizadeh, F., Karp, R.M., Newberg, L.A., Weisser, D.K.: Physical mapping of chromosomes: A combinatorial problem in molecular biology. Algorithmica 13, 52–76 (1995)
4. Atkins, J., Boman, E., Hendrickson, B.: Spectral algorithm for seriation and the consecutive ones problem. SIAM J. Comput. 28, 297–310 (1999)
5. Aykanat, C., Pinar, A., Catalyurek, U.: Permuting sparse rectangular matrices into block-diagonal form. SIAM Journal on Scientific Computing 25, 1860–1879 (2004)
6. Baeza-Yates, R., RibeiroNeto., B.: Modern Information Retrieval. Addison-Wesley (1999)
7. Banerjee, A., Krumpelman, C., Ghosh, J., Basu, S., Mooney, R.: Model-based overlapping clustering. In: Proceedings of Knowledge Discovery and DataMining, pp. 532–537 (2005)
8. Blake, C.L, Merz, C.J.: Uci repository of machine learning databases (1998), http://www.ics.uci.edu/~mlearn/MLRepository.htm
9. Coenen, F.P., Goulbourne, G., Leng, P.: Computing association rules using partial totals. In: Siebes, A., De Raedt, L. (eds.) PKDD 2001. LNCS (LNAI), vol. 2168, pp. 54–66. Springer, Heidelberg (2001)
10. Cuthill, A.E., McKee, J.: Reducing bandwidth of sparse symmentric matrices. In: Proceedings of the 1969 29th ACM National Conference, pp. 157–172 (1969)
11. Fortelius, M., Puolamaki, M.F.K., Mannila, H.: Seriation in paleontological data using markov chain monte method. PLoS Computational Biology, 2 (2006)
12. Garriga, G.C., Junttila, E., Mannila, H.: Banded structures in binary matrices. Knowledge Discovery and Information System 28, 197–226 (2011)
13. Junttila, E.: Pattern in Permuted Binary Matrices. Ph.D. thesis (2011)
14. Von Luxburg, U.A.: A tutorial on spectral clustering. Statistical Computation 17, 395–416 (2007)
15. Mueller, C.: Sparse matrix reordering algorithms for cluster identification. Machune Learning in Bioinformatics (2004)
16. Mäkinen, E., Siirtola, H.: The barycenter heuristic and the reorderable matrix. Informatica 29, 357–363 (2005)
17. Rosen, R.: Matrix bandwidth minimisation. In: ACM National conference Proceedings, pp. 585–595 (1968)

Approximation of Frequent Itemset Border by Computing Approximate Minimal Hypergraph Transversals

Nicolas Durand and Mohamed Quafafou

Aix-Marseille Université, CNRS, LSIS UMR 7296, 13397, Marseille, France
{nicolas.durand,mohamed.quafafou}@univ-amu.fr
http://www.lsis.org

Abstract. In this paper, we present a new approach to approximate the negative border and the positive border of frequent itemsets. This approach is based on the transition from a border to the other one by computing the minimal transversals of a hypergraph. We also propose a new method to compute approximate minimal hypergraph transversals based on hypergraph reduction. The experiments realized on different data sets show that our propositions to approximate frequent itemset borders produce good results.

Keywords: frequent itemsets, borders, hypergraph transversals, approximation.

1 Introduction

The discovery of frequent itemsets has quickly become an important task of data mining [1]. This corresponds to find the sets of items (i.e. attribute values) which appear together in at least a certain number of transactions (i.e. objects) recorded in a database. These sets of items are called frequent itemsets. The main use of the frequent itemsets is the generation of association rules. Nevertheless, the uses have been extended to other tasks of data mining such as supervised classification and clustering [1]. Two points are important in the discovery of frequent itemsets. The first point is the reduction of the search space due to combinatorial explosion. The second point is the reduction of the number of generated itemsets to make easier the exploitation. In this paper, we focus on the second point. To reduce the number of itemsets, some condensed representations of frequent itemsets, such as the frequent closed itemsets, have been proposed [1]. The maximal frequent itemsets (w.r.t. set inclusion) also represent a reduced collection of itemsets. They correspond to a subset of the set of the frequent closed itemsets. The regeneration of all the frequent itemsets is possible from the maximal frequent itemsets but they are not considered as a condensed representation of frequent itemsets because the database must be read to compute the frequencies. The maximal frequent itemsets and the minimal infrequent itemsets correspond respectively to the positive border and the negative border

---

L. Bellatreche and M.K. Mohania (Eds.): DaWaK 2014, LNCS 8646, pp. 357–368, 2014.

of the set of the frequent itemsets [2]. These two borders are linked together by the computation of minimal hypergraph transversals (also called "minimal hitting sets") [2,3]. Thus, it is possible to switch to a border from the other one. The number of itemsets of the borders can still be huge.

In this paper, we propose a new approach to approximate the positive border of frequent itemsets in order to reduce the size of the border. This approach is based on the transition from a border to the other one by computing minimal hypergraph transversals. The approximation is obtained by the computation of approximate minimal transversals. Through the approximation, we also want to find new items which could be interesting for some applications like document recommendation. Another contribution is the proposition of a new method to approximate the minimal transversals of a hypergraph by reducing the hypergraph. To the best of our knowledge, this is the first time that such approaches are proposed. Some experiments have been performed on different data sets in order to evaluate the number of generated itemsets and the distance between the computed approximate borders and the exact borders. We focus on the comparison between our method and two other algorithms which compute approximate minimal transversals, in considering our approach of border approximation.

The rest of this paper is organized as follows. Section 2 defines the notations and the notions necessary for understanding the paper. The proposed approach of border approximation is detailed in Section 3. In Section 4, we present our method to compute approximate minimal hypergraph transversals. Related works are discussed in Section 5. The experiments and the results are presented in Section 6. We conclude in Section 7.

2   Preliminaries

Let $\mathcal{D} = (\mathcal{T}, \mathcal{I}, \mathcal{R})$ be a data mining context, $\mathcal{T}$ a set of transactions, $\mathcal{I}$ a set of items (denoted by capital letters), and $\mathcal{R} \subseteq \mathcal{T} \times \mathcal{I}$ is a binary relation between transactions and items. Each couple $(t, i) \in \mathcal{R}$ denotes the fact that the transaction t is related to the item i. A transactional database is a finite and nonempty multi-set of transactions. Table 1 provides an example of a transactional database consisting of 6 transactions (each one identified by its "Id") and 8 items (denoted $A \ldots H$).

Table 1. Example of transactional database

Id	Items					
t_1	A		C	E		G	
t_2		B	C	E		G	
t_3	A		C	E			H
t_4	A			D	F		H
t_5		B	C		F		H
t_6		B	C	E	F		H

An *itemset* is a subset of $\mathcal{I}$ (note that we use a string notation for sets, e.g., AB for $\{A, B\}$). The complement of an itemset X (according to $\mathcal{I}$) is noted $\overline{X}$. A transaction t supports an itemset X iff $\forall i \in X, (t, i) \in \mathcal{R}$. An itemset X is *frequent* if the number of transactions which support it, is greater than (or is equal to) a miminum threshold value, noted *minsup*. The set of all-frequent itemsets is noted S. Let us take the example of Table 1, if *minsup*=3 then the itemset H is frequent because 4 transactions support it (t_3, t_4, t_5 and t_6). AE is not frequent because only t_1 and t_3 support it.

The set of all maximal frequent itemsets (resp. minimal infrequent itemsets), w.r.t. set inclusion, in $\mathcal{D}$ is the *positive border* (resp. *negative border*) [2] of S and is noted $Bd^+(S)$ (resp. $Bd^-(S)$): $Bd^+(S) = \{X \in S \mid \forall Y \ tq \ X \subset Y, Y \notin S\}$ and $Bd^-(S) = \{X \in 2^\mathcal{I} \setminus S \mid \forall Y \ tq \ Y \subset X, Y \in S\}$. Let us take the example of Table 1, if *minsup*=3, $Bd^+(S) = \{A, BC, CE, CH, FH\}$ and $Bd^-(S) = \{D, G, AB, AC, AE, AF, AH, BE, BF, BH, CF, EF, EH\}$.

Before the presentation of the relationship between the positive border and the negative border of frequent itemsets, we need to introduce the notion of minimal transversals of a hypergraph. A *hypergraph* $\mathcal{H} = (V, E)$ is composed of a set V of vertices and a set E of hyperedges [4]. Each hyperedge $e \in E$ is a set of vertices included or equal to V. The *degree* of a vertex v in $\mathcal{H}$, denoted $deg_\mathcal{H}(v)$, is the number of hyperedges of $\mathcal{H}$ containing v. Let τ be a set of vertices ($\tau \subseteq V$). τ is a *transversal* of $\mathcal{H}$ if it intersects all the hyperedges of $\mathcal{H}$. A transversal is also called a "hitting set". The set of all the transversals of $\mathcal{H}$ is: $Tr(\mathcal{H}) = \{\tau \subseteq V \mid \forall e_i \in E, \tau \cap e_i \neq \emptyset\}$. A transversal τ of $\mathcal{H}$ is *minimal* if no proper subset is a transversal of $\mathcal{H}$. The set of all minimal transversals of $\mathcal{H}$ is noted $MinTr(\mathcal{H})$. The relationship between the notion of borders and minimal transversals has been presented in [2] and [3].

In [2], the following property has been showed:
$$Bd^-(S) = MinTr(\overline{Bd^+(S)})$$
where $\overline{Bd^+(S)}$ is the hypergraph formed by the items of $\mathcal{I}$ (i.e. the vertices) and the complement of the itemsets of the positive border of S (i.e. the hyperedges).

In [3], the following property has been showed:
$$Bd^+(S) = \overline{MinTr(Bd^-(S))}$$
where $Bd^-(S)$ is the hypergraph formed by the items of $\mathcal{I}$ (i.e. the vertices) and the itemsets of the negative border of S (i.e. the hyperedges).

The term *dualization* refers to the use of the previous properties to compute the negative border from the positive border, and vice versa. The size of the borders can be huge according to *minsup*. In the two next sections, we propose a new approach to approximate the borders and to reduce their size. In this way, the exploitation of the itemsets of the borders will be easier.

3   Proposed Approach of Border Approximation

The proposed approach of border approximation exploits the dualizations between the positive border and the negative border. Let f and g be the functions that allow to compute respectively the negative border from the positive border and the positive border from the negative border:

$$f : \begin{vmatrix} 2^{\mathcal{I}} \to 2^{\mathcal{I}} \\ x \mapsto MinTr(\overline{x}) \end{vmatrix} \qquad\qquad g : \begin{vmatrix} 2^{\mathcal{I}} \to 2^{\mathcal{I}} \\ x \mapsto \overline{MinTr(x)} \end{vmatrix}$$

The principle of the proposed approach is to replace the function f by a function $\widetilde{f}$ which performs an approximate computation of the negative border. The new function $\widetilde{f}$ uses an approximate minimal transversals computation, noted $\widetilde{MinTr}$:

$$\widetilde{f} : \begin{vmatrix} 2^{\mathcal{I}} \to 2^{\mathcal{I}} \\ x \mapsto \widetilde{MinTr}(\overline{x}) \end{vmatrix}$$

From the positive border, the approach computes an approximate negative border (noted $\widetilde{Bd^-}(S)$): $\widetilde{f}(Bd^+(S)) = \widetilde{MinTr}(\overline{Bd^+(S)}) = \widetilde{Bd^-}(S)$. The return to a positive border (via the function g) allows to obtain an approximate positive border (noted $\widetilde{Bd^+}(S)$): $g(\widetilde{Bd^-}(S)) = \overline{\widetilde{MinTr}(\widetilde{Bd^-}(S))} = \widetilde{Bd^+}(S)$. Thus, our approach produces the approximate negative border $\widetilde{Bd^-}(S)$ and the corresponding approximate positive border $\widetilde{Bd^+}(S)$. Let us take the example of Table 1 and let us compute the approximate border: $\widetilde{Bd^-}(S) = \widetilde{f}(Bd^+(S)) = \widetilde{MinTr}(\overline{Bd^+(S)}) = \widetilde{MinTr}(\{\overline{A}, \overline{BC}, \overline{CE}, \overline{CH}, \overline{FH}\} = \widetilde{MinTr}(\{BCDEFGH, ADEFGH, ABDFGH, ABDEFG, ABCDEG\})$. Let us assume that the approximate minimal transversals computation provides the following result: $\widetilde{Bd^-}(S) = \{D, E, G, AF, AH, BF, BH\}$. The approximate positive border is obtained by dualization: $\widetilde{Bd^+}(S) = g(\widetilde{Bd^-}(S)) = \overline{\widetilde{MinTr}(\widetilde{Bd^-}(S))} = \{\overline{ABDEG}, \overline{DEFGH}\} = \{CFH, ABC\}$. We can remark that A, B, C and BC are frequent itemsets (according to $minsup = 3$) and here ABC is considered as a frequent itemset. CFH is not frequent (its support is equal to 2) but it is almost frequent. These two itemsets can be interesting for applications like document recommendation. For instance, without our approach, CF is frequent and CFH is not frequent. The item H is potentially interesting. If the items are documents, with our approach, the item H can be recommended to a user.

4   Computation of Approximate Minimal Transversals

In order to complete the approach presented in the previous section, we proposed a method to compute the approximated minimal transversals of a hypergraph. The method is based on the reduction of the initial hypergraph. The aim is to compute the minimal transversals on the reduced hypergraph (smaller than the initial hypergraph). The proposed algorithm of reduction is specially designed to compute minimal transversals. It exploits the fact that the hyperedges formed by the complements of the itemsets of the positive border, strongly intersect (i.e. the average degree of a vertex is high). Indeed, in the example this hypergraph is: $\{BCDEFGH, ADEFGH, ABDFGH, ABDEFG, ABCDEG\}$. The proposed method is composed of two steps: (1) Reduction of the hypergraph, (2) Computation of the (exact) minimal transversals of the reduced hypergraph. At the end, the minimal transversals obtained from the reduced hypergraph are declared as the approximate minimal transversals of the initial hypergraph.

4.1   Reduction of the Hypergraph

The hypergraph reduction of the initial hypergraph is based on the intersections of its hyperedges and on the degree of each vertex. The representative graph [4] (also called "line-graph") of the hypergraph is thus generated. Let us recall that the representative graph of the hypergraph $\mathcal{H}$ is a graph whose vertices represent the hyperedges of $\mathcal{H}$ and two vertices are adjacent if and only if the corresponding hyperedges in $\mathcal{H}$ intersect. In our algorithm, we add values to the edges of the representative graph. Algorithm 1 presents the reduction of a hypergraph $\mathcal{H}$. The algorithm is composed of three steps: (1) Computation of the degree of each vertex in $\mathcal{H}$ (lines 1-3), (2) Generation of the valued representative graph of $\mathcal{H}$ (lines 4-9), (3) Generation of the reduced hypergraph from the valued representative graph (lines 10-17). The complexity of the algorithm is in $O(m^2)$ where m is the number of hyperedges of the initial hypergraph.

Algorithm 1. HR (Hypergraph Reduction)

Require: a hypergraph $\mathcal{H}=(V, E)$ where $|V|=n$ and $|E|=m$
Ensure: the reduced hypergraph $\mathcal{H}_R$
1: **for all** $v \in V$ **do**
2:    Compute $deg_{\mathcal{H}}(v)$
3: **end for**
4: $V' \leftarrow \{v_i'\}$ $i = 1,\dots,m$; {each $v_i' \in V'$ represents $e_i \in E$}
5: $E' \leftarrow \{\}$;
6: **for all** $v_i' \cap v_j' \neq \emptyset$ **do**
7:    $E' \leftarrow E' \cup \{(v_i', v_j')\}$;
8:    $w_{(v_i', v_j')} \leftarrow \displaystyle\sum_{v \in \{\psi^{-1}(v_i') \cap \psi^{-1}(v_j')\}} deg_{\mathcal{H}}(v)$;
9: **end for**
10: $V_R \leftarrow \{\}$;
11: $E_R \leftarrow \{\}$;
12: **while** $E' \neq \emptyset$ **do**
13:    Select $e'max = (v'max_i, v'max_j)$ having the maximal weight value
14:    $V_R \leftarrow V_R \cup \{\psi^{-1}(v'max_i) \cap \psi^{-1}(v'max_j)\}$;
15:    $E_R \leftarrow E_R \cup \{\{\psi^{-1}(v'max_i) \cap \psi^{-1}(v'max_j)\}\}$;
16:    Delete the edges $e' \in E'$ where $v'max_i$ or $v'max_j$ is present
17: **end while**
18: **return** $\mathcal{H}_R$;

Valued Representative Graph Generation (lines 1-9). Let be $\mathcal{H} = (V, E)$ a hypergraph ($|V| = n$ and $|E| = m$). The algorithm constructs a valued graph $G=(V', E')$ where $V' = \{v_i'\}$ ($i = 1,\dots,m$) and $E' = \{e_k'\}$ ($k = 1,\dots,l$). A vertex v_i' represents a hyperedge e_i from $\mathcal{H}$. Let be $\psi : E \to V'$ the bijective function who associates a hyperedge e_i to a vertex v_i'. A hyperedge between v_i' and v_j' shows that the intersection between the hyperedges $\psi^{-1}(v_i')$ and $\psi^{-1}(v_j')$ (e_i and e_j from $\mathcal{H}$) is not empty. The weight of an edge is based on the degree of each vertex in the corresponding intersection. To evaluate the weight of a generated edge, we use the degree of each vertex from the initial hypergraph.

The idea is that a vertex very present has a good chance to be in a minimal transversal. This expresses a "degree" of transversality. If the degree of a vertex is equal to the number of hyperedges then this vertex is a minimal tranversal. Let us note that this heuristic is used by several algorithms that compute transversals [5,6]. The weight of an edge $e'_k = (v'_i, v'_j)$, noted $w_{e'_k}$, is the sum of the degree of the vertices present in the intersection which has led to create this edge (see (1)).

$$w_{e'_k} = \sum_{v \in \{\psi^{-1}(v'_i) \cap \psi^{-1}(v'_j)\}} deg_{\mathcal{H}}(v).$$ (1)

Generation of the Reduced Hypergraph (lines 10-17). After the creation of the valued representative graph, the algorithm performs a selection of edges with a greedy approach. It selects the edge having the higher weight value while there are edges left in G. Each selected edge is transformed to a hyperedge of the reduced hypergraph. This hyperedge contains the vertices from $\mathcal{H}$ corresponding to the intersection of the two vertices of the edge. We obtain, at the end, a set of hyperedges corresponding to the reduced hypergraph $\mathcal{H}_R = (V_R, E_R)$. Let us remark that if several edges have the same weight, the first found edge is selected.

Let us consider the example of Table 1 as a hypergraph $\mathcal{H}$ (6 hyperedges, 8 vertices). The reduced hypergraph is $\mathcal{H}_R = (V_R, E_R)$ where $V_R = \{A, B, C, E, F, H\}$ and $E_R = \{\{A, C, E\}, \{B, C, F, H\}\})$.

4.2   Minimal Transversal Computation

The last step is the computation of the (exact) minimal transversals of the reduced hypergraph. These transversals correspond to the approximate minimal transversals of the initial hypergraph: $\widetilde{MinTr}(\mathcal{H}) = MinTr(\mathcal{H}_R)$.

Let us take our example, the minimal transversals of $\mathcal{H}_R$ are: $\{C, AB, AF, AH, BE, EF, EH\}$. We consider them as the approximate minimal transversals of $\mathcal{H}$. Let us remark that the (exact) minimal transversals of $\mathcal{H}$ are: $\{AB, AC, CD, CF, CH, EF, EH, GH, AFG, BDE\}$.

5   Related Works

Numerous methods have been proposed to reduce the number of itemsets. In [7], the authors have studied the problem of randomly sampling maximal itemsets without explicit enumeration of the complete search space. They have employed a simple random walk that only allows additions of singletons to the current set untill a maximal set is found. In [8], the authors have used the Minimum Description Length (MDL) principle: the best set of itemsets is that set that compresses the database best. In [9], the approximation of a collection of frequent itemsets by the k best covering sets has been studied. The proposed algorithm takes in input the whole collection of the frequent itemsets or the positive border. The authors have explained the difficulties to use a greedy algorithm to obtain, from the positive border, k covering sets belonging to the initial collection. Our approach computes an approximate border from a border given completely in input. In that respect, we are close to the works presented in [9]. The exact positive border is the algorithm's input. Nevertheless, we do not want to find some covering sets and the itemsets of the approximate border do not necessarily belong to the initial collection. Our approach is more controllable than MDL used in [8]. We have the possibility to have several different methods to approximate the negative border ($\widetilde{f}$). Moreover, we have an understanding mapping between the exact border and the approximation border.

The computation of minimal transversals is a central point in hypergraph theory [4]. The algorithms to compute the minimal transversals come from different domains: graph theory, logic and data mining [10]. This is a NP-hard problem. The algorithms of approximation of minimal transversals are rare. Some works approximate the minimal transversals in order to obtain some ones or only one [6]. Some works are based on an evolutionary computation [11] where the transversality and the minimality are transcribed in a fitness function where a parameter, noted ϵ, is the fraction of hyperedges needed to intersect by any generated transversals. In [5], the *Staccato* algorithm computes low-cost approximate minimal transversals with a depth-first search strategy. It has been designed for model-based diagnosis. We have adapted *Staccato* in order to compute approximated minimal transversals in general. *Staccato* sorts the vertices according to their degree in increasing order. At each step, only the first λ (%) vertices of the remaining hypergraph are used. The more the λ value is high, the more the result is close to the set of the minimal transversals. The algorithm presented in [12], that we call δ-*MTminer*, produces minimal transversals which can miss at most δ hyperedges. It uses a breadth-first search strategy and several itemset discovery techniques. We have a different approach to compute approximate minimal transversals. We propose to apply a hypergraph reduction and then to compute the minimal transversals of the reduced hypergraph. These transversals are considered as the approximate minimal transversals of the initial hypergraph. Moreover, we don't need to set any parameters.

6   Experiments

6.1   Data and Protocol

Four data sets have been used: Mushroom, Chess, Connect and Kosarak. They have been downloaded from the FIMI web site[1]. Mushroom contains data on 23 species of gilled mushrooms. Chess contains some strategies for chess sets. Connect contains strategies for the game of connect-4. Kosarak contains anonymized click-stream data of a hungarian on-line news portal. The data sets (see Table 2) have been chosen to cover the different types of existing data sets according to the classification proposed by Gouda & Zaki [13].

The protocol of the experiments is as follows: For each data set and for some minimum support threshold values, (1) Compute the positive border according the minimum support threshold value, with IBE [14], (2) Compute the (exact)

---

[1] Frequent Itemset Mining Implementations, http://fimi.ua.ac.be/data/

Table 2. Data sets used in the experiments

Dataset	#transactions	#items	Avg. size of a trans.	Gouda & Zaki
Mushroom	8124	119	23	type 4
Chess	3196	75	37	type 1
Connect	67557	129	43	type 2
Kosarak	990002	41270	8,1	type 3

negative border with *Border-Diff* [15], the approximate negative border with 1-*MTminer* [12], the approximate negative border with *Staccato* [5], and the approximate negative border with our method (noted *AMTHR - Approximate Minimal Transversals by Hypergraph Reduction*), (3) Dualize to the positive borders (1 exact border and 3 approximate borders) with the *Border-Diff* algorithm which computes minimal transversals. For δ-*MTminer* (cf. Section 5), we have set δ to 1 because this value has produced the best results for δ-*MTminer*. For *Staccato* (cf. Section 5), we have chosen the highest values of λ before being impraticable: λ=0.8 for Mushroom, λ=0.65 for Chess, λ=0.7 for Connect, and λ=0.95 for Kosarak. In Steps 2 and 3, some statistics are computed: the number of itemsets of the computed border, the average size of the itemsets of the computed border, and the distance between the set of the itemsets of the computed border and the set of itemsets of the exact border. To evaluate the distance between two borders, we have used the distance of Karonski & Palka based on the Hausdorff distance. The cosine distance (see (2)) have been chosen to compute the distance between two elements (i.e. two itemsets). The distance D between two set of itemsets $\mathcal{X}$ and $\mathcal{Y}$ is defined in (3).

$$d(X,Y) = 1 - \frac{|X \cap Y|}{\sqrt{|X| \times |Y|}}. \qquad (2)$$

$$D(\mathcal{X},\mathcal{Y}) = \frac{1}{2}\{h(\mathcal{X},\mathcal{Y}), h(\mathcal{Y},\mathcal{X})\} \text{ where } h(\mathcal{X},\mathcal{Y}) = \max_{X \in \mathcal{X}} \{\min_{Y \in \mathcal{Y}} d(X,Y)\}. \qquad (3)$$

6.2 Results and Discussion

Due to space constraints, Figures about the average size of an itemset of the computed borders, are not presented in the paper. All the figures are available online[2]. Moreover, we do not present the execution times because this is not our main objective. For information, the computation of $\widetilde{Bd^-}(S)$ with $AMTHR$ is longer than with the other algorithms. The computation of $Bd^+(S)$ is the fastest with $AMTHR$.

Fig. 1, 2, 3 and 4 present, for each data sets, the number of itemsets of the computed negative borders and the distance between the computed negative borders and the exact negative borders. We can observe that the cardinality of $\widetilde{Bd^-}(S)$ is lower than the cardinality of $Bd^-(S)$ for each data sets. For information, the itemsets of $\widetilde{Bd^-}(S)$ are shorter than the itemsets of $Bd^-(S)$. They

---

[2] http://nicolas.durand.perso.luminy.univ-amu.fr/amthr/

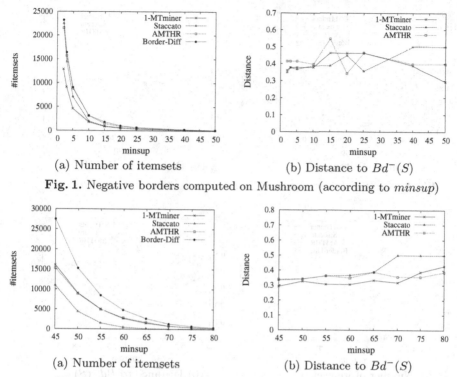

(a) Number of itemsets   (b) Distance to $Bd^-(S)$

Fig. 1. Negative borders computed on Mushroom (according to *minsup*)

(a) Number of itemsets   (b) Distance to $Bd^-(S)$

Fig. 2. Negative borders computed on Chess (according to *minsup*)

are the shortest with *Staccato* for each data sets. On Mushroom and Kosarak, the number of itemsets of $\widetilde{Bd^-(S)}$ produced by *AMTHR* is very close to the number of itemsets of $Bd^-(S)$. The generated itemsets with *AMTHR* are a little shorter than for the exact borders on Mushroom and Kosarak. Nevertheless, the itemsets of $\widetilde{Bd^-(S)}$ are different in view of the observed distances. This is an interesting remark. Some itemsets have been changed and they can be potentially interesting items. On Chess and Connect, *AMTHR* and 1-*MTminer* have produced a similar number of itemsets. These itemsets have a very close average size. Regarding the distance (between $\widetilde{Bd^-(S)}$ and $Bd^-(S)$), *Staccato* has obtained the closest borders on Mushroom and Kosarak. 1-*MTminer* has produced the closest borders on Chess and Connect. Nevertheless, we can observe that *AMTHR* is close to the best algorithm for each data sets.

Fig. 5, 6, 7 and 8 present, for each data sets, the number of itemsets of the computed positive borders and the distance between the computed positive borders and the exact positive borders. For information, the itemsets of $\widetilde{Bd^+(S)}$ are longer than the itemsets of $Bd^+(S)$. They are the longest with *Staccato* or *AMTHR* on each data sets. The number of itemsets of $\widetilde{Bd^+(S)}$ with *AMTHR* is the lowest on Mushroom. On the other data sets, *Staccato* has generated

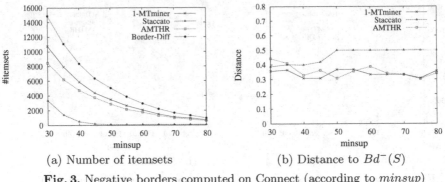

(a) Number of itemsets

(b) Distance to $Bd^-(S)$

Fig. 3. Negative borders computed on Connect (according to *minsup*)

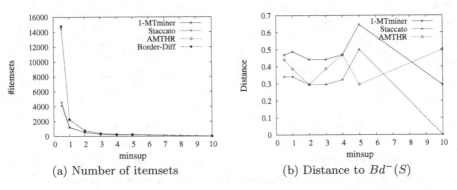

(a) Number of itemsets

(b) Distance to $Bd^-(S)$

Fig. 4. Negative borders computed on Kosarak (according to *minsup*)

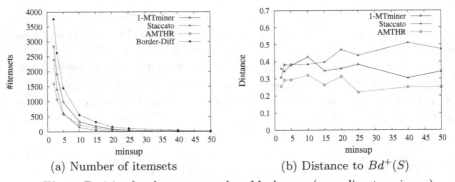

(a) Number of itemsets

(b) Distance to $Bd^+(S)$

Fig. 5. Positive borders computed on Mushroom (according to *minsup*)

the lowest number of itemsets. $1\text{-}MTminer$ have produced more itemsets than $AMTHR$, except for Kosarak. $AMTHR$ has obtained the closest $\widetilde{Bd^+(S)}$ to $Bd^+(S)$ on Mushroom, Chess and Kosarak. On Connect, $1\text{-}MTminer$ has also obtained good results. On Kosarak, *Staccato* and $\delta\text{-}MTminer$ have produced bad results.

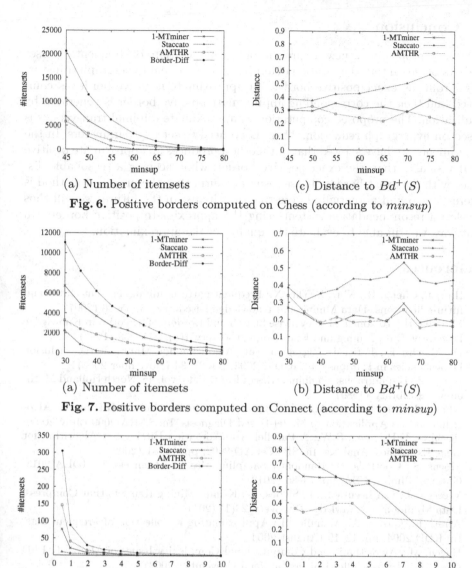

(a) Number of itemsets

(c) Distance to $Bd^+(S)$

Fig. 6. Positive borders computed on Chess (according to $minsup$)

(a) Number of itemsets

(b) Distance to $Bd^+(S)$

Fig. 7. Positive borders computed on Connect (according to $minsup$)

(a) Number of itemsets

(b) Distance to $Bd^+(S)$

Fig. 8. Positive borders computed on Kosarak (according to $minsup$)

To resume, we can note that the proposed method, $AMTHR$, has reduced the number of itemsets of the generated positive borders, while keeping a reasonable distance to the exact positive borders. Moreover, our method seems to be robust according to the different types of data sets.

7   Conclusion

We have proposed a new approach of approximation of frequent itemset borders based on the computation of approximate minimal hypergraph transversals. From the exact positive border, an approximate negative border is computed and then the corresponding approximate positive border is generated by dualization. The proposed computation of approximate minimal transversals is based on hypergraph reduction. There is no need to set any parameters. In the experiments, we have showed that our method produces an approximate positive border smaller than the exact positive border, while keeping a reasonable distance with the exact border. These results confirm that the proposed method is interesting to find potentially interesting new items. In the future, we will thus develop a recommendation system using the approximate positive borders. In that way, we will able to evaluate the quality of the approximation.

References

1. Han, J., Cheng, H., Xin, D., Yan, X.: Frequent pattern mining: current status and future directions. Data Mining and Knowledge Discovery 15, 55–86 (2007)
2. Mannila, H., Toivonen, H.: Levelwise Search and Borders of Theories in Knowledge Discovery. Data Mining and Knowledge Discovery 1(3), 241–258 (1997)
3. De Marchi, F., Petit, J.: Zigzag: a New Algorithm for Mining Large Inclusion Dependencies in Database. In: ICDM 2003, pp. 27–34 (November 2003)
4. Berge, C.: Hypergraphs: Combinatorics of Finite Sets, vol. 45. North Holland Mathematical Library (1989)
5. Abreu, R., van Gemund, A.: A Low-Cost Approximate Minimal Hitting Set Algorithm and its Application to Model-Based Diagnosis. In: SARA 2009 (July 2009)
6. Ruchkys, D.P., Song, S.W.: A Parallel Approximation Hitting Set Algorithm for Gene Expression Analysis. In: SBAC-PAD 2002, pp. 75–81 (2002)
7. Moens, S., Goethals, B.: Randomly Sampling Maximal Itemsets. In: IDEA 2013, Chicago, Illinois, USA, pp. 79–86 (2013)
8. Vreeken, J., van Leeuwen, M., Siebes, A.: Krimp: Mining Itemsets that Compress. Data Mining and Knowledge Discovery 23(1) (2011)
9. Afrati, F., Gionis, A., Mannila, H.: Approximating a Collection of Frequent Sets. In: KDD 2004, pp. 12–19 (August 2004)
10. Hagen, M.: Algorithmic and Computational Complexity Issues of MONET. PhD thesis, Friedrich-Schiller-Universitat Jena (November 2008)
11. Vinterbo, S., Øhrn, A.: Minimal Approximate Hitting Sets and Rule Templates. Approximate Reasoning 25, 123–143 (2000)
12. Rioult, F., Zanuttini, B., Crémilleux, B.: Nonredundant Generalized Rules and Their Impact in Classification. In: Ras, Z.W., Tsay, L.-S. (eds.) Advances in Intelligent Information Systems. SCI, vol. 265, pp. 3–25. Springer, Heidelberg (2010)
13. Gouda, K., Zaki, M.J.: Efficiently Mining Maximal Frequent Itemsets. In: ICDM 2001, pp. 163–170 (November 2001)
14. Satoh, K., Uno, T.: Enumerating Maximal Frequent Sets Using Irredundant Dualization. In: Grieser, G., Tanaka, Y., Yamamoto, A. (eds.) DS 2003. LNCS (LNAI), vol. 2843, pp. 256–268. Springer, Heidelberg (2003)
15. Dong, G., Li, J.: Mining Border Descriptions of Emerging Patterns from Dataset-Pairs. Knowledge and Information Systems 8(2), 178–202 (2005)

Clustering Based on Sequential Multi-Objective Games

Imen Heloulou[1], Mohammed Said Radjef[1], and Mohand Tahar Kechadi[2]

[1] University A. MIRA of Bejaia, Road Targua Ouzemour 06000, Bejaia, Algeria
[2] School of Computer Science and Informatics, University College Dublin (UCD), Belfield,
Dublin 4, Ireland
heloulou.imen@gmail.com

Abstract. We propose a novel approach for data clustering based on sequential multi-objective multi-act games (ClusSMOG). It automatically determines the number of clusters and optimises simultaneously the inertia and the connectivity objectives. The approach consists of three structured steps. The first step identifies initial clusters and calculates a set of conflict-clusters. In the second step, for each conflict-cluster, we construct a sequence of multi-objective multi-act sequential two-player games. In the third step, we develop a sequential two-player game between each cluster representative and its nearest neighbour. For each game, payoff functions corresponding to the objectives were defined. We use a backward induction method to calculate Nash equilibrium for each game. Experimental results confirm the effectiveness of the proposed approach over state-of-the-art clustering algorithms.

Keywords: clustering, multi-objective, sequential game, multi-act, inertia, pay-off functions, connectivity, backward induction, Nash equilibrium.

1   Introduction

Nowadays, data clustering is a well-established field, which is growing rapidly in many domains such as pattern-analysis and grouping, and decision-making [1]. Many clustering methods have been proposed to satisfy these application requirements. However, in many real-world problems, more than one objective are needed to be optimised. The multi-objective clustering methods attempt to identify clusters in such a manner that several objectives are optimised during the procedure [1]. Traditionally, multi-objective methods have been categorised as ensemble, evolutionary and micro-economic methods. Cluster ensemble frameworks combine different partitions of data using consensus functions [2]. Ensemble methods based clustering have been proven to more powerful method than individual clustering methods. However, they are not able to deal with multi-objective optimisation [3]. On the other hand, evolutionary algorithms such as MOEA [4], PESA-II [3] and MOCK [4] identify better clusters than ensemble clustering methods, as they optimise many objectives concurrently [5]. Microeconomic models naturally analyse the situations of conflicting objectives in a game theoretic setting [6].

L. Bellatreche and M.K. Mohania (Eds.): DaWaK 2014, LNCS 8646, pp. 369–381, 2014.

Gupta et al [7] and Badami et al [8] used a microeconomic game theoretic approach for clustering, which simultaneously optimises compaction and equi-partitioning. Garg et al [9] proposed the use of Shapley value to give a good start to K-means. Bulo and Pelillo [10] used the concept of evolutionary games for hyper-graph clustering.

Despite the large number of algorithms, the priority is always given to partitioning algorithms. They are attractive as they lead to elegant mathematical and algorithmic proofs and settings. However, there are several limitations with this oversimplified formulation. Probably the best-known limitation of the partitioning approaches is the typical requirement of the number of clusters to be known in advance, which is not always easy, especially when there is no sufficient information on the data set. The choice of the clusters' centres represents another major problem. Moreover, the partitioning techniques were only good in minimising the criterion of compactness and in detecting clusters of spherical form. In our endeavour to provide answers to the questions raised above, we found that game theory offers a very elegant and general perspective that serve well our purposes and which has found applications in diverse fields. Specifically, in this paper we have developed a clustering technique based on non-cooperative games theory in sequential form. This novel approach performs the optimisation on the basis of two conflicting objectives, inertia and connectivity in a simultaneous manner. We use a backward induction method to derive the right number of the clusters. In this way, not only the number of clusters is determined dynamically, but also the distribution of objects to clusters will be also done by negotiation.

2   Multi-Objective Clustering Game

Before describing our clustering approach, lets define the fundamentals of every clustering algorithm, which is the similarity measure. The similarity measure allows us to evaluate how much clustering is good or bad by evaluating either intra-cluster or inter-cluster inertia or both. In our case we attempt to optimise intra-cluster/inter-cluster inertia and the connectivity objectives. The intra-cluster inertia should be as small as possible in order to have a set of homogeneous clusters. Let ξ a set of clusters of the initial data set D. The inter-cluster inertia is given by [11]:

$$I_A(\xi) = \frac{1}{n}\sum_{C_i \in \xi} \sum_{j \in C_i} d(j, ch_i) \tag{1}$$

where d is the Euclidean distance between object j and the centre ch_i of the cluster C_i. A larger value of the inter-cluster inertia leads to a good separation of the clusters. It can be calculated using the following relationship [11]:

$$I_R(\xi) = \frac{1}{n}\sum_{i=1}^{K} |C_i| \ d(ch_i, g) \tag{2}$$

where $g = (g_1, ..., g_j, ..., g_d)$ and g_j is the gravity centre of D along the jth dimension.

$$g_j = \frac{1}{n} \sum_{i=1}^{n} i_j \tag{3}$$

The connectivity measure evaluates which neighbouring data objects are placed in the same cluster. It is computed as follows [12]:

$$Connc(\xi) = \frac{1}{n}\sum_{i=1}^{n}\frac{\sum_{j=1}^{L}x_{i,nn_{ij}}}{L} \qquad (4)$$

where:

$$x_{rs} = \begin{cases} 1 \ if \ \exists \ C_l : r,s \ \in C_l \\ 0, \qquad otherwise \end{cases}$$

nn_{ij} is the jth nearest neighbour of object i and L, a parameter, is the number of neighbours that contribute to the connectivity measure.

Table 1. Notations and Terminology.

D	Data set; $D = \{1,2,\dots i,\dots,n\}$ each object i is described by a set of δ attributes $i = (i_1, i_2, \dots, i_\delta)^T \in \Re^\delta$
$C_i^{(t)}$	Cluster C_i at time t
$K^{(t)}$	Total number of clusters at time t
$\xi^{(t)}$	Set of clusters at time t; $\xi^{(t)} = \{C_1^{(t)}, \dots, C_{K^{(t)}}^{(t)} (K^{(t)} \leq n)\}$
ch_i	Center of cluster $C_i^{(t)}$; $Argmin_{m \in C_i^{(t)}} \frac{1}{|C_i^{(t)}|}\sum_{j \in C_i^{(t)}} d(m,j)$
h_D	h measurement calculates the dissimilarity of object i with respect to all dataset's objects; $h_D(i) = \frac{1}{n}\sum_{\substack{m \in D \\ l \neq m}} d(i,m)$
$I_A(C_i^{(t)})$	Intra-cluster inertia for cluster C_i at time t; $I_A(C_i^{(t)}) = \sum_{j \in C_i^{(t)}} d(j, ch_i)$
$Connc(C_i^{(t)})$	Connectivity of cluster C_i at time t; $Connc(C_i^{(t)}) = \frac{1}{|C_i^{(t)}|}\sum_{l=1}^{|C_i^{(t)}|}\frac{\sum_{j=1}^{L}x_{i,nn_{ij}}}{L}$
$neig(i)$	Set of L nearest neighbours of object i arranged in ascending order according to theirs Euclidean distance
ALC	Average linkage clustering measurement; $\frac{1}{|C_l^{(t)}| \cdot |C_m^{(t)}|}\sum_{l \in C_l^{(t)}}\sum_{h \in C_m^{(t)}} d(l,h)$
$neig(C_i^{(t)})$	Nearest neighbour of cluster $C_i^{(t)}$ according to ALC; $neig(C_i^{(t)}) = Argmin_{C_m^{(t)} \in \xi^{(t)}\setminus C_i^{(t)}} \frac{1}{|C_i^{(t)}| \cdot |C_m^{(t)}|}\sum_{l \in C_i^{(t)}}\sum_{j \in C_m^{(t)}} d(l,j)$
$E(i)$	Set of i's neighbours belonging to nearest neighbour of cluster $C_j^{(t)}$; $E(i) = neig(i) \cap neig(C_j^{(t)})$, $\forall i \in C_j^{(t)}$
$E(C_j^{(t)})$	Set of all $E(i)$, $\forall i \in C_j^{(t)}$; $E(C_j^{(t)}) = \cup_{i \in C_j^{(t)}} E(i)$
$E(C_i^{(t)}, C_j^{(t)})$	Set of all neighbours in conflict between ch_i and ch_j; $E(C_i^{(t)}) \cap E(C_j^{(t)})$
$E_i(C_i^{(t)}, C_j^{(t)})$	For all $m \in E(C_i^{(t)}, C_j^{(t)})$, assuming: $C_i^{(t)} = C_i^{(t)} \cup \{m\}$ and calculating $d(m, ch_i)$. Then, ch_i constructs $E_i(C_i^{(t)}, C_j^{(t)})$ containing elements of $E(C_i^{(t)}, C_j^{(t)})$ arranged in ascending order based to Euclidean distance according to ch_i
$B(C_i^{(t)}, C_j^{(t)})$	Set of objects which ch_i would to exchange with ch_j arranged in ascending order according to theirs Euclidean distance compared to ch_j; $B(C_i^{(t)}, C_j^{(t)}) = \{m \in C_i^{(t)}/d(ch_i, m) > d(ch_j, m) \ and \ C_j^{(t)} = neig(C_i^{(t)})\}$
$conflict^{(t)}$	Set of conflict-clusters; $C_i^{(t)} \in conflict^{(t)} \Leftrightarrow \exists m,j \in \{1, \dots, K^{(t)}\}/ C_m^{(t)}, C_j^{(t)} \in \xi^{(t)} \ and \ C_i^{(t)} = neig(C_j^{(t)}) = neig(C_m^{(t)})$
$conflict(C_i^{(t)})$	Set of clusters which for them $C_i^{(t)}$ is the nearest neighbour; $conflict(C_i^{(t)}) = \{C \in \xi^{(t)}/C_i^{(t)} = neig(C)\}$

R-Square (R^2) is used to estimate the number of clusters. It is defined by [13]:

$$R^2(\xi) = \frac{I_R(\xi)}{I_A(\xi) + I_R(\xi)} \qquad (5)$$

The more close to 1 it is, the better is the classification. It should not be maximised at all costs, since it would lead thus to very large number of clusters [13]. By combining the connectivity and R^2 objectives, a trade-off is required to determine the appropriate value of K. We expect this product to be large:

$$\varphi(\xi) = R^2(\xi) * Connc(\xi) \qquad (6)$$

When we go beyond the right number of clusters, φ will decrease: the decrease in R^2 will be less significant but comes at a high cost in terms of connectivity (because a true cluster is being split). ClusSMOG consists of three components briefly explained in the following subsections. The notations and terminology used in the rest of the paper are given in Table 1.

2.1 Step 1: Initialisation

This step consists of initialising the primary $K^{(0)}$ clusters with their cluster representatives. Initially, we construct for each object i its **neig(i)** set and the similarity matrix. Then, we calculate for each object i the $h_D(i)$ value. Objects with a minimum value of h_D means that they have high density around them and we could consider them as initial cluster representatives. A cluster representative should not be among the 5% L first nearest neighbours of clusters representatives previously selected. Thereafter, the remaining objects are assigned to the nearest clusters according to d and then the cluster representatives are updated. After, each $C_i^{(0)}$ seeks its nearest neighbour **neig($C_i^{(0)}$)**, which is shown like resource to receive neighbouring objects. Due to this, a cluster may be the nearest neighbour of several clusters; it is called conflict-cluster, so the $conflict^{(0)}$ set is constructed containing conflict-clusters.

2.2 Step 2: Sequential Games for Conflict-Clusters Objects

The purpose of this step is to maximise φ value in order to achieve the correct number of clusters. At instant t, each $C_i^{(t)}$ seeks its nearest neighbour $neig(C_i^{(t)})$, which contains the neighbours of its objects, to integrate them in its cluster in order to increase its connectivity and maximise R^2. An issue may arise when a cluster is the nearest neighbour for several clusters, which at the same time trying to attract objects of this cluster, (**conflict-cluster**). However, the real competition will be on the objects coveted by different clusters. To analyse and find a solution to this competition, we have modelled it as a multi-objective multi-act sequential non-zero-sum game with perfect information. A game consists of a set of players, a set of moves (or strategies) available to those players, and a specification of payoffs for each combination of strategies. Sequential games are games where later players have some knowledge about earlier actions. The game has *perfect information* if each player, when making any decision, is perfectly informed of all the events that have previously occurred. Consequently, a problem may occur when conflict-cluster allocates all its objects. This cluster will be removed if its deletion improves the global objectives simultaneously. This step is, therefore, responsible for reducing the number of clusters.

The step starts by constructing the $conflict^{(t)}$ set, which contains all conflict-clusters. We define for each $C_l^{(t)} \in conflict^{(t)}$ the $conflict(C_l^{(t)})$ set of clusters for which $C_l^{(t)}$ is their nearest neighbour. The $conflict^{(t)}$ set is arranged in descending order according to the clusters cardinality, $|conflict(C_l^{(t)})|$ and the elements of each $conflict(C_l^{(t)})$ set are ordered in ascending order according to ALC.

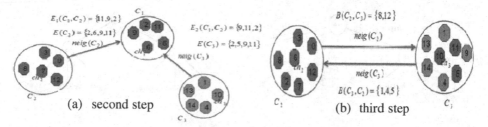

Fig. 1. An example of a clustering game

Starting by $C_l^{(t)} \in conflict^{(t)}$, which has the highest cardinality, a game is formulated between ch_i and ch_j, the representatives of both nearest clusters of $C_l^{(t)}$. So, the set of players competing for $C_l^{(t)}$ objects at time t is defined as follow:

$$N^{(t)} = \{ch_i, ch_j/neig(C_i^{(t)}) = neig(C_j^{(t)}) = C_l^{(t)}\} \tag{7}$$

As shown in Figure 1a, a game is initiated between ch_2 and ch_3, where their clusters have the same neighbour, C_1. They try to attract their neighbouring objects in order to maximise their payoffs. To do this, each player constructs its $E(C_i^{(t)})$ set, which contains its neighbours belonging to the conflict-cluster. Each ch_i begins by integrating the objects that are not coveted by the opponent player; $E(C_i^{(t)}) \setminus E(C_i^{(t)}, C_j^{(t)})$ set. They affect objects that best improve the overall objectives, i.e. the φ value. However, the major issue that needs to be solved is when the objects are covered by both clusters, $E(C_i^{(t)}, C_j^{(t)})$. For example, object 11 is in conflict set because it is a neighbour of object $10 \in C_3$ and a neighbour of object $12 \in C_2$. The solution is to first affect the object to the player with higher connectivity degree; $Connc(C_i)^{(t)} > Connc(C_j)^{(t)}$. Thereafter, each ch_i arranges the elements of $E(C_i^{(t)}, C_j^{(t)})$ set in ascending order according to their distance to ch_i to construct the $E_i(C_i^{(t)}, C_j^{(t)})$ set. Sequential game is presented as tree (as shown in Figure 2). Each node represents a choice for a player. The links represent a possible action for that player. The payoffs are specified at the bottom of the tree. ch_2 moves first and chooses either to integrate in its cluster $\{11\}$, $\{11,9\}$ or $\{11,9,2\}$ objects. ch_3 sees ch_2's move and then chooses its action. It has the possibility to choose among remaining objects. Then, ch_2 is called again to choose its action. The same process is repeated until the end of conflict-objects (see Figure 2a). This game is called multi-act game; a player is allowed to act more than once. So, the actions set for a player ch_i at time t and the level p is defined as follows:

$$X_i^{(t,p)} = \{M_l, l = 1, .., |E_i(C_i^{(t)}, C_j^{(t)}) \setminus (\cup^{\;P} < p \;(X_i^{(t,p)} \cup X_j^{(t,p)}))|\} \tag{8}$$

$$M_l = \{m_r \in E_i(C_i^{(t)}, C_j^{(t)}) \setminus (\cup^{\;P} < p \;(X_i^{(t,p)} \cup X_j^{(t,n)})), r \le l\} \tag{9}$$

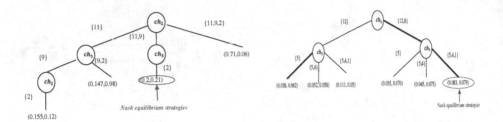

(a) Two-act sequential non-zero-sum two-player game

(b) Single-act sequential game non-zero-sum two-player game

Fig. 2. Sequential game Models

r is the rank of the object m in $E_i(C_i^{(t)}, C_j^{(t)}) \setminus (\cup \; p < p \; (X_i^{(t,p)} \cup X_j^{(t,p)}))$. This allows ch_i to assign neighbours that maximise its payoff; provided that the chosen objects are not selected by any of the two players in the previous steps. We do not take all the possible objects coalitions but only the closest ones, as illustrated in Figure 2a. This definition reduces the complexity of the game.

The game performance is extremely restricted to proper definition of the payoffs function. The players choose their actions in order to maximise the connectivity of their clusters and maximise R^2. So the connectivity objective can be seen as a private objective and R^2 as a public objective (collective). So each cluster representative has a vector function (bi-criteria) of the payoffs:

$$f_i^{(t)}(x_1, x_2) = \left(Connc(C_i^{(t)}), R^2(\xi^{(t)})\right) \tag{10}$$

Where $f_i^{(t)}(.,.): X_1^{(t)} * X_2^{(t)} \xrightarrow{yields} \mathbb{R}^2$. The two objectives are conflicting, because connectivity's improvement can lead to the decrease of R^2 by reducing the number of the clusters. Every time we get rid of a conflict-cluster we increase the intra-cluster inertia. Decreasing R^2 will be less significant as each player competes for objects (resources) to improve their connectivity by trying to avoid a big loss in intra-cluster inertia.

After, we analyse the game using backward induction methodology to calculate the Nash equilibrium strategies, which represent the best structure of clustered data. A temporary reallocation of objects is performed according to the chosen actions. If the reallocations improve the overall objective according to φ, the allocations are committed, the clusters representatives are then updated and that conflict-cluster is removed. If the played game did not improve the system's objectives, a sequence of sequential games is formed for this conflict-cluster between all possible pairs of players starting by the nearest clusters. We have so a maximum $\sum_{C_i^{(t)} \in conflict^{(t)}} \frac{|conflict(C_i^{(t)})| * (conflict(C_i^{(t)}) - 1)}{2}$ games at time t. While $conflict^{(t)}$ set is not empty, a game is formulated for another conflict-cluster between two clusters representatives if theirs clusters are not changed in previous steps. If the $conflict^{(t)}$ set is empty and no improvement for the system thus go to third step, else restart again step2.

2.3 Step 3: Sequential Games for Clusters Neighbours

This step deals with intra-cluster inertia optimisation, i.e. construct homogeneous clusters. Each ch_i engages in exchanging objects with ch_j of its nearest neighbour cluster. This exchange between the two clusters is modelled as a single-act sequential non-zero-sum two-player game with perfect information, which we will identify its factors below. In a single-act game, each player makes a decision only once (as shown in Figure 2.b). The difference between this step and the previous step lies in the existence of common objects between the players, so the necessity to use multi-act in order to explore the effect of possible combinations of objects. Unlike Step 2, the set of exchanged objects are distinct. The players' set at time t is given by:

$$N^{(t)} = \{ch_i, ch_j / C_j^{(t)} = neig(C_i^{(t)})\} \tag{11}$$

Before starting the game, each ch_i constructs the $B(C_i^{(t)}, C_j^{(t)})$ set. This set consists of objects that maximise its intra-cluster inertia and they will be ordered in ascending order according to their distance to the opponent player ch_j, since they will be transferred to him. The objects concerned are whose having minimal distance compared to the second player. As shown in Figure 1b, a game is formulated between ch_2 and ch_3, where ch_2 wants to exchange with ch_3 the $\{8,12\}$ objects because they are close to ch_3 = object 10 rather than ch_2 = object 6. This is the same for second player. Thus, the actions set for each player at time t is as follow:

$$X_i^{(t)} = \{M_l, l = 1, ..., |B(C_i^{(t)}, C_j^{(t)})|\} \tag{12}$$

$$M_l = \{m_r \in B(C_i^{(t)}, C_j^{(t)}), r \le l\} \tag{13}$$

r is the rank of the object m in $B(C_i^{(t)}, C_j^{(t)})$. The player having the smallest number of objects to exchange with the second player; $Argmin_{i,j}\{|B(C_i^{(t)}, C_j^{(t)})|, |B(C_j^{(t)}, C_i^{(t)})|\}$ is called leader and will play first.

It remains to define the payoffs function of the players. As we are interested in intra-cluster inertia; weaker intra-cluster inertia better is the homogeneity of objects, the payoffs function of each player is given by:

$$f_i^{(t)}(x_1, x_2) = \frac{1}{I_A(C_i^{(t)})} \tag{14}$$

Now, all elements characterising the game are well defined, so we can construct the tree representing the sequential form, as illustrated in Figure. 2b. To minimise the complexity of the tree, the number of combinations is reduced according to the order of objects in $B(C_i^{(t)}, C_j^{(t)})$, as shown in Figure. 2b. If a cluster is singleton, its object will be assigned immediately to the second player without playing the game.

After game's resolution by application of the backward induction, temporary reallocation of objects is performed according to the Nash equilibrium strategies representing the optimal intra-cluster inertia for both players. If the reallocations improve the overall objective; minimise the system's intra-cluster inertia $I_A(\xi^{(t)})$, the allocations are committed and the clusters representatives are then updated. This process is repeated at maximum $K^{(t)}$ times if the concerned clusters are not changed in the previous steps, because the neighbour of each cluster may change if the content of cluster is changed.

2.4   Solution Concept: Backward Induction

The backward induction is the most common solution concept for sequential games. Taking the example described at Fig. 2b, player 1 makes the first decision x_1 from its actions set $X_1 = \{\{8\}, \{8,12\}\}$ and player 2 makes its decision x_2 from its actions set $X_2 = \{\{1\}, \{1,4\}, \{1,4,5\}\}$ after player 1. The payoffs functions of both players are given by $f_1(x_1, x_2)$ and $f_2(x_1, x_2)$ respectively. Backward induction uses the assumption of rationality, meaning that player 2 will maximise its payoff in any given situation. Player 2 chooses its best response x_2^* to the actions of player 1 which is the solution of this program:

$$x_2^* = Argmax_{x_2 \in X_2} f_2(x_1, x_2) \tag{15}$$

By anticipating the reaction x_2^* of player 2, we can reduce the size of our tree by eliminating the choices that player 2 will not choose. In this way, the links that maximise the player's payoff at the given information set are in bold. After this reduction, player 1 can maximise its payoffs once the player 2 choices are known. His best response is the solution of this program:

$$x_1^* = Argmax_{x_1 \in X_1} f_1(x_1, x_2^*) \tag{16}$$

The result (x_1^*, x_2^*) is a Nash equilibrium found by backward induction of player 1 choosing $\{12,8\}$ and player 2 choosing $\{5,4,1\}$.

2.5   Stopping Criterion

If the overall objectives of the system are improved in the antecedent steps, this process starts again (step 2 and step 3) until no further improvement is possible.

3   Experimental Setup

We carried out extensive experimentations to compare ClusSMOG with state-of-the-art algorithms on several artificial datasets (Square1, Square4 [14], Ellipse [15], Dataset_9_2, Dataset_3_2, Dataset_4_3 [16]) and real-life datasets [17]. All experiments are implemented in Java and run on 2.20 GHz Intel core 2 Duo CPU with 3 GB RAM. Single-objective algorithms are performed using Rapid Miner [18] tool.

3.1   Parameters Settings

In this subsection, we discuss the specification of parameters for ClusSMOG and MOCK algorithms. For ClusSMOG, the initial number of clusters $K^{(0)} = \sqrt{n}$. While the choice of a reasonably large value of L is necessary to prevent outliers from being classified as individual clusters. However, if L is too large this may result in large number of small clusters. In our experiments, we chose $L = 10\%n$, which allows robust detection of clusters and better performance on all studied datasets. MOCK was performed with the source codes available from [14].

3.2   Results and Discussion

In order to evaluate the performance of our algorithm, we have compared ClusSMOG to both single and multi-objective algorithms: MOCK, K-means, K-medoid, Dbscan and X-means based on different evaluation measure. Specifically we choose for internal measures the Purity, the Rand Index [19], the adjusted Rand index (ARI) [20] and the F-measure [21] and the Silhouette Index for external measures [22]. The results are summarised in Table 2 and show the robustness of ClusSMOG from the simultaneous optimisation of system's objectives: intra-cluster/inter-cluster inertia and connectivity. The best entries have been marked in bold in each row. While it may be marginally beaten by MOCK algorithm on Iris and Wine datasets and X-means on Suqare1 and Suqare4 datasets, it shows an impressive performance across the entire datasets. This is not only reflected in the high values of the Adjusted Rand Index, but also in the close agreement between the number of clusters in the generated solution, and the correct K on all datasets. It is clear that ClusSMOG and MOCK exceed X-means in the detection of the adequate number of clusters. ClusSMOG and X-means outperform MOCK on the Adjusted Rand Index value on Square4, Square1, Dataset_3_2 and Dataset_4_3 datasets. Our algorithm is better than MOCK and X-means on Glass, Dataset_9_2 and Ellipse datasets. In conclusion, our algorithm is able to detect clusters of arbitrary shapes; this is due to the simultaneous optimisation of connectivity and inertia objectives.

Table 2. Adjusted Rand Index value comparison between ClusSMOG, MOCK and X-means

Dataset				ClusSMOG		MOCK		X-means	
Name	K	n	δ	K	ARI	K	ARI	K	ARI
Square1	4	1000	2	4	0.963	4.22	0.9622	4	**0.9735**
Square4	4	1000	2	4	0.8274	4.32	0.7729	4	**0.8348**
Ellipse	2	400	2	5	**0.3017**	7.8	0.2357	4	0.2263
Dtaset_9_2	9	900	2	9	**0.8233**	8.52	0.8109	4	0.3353
Dtaset_3_2	3	76	2	3	**1**	3.33	0.9465	3	**1**
Dataset_4_3	4	400	3	4	**1**	3.78	0.8787	4	**1**
Iris	3	150	4	3	0.5962	3.05	**0.7287**	4	0.6744
Glass	6	214	9	6	**0.2449**	6.18	0.1677	4	0.2391
Wine	3	178	13	3	0.4331	3.59	**0.8647**	4	0.3034

We also evaluated and compared ClusSMOG with several others state-of-the-art algorithms like K-means, Dbscan and K-medoid using other internal measures evaluation. Figure. 3 shows the results of the value of purity on several datasets. ClusSMOG gives higher values for purity on the majority of datasets. Indeed, the obtained clusters have a better homogeneity than K-medoid and Dbscan and K-means. It is obvious in Figure. 4 and Figure. 5 that ClusSMOG is able to make good decisions in the assignment of objects, which leads to high values of F-measure and Rand Index obtained by our algorithm in all datasets. Thus, overall, ClusSMOG presents definitely better performance than others algorithms and similar to K-means, in addition our technique automatically calculates the number of the clusters while K-means requires the desired number of clusters as input.

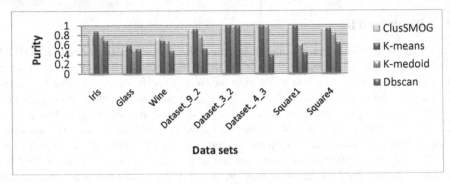

Fig. 3. Purity value comparison between different algorithms

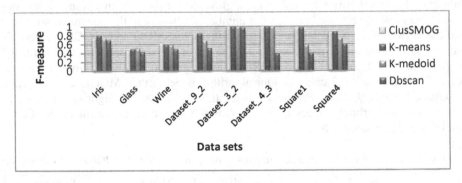

Fig. 4. F-measure value comparison between different algorithms

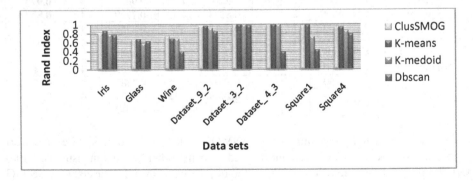

Fig. 5. Rand Index value comparison between different algorithms

In order to analysis the proposed clustering technique more accurately, we use silhouette metric as a method of interpretation and validation of clustered data as it is shown in Figure. 6. ClusSMOG clusters the data objects with high inter-cluster and low intra-cluster. One may have high silhouette value for K-means in some cases, this happens due to fact that the silhouette metric considers only intra-cluster inertia. Since K-means is a single objective clustering method, it optimises intra-cluster inertia effectively, especially when it is not easy to consider both objectives. However, in most cases, the presented algorithm gives higher silhouette metric, which indicates the effectiveness of the ClusSMOG.

From these results, ClusSMOG has proven its robustness compared to other mono/multi objectives algorithms using various evaluation measures. It gives solutions to clustering's problems by using the concepts of sequential games theory with a cluster initialisation mechanism, which plays a very important role on the final result of clustering.

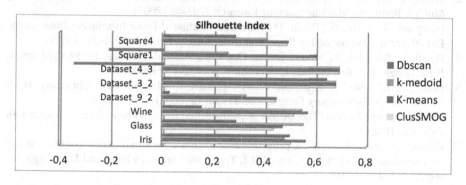

Fig. 6. Silhouette Index comparison between different algorithms

The proposed approach is also analysed on the basis of complexity metric. As sequential games are represented in a tree form, the complexity of a tree browsing is $O(m)$, where m is the number of nodes. In the second step where we have formulated multi-act sequential two-player games, the size of a player's actions is limited to $|X_i| = L * |C_i|$. Assuming that the average size of each cluster $|C_i| = \left\lfloor \frac{n}{K^{(0)}} \right\rfloor$ and $conflict^{(t)} = K^{(0)}$. Hence, in the worst-case, the complexity is $O(m * n * \sqrt{n})$, where $m \leq n^2$. Assuming that the third step is carried out also in the worst-case, with $K^{(0)}$ single-act sequential games, the complexity is $O(n * \sqrt{n})$.

4   Conclusion

We proposed a novel microeconomic-theory-based technique for simultaneous multi-objective clustering based on conflicting objectives, intra-cluster / inter-cluster inertia and connectivity with automatic K-determination. Our methodology is based on the backward induction in order to derive a desirable fairness in the final clustering results. The proposed technique is also able to determine the appropriate number of cluster dynamically. For this, we developed an interesting and very important cluster initialisation mechanism that has direct impact on the final clustering results. The experimental study conducted on some well-known benchmarks datasets provided important insights on the performance of the game theoretic algorithms. Experimental results confirm the effectiveness of the proposed approach over state-of-the-art clustering algorithms (including single and multi-objective techniques. As future work, we will look at some parts of the technique where we can use parallelism or concurrent actions.

References

1. Law, M.H.C., Topchy, A.P., Jain, A.K.: Multiobjective Data Clustering. In: Computer Society Conference on Computer Vision and Pattern Recognition (2004)
2. Strehl, A., Ghosh, J.: Cluster Ensembles: A Knowledge Reuse Framework for Combining Multiple Partitions. Machine Learning Research 3, 93–98 (2003)
3. Dempster, A., Laird, N., Rubin, D.: Maximum Liklihood From Incomplete Data via the EM Algorithm. Journal of The Royal Statistical Society, Series B 39, 1–38 (1977)
4. Handl, J., Knowles, J.: MultiObjective Clustering Around Medoids. In: Proceedings of IEEE Congress on Evolutionary Computation, pp. 632–639 (2005)
5. Handl, J., Knowles, J.: An Evolutionary Approach to Multiobjective Clustering. IEEE Transaction on Evolutionary Computation 11, 56–76 (2007)
6. Von Neumann, J.: Zur Theorie der Gesellschaftsspiele. Mathematische Annalen 100, 295–320 (1928)
7. Gupta, U., Ranganathan, N.: A game theoretic approach for simultaneous compaction and equipartitioning of spatial data sets. IEEE Transcation on Knowledge and Data Engineering 22, 465–478 (2010)
8. Badami, M., Hamzeh, A., Hashemi, S.: An enriched game-theoretic framework for multiobjective clustering. Applied Soft Computing 13, 1853–1868 (2013)
9. Garg, V.K., Narahari, Y., Murty, M.N.: Novel Biobjective Clustering (BiGC) based on Cooperative Game Theory. IEEE Transactions on Knowledge and Data Engineering (2013)
10. Bulo, S.R., Pelillo, M.: A game-theoretic approach to hypergraph clustering. Advances in Neural Information Processing Systems (2009)
11. Goutte, C., Toft, P., Rostrup, E., Nielsen, F.Å., Hansen, L.K.: On Clustering fMRI Time Series. NeuroImage 9, 298–310 (1999)
12. Handl, J., Knowles, J.D.: Evolutionary Multiobjective Clustering. In: Yao, X., et al. (eds.) PPSN 2004. LNCS, vol. 3242, pp. 1081–1091. Springer, Heidelberg (2004)
13. Tufféry, S.: Data Mining et Statistique Décisionnelle: L'intelligence des données (2010)
14. http://personalpages.manchester.ac.uk/mbs/Julia.Handl/mock.html
15. Bandyopadhyay, S., Saha, S.: GAPS: A Clustering Method Using a New Point Symmetry-Based Distance Measure. Pattern Recognition 40, 3430–3451 (2007)
16. http://www.isical.ac.in/~sanghami/data.html

17. http://archive.ics.uci.edu/ml/datasets.html
18. RapidMiner website, http://www.rapidminer.com
19. Rand, W.: Objective Criteria for the Evaluation of Clustering Methods. J. Amer. Statist. Assoc. 66, 846–850 (1971)
20. Hubert, A.: Comparing partitions. J. Classification 2, 193–198 (1985)
21. van Rijsbergen, C.: Information Retrieval, 2nd edn. Butterworths (1979)
22. Rousseeuw, P.J.: Silhouettes: A Graphical Aid to the Interpretation and Validation of Cluster Analysis. J. Comput. Appl. Math. 20, 53–65 (1987)

Opening up Data Analysis for Medical Health Services: Cancer Survival Analysis with CARESS

David Korfkamp[1], Stefan Gudenkauf[1], Martin Rohde[1], Eunice Sirri[2], Joachim Kieschke[2], and H.-Jürgen Appelrath[1]

[1] OFFIS - Institute for Computer Science, Escherweg 2, Oldenburg, Germany
{david.korfkamp,stefan.gudenkauf,martin.rohde}@offis.de,
appelrath@informatik.uni-oldenburg.de
[2] Epidemiological Cancer Registry Lower Saxony, Oldenburg, Germany
{eunice.sirri,kieschke}@offis-care.de

Abstract. Dealing with cancer is one of the big challenges of the German healthcare system. Originally, efforts regarding the analysis of cancer data focused on the detection of spatial clusters of cancer incidences. Nowadays, the emphasis also incorporates complex health services research and quality assurance. In 2013, a law was enacted in Germany forcing the spatially all-encompassing expansion of clinical cancer registries, each of them covering a commuting area of about 1 to 2 million inhabitants [1]. Guidelines for a unified evaluation of data are currently in development, and it is very probable that these guidelines will demand the execution of comparative survival analyses.

In this paper, we present how the CARLOS Epidemiological and Statistical Data Exploration System (CARESS), a sophisticated data warehouse system that is used by epidemiological cancer registries (ECRs) in several German federal states, opens up data analysis for a wider audience. We show that by applying the principles of integration and abstraction, CARESS copes with the challenges posed by the diversity of the cancer registry landscape in Germany. Survival estimates are calculated by the software package periodR seamlessly integrated in CARESS. We also discuss several performance optimizations for survival estimation, and illustrate the feasibility of our approach by an experiment on cancer survival estimation performance and by an example on the application of cancer survival analysis with CARESS.

Keywords: Data analytics, cancer survival, CARESS, periodR.

1 Introduction

With an estimated annual number of 470,000 incident cases and nearly 215,000 deaths, dealing with cancer is one of the big challenges of the German healthcare system [2,3]. The analysis of cancer data can provide valuable insights on oncological care. Typical analyses of interest are, for example: detecting region-specific changes in the survival of cancer patients which may be attributable

L. Bellatreche and M.K. Mohania (Eds.): DaWaK 2014, LNCS 8646, pp. 382–393, 2014.

to improvements in diagnostics, therapeutics and secondary prevention, and detecting regional and international differences in the survival of cancer patients.

Originally, efforts regarding the analysis of cancer data focused on the detection of spatial clusters of cancer incidences, for example, finding bursts of leukemia in the proximity of nuclear power plants. In 1995, a national law was enacted directing the establishment of population-based cancer registries in all German federal states [4]. However, up until now, federal activities are still isolated and the landscape of cancer survival analysis is still diverse. This also applies to the regional level. For example, certified organ cancer centers,[1] oncology centers and clinics, which treat the majority of cancer patients in Germany, rely on heavily customized software systems with heterogeneous data storage systems, making it even harder to obtain comparable data bases for analysis. Although the Association of Population-based Cancer Registries in Germany (GEKID) provides a coordinated effort to harmonize cancer incidence data collection since 2006, their recommendations are still to be widely implemented and data acquisition, reporting activities, as well as the regulatory frameworks remain inconsistent.

Nowadays, the emphasis also incorporates complex health services research and quality assurance. Additionally, in 2013 a law was enacted forcing the spatially all-encompassing expansion of clinical cancer registries, each of them covering a commuting area of about 1 to 2 million inhabitants [1]. Guidelines for a unified evaluation of data are currently in development, and it is very probable that these guidelines will demand the execution of comparative survival analyses.

In this context of heterogeneity, cancer registries represent a necessity. As data warehouse (DWH) systems [5], they physically integrate cancer data of various formats and stemming from various sources into a single system. They provide an integrated view on population-based cancer data confined to a specific region and appropriate tools to enable their analysis.

There are several software tools for cancer survival analysis, for example SURV3/4 and periodR [6, p. 527ff]. Although proven to be practical regarding applicability [2], most lack in accessibility: The tools are isolated, meaning that the user must provide a prepared dataset of previously selected cancer data beforehand – a task that is notoriously time-consuming and error-prone, requires extensive technical skills, and represents a recurring discontinuity in the digital workflow. Moreover, none of them is particularly suited to generate and publish end-user-friendly reports on-the-fly.

In this paper, we show how specific data warehouse systems can open up data analysis for a wider audience. In an example we show survival analysis on cancer data with CARESS, an epidemiologic cancer registry (ECR) system that is utilized in several federal states in Germany. First, we introduce several methods for the computation of cancer survival estimates. Second, we introduce the CARESS system and its conceptual architecture, including the integration of cancer survival analysis. Next we highlight technical challenges of the implementation

---

[1] See http://www.onkozert.de/ [last visited 2014/03/27]

and how we overcame them, especially regarding performance optimization, and present an example on the application of cancer survival analysis with CARESS.

2   Methods of Cancer Survival Analysis

Cancer survival analysis employs statistical methods to analyze cancer data by considering the time period between a defined starting point (e.g., the documented date of initial diagnosis) and the occurrence of an event of interest (e.g., patient death) [7]. Cancer survival estimates can be computed by a variety of methods, and the computation itself can be executed by a variety of software tools. This section presents an overview of the different types of cancer survival analyses and selected tools to perform such analyses.

The first dimension is whether the computation is cohort-based or period-based. The traditional cohort-based approach includes a group of patients in the analysis (i.e., the cohort) by considering a defined period of diagnosis (i.e., years of diagnosis), with all follow-up diagnoses within a defined timeframe [2,6]. Although this approach is considered limited regarding the reflection of recent progress made on cancer care, this shortcoming can be mitigated with complete analysis, a variant of the cohort-based analysis that additionally considers more recently diagnosed patients regardless of the length of follow-up [8]. Period-based analysis, in contrast, is an approach that focuses on information of recently departed patients by applying a survival function to the observed survival experience within a defined timeframe (i.e., the period) to estimate the survival of the patients within this timeframe of follow-up years [2]. In several experiments using historical data, period-based analysis has proven to be more accurate than cohort-based analysis in estimating the chance of survival of more recently diagnosed patients. [2,9,10,11]

The second dimension of cancer survival analysis is whether the computation is absolute or relative. According to Holleczek et al., absolute computation calculates survival in terms of proportions of patients still alive after a given time span after diagnosis, typically reported in 5 or 10-year survival [8]. Relative computation of cancer survival is instead calculated as the ratio of the observed survival in a group of patients and the expected survival of a comparable group considered not to have the cancer of interest in terms of age, sex and calendar period as obtained from population life tables [8]. Thus, the reported survival is corrected for other causes of death than the cancer without requiring detailed information on the exact cause of death.

There exist several methods and tools for estimating survival. Widely accepted methods are Ederer I, Ederer II, Hakulinen and Kaplan-Meier. Tool support can be differentiated into openly accessible software such as SURV-4, or periodR, and proprietary tooling that is directly integrated into ECR specific database systems. For example, Table 1 presents a categorization of the software tool periodR according to Holleczek et al., an open source add-on package to the R programming language and environment for statistical computation. As the table shows, periodR covers the whole range of cancer survival analyses and employs

Table 1. Categorization of the software tool periodR. X = supported. (a) = Ederer II, Hakulinen. (b) = Greenwoods method.

	Absolute	Relative
Period-based analysis	X	X (a)
Complete analysis	X	X (a)
Cohort-based analysis	X	X (a)
Standard-error detection	X (b)	X (b)

widely accepted methods to do so. As it provides an Application Programming Interface (API) naturally, we deem periodR a fit choice to integrate in an ECR system.

3 The CARESS System

The *CARLOS Epidemiological and Statistical Data Exploration System* (CA-RESS) is a sophisticated data warehouse system that is used by the ECRs in several German federal states. Originally developed in 1993 in the pilot project Cancer Registry Lower-Saxony (CARLOS) serving as a geographic information system (GIS) tool for analyzing clusters of cancer incidences [12,13], the CA-RESS system was subsequently extended into a full-fledged ECR data warehouse system and adopted by the federal states of Hamburg, Schleswig-Holstein and North Rhine-Westphalia and the center for cancer registry data (ZFKD) at the Robert-Koch-Institute, which pools data from all federal states in Germany. The system supports several stakeholders in medical health services such as doctors and epidemiologists by providing sophisticated tools for data analysis in a highly accessible user interface, enabling them to carry out explorative analyses, ad-hoc queries and reporting activities without extensive technical skills.

CARESS consists of three layers. The *data source integration layer* provides a unified physical integration of various heterogeneous data sources. CARESS supports DWH products from different vendors such as Microsoft Analysis Services, Pentaho Mondrian and Jedox Palo.

The *component integration layer* provides a service facade to client applications in order to invoke the services supported by CARESS, see Figure 1. Complex statistical queries are executed by outsourcing requests, for example, to automate the calculation of cancer survival estimates using the R programming language and environment. Requests to the underlying data source integration layer are encapsulated as services as well. Additionally, the component integration layer provides access to the system's metadata repository containing complex analysis configurations. As the service factory only provide access to service interfaces instead of concrete services, the respective service implementations can be exchanged easily. For example, we introduced an optimized CachedSurvivalAnalysisDataService (see Section 4) that maintains an instance of the original SurvivalAnalysisDataService to forward any request not previously cached to this instance.

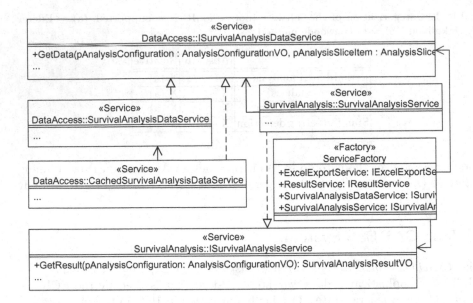

Fig. 1. Service classes of the CARESS component integration layer

The *client layer* provides a convenient graphical user interface (GUI) to access the service façade of the component integration layer. The GUI is realized using Microsoft Windows Presentation Foundation (WPF) and the Prism framework for modular applications. [14] Also, the GUI supports arbitrary inquiries (e.g., constructing a pivot table by combining arbitrary summary attributes with arbitrary compatible dimensions) as well as topic-oriented analyses with predefined statistical methods such as cancer survival analysis. This enables epidemiologists to carry out semi-automated explorative analyses and ad-hoc queries with minimal effort. In addition, CARESS offers a rich set of reporting functions for end users such as doctors and nurses, and supports different export formats such as XML, CSV, Excel and PNG image files.

In contrast to the use of individual tools for estimating survival analyses, requiring the user to provide a prepared dataset of previously selected cancer data beforehand, CARESS provides an experience that enables the user to focus on the task at hand without distraction. Setting up CARESS is a one-time-effort of DWH engineers, while operation is guaranteed by DWH and software engineers, depending on the question which architecture layer has to be adapted or extended with additional functionality.

4   Optimizing CARESS for Survival Analyses

Survival analyses impose particular challenges regarding data acquisition. First, data required to run survival analyses should be highly detailed. For example, the diagnosis date of cancer cases should be available at least in terms of months to achieve a high precision of survival estimates.

Second, CARESS uses a single database as its *single point of truth* that conforms to the Online Analytical Processing (OLAP) cube paradigm. While the OLAP paradigm proves powerful when navigating through multidimensional data spaces, it is rather limited regarding the acquisition of heterogeneous data at high levels of detail compared to the conventional relational database model. For example, a typical 3-year period analysis with 5 years of follow-up diagnoses results in a 5-dimensional cube, consisting of approximately 1,400,000 cells (100 distinct age categories $\times (3 + 5) \times 12$ diagnosis months $\times 3 \times 12$ death months $\times 2$ gender categories $\times 2$ vital status), resulting in comparatively large requests in contrast to an equivalent relational database request.[2] As a consequence, OLAP result sets for highly detailed survival analysis are typically large and require much time and memory for processing.

To address these challenges we enhanced CARESS' *SurvivalAnalysisDataService*, the component responsible for retrieving data from the underlying multidimensional database, and integrated the following adaptations for survival analyses in contrast to the regular DataService.

First, we optimized the Multidimensional Expressions (MDX) query used to retrieve data from the OLAP database by applying the *NON EMPTY* keyword to each axis. Axes tagged *NON EMPTY* contain only classification nodes with actual values with respect to the other axes. For example, if a query result contains no single cancer case for a patient who at the time of diagnosis was 40 years old, the result will not include the 40 years node from the age dimension at all, although this age was requested in the analysis. Depending on the actual distribution of cases this measure can significantly decrease the size of the returned result. However, the optimization is likely to go unnoticed for analyses that are performed on large areas, since most combinations of age, gender, date of diagnosis, date of death, and vital status include at least one case.

Second, we extended the SurvivalAnalysisDataService with functions to split, parallelize and merge query requests. The actual implementation operates as follows: (1) Candidate dimensions are identified for splitting the request into several smaller requests. In general, all classifying dimensions of an analysis (age, gender, vital status, date of diagnosis, and date of death) are considered as to be candidates. However, in certain situations some of those dimensions can not be used for splitting. For example, this is the case when a classifying dimension was selected as a *slicer dimension* (e.g., for retrieving age- or gender-specific survival estimates – in this case we only retrieve the data of the selected slice). (2) Of the remaining candidates up to two dimensions are selected automatically as *split dimensions*. In case of two split dimensions, partial cubes are retrieved based on the cross product of each classification node of the two dimensions.

---

[2] A relational database request would result in only a few hundred rows, when, for example, a rare diagnosis or a specific regional area is analyzed.

The resulting partial cubes are then being requested in parallel from the underlying multidimensional database in order to reduce the overall request time. Once all partial cubes are retrieved, they are merged into a single result cube available for further computations.

Third we introduced a caching algorithm in order to reduce the speed of subsequent survival analysis requests that are based on the same data. For example, these can occur when the statistical method for survival analyses is changed, the life table is exchanged, or a previously rendered survival analysis is exported to Microsoft Excel.[3] To do so, we used the Microsoft .Net Framework's native MemoryCache class to store results from the SurvivalAnalysisDataService, since it already provides functionalities for caching – for example, the ability to let cache values expire after a predefined period. To store and access the cache we derived a normalized key from the request object used to interact with the SurvivalAnalysisDataService. This key consists of sorted lists of the classifying and restricting dimensions containing the selected classification nodes (also sorted). As a result, small changes to queries such as the rearrangement of classification nodes or slicer axes do not result in new database requests.

In the following, we describe an experiment on the response-time of cancer survival analysis to illustrate the effectiveness of our optimizations.

4.1   Experimental Setup

The experiment was conducted on a single machine running the complete CARESS stack, including the database server, to minimize external effects. The machine was equipped with a Dual-Core Opteron 2220 processor clocked at 2,6 GHz and with 8 GB RAM. Microsoft SQL Server 2012 Analysis Services were used as the database backend. We ran several tests in which always the same realistic example configuration was computed. After every computation we restarted CARESS to avoid interfering effects introduced by caching. For each run we recorded the response-time using CARESS' builtin logging mechanism. Thereby, the measured response-time represents the effects notable by the user. It includes the time required to retrieve the data and subsequent activities such as statistical calculations in the R component and client-side rendering.

We conducted 15 survival analyses for the original and for the optimized survival analysis each. We used the arithmetic mean and the 95% confidence interval (CI) for the results of the analyses. The CIs allow us to address two questions: (1) Do the different implementations show significantly different behavior or not? (2) How large is the performance variation of the individual measurements for a single implementation? The CIs are computed using the Students t-distribution, as the number of measurements is small ($n < 30$). We refer to the work of Georges et al. as an excellent reading on the importance of confidence intervals for statistically rigorous performance evaluation. [15]

---

[3] Exporting a survival analysis to Excel is performed by the *ExcelExportService*, which uses the *CachedSurvivalAnalysisDataService* via the *ResultService*.

4.2 Results

As Figure 2 shows, the mean response-times of both implementations differ and the 95% confidence intervals (CI) do not overlap, showing that the difference between the measurements is statistically significant (optimized analysis: CI 95% max. 10.493, original analysis: CI 95% min. 381.567), indicating that the optimized version is much faster. Furthermore, the CI show that the variance of the measurements of the original implementations is much greater than the variance of the optimized version, indicating a more stable behavior.

However, any empirical study like ours is vulnerable to certain threats to validity. For example, the experiment has only been executed on a single machine and wider applicability is yet to prove. Also, the optimized survival analysis is only compared to its functionally equivalent original version. Comparative analyses to other survival analysis tools are desirable. On the other hand, to the best of our knowledge, there are currently no other implementations for survival analysis that comprise both the actual analysis as well as automatized data preparation and retrieval. We leave these topics amongst others for future work (i.e., more experiments).

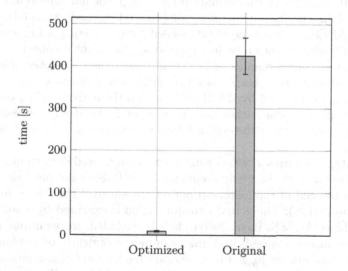

Fig. 2. Response-time of the optimized and the original survival analysis implementation. The optimized implementation required 9.07 seconds in average (CI 95% 1.427), the original approach required 424.67 seconds in average (CI 95% 43.100).

5 Application of Cancer Survival Analysis with CARESS

In the following, we describe how cancer survival estimates are computed with CARESS by the example of the ECR in Lower Saxony (Epidemiologisches Krebsregister Niedersachsen (EKN)). The system integrates data from various data sources into a data warehouse, including cancer incidence data along with date of diagnosis, date of death (if applicable), vital status, gender, diagnosis, death certificate only (DCO) cases etc. gathered regularly at EKN. Typical stakeholders to the system are epidemiologists that prepare mandatory annual reports on cancer and ad-hoc reports for governmental requests and journalistic inquiries. Since 2003, EKN has comprehensively covered the state of Lower Saxony in Germany, inhabited by approximately eight million people. The completeness of registration was reported to be over 95% in 2010 [16].

The minimum data requirements for survival analysis include sex, month and year of diagnosis (dm and dy), age at diagnosis, month and year of end of follow-up (fm and fy) and vital status at the end of follow-up. A detailed specification of the minimum data requirements and the concrete periodR functions are described by Holleczek et al. [8]

CARESS calculates cancer survival estimates by performing a three-stage procedure. The software component that manages these three stages was extensively empirically evaluated to guarantee that survival estimates and the corresponding plotted survival curves are correct. Stage one includes querying the required data from the data warehouse by narrowing the data space according to the user input. In general, cancer survival estimation requires particular attention to define the temporal dimensions date of diagnosis (dm and dy) and end of follow-up (fm and fy). CARESS reduces the effort needed by considering a higher level of abstraction: the user defines a period (period approach) or a cohort (cohort approach) of interest, and a number of follow-up years. The parameters dm and dy are then constrained to a range based on either the cohort or the period, while fm and fy are calculated by CARESS based on the corresponding date of death, including all cases that either have no date of death at all (representing patients that were still alive at the end of follow-up) or died in the course of the follow-up years.

The second stage performs a transformation of the retrieved data to meet the aforementioned minimum data requirements for periodR. For example, cases that are still alive at the end of the follow-up period are assigned the end of follow-up dates (i.e., fm and fy). The actual transformation is executed by a software component within CARESS. Inconclusive data is excluded, for example, cases with unknown month of diagnosis and unknown month of death, or implausible dates. Excluded datasets are logged in a separate file for later examination.

In the final stage the prepared dataset is handed to the periodR component. The results returned include a chart showing the absolute and relative survival rates by follow-up years as well as tables that show the survival estimates along with 95% confidence intervals and standard errors. Both are visualized directly in the CARESS client. The chart is illustrated in Figure 3. For convenient reporting, results can be exported to different formats such as Excel and PNG image files.

As an example, we illustrate cancer survival analysis with CARESS using a dataset that includes 33,611 records of lung cancer patients aged 15-99 years, diagnosed in 2003-2010 with passive mortality follow-up until December 2010 and for the period 2008-2010. The event of interest considered for survival estimates

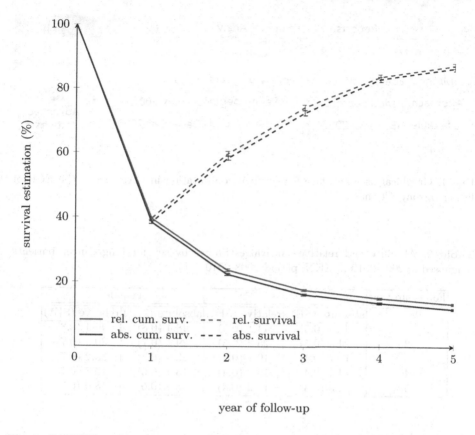

Fig. 3. CARESS analysis report on cumulative absolute and relative survival estimates of breast cancer patients grouped into three age categories for five years of follow-up for patients diagnosed in 2003-2010, with mortality follow-up until 2010 and for period 2008-2010.

was the death of the patient. Therefore, patients still alive at the end of 2010 were right censored: For these patients, fm and fy were imputed as 12 and 2010, respectively, automatically by the CARESS system. Addressing further data quality concerns such as the exclusion of DCO cases can be performed by the user by simply removing the corresponding data entries from a filter view within the user interface of CARESS. The user interface also lets the user choose the actual analysis methods supported by periodR (see Figure 4).

A particular advantage of CARESS for survival analysis is the high degree of automation. For example, computing survival estimates by prognostic variables (e.g., sex, age groups, stage of disease at diagnosis, histology, anatomic subsite) is executed automatically once the user has selected the respective variables in the user interface. Table 2 illustrates the results for absolute and relative survival estimates of the cohort of patients diagnosed in 2003-2010 and of the period 2008-2010, as analyzed with CARESS. The estimation is ordered by sex.

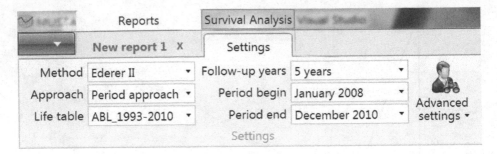

Fig. 4. Graphical user interface for survival computation in CARESS at the ECR of Lower Saxony, Germany

Table 2. Absolute and relative survival estimates by sex for lung cancer patients diagnosed in 2003-2010 in EKN, period 2008-2010

Follow-up year	Male		Female	
	absolute (SE*)	relative (SE*)	absolute (SE*)	relative (SE*)
1	42.9 (0.5)	44.1 (0.5)	48.4 (0.8)	49.4 (0.8)
2	24.5 (0.5)	25.9 (0.5)	29.5 (0.7)	30.6 (0.7)
3	17.7 (0.4)	19.1 (0.4)	22.0 (0.7)	23.2 (0.7)
4	14.4 (0.4)	16.2 (0.4)	18.4 (0.6)	19.7 (0.7)
5	12.3 (0.3)	14.2 (0.4)	16.5 (0.6)	18.0 (0.7)

6   Conclusions and Outlook

Cancer epidemiology is an explorative art on the one hand and uses complex statistical methods like survival analyses on the other hand. A sophisticated multidimensional data model for data warehouse systems in health care must provide integration of statistical methods and definition of ad-hoc aggregations at run-time. In contrast to standard OLAP tools and standard statistical tools the CARESS system provides sophisticated mechanisms to integrate domain-specific statistical methods into the multidimensional data model and makes them available for epidemiologists and scientists via a convenient graphical user interface. Additionally, CARESS is specifically optimized for OLAP-based survival analysis. This is illustrated by experiments on the response-time of cancer survival analyses.

Further extension of CARESS comprise the implementation of Cox regression methods and more experiments. In addition, the EKN reckons with being assigned the task of analyzing data from all future clinical cancer registries in Lower Saxony, comprising a population of about 8 million people. An corresponding extension of the CARESS system is subject of our current efforts in anticipation of this task. With this extension, we deem CARESS an appropriate candidate as the DWH system of choice for regional clinical cancer registries.

References

1. Gesetz zur Weiterentwicklung der Krebsfrüherkennung und zur Qualitätssicherung durch klinische Krebsregister (Krebsfrüherkennungs- und -registergesetz – KFRG) of 2013, BGBl. I, pp. 617–623 (April 3, 2013) (German)
2. Nennecke, A., Brenner, H., Eberle, A., Geiss, K., Holleczek, B., Kieschke, J., Kraywinkel, K.: Cancer survival analysis in Germany – Heading towards representative and comparable findings. Gesundheitswesen 72, 692–699 (2010) (German)
3. Robert Koch-Institut und die Gesellschaft der epidemiologischen Krebsregister in Deutschland e. V., editors. Krebs in Deutschland 2007/2008, Berlin (2012) (German)
4. Gesetz über Krebsregister (Krebsregistergesetz – KRG) of 1994, BGBl. I pp. 3351–3355 (November 4, 1994) (German)
5. Bauer, A., Günzel, H.: Data Warehouse Systeme, 3rd edn. dpunkt.verlag, Heidelberg (2009) (German)
6. Kleinbaum, D.G., Klein, M.: Survival Analysis, 3rd edn. Springer, New York (2012)
7. Altman, D.G., Bland, J.M.: Time to event (survival) data. British Medical Journal 317, 468–469 (1998)
8. Holleczek, B., Gondos, A., Brenner, H.: periodR – an R Package to Calculate Long-term Cancer Survival Estimates Using Period Analysis. Methods Inf. Med. 48, 123–128 (2009)
9. Brenner, H., Söderman, B., Hakulinen, T.: Use of period analysis for providing more up-to-date estimates of long-term survival rates: Empirical evaluation among 370 000 cancer patients in Finland. Int. J. Epidemiol. 31, 456–462 (2002)
10. Talbäck, M., Stenbeck, M., Rosén, M.: Up-to-date long-term survival of cancer patients: An evaluation of period analysis on Swedish Cancer Registry data. Eur. J. Cancer 40, 1361–1372 (2004)
11. Ellison, L.F.: An empirical evaluation of period survival analysis using data from the Canadian Cancer Registry. Ann. Epidemiol. 16, 191–196 (2006)
12. Wietek, F.: Spatial Statistics for Cancer Epidemiology – the Cancer Registry's Epidemiological and Statistical Data Exploration System (CARESS). In: Fehr, R., Berger, J., Ranft, U. (eds.) Environmental Health Surveillance. Fortschritte in der Umweltmedizin, pp. 157–171. ecomed-Verlag, Landsberg (1999)
13. Kamp, V., Sitzmann, L., Wietek, F.: A spatial data cube concept to support data analysis in environmental epidemiology. In: Proceedings of the 9th International Conference on Scientific and Statistical Database Management, August 11-13. IEEE, Olympia (1997)
14. Brumfield, B., Cox, G., Hill, D., Noyes, B., Puleo, M., Shifflet, K.: Developer's Guide to Microsoft Prism 4: Building Modulare MVVM Applications with Windows Presentation Foundation and Microsoft Silverlight. Microsoft Press (2010)
15. Georges, A., Buytaert, D., Eeckhout, L.: Statistically Rigorous Java Performance Evaluation. ACM SIGPLAN Notices - Proceedings of the 2007 OOPSLA Conference 42(10), 57–76 (2007)
16. Registerstelle des Epidemiologischen Krebsregisters Niedersachsen, editor. Krebs in Niedersachsen, Oldenburg, 117 p. (2010) (German)

"Real-time" Instance Selection
for Biomedical Data Classification

Chongsheng Zhang[1], Roberto D'Ambrosio[2], and Paolo Soda[2]

[1] Henan University, China
chongsheng.zhang@yahoo.com
[2] Università Campus Bio-Medico di Roma, Italy
{r.dambrosio,p.soda}@unicampus.it

Abstract. Computer-based medical systems play a very important role in medical applications because they can strongly support the physicians in the decision making process. Several existing methods infer a classification function from labeled training data. The large amount of data nowadays available, although collected from high quality sources, usually contain irrelevant, redundant, or noisy information, suggesting that not all the training instances are useful for the classification task. To address this issue, we present here an instance selection method that, different from the existing approaches, selects in "real-time" a subset of instances from the original training set on the basis of the information derived from each test instance to be classified. We apply our method to seven public benchmark datasets, showing that the recognition performances are improved. We will also discuss how method parameters affect the experimental results.

1 Introduction

The use of clinical decision support systems, also known as computer-based medical systems, is spreading over several fields of medicine and biology. Among the large variety computer-based medical systems proposed to date, in this paper we focus on those classification based systems belonging to supervised learning. The labeled datasets collected in such fields are very large in terms of both the number of features and the number of samples, but not all the information is helpful for the classification tasks. Although the data are collected from high quality sources, they could contain irrelevant, redundant, noisy or unreliable information [16,9].

In this respect, several machine learning and data mining work focus on how to select relevant data [3]. Existing preprocessing techniques typically reduce the number of features, or discretize the feature space, or select the instances. While next section briefly summarizes these approaches, in this paper we focus on instance selection methods. They select a subset of examples from the original training data, this subset is then used to train a classifier.

Within this framework we notice that, to the best of our knowledge, existing work reduces the size of the training set once for all test samples. Moreover, such algorithms compute the similarities between training instances in the whole feature space. Our approach, however, selects on the fly a subset of training samples for each test instance to be classified. To this aim, for each feature we first determine the top-k training samples

L. Bellatreche and M.K. Mohania (Eds.): DaWaK 2014, LNCS 8646, pp. 394–404, 2014.

that are most similar to the test instance. We then remove those samples whose occurrences in the top-k lists are lower than a threshold. The remaining instances will be used to train a classification model that will predict the label of the test instance.

Our instance selection method has been applied to seven medical and biological benchmark datasets. The recognition performances achieved by three different classifiers trained on the reduced training set using our proposal have been significantly improved over those attained by the same classifiers trained on the original training set. We also experimentally investigate how the parameters of the algorithm as well as the dimensionality of the feature space affect the final performances.

2 Background and Motivations

Most algorithms train a classifier using all the samples from the training data. However, there are several reasons, e.g., the noise and the curse of dimensionality, requiring us to reduce the original training set to a smaller one. This phenomenon is particularly evident for the k-Nearest Neighbour (kNN) algorithm [6], which presents low tolerance to noise since it considers all data as relevant, even when they are incorrect. This drawback is common for all learners employing a classification scheme similar to kNN, which are usually named as instance-based learners, lazy learners, or case-based learners. Besides these classifiers, it is well-known that noisy, harmful and superfluous training instances can negatively affect the performances of other classification architectures, such Support Vector Machines and Neural Networks [4].

Therefore, a very active line of research in machine learning and data mining deals with scaling down data. Existing preprocessing techniques can be traced back to *data reduction, data cleaning, data integration* and *data transformation* [3]. While a detailed review of each of these categories is out of the scope of the manuscript, we now focus on data reduction methods. These methods, further to be able to face with the computational complexity, storage requirements, and noise tolerance of kNN [6], can be successfully applied to preprocess the training set. These reduction techniques aim to obtain a representative training set with a lower size than the original one and with a similar, or even higher, classification accuracy for the new incoming data. This goal can be achieved by reducing the number of columns in a data set, i.e., selecting features, reducing the number of possible values of discretized features, i.e., discretizing feature-values, and reducing the number of rows in a data set, i.e., selecting instances [10]. In the literature, the latter approaches are known as data reduction techniques, prototype selection or instance selection (IS) methods. They select a subset of examples from the original training data, and then a classifier D is trained on this subset. Next, each new instance is classified by D. It is worth noting that, depending on the strategy adopted, IS methods can remove noisy, redundant, and both kinds of examples, without generating new artificial data [6].

A widely used categorization of IS methods divide them into *edition* methods, *condensation* methods, and *hybrid* methods. Edition methods aim to remove noisy instances to increase classifier accuracy. Condensation methods reduce the size the training set by removing superfluous instances that will not affect the classification accuracy of the training set itself. Hybrid methods search for a small subset of the training set that simultaneously achieves the elimination of both noisy and superfluous instances. The interested readers can find further details in the reviews of IS methods reported in the literature [6,8,13].

We note that in several different application fields, the application of IS methods enables experts to derive classifiers with reasonable performances, even when they select a small number of samples. This observation also holds in the biological and medical areas, e.g., [12], and it becomes very interesting since biomedical devices nowadays produce a vast amount of data. Furthermore, humans are limited in their ability to detect abnormal cases due to their non-systematic search patterns and in the presence of noisy, irrelevant, redundant, or unreliable data. Therefore, in these fields IS methods can play a pivotal role to develop decision support systems.

To the best of our knowledge, IS methods *a-priori* reduce the size of training set, i.e., they remove instances once and without considering any information related to the test sample. For example, in [6] the kNN method classifies a test instance on a reduced reference set, whereas in [4] the authors extract from the training set a minimum number of points completely characterizing the separating plane of a Support Vector Machine (SVM). On the contrary, in this work we propose an IS method which reduces the training set on the basis of information directly derived from each test sample to be classified. Therefore, we do not remove *a-priori* the irrelevant, redundant, noisy or unreliable data contained in the dataset, but do "real-time" instance selection for each test instance to be classified.

Furthermore, existing algorithms pick the training instances using all dimensions of the feature space, or even a reduced space dimensionality given by a feature selection algorithm. In [14], we presented a preliminary work on biomedical instance selection. In this work, we measure the similarity between the training samples and the new instance to be classified by considering each feature independently and selecting separately the top-k training samples that are most similar to the new instance on each feature. Only the training samples with occurrences in the top-k lists above a certain threshold will be chosen in the final reduced training set. Details of our proposal will be presented in the next section.

3   Methods

Our proposal belongs to the wide domain of IS methods: it selects training instances from the knowledge base to build a classification model. Existing approaches reduce the set of training instances once and then build a single classification model, which is therefore equal for all test instances. In contrast, our method reduces the training set on the basis of information related to each test instance. For this reason, in the following we refer to this approach as *Real-time Instance Selection* (RtIS) method. This implies that each test sample has a different training set: we deem that building a specific classification model for each test sample can be beneficial in terms of classification performances. On the one side, a computer based medical system relying on a general classification model built upon all the historical data is not always the more reasonable approach. On the other side, our proposal should permit to remove the training samples which are irrelevant, redundant, noisy or unreliable with respect to the specific test sample. Although this approach can introduce a computation burden, we deem that it is likely to become less severe because of a smaller training dataset and a simpler prediction model, and the development of modern computational technology, e.g., faster hardware, higher memory capacity, and parallel programming exploiting multi-cores machines.

Let $Z^{tr} = \{x_i^{tr}, \omega_i\}_{i=1}^N$ be the training set, with N the number of training instances, $x_i \in \Re^n$ the n-dimensional feature vector, i.e., $x_i^{tr} = \{x_{1i}^{tr}, \ldots, x_{ji}^{tr}, \ldots x_{ni}^{tr}\}$, and $\omega_i \in \Omega = \{\omega_1, \ldots, \omega_c\}$ the label class. Therefore, Z^{tr} can be represented as a matrix with size $N \times n$. We also denote with Z^{te} the test set, and a test instance is represented by $x^{te} = \{x_1^{te}, \ldots, x_j^{te}, \ldots x_n^{te}\}$.

A formal specification of the traditional IS approach is the following: let $\hat{Z}^{tr} \subseteq Z^{tr}$ be the subset of selected samples resulting from the execution of an IS algorithm; then a classifier D trained over $\hat{Z}^{tr}$ instead of Z^{tr} assigns to x^{te} a class label from Ω. Different from this traditional approach, here we determine $\hat{Z}^{tr}$ for each x^{te} as follows.

Given the test instance, we consider the j-th dimension of the feature space as an independent feature space and determine the set $s_j(x^{te})$ of top-k training patterns similar to x^{te}. Hence:

$$s_j(x^{te}) = \{x_i^{tr}(\epsilon_j^k), \omega_i\}_{i=1}^k \tag{1}$$

with

$$x_i^{tr}(\epsilon_j^k) = \{x_{ji}^{tr} \in \Re : |x_{ji}^{tr} - x_j^{te}| \le \epsilon_j^k\} \tag{2}$$

where ϵ_j^k is the mono-dimensional euclidean distance of further training sample among the k neighborhoods considered. Notice that, according to equation 1, $s_j(x^{te})$ can be represented as a matrix with size $k \times n$. Therefore, in the collection of $s_j(x^{te})$, computed for all $j = 1, \ldots, n$, the same training instance can appear r times, with $r \ge 1$. In this respect, we rewrite the collection of $s_j(x^{te})$ as:

$$\overline{s}(x^{te}) = \{(x_i^{tr}(\epsilon_j^k), \omega_i), r\} \tag{3}$$

On this basis, we get $\hat{Z}^{tr}$ analyzing $\overline{s}(x^{te})$ as follows:

$$\hat{Z}^{tr} = \{x_i^{tr}(\epsilon_j^k), \omega_i\}_{r \ge R} \tag{4}$$

where $R \le N$, with $R \in \mathbb{N}^+$. Therefore, $\hat{Z}^{tr}$ collects only those training samples satisfying two conditions: they are the most similar to the test instance and they most frequently occur among the top-k lists of most similar training samples.

In summary, RtIS determines the reduced training set associated to a test sample, i.e., $\hat{Z}^{tr}$, in two steps. First, in the linear space associated to the j-th descriptor it looks for the k training samples closest to the test instances. To the best of our knowledge, this is one of the differences between our proposal and what reported in the literature: exiting works consider the sample representation in the whole feature space, while we work with one feature at a time. In the second step RtIS removes the training samples whose number of occurrences in the top-k lists of most similar training samples on each feature are lower than the threshold R. Therefore, the other novelty of our approach is the application of the "bag-of-samples" concept that, using the frequency of occurrence of each training sample in the set of top-k similar instances, enables us to determine the final training set.

4   Experimental Setup

In this section we describe the datasets, the classifiers, the performance metrics and the configuration setups of the tests.

We use the seven public medical and biological datasets summarized in Table 1, which have a different number of samples, features, classes and sample distribution among classes [1,2].

Table 1. Summary of the datasets used

Dataset	Number of samples	Number of classes	Majority class	Minority class	Number of features
ABALONE	4138	18	16.6%	0.34%	8
CARDIO	2126	3	77.9%	8,3 %	22
FER	876	6	28.1%	7.5%	50
GLASS	205	5	37.0%	6.3%	9
NTHYR	215	3	69.7%	13.95%	5
THYR	720	3	92.5%	2.4%	21
WINE	178	3	39.9%	27.0%	13

On these datasets we apply three different classifiers, namely a Support Vector Machine (SVM) as a kernel machine, a k-Nearest Neighbor (kNN) as a non-parametric classifier, and a Naïve Bayes (NB) as a statistical classifier. We use the default values of classifier parameters included in the Weka software package [7], as suggested in [5]. In case of SVM with the intersection kernel we use $C = 1$, whereas for kNN we use $k = 1$ and the euclidean metric. Although we acknowledge that the tuning of the parameters for each method on each particular problem could lead to better results, we prefer maintaining a baseline performance of each method as the basis for comparison. Since we are not comparing base classifiers among them, our hypothesis is that the methods that win on average on all problems would also win if a better setting was performed. Furthermore, in a framework where no method is tuned, winner method tends to correspond to the most robust, which is also a desirable characteristic.

The performances of these classifiers are estimated using the confusion matrix where it is possibile to derive the following measures. The first estimator is the accuracy. However, this metric is not sufficient to assess the performances when the dataset exhibits a skewness between classes, which happens in most of real datasets since it is well known that recognition performances on the minority class can be low although the accuracy is high [11,15]. To overcome this problem we estimate the recall per class, which measures the proportion of actual samples of a class which are correctly identified.

Another relevant metric is the precision, which is the number of correct results divided by the number of all returned results. For each class, these two metrics are then summarized by the f-measure.

For each dataset, the experimental protocol consists of training one of the aforementioned classifiers using the whole feature spaces and all samples. The corresponding performances achieved are referred to as *baseline* in the following. On the same dataset we apply the RtIS method performing grid-based tests, where the values of the grid are the parameters of our method, i.e., k and R (see section 3). Given the number N of training instances, the array of k values were computed applying the following formula for different values of m, with $m \in \mathbb{N}_0$:

$$k = \begin{cases} \lfloor 0.75 \cdot N \rceil & if\ m = 0 \\ \lfloor \frac{N}{2^m} \rceil & if\ m \geq 1 \wedge \lfloor \frac{N}{2^m} \rceil > 10 \\ 10 & if\ \lfloor \frac{N}{2^m} \rceil \leq 10 \end{cases} \tag{5}$$

where $\lfloor \cdot \rceil$ denotes the round operator. For R we use the integer values in [1;5].

We are also interested in analyzing RtIS performances when the number of features describing the samples vary. For this reason, we ran the experimental tests also as follows. Using the Principal Component Analysis (PCA) we find out the principal components and their variances. Then, we eliminated those principal components that contribute less than 5% to the total variation in each dataset so that we derived a features space with lower dimensions. Finally, in this reduced feature space we repeated the aforementioned experimental procedure consisting of computing the baseline performances and running the grid tests.

5 Results

The results of the experiments validating the RtIS method were carried out using the ten fold cross validation and then averaging out the results. As presented in section 4, we use three different classification schemes, seven datasets and two different test configurations, i.e., with and without PCA, resulting in more than 700 independent experiments. We summarize these results in Table 2, which is composed of three panels showing the accuracy, the average recall per class and the average f-measure per class, respectively. Hereinafter, for brevity, the average recall per class and the average f-measure per class are referred to as recall and f-measure. Rows reporting results of each classifier are divided into two groups marked as "noPCA" and "PCA". The former reports the results achieved when we maintain the dimensionality of the original feature space, whereas the latter presents the results attained when we reduce the dimensionally of the feature space by selecting a subset of the principal components. In each group we report the following results: i) *baseline*: the results obtained when the classification algorithm is trained using all training instances; ii) δ: the largest performance improvement achieved by RtIS in comparison with the baseline; iii) R-m: the pair of RtIS parameters used to achieve the largest performance. The tabular "-" means that RtIS does not improve the baseline performances. The last column of the table reports the average performances among the datasets.

Focusing on the results measured in terms of accuracy, we observe that the proposed method improves the classification performance in the 76% of the cases, whereas in terms of recall and f-measure the improvements occur in the 71% and 78% of cases, respectively. The largest improvements are achieved by the SVM+PCA scheme on the FER dataset where, in terms of accuracy, recall and f-measure the values of δ are equal to 29.37%, 35.38% and 35.18%, respectively.

Table 2. Results reported in terms of accuracy (%) (top panel), average recall per class % (middle panel), and average f-measure per class (%) (bottom panel).

| | | | | Datasets | | | | | | | |
				ABALONE	CARDIO	FER	GLASS	NTHYR	THYR	WINE	μ
Accuracy	SVM	noPCA	Baseline	27.10	85.23	89.69	73.77	99.07	92.51	92.21	79.94
			δ	0.66	3.57	7.45	5.56	-	0.55	3.41	3.03
			R-m	1-4	4-3	3-6	5-2	-	1-6	5-3	
		PCA	Baseline	26.40	69.86	61.38	72.29	96.26	93.34	96.10	73.66
			δ	0.12	3.44	29.37	1.60	0.50	0.14	1.61	5.25
			R-m	1-2	2-3	4-2	3-2	2-3	3-2	1-5	
	kNN	noPCA	Baseline	21.03	76.68	97.82	73.69	93.96	89.85	75.97	75.57
			δ	0.22	0.09	-	2.22	1.84	1.95	19.09	3.63
			R-m	3-4	5-3	-	3-4	2-5	4-4	5-3	
		PCA	Baseline	20.47	71.78	94.63	68.57	96.73	89.85	94.34	76.62
			δ	-	-	-	1.23	0.50	0.14	1.12	0.43
			R-m	-	-	-	4-2	1-5	2-4	2-3	
	NB	noPCA	Baseline	6.00	20.09	39.59	49.48	96.73	95.13	97.22	57.75
			δ	3.74	4.70	0.34	3.09	-	-	-	1.70
			R-m	4-1	2-8	2-1	5-3	-	-	-	
		PCA	Baseline	7.47	23.93	45.89	50.63	93.98	76.42	97.80	56.59
			δ	5.39	14.03	14.83	2.52	-	15.95	-	7.53
			R-m	1-0	4-3	4-3	1-0	-	3-4	-	
Average recall per class	SVM	noPCA	Baseline	16.27	82.97	88.30	64.11	98.67	34.95	92.83	68.30
			δ	0.54	2.93	8.49	7.42	-	8.33	3.21	4.42
			R-m	4-5	4-3	3-6	5-2	-	1-6	5-3	
		PCA	Baseline	15.91	60.18	53.67	66.08	92.89	55.94	96.32	63.00
			δ	0.05	2.72	35.38	0.81	1.11	1.67	1.59	6.19
			R-m	1-1	2-3	4-2	2-5	1-5	3-2	1-5	
	kNN	noPCA	Baseline	15.74	72.40	97.72	72.67	90.11	37.41	75.51	65.94
			δ	0.49	0.01	-	0.63	3.11	9.47	20.05	4.82
			R-m	3-4	5-3	-	3-3	2-3	2-6	5-3	
		PCA	Baseline	14.57	66.24	93.43	64.08	96.06	51.68	95.05	68.73
			δ	-	0.21	-	2.34	0.22	-	0.87	0.52
			R-m	-	2-4	-	2-5	2-5	-	2-3	
	NB	noPCA	Baseline	5.04	8.23	37.64	48.31	93.11	66.52	97.54	50.91
			δ	2.39	5.63	0.38	0.63	-	-	-	1.29
			R-m	5-1	2-8	1-1	2-0	-	-	-	
		PCA	Baseline	6.15	10.19	35.73	47.17	88.50	46.38	97.83	47.42
			δ	1.67	21.02	20.98	-	-	-	-	6.24
			R-m	1-5	4-3	4-3	-	2-4	-	-	
Average f-measure per class	SVM	noPCA	Baseline	15.03	83.40	88.29	61.13	98.63	34.26	92.47	67.60
			δ	1.73	2.94	8.67	8.96	-	10.09	3.33	5.10
			R-m	4-5	5-3	5-5	5-2	-	1-6	5-3	
		PCA	Baseline	14.67	62.42	53.89	63.60	93.85	56.05	96.18	62.95
			δ	0.07	3.02	35.18	1.24	0.79	1.20	1.52	6.15
			R-m	1-1	2-3	4-2	2-5	2-3	3-2	1-5	
	kNN	noPCA	Baseline	15.61	73.39	97.71	69.99	91.38	37.39	74.64	65.73
			δ	0.51	0.05	-	1.10	2.54	8.27	20.56	4.72
			R-m	3-4	5-3	-	3-3	2-3	2-6	5-3	
		PCA	Baseline	14.56	66.40	93.25	62.29	95.51	49.28	94.25	67.93
			δ	-	0.21	0.08	1.15	0.59	0.49	1.25	0.54
			R-m	-	2-4	2-3	2-5	2-5	1-6	2-3	
	NB	noPCA	Baseline	2.37	9.54	34.56	43.62	94.44	67.78	97.29	49.94
			δ	2.38	1.69	0.89	5.78	-	-	-	1.53
			R-m	1-6	4-2	2-1	4-4	-	-	-	
		PCA	Baseline	2.42	10.05	34.21	43.67	89.62	38.56	97.85	45.20
			δ	1.94	25.20	28.02	1.08	-	-	-	8.04
			R-m	1-5	4-3	4-3	4-3	-	-	-	

Average results reported in the last column of Table 2 show that the proposed method improves the performances of classification algorithms trained on a set containing all available instances. Indeed, accuracy improvements achieved by RtIS range in [0.43%; 7.53%], whereas improvements in terms of recall and f-measure range in [0.52%; 6.19%] and [0.54%; 8.04%], respectively. To quantitatively asses the improvements provided by RtIS, we focus on the SVM+noPCA scheme, which shows the largest average performances. When used as baseline, it achieves values of accuracy, recall and f-measure equal to 79.94%, 68.30%, and 67.60%, respectively. When this classifier is trained with a reduced set given by RtIS method these values become 82.97%, 72.72%, and 72.70%, respectively. Furthermore, last column suggests that our method provides larger classification performances when the feature space dimensionality has been preliminary reduced by the PCA. Indeed in the 66% of cases, the values of δ are larger when the PCA is employed.

We now report how the values of k and R affect the classification performances (Fig. 1). Each row of the figure shows results achieved by a classification scheme in terms of accuracy (left), recall (middle) and f-measure (right). Each plot reports the differences between the results achieved by a classifier trained on a reduced training set computed by RtIS and a classifier trained on the original training set (i.e. the baseline), where these values of these differences were normalized in [-1;1] and averaged out among the datasets. The x and y-axes shows the values of m (which determines the value of k according to equation 5) and R, respectively. In this figure, the pairs *color - value* ranges for performance differences are: purple [0.5;1], green [0;0.5], red [-0.5;0] and blue [-1;-0.5].

The top-right corners of all plots show that the application of the RtIS method with the largest values of parameters reduces the classification performances. This result is expected since large values of m and R significantly reduce the number of training samples, thus resulting in low generalization ability of the classifier. This happens because when m increases the number of top-k neighbors decreases; furthermore, when R is large, a training instance to be selected has to occur at several times in the neighborhood of the test sample.

Focus now the attention on results when $R = 1$, i.e., when all top-k samples are included in the reduced training set. When m increases, RtIS reduces the number of samples retained for training, thus removing noisy and unreliable sample. Most of the charts show that RtIS favorably compares with the baseline classifier.

When R increases the number of disregarded samples increases as well. Although in some cases the performances decrease, in others RtIS continues to outperform the baseline approach. While this figure displays average results computed over all datasets, interestingly Table 2 shows that best RtIS results are achieved also using a large value of R, which corresponds to remove several training samples.

When the feature space dimensionality is reduced by the PCA, we note that best results are achieved with lower values of the pair (m, R): this result could be expected since the PCA reduces the information stored in the datasets, thus requiring to select more samples for the training set. This also explains why the charts of tests where we ran the PCA show smaller purple areas than those without the PCA.

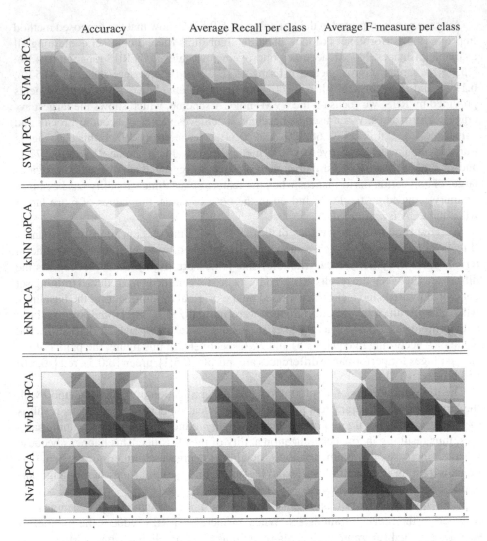

Fig. 1. Each plot reports the difference between the results achieved by a classifier trained on a reduced training set computed by RtIS and a classifier trained on the original training set, where these values of these differences were normalized in [-1;1] and averaged out among the datasets. The x and y-axes shows the values of m (which determines the value of k) and R, respectively. The pairs *color - value range* are: purple [0.5;1], green [0;0.5], red [-0.5;0] and blue [-1;-0.5]. The charts in the same row show the results of a classification scheme, whereas charts in the same column report the results estimated using the same performance metric.

Furthermore, looking at the last column of Table 2, and comparing the performances of our proposal with the baseline ones, we observe larger values of δ when using the PCA rather than when we keep the original feature space. This result can be explained considering two issues. First, the application of the PCA averagely reduces the performances of the baseline classifier. Second, the application of the PCA to the RtIS method does not significantly reduces its performances in comparison to the RtIS method applied to the original feature space. This suggests us that RtIS method successfully remove training samples that noisy and unreliable with respect to the test instance at hand.

6 Conclusion

The large amount of biomedical data nowadays available, although collected from quality sources, may contain irrelevant or redundant information, or they might be noisy and unreliable. To improve the classification accuracy, we propose an instance selection method that, different from the existing approaches, selects a subset of the examples from the original training set on the basis of the information derived from the test sample at hand. We compared its performances with a classifier trained on the whole training set. Future works are directed towards applying our proposal to medical data stream classification.

Acknowledgement. This work was partially supported by the National Natural Science Foundation of China under Grants no. 61300215 and 61272545, the Key Scientific and Technological Project of Henan Province under Grants no. 122102210053 and 132102210188, Henan University under Grants no. 2013YBZR014 and 0000A40442, and China-Italy Scientific and Technological Cooperation Programme for the years 2013-2015 under Grant no. CN13MO4.

References

1. Asuncion, A., Newman, D.J.: UCI machine learning repository (2007)
2. D'Ambrosio, R., Iannello, G., Soda, P.: Automatic facial expression recognition using statistical-like moments. In: Maino, G., Foresti, G.L. (eds.) ICIAP 2011, Part I. LNCS, vol. 6978, pp. 585–594. Springer, Heidelberg (2011)
3. Fernández, A., Duarte, A., Hernández, R., Sánchez, Á.: GRASP for instance selection in medical data sets. In: Rocha, M.P., Riverola, F.F., Shatkay, H., Corchado, J.M. (eds.) IW-PACBB 2010. AISC, vol. 74, pp. 53–60. Springer, Heidelberg (2010)
4. Fung, G., Mangasarian, O.L.: Data selection for support vector machine classifiers. In: ACM SIGKDD, pp. 64–70 (2000)
5. Galar, M., Fernández, A., Barrenechea, E., Bustince, H., Herrera, F.: An overview of ensemble methods for binary classifiers in multi-class problems: Experimental study on one-vs-one and one-vs-all schemes. Pattern Recognition 44(8), 1761–1776 (2011)
6. Garcia, S., Derrac, J., Cano, J.R., Herrera, F.: Prototype selection for nearest neighbor classification: Taxonomy and empirical study. IEEE Transactions on PAMI 34(3), 417–435 (2012)
7. Hall, M., Frank, E., Holmes, G., Pfahringer, B., Reutemann, P., Witten, I.H.: The WEKA data mining software: an update. ACM SIGKDD Explorations Newsletter 11(1), 10–18 (2009)
8. Kim, S., Oommen, B.J.: A brief taxonomy and ranking of creative prototype reduction schemes. Pattern Analysis & Applications 6(3), 232–244 (2003)
9. Kuncheva, L.I.: Editing for the k-nearest neighbors rule by a genetic algorithm. Pattern Recognition Letters 16, 809–814 (1995)
10. Liu, H., Motoda, H.: On issues of instance selection. Data Mining and Knowledge Discovery 6(2), 115–130 (2002)

11. Soda, P.: A multi-objective optimisation approach for class-imbalance learning. Pattern Recognition 44, 1801–1810 (2011)
12. Sordo, M., Zeng, Q.: On sample size and classification accuracy: A performance comparison. In: Oliveira, J.L., Maojo, V., Martín-Sánchez, F., Pereira, A.S. (eds.) ISBMDA 2005. LNCS (LNBI), vol. 3745, pp. 193–201. Springer, Heidelberg (2005)
13. Wilson, D.R., Martinez, T.R.: Reduction techniques for instance-based learning algorithms. Machine Learning 38(3), 257–286 (2000)
14. Zhang, C., D'Ambrosio, R., Soda, P.: Real-time biomedical instance selection. In: The 27th IEEE International Symposium on Computer-Based Medical Systems, CBMS 2014 (2014)
15. Zhang, C., Soda, P.: A double-ensemble approach for classifying skewed data streams. In: Tan, P.-N., Chawla, S., Ho, C.K., Bailey, J. (eds.) PAKDD 2012, Part I. LNCS, vol. 7301, pp. 254–265. Springer, Heidelberg (2012)
16. Zhu, X., Wu, X.: Scalable representative instance selection and ranking. In: Proceedings of the 18th International Conference on Pattern Recognition, pp. 352–355 (2006)

Mining Churning Factors in Indian Telecommunication Sector Using Social Media Analytics

Nitish Varshney and S.K. Gupta

Department of Computer Science and Engineering,
Indian Institute of Technology Delhi, India
{nitish.mcs12,skg}@cse.iitd.ac.in

Abstract. In this paper we address the problem of churning in the telecommunication sector in Indian context. Churning becomes a challenging problem for telecom industries especially when the subscriber base almost reaches saturation level. It directly affect the revenue of the telecom companies. A proper analysis of factors affecting churning can help the telecom service providers to reduce churning, satisfy their customers and may be design new products to reduce churning. We use social media analytics, in particular twitter feeds, to get opinion of the users. The main contribution of the paper is feasibility of data mining tools, in particular association rules, to determine factors affecting churning.

Keywords: Sentiment analysis, Social data analysis, Data mining applications: telecommunication, Churn pertaining factors.

1 Introduction

Mobile service usage in India has increased rapidly following the reduction in call cost and emerging use of new mobile phone technologies. Currently India's telecommunication network is the second largest in the world in terms of total number of users (both fixed and mobile phone) [14] and it has one of the lowest call tariffs enabled by the mega telephone network operators and hyper-competition among them. On 30th September, 2013, country's telecom subscriber's base was as huge as 899.86 million [13] and penetration rate was about 71%. Out of these 899.86 million total telecom subscribers, about 97 % utilize wireless services. Due to such large number, our focus is on wireless telecom services and mobile service providers as they are predominant in numbers.

There are about 15 mobile carriers in India. Most of them are quite stable and have pan India presence. It is estimated that about 96 % of all mobile subscribers opt for a prepaid service and there is a fiercely competitive dynamic environment leading to luring customers from one service provider to another.

Above situation depicts a condition where market is almost saturated and telecom service providers are stable. It leads to intensification of competition among existing mobile service providers in order to maintain their subscriber base. In such a situation, the significant business drivers would be customer subscriber base retention and increase in average revenue per customer [5].

L. Bellatreche and M.K. Mohania (Eds.): DaWaK 2014, LNCS 8646, pp. 405–413, 2014.
© Springer International Publishing Switzerland 2014

There is a trade off in between these business drivers. Customer retention depends on factors like call rate and quality of services provided by a service provider. A superior quality of service imposes heavy implementation cost which has to be passed on to the customer. This has twin fall back of either higher call rate charges or subscriber churning. Hence, telecom service providers would want optimal values for both.

However, in Indian market most of the devices have multi-SIM card capability and it is easier to switch the service provider by getting a new SIM or using Mobile Number Portability (MNP). MNP enabled subscribers to retain their telephone numbers when switching from one service provider to another. It made it difficult for service providers to retain customers.

In telecommunication, customer movement from one service provider to other service provider is termed as churning. Churn rate is the percentage of subscribers who discontinue services with a service provider and change their service provider as per their choice. Customer churning is of great concern for any service provider. However, according to statistical information provided by Telecom Regulatory Authority of India (TRAI) already 100+ million users have utilized mobile number portability service [13]. This is relatively very high, specially when the aim is to retain the existing customers. Companies need to fully understand the factors leading to customer churn. These problems have not been fully addressed in the literature. In this paper, we present a data mining based approach to determine factors affecting churn in the Indian telecom sector.

2 Related Work

Existing literature on churn factors can be classified into two categories. Both categories attempt to predict customers getting ready to switch, understand why and connect with them to offer incentives to mitigate churn [9,15].

The first approach is based on large-scale actual customer transaction and billing data. This is proprietary data of a service provider. Such studies use various machine learning techniques like Support Vector Machine [2], Regression [6], Decision Trees [6,9] etc. They rely on the rich subscriber call data records which are available inhouse. Such dataset describe calling behavior of a customer by providing information such as their voicemail plan, call lengths and usage patterns. This leads to determination of patterns like:
- Patterns followed by churning customers.
  - If I am calling more than X minutes, then I will churn.
  - If I am calling to customer care more than Y times, then I will churn.
  - If my in net call duration is low, then I will churn.
- Potential value of a customer.

The second approach avoids the proprietary nature of actual customer call record data, and deals with consumer survey data [7,11] avoiding privacy issues. These consider consumers perceptions of service experiences and intention to churn. However, the survey data may not fully represent the customers actual future continued patronage decision.

The above two approaches focus on predictive accuracy rather than descriptive explanation or reasons thereof. Also such studies can not be directly useful in making a decision support system. To illustrate it, suppose if Vodafone reduces 3G usage charges to 2paisa/10KB from 10paisa/10KB, then should Idea also reduce their 3G usage charges? Here mining the impact of not reducing the charges is one of the most important parameters in decision making. The above approaches would not provide any information about such queries. So there were problems with such works as they can not pinpoint driving forces and used to measure their impact. We therefore focus our attention to know what went wrong and what would be the impact.

This paper has two distinct research objectives compared with the previous approaches. The first objective is to identify factors pertaining to churn which may help decision makers improve operations in terms of their marketing strategy, specifically customer churn prevention programs. The second objective is to develop a comprehensive model which can help in taking business decisions. Specifically, this research uses data mining techniques to find a model to achieve above two objectives.

3 Methodology

Figure 1 shows methodology applied to find dominant churning factor.

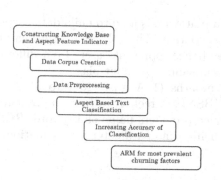

Fig. 1. Methodology applied to find dominant churning factor

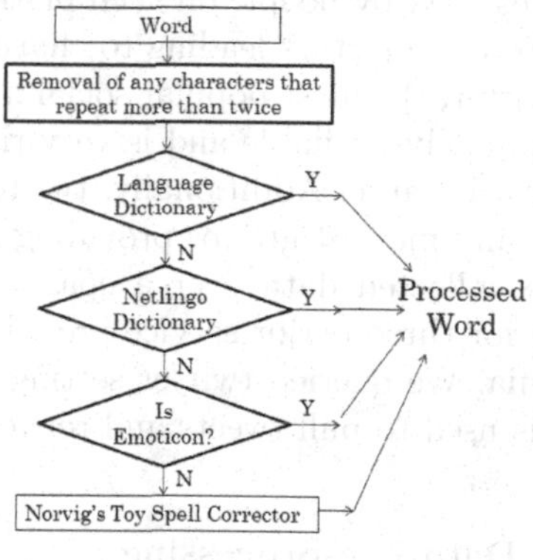

Fig. 2. NLP based spell corrector which takes a word as input and tries to correct spell errors if any.

3.1 Constructing Knowledge Base and Aspects Feature Indicators

An aspect (target) based extraction model has been developed, which looks for certain words in tweets using lexicon matching approach. Three major aspects (Price, Service and Satisfaction) in telecom business have been considered, which can influence a customer to churn [8]. Price aspect incorporates call rates and pricing options. Service aspect incorporates call quality, coverage, billing and customer service. Satisfaction aspect incorporates customers who have already churned. Every other churn indicators have been put up in miscellaneous aspect. Miscellaneous aspect also incorporates insulting effects on the reputation of the company. Each aspect has some feature indicators. For example, price aspect has rate, price, charge and tariff as feature indicators.

A knowledge base having unigram, bigram and trigram features are created manually. Bigram features are of the form string-string like tariff slashed, too-much interrupt. Trigram features are constructed to capture cases like ported to bsnl. In our approach, bigram and trigram features are not required to co-occur consecutively. Additionally, our approach partly utilizes a list of domain inde-pendent strongly positive and negative words provided by Hu and Liu [4], while building knowledge base. All words which are not strongly positive / negative in context of telecom business, are removed from the list. Few domain depen-dent strongly positive / negative word are also added to above list for example highspeed, slashes, flop etc. Features in the knowledge base are termed as senti-ment describing terms and aspect feature indicators are termed as topical terms. These features are manually identified.

3.2   Data Corpus Creation

People often share and exchange ideas on social media platform. Also, there are a lot of personal thoughts to public statements about telecommunication services used by people on such platforms. The key idea is to utilize these feeds to determine factors leading to churn.

Twitter, the most popular social media platform, is used to build data corpus. It is publicly available and is very rich in content. The collected corpus can be arbitrarily large. Additionally, twitter audience represents users from different social and interest groups providing a good sample without bias.

We collected data over a span of 9 months (1 August, 2012 to 31 April, 2013) for three major service providers (BSNL, Aircel, Tata Indicom/Docomo) in India. We queried twitter setting service providers name as keyword. REST API is used to pull tweets and retweets using time parameter for pagination.

3.3   Data Pre-processing

Tweets often have lots of grammatically misspelled words due to 140 character limit. We propose to apply lexicon based text categorization method on such feeds which require us to overcome such errors. However, most popular Norvigs toy spell corrector [10] cannot be applied in its raw form in our corpus as feeds often incorporate internet lingo, colloquial expressions and emoticons (as toy spell correctors base file big.txt rarely contains any such words). It leads us to build a NLP based spell corrector, as described in Figure 2. It first tries to remove any characters that repeat more than twice making a word cool to appear as coooool as people sometimes repeat characters for added emphasis. Second it matches words in tweet with an English dictionary. Then to handle netlingo words, slangs and abbreviations our model incorporated words taken

Tweet 1: aircel i *hate* u, no **wifi** since morning.
Tweet 2: idea and bsnl both are *shit* i am pretty sure about this need to **shift to** virgin.
Tweet 3: @yearning4d_sky airtel is also *crap* i **mnped from** airtel to bsnl one of the best decisions taken in telecom industry.
Tweet 4: bsnl *launches* new call **tariff** *reduction* stvs for **2g 3g** prepaid customers bsnl telecom.
Tweet 5: bsnl suck. no line never work uncouth staff.

Fig. 3. Sample Tweets

from Netlingo[1], NoSlang[2] and Webopedia[3] for matching. Further, emoticon list provided by DataGenetics[4] is used. If no match is found for the word then big.txt is used to calculate edit distance as explained by Norvig [10]. Still our approach is not able to capture phenomena such as sarcasm, irony, humor etc., but overall, data captured in such a manner is quite reasonable.

In addition to these, NLTK wordnet lemmatizer was used to lemmatize each word. Further to structure text and handle moderately-sized dataset, feeds are pushed into a MySQL RDBMS. We retrieved around 75K feeds in such a way.

3.4   Aspect Based Text Classification

The collected dataset is cross-checked against knowledge base to extract aspect and sentiment. In sentiment extraction, ternary classification task were specifically designed to classify a tweet x^i, to a class $y^i \in$ {NEGATIVE, NEUTRAL, POSITIVE}. Neutral class has been incorporated in the model to classify tweets that are not assigned a positive / negative class.

Classification task can be broadly broken into two phases. In the first phase, sentiment describing terms and topical terms are matched with words appearing in a tweet. To collect these sentiments, we followed the same procedure as described in Taboada et al. [12]. The second phase calculates overall sentiment score based on summation of individual lexicon sentiment retrieved in the first phase. Our model assumes that customers providing negative tweets would lead to churn in few months. Figure 3 shows some examples of tweets fetched, sentiment describing terms are *italicized* and topical terms are in **bold**. A tweet can have multiple aspects or it may have none like Figure 3, tweet 4 has multiple aspects tariff, 2g and 3g. Tweets having positive or negative polarity and having price, service, satisfaction or miscellaneous aspect are used, other tweets are ignored.

---

[1] http://www.netlingo.com
[2] http://www.noslang.com/spolling.php
[3] http://www.webopedia.com/quick_ref/textmessageabbreviations.asp
    retrieved on February 10th 2013
[4] http://datagenetics.com/blog/october52012/index.html

3.5   Increasing Accuracy of Classification

Aspect based text classifier is not able to properly assign sentiment to tweets having multiple service providers because inclusion of multiple service providers in a tweet presents semantic issues. For example, tweet 2 and tweet 3 in figure 3, would lead to misclassification due to <shit, virgin> and <crap, bsnl> matching as a bigram feature. To tackle the issue, the classifier has been modified to use a simple neighbourhood proximity based approach. Sentiment describing terms are attached to those aspect feature indicators that occur closest to them. Experiments show that utilizing proximity improved accuracy of our model from 69% to 79.5%.

Negating words are capable of reversing polarity of sentiment again. Experiments show that if negative context lies in proximity of antonym of sentiment term and topical term, it does not change polarity of a tweet. For example tweet1 and tweet 5 in figure 3, should be classified as negative due to presence of bigram features <hate, wifi>, <suck, staff>, <uncouth, staff>. However our classifier would end up classifying such tweets as positive due to presence of negative words <no>, <never>. In an experiment we have manually annotated 200 tweets having negative words in the proximity of sentiment word and topical term, 85.5% of them are found to be wrongly classified using aspect based text categorization approach. Therefore we modified our approach to classify a tweet negative even in presence of negating words.

3.6   Association Rule Mining for Most Prevalent Churning Factors

Association rule mining (ARM)[1] is one of the most common data mining techniques, which is used to discover rules among multiple independent elements that co-occur frequently. We found out that it can be used to mine out co-occurrence between set of aspects and sentiments. Previous steps are considered as preprocessing steps for mining association rule with maximum confidence within a time window, say month. Top most of all such rules can be used to denote most prevailing churning factor within that time frame.

Association rules mined in such a manner are implication of the form $A \rightarrow B$, where $A \subseteq I$, $B \subseteq I$, $A \cap B = \Theta$ and I be a set of literals called items. For our analysis, I = {positive, negative, price, service, satisfaction, miscellaneous}. Classification output obtained after applying aspect based text classification has been used to fill positive / negative item. Hence, in a tuple, either 'positive' or 'negative' item can have value 1. Aspect feature indicators obtained in the process are used to fill price, service, satisfaction, miscellaneous. In such a way, a binary table is obtained, similar to to transaction table generally used for ARM. Such a table for tweets in Figure 3 is shown in Table 1. We use Top-K Association Rule mining technique [3] as otherwise we are flooded with many association rules. In particular, we find that K=2 serves our purpose.

Top-K rules obtained after applying ARM algorithm on this dataset denote most pertaining churning factor, if {negative} $\subseteq A$ in association rule found. To illustrate it, suppose we obtain a rule negative $\rightarrow$ price in a time frame for

Table 1. Tweets in relational format for Binary Association Rule Mining

TID	Positive	Negative	Price	Service	Satisfaction	Miscellaneous
1	0	1	0	1	0	0
2	0	1	0	0	1	0
3	1	0	0	0	1	0
4	1	0	1	1	0	1
5	0	1	0	1	0	0

a service provider, it means within that time frame tariff charges of services provided by the service provider are not competitive and high in comparision to other service providers. Service provider needs to reduce tariff of the service otherwise customers are going to churn from it. All such rules obtained for Aircel service provider have been shown in table 2.

4   Experimental Evaluations

In this section we will be looking at the results obtained after applying above approach on data collected within a time frame. Table 2 shows top-k association rules mined, with k=2 and total customers for aircel service provider. We could interpret these results in the light of sentiments derived from tweets. Top-k association rules obtained are given in first row. Rules are presented in the form {A,B x} A→B with x confidence. Pos, Neg represents positive and negative sentiment respectively. Details of total customers of the service provider provided by

Table 2. Association Rules Mined for Aircel service provider with actual customers Aircel had during the period

	Aug, 2012	Sep, 2012	Oct, 2012	Nov, 2012	Dec, 2012
Top-K Rules	{Pos,Mis 54.1}, {Pos,Ser 36.67}	{Pos,Mis 47.4}, {Pos,Ser 34.54}	{Pos,Mis 48.5}, {Pos,Ser 38.37}	{Neg,Ser 38.2} ,{Pos,Mis 34.15}	{Neg,Ser 50.3}, {Pos,Mis 44.18}
Total Customers	65952244	66607361	66786295	65323317	63347284
Change in Customers	793717	655117	178934	**-1462978**	**-1976033**

	Jan, 2013	Feb, 2013	Mar, 2013	Apr, 2013	
Top-K Rules	{Neg,Ser 47.6}, {Pos,Ser 46.86}	{Neg,Ser 57.96}, {Neg,Sat 42.7}	{Pos,Mis 38.1}, {Pos,Ser 30.51}	{Pos,Mis 68}, {Neg,Sat 35.32}	
Total Customers	61571291	60872785	60071967	60080216	
Change in Customers	**-1775993**	**-698506**	**-800818**	**8249**	

TRAI [13] are shown in next line. Few interesting rules obtained are in **bold**. For example, Top-K rule for Aircel in Nov,2012 , Dec,2012 , Jan,2013 and Feb,2013 months have higher negative confidence, due to which during these months total customers of Aircel might be dropped. Hence, churning factors during these months are {Service}, {Service}, {Service, Satisfaction} and {Satisfaction}. Similar, results are obtained for other service providers.

5 Conclusion and Future Work

In this paper we presented a technique which attempts to mine churning factors in telecommunication sector of India using social media analytics. In the preprocessing stage telecom specific tweets are pulled, cleaned for misspelled words, stemming is performed and data is translated into relational format. Further tweets are classified into three categories using lexicon based classifier. Finally ARM is applied to find the dominant churn factor out of a selected few factors as determined by domain expert. Results obtained are helpful in interpreting the customer satisfaction and also knowing the reason of customer dissatisfaction.

The results can be improved further by considering availability of inhouse data and performing deeper analysis of tweets for genuineness.

References

1. Agrawal, R., Imieliski, T., Swami, A.: Mining association rules between sets of items in large databases. ACM SIGMOD Record 22(2), 207–216 (1993)
2. Cheung, K.W., Kwok, J.T., Law, M.H., Tsui, K.C.: Mining customer product ratings for personalized marketing. Decision Support Systems 35, 231–243 (2003)
3. Fournier-Viger, P., Wu, C.-W., Tseng, V.S.: Mining top-k association rules. In: Kosseim, L., Inkpen, D. (eds.) Canadian AI 2012. LNCS, vol. 7310, pp. 61–73. Springer, Heidelberg (2012)
4. Hu, M., Liu, B.: Mining and summarizing customer reviews. In: Proceedings of the Tenth ACM SIGKDD International Conference on Knowledge Discovery and Data Mining. ACM (2004)
5. Hung, S.Y., Yen, D.C., Wang, H.Y.: Applying data mining to telecom churn management. Expert Systems with Applications 31(3), 515–524 (2006)
6. Hwang, H., Jung, T., Suh, E.: An LTV model and customer segmentation based on customer value: a case study on the wireless telecommunication industry. Expert Systems with Applications 26(2), 181–188 (2004)
7. Keaveney, S.M.: Customer switching behavior in service industries: An exploratory study. Journal of Marketing 59(2), 71–82 (1995)
8. Kim, H.S., Yoon, C.H.: Determinants of subscriber churn and customer loyalty in the Korean mobile telephony market. Telecommunications Policy 28(9/10), 751–765 (2004)
9. Kim, S.Y., Jung, T.S., Suh, E.H., Hwang, H.S.: Customer segmentation and strategy development based on customer lifetime value: A case study. Expert Systems with Applications 31, 101–107 (2006)
10. Norvig, P.: How to write a spelling corrector, http://norvig.com/spell-correct.html (visited February 8, 2013)

11. Oghojafor, B., et al.: Discriminant Analysis of Factors Affecting Telecoms Customer Churn. International Journal of Business Administration 3(2) (2012)
12. Taboada, M., et al.: Lexicon-based methods for sentiment analysis. Computational Linguistics 37(2), 267–307 (2011)
13. Telecom Regulatory Authority of India, Telecom Subscription Data as on 30th September, Press Release No. 78/2013
14. Telecommunications in India, In Wikipedia, http://en.wikipedia.org/wiki/Telecommunications_in_India (retrieved January 24, 2014)
15. Wei, C.P., Chiu, I.T.: Turning telecommunications call details to churn prediction: A data mining approach. Expert Systems with Applications 23(2), 103–112 (2002)

Drift Detector for Memory-Constrained Environments

Timothy D. Robinson, David Tse Jung Huang, Yun Sing Koh, and Gillian Dobbie

Dept. of Computer Science, The University of Auckland, New Zealand
trob807@aucklanduni.ac.nz, {dtjh,ykoh,gill}@cs.auckland.ac.nz

Abstract. Current approaches to drift detection assume that stable memory consumption with slight variations with each stream is suitable for all programs. This is not always the case and there are situations where small variations in memory are undesirable such as drift detectors on medical vital sign monitoring systems. Under these circumstances, it is not sufficient to have a memory use that is predictable on average, but instead memory use must be fixed. To detect drift using fixed memory in a stream, we propose DualWin: a technique that keeps two samples of controllable size, one is stored in a sliding window, which represents the most recent stream elements, and the other is stored in a reservoir, which uses reservoir sampling to maintain an image of the stream since the previous drift was detected. Through experimentation, we find that DualWin obtains a rate of true positive detection which is comparable to ADWIN2, a rate of false positive detection which is much lower, an execution time which is faster, and a fixed memory consumption.

Keywords: Data Streams, Drift Detection, Fixed Memory, Reservoir Sampling.

1 Introduction

With the proliferation of handheld devices and small sensors equipped with high tech-specs, we can now perform tasks that previously required high performance systems on these small devices. Data streams in and/or produced on such small devices can serve a number of important applications in areas such as medical, stock market analysis and social networks. This requires the processing of data streams to be performed locally on-board small devices with low computational power, which are also known as resource-constrained environments.

In many applications, learning algorithms act in dynamic environments where the data flows continuously. If the process is not strictly stationary, which is the norm in most real-world applications, the target concept could change over time. Detecting concept changes, such as a change of customer preference for telecom services, is very important in terms of prediction and decision applications. A concept change, also known as concept drift, in the distribution of the stream will be detected if there is a sufficiently large difference in the means of two windows or subwindows. Drift can have different gradients over which the change in distribution occurs. When analyzing two stream samples for concept drift, one sample will be weighted towards representing older elements in the stream, and the other weighted towards newer elements. When drift occurs, one sample will represent the most recent distribution, and the other will represent the previous distribution.

L. Bellatreche and M.K. Mohania (Eds.): DaWaK 2014, LNCS 8646, pp. 414–425, 2014.

Some current strategies for drift detection often implement a variation of the *sliding window*. A sliding window is a window which stores new elements arriving from the stream and drops older elements according to some rule. Traditionally the window is implemented with a fixed size. Typically we require a drift detector that provides good accuracy whilst balancing between other important performance measures such as false positives, delay, and memory use.

We propose a new method for drift detection in data streams, DualWin, which uses a fixed, controllable amount of memory while remaining comparable in performance measures to the current state-of-the-art technique, ADWIN2 [2] . Through our experiments we show that DualWin has comparable accuracy (detection rate, false negative rate) to ADWIN2, a faster execution time, and a lower rate of false positives. DualWin has a slightly higher but competitive delay. Our method is suitable for situations where memory is limited, or situations where it is vital that the program uses a fixed amount of memory. In our technique, we keep two samples of the stream: one a sliding window which reflects the most recent stream items to arrive, and the other a reservoir which represents the stream since the previous drift was detected. Every new element from the stream is added to the sliding window. The oldest element in the sliding window is removed and replaced with each new element, so the window 'slides' across the most recently seen elements from the stream. The removed elements are inserted into the reservoir based on a probability that decreases over time. This technique of reservoir sampling [10] ensures the reservoir is representative of the stream since the previous drift was detected. Our technique will signal a drift if there is a large enough difference in the mean of the window and the reservoir.

An example where DualWin is useful is in an in-home health monitoring system, such as a machine that monitors vital signs such as the heart rate or the amount of a chemical substance in the blood. In these applications it is important that a drift can be detected in a timely fashion to signal an emergency alert if change has occurred. These machines need to be small, cheap, or widely produced, which would limit the memory available to the program. Small uncertainty in memory use, especially when used over many iterations of a possibly infinite stream, may exceed the total memory available. Drift detectors that run in these situations are therefore required to (1) use a fixed memory size; (2) have a fast execution time; (3) have a small delay; and (4) maintain high accuracy. Constraints (2 - 4) have mostly been addressed in previous research; in this paper we focus on constraint (1).

The structure of the paper is as follows. In the next section we review related work in the area of drift detection. Section 3 gives a formal definition of the problem undertaken in this research. In Section 4 we outline our approach and describe its details. Section 5 presents the experimental evaluation of our work against ADWIN2 and Section 6 concludes the paper.

2   Related Work

The concept change detection problem has a classic statistical interpretation: given a sample of data, we would like to determine whether the sample represent a single homogeneous distribution or whether at some point in the data (i.e. the concept drift point) that data distribution has undergone a significant shift. Current concept drift detection approaches in the literature formulate the problem from this viewpoint but the models and the algorithms used to solve this problem differ greatly in their detail. Current methods used fall into four basic categories [9]: Statistical Process Control (SPC) [3,5], Adaptive Windowing [2], Fixed Cumulative Windowing Schemes [6] and finally other statistical change detection methods such as the Page-Hinkley test [8], Martingale frameworks [4], kernel density methods [1], and support vector machines [7].

ADWIN2 [2] is an adaptive windowing scheme which is based on the use of the Hoeffding bound to detect concept change. The ADWIN2 algorithm was shown to outperform the SPC approach and has the attractive property of providing rigorous guarantees on false positive and false negative rates. The initial version, called ADWIN0, maintains a window (W) of instances at a given time and compares the mean difference of any two sub windows (W_0 of older instances and W_1 of recent instances) from W. If the mean difference is statistically significant, then ADWIN0 removes all instances of W_0 considered to represent the old concept and only carries W_1 forward to the next test. ADWIN used a variation of exponential histograms and a memory parameter to limit the number of hypothesis tests done on a given window. ADWIN2 was shown to be superior to Gama's method and fixed size window with flushing [6] on almost all performance measures such as the false positive rate, false negative rate and sensitivity to slow gradual changes [2]. Due to these reasons we chose to compare our technique against ADWIN2.

3   Preliminaries

Let us describe a data stream and drift detection more formally. Let there be a (possibly infinite) stream of real values $S = \{x_1, x_2, x_3, ..., x_t\}$. The value of x_t is only avaliable at time t; if x_t is not stored at time t, it cannot be retrieved past time t. Each x_t is generated according to some distribution D_t, independently for every t. We assume x_t is always in [0, 1].

Let $S_1 = (x_1, x_2, ..., x_m)$ and $S_2 = (x_{m+1}, ..., x_n)$, with $0 < m < n$, represent two samples from the stream with means μ_1 and μ_2 respectively. The problem of drift detection can be expressed as testing whether the samples are from the same distribution, where the null hypothesis H_0 that $\mu_1 = \mu_2$, versus samples from different distributions, where the alternate hypothesis, H_1, that $\mu_1 \neq \mu_2$. If H_0 is accepted when a change has taken place, then a false negative is said to have occured, and the drift detector has failed to detect a drift. If H_1 is accepted when no change has taken place, then a false positive is said to have occurred, and the drift detector has detected a drift that has not taken place.

To determine if the sample means μ_1 and μ_2 are from the same distribution we formally describe H_1 as:

$$Pr((|\mu_1 - \mu_2|) \geq \epsilon_{cut}) > \delta$$

where δ lies on the interval (0,1) and is a parameter that controls the maximum false positive rate allowed, and ϵ_{cut} is a function of δ and is the test statistic used to model the difference between sample means. A drift is signalled when the difference in sample means is greater than the ϵ_{cut} threshold.

4 Dual Window Drift Detection

Our technique keeps two samples from the stream: a sliding window, W, which contains the most recent elements to arrive from the stream, and a reservoir, R, which represents the stream since the previous drift was detected. At specific check point intervals we compare the distribution between the samples in W to R. If $|\mu_W - \mu_R| > \epsilon_{cut}$ then we say that a drift has occurred. In Section 4.1 and Section 4.2 we discuss the implementation details of the sliding window and reservoir.

4.1 Sliding Window

Intuitively the sliding window reflects the most recently seen stream elements by the drift detector, and describes the current state of the stream. Our algorithm keeps sliding window W and stores the most recently read x_t. Here W has a fixed size n_W and is implemented using a circular array. Once full, each new element added will replace the oldest element in W; so at any time t, the oldest element in W is $x_{(t-n_W)}$. For example, assume pointer | dictates the oldest element in the circular array (older elements are on the right):

$$101101011|000100$$

If a new bit, 1, enters the drift detector and is to be stored in W.

$$1011010111|00100$$

1 is stored in place of the oldest element, 0. The pointer moves one place to the right.

We chose to use a sliding window to represent the current state of the stream, as it allows us to maintain the time sequence of when the items appeared. This is crucial if we want to detect drift with low delay. The total memory cost of the circular array can be calculated to be the size of the window n_W multiplied by 8 bytes.

4.2 Reservoir

The second component is a reservoir R, with size n_R. The reservoir is designed to reflect the stream since the previous drift was detected, and uses random reservoir sampling to maintain a uniformly chosen sample of elements [10]. Reservoir sampling adds to the reservoir with a decreasing probability (to reflect an increase in stream size seen so far) and removes elements from the reservoir at random. Each element in the stream has comparable probability of contributing to the sample the reservoir holds.

With each add to our sliding window W, the oldest element $x_{(t-n_W)}$ is removed. This is then added to the reservoir with the probability $P(insert) = \frac{1}{c}$. c is a natural number that starts at one and is incremented by one every time an insertion to the reservoir takes place. For each insertion, the reservoir first removes an element at random. This technique ensures that neither elements that arrive earlier or later in the stream have an advantage chance of contributing to the reservoir; the reservoir will reflect the since the previous drift was detected. As the removal rule of R is random (and not a queue like our sliding window), it can be implemented simply as a single real integer

r, where r is the sum of every element in R that has a value of 1. To remove from R, it must first be decided if the bit is a 1 or a 0; there is a $\frac{r}{n_R}$ chance it will be a 1, else it will be a 0. To insert an element that is 1 in R, replacing a 0 bit, it is the simple case of increasing r by 1; whereas to insert an element that is 0 in R, replacing a 1 bit, it is the simple case of decreasing r by 1. This implementation reduces reservoir memory use to a relatively negligible amount, and allows for a large n_R value and subsequently, greater representation of the stream as a whole.

Since the reservoir uses constant memory, the only varying factor for memory use is the size of the sliding window. Given a device of a fixed memory size, we can calculate the suitable size of the sliding window based on the following formula:

$$n_{W(calculated)} = \frac{AvailableMemory - Overhead}{8} \qquad (1)$$

where $n_{W(calculated)}$ is the size of calculated window, $AvailableMemory$ is the size of the available memory on the device, and $Overhead$ is a constant value including the size of the reservoir. The denominator is the size of a single element in the sliding window, which is 8 bytes. A calculated example will be given in the experimental section which shows the size of the overhead.

4.3   Drift Detection

In our technique updated mean statistics are kept for both the sliding window and reservoir determined as μ_W and μ_R respectively. To determine if a drift has occurred, we take an established approach in drift detection and compare the mean of each sample [2]. If there is a difference which satifies the ϵ_{cut} threshold, a drift is considered to have occurred. If $|\mu_W - \mu_R| > \epsilon_{cut}$ then a drift has occurred, otherwise if $|\mu_W - \mu_R| \le \epsilon_{cut}$ then a drift has not occurred. There are many statistical tests which could be used to calculate ϵ_{cut}. Here ϵ_{cut} is determined by using Hoeffding's Inequality with Bonferroni correction [2]:

$$\epsilon_{cut} = \sqrt{\frac{2\sigma_w^2}{m} \ln \frac{2}{\delta'} + \frac{2}{3m} \ln \frac{2}{\delta'}}$$

where σ_w^2 is the observed variance of the elements in W and R, $m = \frac{1}{(1/|n_W|+1/|n_R|)}$, and $\delta' = \frac{\delta}{\ln(n_W+n_R)}$.

It is not necessary to check for drift after every insertion of a new stream element into our window. We limit the number of checks taken with a *check point* variable, α, which is the number of insertions that take place before each check. Increasing α will reduce execution time as drift will be checked for less frequently, but it will also increase the delay time and may affect accuracy. We examine effects of varying α in Section 5.3.

Figure 1 shows an example of DualWin handling three insertions of stream elements. In our example, the reservoir is represented by an integer r value of the number of stars it holds, just as we do for positive bits in our actual implementation. In Figure 1(b), a triangle is passed from the stream to the drift detector. It is stored in place of the oldest element in the sliding window, a star. The star is then passed to the reservoir with the probability of $\frac{1}{c}$ (which we assume occurs). The reservoir removes an element at random in Figure 1(c); in this case it is a triangle. The star is added, and the r count of the

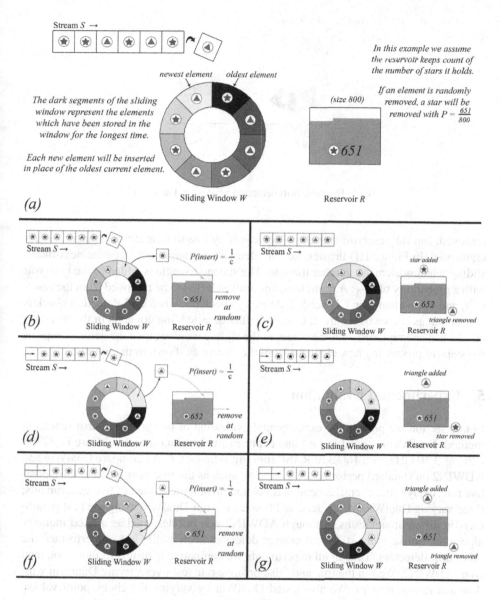

Fig. 1. Example insertion of three stream elements into DualWin

reservoir is increased. c is also incremented, as an insertion into the reservoir has taken place. In Figure 1(d) the next stream element, a star, replaces the next-oldest sliding window element, a triangle. The triangle is added to the reservoir with the probability of $\frac{1}{c+1}$. This is a slightly smaller probability than before, which reflects an increase in the size of the stream seen by our drift detector so far. The current element occupies a smaller proportion of the entire stream than previous elements, and so has a smaller chance of being added to the reservoir. The triangle is added in (e), a star is randomly

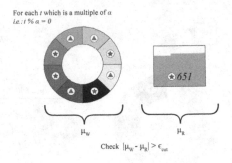

Fig. 2. Example drift detection process in DualWin

removed, and the reservoir star count decreases by one to reflect this change. c is incremented. In Figure 1(f) the next stream element, a triangle, replaces the next-oldest sliding window element, another triangle. The second triangle is added to the reservoir with a probability of $\frac{1}{c+2}$. A triangle is randomly selected to be removed from the reservoir, and so no change in r is needed. At every check point α a drift detection check is carried out as shown in Figure 2. Once a drift is detected, the contents in the reservoir are flushed and replaced with the values in the sliding window, then the detector repeats the steps of processing new elements from the stream as shown in the figures.

5   Experimental Evaluation

In this section we present the experimental evaluation of our proposed drift detection method. DualWin is coded in Java and experiments are run on an Intel Core i7-3770S CPU @ 3.10 GHz with 16GB of RAM, running Windows 7. We compare DualWin and ADWIN2 on standard performance measures, such as the true positive rate, false positive rate, delay time, overall execution time, and memory use. We ran five experiments; three varying DualWin parameters, and two testing our DualWin using the best parameters in different situations. Although ADWIN2 was not designed as a fixed memory algorithm, unlike other statistical change detectors such as SPC, it outperformed the other drift detectors. Thus in all experiments we compare our results to those obtained from ADWIN2. We ran the true and false positive rate test over varying DualWin window and reservoir sizes. We also tested DualWin by varying the check point values, varying stream size and varying types of input stream.

5.1   Performance Comparison of DualWin and ADWIN2

We test DualWin with different sliding window and reservoir size combinations. Our experiment is conducted using a stream size of $L = 1,000,000$, generated from a stationary Bernoulli distribution of parameter μ. We keep $\mu = 0.2$ for the first L-5000 time steps and then induce a drift by increasing μ linearly during the final 5000 steps.

We use 4 different slopes; 1.00×10^{-4}, 2.00×10^{-4}, 3.00×10^{-4}, 4.00×10^{-4} as well as a slope of 0 to check for false positive detection. We use a check point value of 32. For both DualWin and ADWIN2, we repeat each of the experiments 100 times, and use $\delta = 0.05$. We test $4 \times 6 = 24$ sets of dataset; containing combinations of sliding window sizes 100, 200, 300, 400 and reservoir sizes: 500, 1000, 1500, 2000, 2500, 3000.

True Positive Rate. In all cases we observe a true positive detection rate of 100%, which is equivalent to the result of ADWIN2 when tested under the same parameters. Given this equivalent result, we compare DualWin closely to ADWIN2 on other performance measures.

Table 1. Average Memory Usage for DualWin and ADWIN2

Stream Size	DualWin				ADWIN2
	Sliding Window Size				
	100	200	300	400	
1.00×10^{-4}	984.00	1784.00	2584.00	3384.00	1932.72
2.00×10^{-4}	984.00	1784.00	2584.00	3384.00	1836.96
3.00×10^{-4}	984.00	1784.00	2584.00	3384.00	1732.80
4.00×10^{-4}	984.00	1784.00	2584.00	3384.00	1670.64

Memory Use. We propose DualWin as a solution for situations where memory is limited, so an important distinguishing factor is memory use. The average memory use of ADWIN2 and DualWin over 100 iterations is displayed in Table 1. Unlike ADWIN2, DualWin uses a fixed amount of memory. This amount is mainly determined by the size of the sliding window and is not affected by the size of the reservoir which is implemented with an integer. As well as using memory that is fixed, DualWin also uses a memory size that is smaller or comparable to that of ADWIN2. A sliding window of size 100 uses much less memory than ADWIN2 and of size 200 has a memory use that is comparable. It can be observed that an increase in sliding window size of 100 will increase the memory use of DualWin by 800 bytes. The reservoir and other variables use a constant memory of 184 bytes. From Equation 1, if we are restricted to 1024 bytes of available memory, we can use this formula and calculate the suitable n_W to be: $n_{W(calculated)} = \frac{1024-184}{8} = 105$ Our paper is aimed at situations where memory is restricted so we will focus our experiments on DualWin with sliding window sizes of 100 and 200 .

Time and Delay. We display time and delay results for DualWin in Table 2 and for ADWIN2 in Table 3. Observing these results, we find that DualWin executes faster than ADWIN2, and has a delay which is competitive, but higher, than ADWIN2. DualWin always executes about five times faster than ADWIN2.

5.2   Comparisions of False Positive Rates

We investigate the rate of false positives detected by DualWin and compare to AD-WIN2. We test the different sizes of DualWin (as tested in Section 5.1) on a stream size

Table 2. Delay and Execution Time for DualWin

Reservoir Size	Slope	Sliding Window Size							
		100				200			
		Avg Delay	Delay Std	Avg Time	Time Std	Avg Delay	Delay Std	Avg Time	Time Std
500	1.00×10^{-4}	1461.32	331.57	129.13	25.93	1300.84	307.16	129.30	26.21
	2.00×10^{-4}	858.76	191.21	127.88	28.01	781.16	150.76	124.75	26.60
	3.00×10^{-4}	647.24	127.11	129.38	25.97	590.12	106.04	124.31	29.90
	4.00×10^{-4}	506.12	93.63	134.23	27.46	462.76	72.59	129.33	22.98
1000	1.00×10^{-4}	1363.56	288.41	150.10	23.63	1181.96	221.52	146.90	25.35
	2.00×10^{-4}	783.40	152.35	142.87	25.50	692.04	125.45	148.96	23.43
	3.00×10^{-4}	601.32	122.16	146.07	23.91	522.44	97.80	147.71	21.78
	4.00×10^{-4}	460.52	96.53	144.79	27.55	420.68	75.66	143.96	23.22
1500	1.00×10^{-4}	1346.76	254.97	129.93	27.01	1143.40	196.55	125.98	24.86
	2.00×10^{-4}	798.60	155.98	134.15	24.57	700.84	135.22	130.51	29.37
	3.00×10^{-4}	594.76	119.33	132.30	24.73	516.84	97.13	132.51	25.21
	4.00×10^{-4}	458.44	97.91	134.27	25.52	417.96	75.13	132.07	24.84
2000	1.00×10^{-4}	1299.88	275.45	143.32	25.69	1113.48	212.45	143.67	21.46
	2.00×10^{-4}	772.20	164.01	142.60	22.29	696.52	137.49	143.14	24.00
	3.00×10^{-4}	594.92	127.82	141.89	26.02	510.60	101.63	139.42	24.35
	4.00×10^{-4}	453.16	97.14	150.07	24.57	405.32	76.55	140.92	20.89
2500	1.00×10^{-4}	1330.76	253.24	128.92	23.75	1086.44	215.67	139.83	22.95
	2.00×10^{-4}	774.92	164.40	129.57	27.13	673.64	140.83	144.10	25.42
	3.00×10^{-4}	589.64	122.52	133.07	26.97	486.44	108.05	149.32	25.33
	4.00×10^{-4}	446.28	98.15	131.10	27.30	395.56	78.52	142.31	24.46
3000	1.00×10^{-4}	1325.16	264.49	121.67	25.35	1091.40	223.33	132.58	26.48
	2.00×10^{-4}	772.52	152.46	124.30	22.98	674.44	134.30	129.86	28.11
	3.00×10^{-4}	575.72	123.10	120.29	26.27	492.04	95.21	132.64	24.64
	4.00×10^{-4}	457.00	93.23	120.18	28.85	405.96	78.57	130.35	25.33

Table 3. Delay and Execution Time for ADWIN2

Slope	Avg Delay	Delay Std	Avg Time	Time Std
1.00×10^{-4}	865.24	187.71	758.66	41.82
2.00×10^{-4}	564.44	121.62	775.44	39.14
3.00×10^{-4}	440.92	84.34	776.11	35.76
4.00×10^{-4}	371.48	73.65	776.23	38.14

$L = 100,000$ generated from a stationary Bernoulli distribution with different parameters of μ and confidence parameters of δ. We test δ values of 0.05, 0.1 and 0.3 against μ values of 0.01, 0.1, 0.3 and 0.5. Due to space limitation, in Table 4 we display the false positive rate for our best-sized DualWin which was 200/2500, the worst rate we attained across all DualWin sizes which was 400/3000, and the rate for ADWIN2. We find that across all sizes of DualWin, there is a significantly lower false positive rate than ADWIN2.

5.3   Varying Check Point Value

For this experiment we test DualWin with different check point, α, values. We use our best DualWin with a sliding window size of 200 and a reservoir size of 2500. The experiment is set up as the performance experiments above. We test over the four slopes used in Section 5.1 as well as slope 0 to check for false positives. We test both ADWIN and DualWin with check point sizes of 8, 16, 24, 32 and 40, shown in Table 5.

Table 4. False positive rates for ADWIN2 and DualWin

μ/δ	DualWin (Best)			DualWin (Worst)			ADWIN2		
	0.05	0.1	0.3	0.05	0.1	0.3	0.05	0.1	0.3
0.01	0.0000	0.0000	0.0003	0.0000	0.0000	0.0003	0.0001	0.0003	0.0014
0.1	0.0001	0.0004	0.0015	0.0003	0.0006	0.0026	0.0008	0.0020	0.0081
0.3	0.0002	0.0005	0.0028	0.0002	0.0007	0.0040	0.0012	0.0030	0.0122
0.5	0.0001	0.0004	0.0025	0.0002	0.0006	0.0036	0.0015	0.0034	0.0136

Table 5. Time and delay for DualWin and ADWIN2 with varying check point sizes

α	Slope	DualWin				ADWIN2			
		Avg Delay	Delay Std	Avg Time	Time Std	Avg Delay	Delay Std	Avg Time	Time Std
8	1.00×10^{-4}	1051.16	230.37	142.55	27.72	846.92	197.06	2398.78	88.71
	2.00×10^{-4}	634.52	146.32	140.69	22.37	547.32	119.34	2388.92	88.45
	3.00×10^{-4}	464.60	105.15	139.50	24.71	426.36	89.65	2402.15	92.19
	4.00×10^{-4}	376.44	75.31	139.54	24.66	356.20	72.49	2387.21	80.86
16	1.00×10^{-4}	1067.56	215.17	141.80	25.44	851.80	191.64	1319.81	57.54
	2.00×10^{-4}	656.20	140.70	144.58	26.44	553.24	122.38	1319.00	53.88
	3.00×10^{-4}	472.84	109.52	145.12	25.67	434.36	85.87	1319.39	50.76
	4.00×10^{-4}	382.44	74.97	144.67	26.00	361.72	72.13	1319.94	55.99
24	1.00×10^{-4}	1075.72	211.95	134.67	25.17	866.28	196.34	948.76	48.82
	2.00×10^{-4}	665.08	145.55	136.03	22.86	561.24	120.53	963.75	50.22
	3.00×10^{-4}	483.40	106.33	136.15	28.76	440.52	85.45	966.61	53.65
	4.00×10^{-4}	392.92	77.99	135.67	25.61	366.60	71.14	975.97	55.51
32	1.00×10^{-4}	1086.44	215.67	127.04	25.56	865.24	187.71	793.30	44.73
	2.00×10^{-4}	673.64	140.83	132.91	24.85	564.44	121.62	796.90	41.95
	3.00×10^{-4}	486.44	108.05	130.66	25.35	440.92	84.34	793.92	51.07
	4.00×10^{-4}	395.56	78.52	124.43	30.54	371.48	73.65	797.17	46.01
40	1.00×10^{-4}	1104.20	222.16	132.64	24.97	871.00	188.04	695.36	41.09
	2.00×10^{-4}	679.80	140.66	131.91	24.79	568.60	123.43	691.01	34.97
	3.00×10^{-4}	501.40	98.94	127.72	24.42	447.80	86.82	705.49	45.64
	4.00×10^{-4}	409.00	76.52	133.04	25.35	375.00	73.10	707.85	44.41

For DualWin, we observe a 100% true positive detection rate and no false positives on the 0 slope. ADWIN2 also achieves a detection rate of 100%, but has noticable false positives. At check point sizes 8 and 16, ADWIN2 has 4 false positives, and at sizes 24, 32 and 40 has 3 false positives. We observe that execution time of DualWin reduces with each increase in check point size, however this reduction plateaus at around 32. The delay time seems to increases at a constant rate, so we choose 32 as our ideal check point size.

5.4 Varying Stream Size

Using DualWin with the best window/reservoir size of 200/2500 was chosen in Section 5.1 and the check point size 32 chosen in Section 5.3, we compare performance of our technique on varying stream sizes. We use sizes of 100,000, 500,000, 1,000,000, 5,000,000 and 10,000,000 bits and continue to locate the drift 5000 bits from the end of the stream and use four slopes, as in Section 5.1 and Section 5.3.

Table 6. Average execution time by slope over different stream size

	Slope	100k	500k	1M	5M	10M
DualWin	1.00×10^{-4}	11.55	64.57	125.9	659.15	1300.48
	2.00×10^{-4}	12.68	66.51	131.41	640.77	1315.14
	3.00×10^{-4}	11.56	64.90	125.59	641.66	1289.17
	4.00×10^{-4}	13.88	64.56	128.71	659.55	1282.64
ADWIN2	1.00×10^{-4}	67.38	374.55	771.58	3845.85	7855.04
	2.00×10^{-4}	67.46	374.48	755.56	3858.54	7919.96
	3.00×10^{-4}	65.46	380.12	755.15	3852.04	7769.87
	4.00×10^{-4}	67.59	374.75	756.64	3934.01	7866.34

We display our results in Table 6. Our experiment shows ADWIN2 execution time increases significantly as the stream size becomes larger. DualWin continues to achieve a similar rate of effectiveness as shown in the other performance metrics as ADWIN2, but for larger streams DualWin uses significantly less time.

5.5 Alternate Stream Generators

In this experiment we compare DualWin and ADWIN2 on the same performance measures as previously, but use different sources of data. We use data sets from four well known data stream generators; RBF generator, Hyperplane generator, LED generator and Waveform generator. Our results are shown in Table 7. ADWIN2 attains a 100% detection rate for all four sets. DualWin attains a 100% detection rate for the first three sets and a 96% detection rate for the Waveform data set. Our results show that DualWin is compatible with different types of data, and will acheive similar results outside of Bernoulli-generated streams.

Table 7. ADWIN2 and DualWin performance on different data sources

	Generator	TP Rate	Avg Delay	Delay Stdev	Avg Time	Time Stdev	Memory Size
ADWIN2	Hyperplane	1.00	1437.72	801.24	370.77	37.30	1860.48
	LED	1.00	39.32	14.04	381.28	39.85	1013.76
	RBF	1.00	160.92	27.46	351.83	41.33	1285.92
DualWin	Hyperplane	1.00	2685.64	2039.93	67.56	26.85	1784.00
	LED	1.00	85.96	21.31	63.62	23.22	1784.00
	RBF	1.00	27.40	23.88	59.42	25.53	1784.00

6   Conclusion and Future Work

In this paper we introduced a novel drift detection technique for data streams, DualWin, which uses a fixed, controllable amount of memory while remaining comparable in other performance measures to ADWIN2. We compare DualWin and ADWIN2 on the essential performance measures: true positive rate, false positive rate, delay time, overall execution time, and memory use. From these experiments we show that even though DualWin is limited by fixed memory, we obtain faster execution time than ADWIN and achieve comparable results in true positive rate, false positive rate, and memory size.

While we have a slightly larger delay, DualWin remained competitive. We showed that our technique was comparable in all accounts whilst maintaining a fixed memory usage.

Our future work involves adapting the reservoir size on-the-fly. Since the size of the reservoir dictates whether the reservoir can provide an honest representation of the stream, having an adaptive size will allow the technique to be more flexible when dealing with different types of data streams. In addition, we would like to experiment on other statistical upper bounds such as the Bernstein bound in place of the current Hoeffding bound.

References

1. Aggarwal, C.C., Han, J., Wang, J., Yu, P.S.: A framework for clustering evolving data streams. In: Proceedings of the 29th International Conference on Very Large Data Bases, VLDB 2003, vol. 29, pp. 81–92. VLDB Endowment (2003)
2. Bifet, A., Gavaldà, R.: Learning from time-changing data with adaptive windowing. In: Proceedings of the Seventh SIAM International Conference on Data Mining. SIAM (2007)
3. Gama, J., Medas, P., Castillo, G., Rodrigues, P.: Learning with drift detection. In: Bazzan, A.L.C., Labidi, S. (eds.) SBIA 2004. LNCS (LNAI), vol. 3171, pp. 286–295. Springer, Heidelberg (2004)
4. Ho, S.S.: A martingale framework for concept change detection in time-varying data streams. In: Proceedings of the 22nd International Conference on Machine Learning, ICML 2005, pp. 321–327. ACM, New York (2005)
5. Jose, M.B., Campo-Ávila, J.D., Fidalgo, R., Bifet, A., Gavaldà, R., Morales-bueno, R.: Early Drift Detection Method. In: Proceedings of the 4th ECML PKDD Int. Workshop on Knowledge Discovery from Data Streams, Berlin, pp. 77–86 (2006)
6. Kifer, D., Ben-David, S., Gehrke, J.: Detecting change in data streams. In: Proceedings of the Thirtieth International Conference on Very Large Data Bases, VLDB 2004, vol. 30, pp. 180–191. VLDB Endowment (2004)
7. Klinkenberg, R., Joachims, T.: Detecting concept drift with support vector machines. In: Proceedings of the Seventeenth International Conference on Machine Learning, ICML 2000, pp. 487–494. Morgan Kaufmann Publishers Inc., San Francisco (2000)
8. Page, E.S.: Continuous inspection schemes. Biometrika 41(1/2), 100–115 (1954)
9. Sebastiao, R., Gama, J.: A study on change detection methods. In: Proceedings of the 14th Portuguese Conference on Artificial Intelligence, EPIA 2009, pp. 353–364. Springer, Heidelberg (2009)
10. Vitter, J.: Random sampling with a reservoir. ACM Transactions on Mathematical Software, 37–57 (1985)

Mean Shift Clustering Algorithm
for Data with Missing Values

Loai AbdAllah[1] and Ilan Shimshoni[2]

[1] Department of Mathematics, University of Haifa, Israel
Department of Mathematics and Computer Science,
The College of Sakhnin for Teacher Education, Israel
[2] Department of Information Systems, University of Haifa, Israel
loai1984@gmail.com, ishimshoni@mis.haifa.ac.il

Abstract. Missing values in data are common in real world applications. There are several methods that deal with this problem. In this research we developed a new version of the *mean shift* clustering algorithm that deals with datasets with missing values. We use a weighted distance function that deals with datasets with missing values, that was defined in our previous work. To compute the distance between two points that may have attributes with missing values, only the *mean* and the *variance* of the distribution of the attribute are required. Thus, after they have been computed, the distance can be computed in $O(1)$. Furthermore, we use this distance to derive a formula for computing the *mean shift* vector for each data point, showing that the *mean shift* runtime complexity is the same as the Euclidian *mean shift* runtime. We experimented on six standard numerical datasets from different fields. On these datasets we simulated missing values and compared the performance of the *mean shift* clustering algorithm using our distance and the suggested mean shift vector to other three basic methods. Our experiments show that *mean shift* using our distance function outperforms *mean shift* using other methods for dealing with missing values.

Keywords: Missing values, Distance metric, Weighted Euclidian distance, Clustering, Mean Shift.

1   Introduction

Mean shift is a non-parametric iterative clustering algorithm. The fact that mean shift does not require prior knowledge of the number of clusters, and does not constrain the shape of the clusters, makes it ideal for handling clusters of arbitrary shape and number. It is also an iterative technique, but instead of the means, it estimates the modes of the multivariate distribution underlying the feature space. The number of clusters is obtained automatically by finding the centers of the densest regions in the space (the modes). The density is evaluated using kernel density estimation which is a non-parametric way to estimate the density function of a random variable. It is also called the Parzen window technique.

L. Bellatreche and M.K. Mohania (Eds.): DaWaK 2014, LNCS 8646, pp. 426–438, 2014.

Mean shift was first proposed by Fukunaga and Hostetler [7]. It was then adapted by Cheng [3] for the purpose of image analysis. Later Comaniciu and Meer [5,4] successfully applied it to image segmentation and tracking. Tao, Jin and Zhang [13] use it in color image segmentation and DeMenthon and Megret [6] employed it for spatio-temporal segmentation of video sequences in a $7D$ feature space.

The problem with this algorithm is that it can not deal with datasets that contain missing values which are common in many real world datasets. Missing values can be caused by human error, equipment failure, system generated errors, and so on. In this paper we have developed a *mean shift* clustering algorithm over datasets with missing values based on the distance function we developed in [1].

Several methods have been proposed to deal with missing data. These methods can be classified into two basic categories: (a) **Case deletion** method, which ignores all the instances with missing values and performs the analysis on the rest. This method has two obvious disadvantages: (1) A substantial decrease in the size of the dataset available for the analysis. (2) The data are not always missing at random that may affect the distribution of the other features [15]. (b) Missing data imputation, which replaces each missing value with a known value according to the dataset distribution. It is important to note that by using this method the mean shift procedure can run like on the complete dataset (each missing replaced with known value). But, as we show in this paper, they perform poorly and our proposed method yields better results.

A common method that imputes missing data is the **Most Common Attribute Value (MCA)** method. The value of the attribute that occurs most often is selected to be the value for all the unknown values of the attribute [9]. One main drawback of this method is that it ignores the other possible values of the attribute and their distribution.

The main idea of the **Mean Imputation (MI)** method is to replace a data point with missing values with the mean of all the instances in the data. However, using a fixed instance to replace all the instances with missing values will change the characteristics of the original dataset. Ignoring the relationship among attributes will bias the performance of subsequent data mining algorithms. A variant of this method is to replace the missing data for a given attribute with the Mean of all known values of that **Attribute-MA** (i.e., the mean of each attribute) in the coordinate where the instance with missing data belongs [10]. This method has the same drawback as the MCA method: both methods represent the missing value with one value and ignore all the other possible values.

The k-**Nearest Neighbor Imputation** method uses the kNN algorithm (using only the known values) to estimate and replace the missing data [16,2] by looking for the most similar instances. Efficiency is the biggest obstacle for this method. Moreover, the selected value of k and the measure of similarity will affect the results greatly.

Finally, the k-means **Imputation** method predicts missing attribute values using simple k-means clustering. This approach deals with labeled datasets. [12].

Again, the main drawbacks of each suggested method can be summarized as inefficiency and inability to approximate the missing value. In our previous work [1] we developed a method to compute the distance function (MD_E) that is not only efficient but also takes into account the distribution of each attribute. In the computation procedure we take into account all the possible values with their probabilities, which are computed according to the attribute's distribution. This is in contrast to the MCA and the MA methods, which replace each missing value only with the mode or the mean of each attribute.

We can summarize our distance for the three possible cases of the two values: (1) Both values are known. In that case, the distance function is identical to the Euclidian distance. (2) One value is missing. In that case, the distance will be computed only according to the two statistics (*mean* and *variance*), where the distance equals the Euclidian distance between the known value and the *mean* plus the *variance* of that attribute (mean squared error). (3) Both values are missing. In that case, the distance will be computed only according to the *variance* of that attribute, and it equals twice the *variance*. Therefore, the runtime of this distance function is the same as for the Euclidian distance.

In order to develop a mean shift algorithm for data with missing values we derived a formula for computing the gradient function of the local estimated density. For this case the runtime complexity of the resulting gradient function using the MD_E distance is the same as the standard one using the Euclidean distance. To measure the ability of the suggested mean shift vector using the MD_E distance to represent the actual mean shift vector when the dataset is complete, we integrated it within the mean shift clustering algorithm on datasets with missing values. The developed algorithm not only yields better results than the other methods, as can be seen in the experiments, but also preserves the runtime of the mean shift clustering algorithms which deals with complete data. We experimented on six standard numerical datasets from different fields from the Speech and Image Processing Unit [14]. Our experiments show that the performance of the mean shift algorithm using our distance function and the proposed mean shift vector were superior to mean shift using other methods.

The paper is organized as follows. A review of our distance function (MD_E) is described in Section 2. An overview of the mean shift clustering algorithm is presented in Section 3. Section 4 describes how to compute and to integrate the (MD_E) distance and the computed mean shift vector within the mean shift clustering algorithm. Experimental results of running the mean shift clustering algorithm on the Speech and Image Processing Unit [14] datasets are presented in Section 5. Finally, our conclusions are presented in Section 6.

2   Our Distance Measure

In this section we give a short overview of our distance function developed in [1]. Let A be a set of points. For the ith coordinate C_i, the conditional probability for C_i will be computed according to the known values for this coordinate from A (i.e., $P(c_i) \sim \chi_i$), where χ_i is the distribution of the ith coordinate.

Our method can be generalized to deal with coordinates whose measurements are dependant, but for simplicity we assume that these measurements are independent. Under these assumptions we will treat each coordinate separately.

Given two sample points X and Y from A, the goal is to compute the distance between them. Let x_i and y_i be the ith coordinate values from points X, Y respectively. There are three possible cases for the values of x_i and y_i: (1) Both values are given. (2) One value is missing. (3) Both values are missing.

Two Values Are known: When the values of x_i and y_i are given, the distance between them will be defined as the Euclidian distance:

$$D_E(x_i, y_i) = (x_i - y_i)^2. \tag{1}$$

One Value Is Missing: Suppose that x_i is missing and the value y_i is given. Since the value of x_i is unknown, we cannot compute its Euclidian distance. Instead we model the distance as a random selection of a point from the distribution of its coordinate χ_i and compute its distance. The mean of this computation is our distance. We will estimate this value as follows: We divide the range of c_i (i.e., $[\min(c_i), \max(c_i)]$) into l equal intervals $(\Delta_1, \ldots, \Delta_l)$. Let m_j be the center point of interval Δ_j. For it we can estimate its probability density $p(m_j)$ using the KDE. The probability for the jth interval Δj is therefore:

$$P(\Delta j) = p(m_j) \cdot \frac{\max(c_i) - \min(c_i)}{l} \approx \int_{x \in \Delta_j} p(x) dx.$$

As a result, we approximate the mean Euclidian distance (MD_E) between y_i and the distribution as:

$$MD_E(\chi_i, y_i) = \sum_{j=1}^{l} P(\Delta j) D_E(m_j, y_i).$$

This metric measures the distance between y_i and each suggested value of x_i and takes into account the probability for this value according to the evaluated probability distribution. Computing the distance in this way is an expensive procedure. We therefore turn to the continuous probability density function $p(x)$ in hope of getting an efficient formula. In this case,

$$MD_E(\chi_i, y_i) = \int p(x)(x - y_i)^2 dx = \int p(x)x^2 + p(x)y_i^2 - p(x)2xy_i) dx$$

$$= \left(E[x^2] + y_i^2 - 2y_i E[x] \right) = \left((y_i^2 - E[x])^2 - E[x]^2 + E[x^2] \right)$$

$$= \left((y_i - \mu_i)^2 + \sigma_i^2 \right), \tag{2}$$

where μ_i, σ_i^2 are the *mean* and the *variance* for all the known values of the attribute. The distance computed according to the last equation has several important properties:

- It is identical to the Euclidian distance, when the dataset is complete. In this case $\mu_i = x_i, \sigma_i = 0$ (*Dirac delta function*).
- It can be applied to any distribution and can be used in any algorithm that uses a distance function.
- It is simple to implement.
- It is very efficient because, to compute the MD_E between two values when one of them is missing, we need only to compute in advance the two statistics (i.e., μ and σ) for each coordinate. After that the runtime is $O(1)$.
- When the variance is small, the real value of the missing value is close to the mean of the attribute and our distance will converge to the Euclidian distance.
- When the variance is large, the uncertainty is high, and as a result the distance should be large.
- It is basically the sum of the bias term $(y_i - \mu_i)^2$ and the *variance* σ_i^2, yielding the mean squared error (MSE).

In contrast to the other imputation methods we do not replace the missing value with any constant value. Moreover, it differs from the **Most Common Attribute Value** method, where the value of the most frequent attribute is selected to be the value for all the unknown values of the attribute, implying that the probability of the most common attribute value is 1 and 0 for all other possible values. Our distance also differs from the **Mean Attribute Value method**, where the mean of a specific attribute is selected to replace the unknown value of the attribute because the dispersion of the values in the distribution is not taken into account.

The Two Values Are Missing: In this case, in order to estimate the mean Euclidian distance, we have to randomly select values for both x_i and y_i. Both these values are selected from distribution χ_i. To compute the distance, the following double sum has to be computed.

$$MD_E(x_i, y_i) = \sum_{q=1}^{l-1} \sum_{j=1}^{l-1} P(\Delta_{1q})P(\Delta_{2j})D_E(m_{1q}, m_{2j}).$$

As we did for one missing value, we turn to the continuous case:

$$MD_E(x_i, y_i) = \int \int p(x)p(y)(x - y)^2 dx dy$$
$$= \left((E[x] - E[y])^2 + \sigma_x^2 + \sigma_y^2 \right) = 2\sigma_i^2. \tag{3}$$

From (3) we can easily see that our metric reflects the similarity between points better than the other methods. In the previous two methods, all the missing values of an attribute are replaced by the mode or the mean of that attribute, and the distance is equal to 0 without paying any attention to the variance of the coordinates. In our metric, however, the distance depends on the variance for each coordinate, which is more logical because if the variance is larger then the distance between the possible values on average is larger.

3 Mean Shift Algorithm

For completeness we will now give a short overview of the mean shift algorithm. Here we only review some of the results described in [4,8] which should be consulted for the details. Assume that each data point $x_i \in \mathbb{R}^d, i = 1, ..., n$ is associated with a bandwidth value $h > 0$. The *sample point* density estimator at point x is

$$\hat{f}(x) = \frac{1}{nh^d} \sum_{i=1}^{n} K\left(\frac{x - x_i}{h}\right). \tag{4}$$

Based on a spherically symmetric kernel K with bounded support satisfying

$$K(x) = c_{k,d}k\left(\|x\|^2\right) \qquad \|x\| \leq 1 \tag{5}$$

is an adaptive nonparametric estimator of the density at location x in the feature space. The function $k(x), 0 \leq x \leq 1$, is called the *profile* of the kernel, and the normalization constant $c_{k,d}$ assures that $K(x)$ integrates to one. Employing the profile notation the density estimator (4) can be rewritten as

$$\hat{f}_{h,k}(x) = \frac{c_{k,d}}{nh^d} \sum_{i=1}^{n} k\left(\left\|\frac{x - x_i}{h}\right\|^2\right). \tag{6}$$

The first step in the analysis of a feature space with the underlying density $f(x)$ is to find the modes of the density. The modes are located among the zeros of the gradient $\nabla f(x) = 0$, and the mean shift procedure is an elegant way to locate these zeros without estimating the density.

The density gradient estimator is obtained as the gradient of the density estimator by exploiting the linearity of (6)

$$\nabla \hat{f}_{h,K}(x) = \frac{2c_{k,d}}{nh^{d+2}} \sum_{i=1}^{n} (x - x_i)k'\left(\left\|\frac{x - x_i}{h}\right\|^2\right). \tag{7}$$

We define the function $g(x) = -k'(x)$ that can always be defined when the derivative of the kernel profile $k(x)$ exists. Using $g(x)$ as the profile, the kernel $G(x)$ is defined as

$$G(x) = c_{g,d}g\left(\|x\|^2\right).$$

Introducing $g(x)$ into (7) yields

$$\nabla \hat{f}_{h,K}(x) = \frac{2c_{k,d}}{nh^{d+2}} \sum_{i=1}^{n} (x_i - x) g\left(\left\|\frac{x - x_i}{h}\right\|^2\right)$$

$$= \frac{2c_{k,d}}{nh^{d+2}} \left[\sum_{i=1}^{n} g\left(\left\|\frac{x - x_i}{h}\right\|^2\right)\right] \left[\frac{\sum_{i=1}^{n} x_i g\left(\left\|\frac{x - x_i}{h}\right\|^2\right)}{\sum_{i=1}^{n} g\left(\left\|\frac{x - x_i}{h}\right\|^2\right)} - x\right], \quad (8)$$

where $\sum_{i=1}^{n} g\left(\left\|\frac{x - x_i}{h}\right\|^2\right)$ is assumed to be a positive number. Both terms of the product in (8) have special significance. The first term is proportional to the density estimate at x computed with the kernel G. The second term

$$m_G(x) = \frac{\sum_{i=1}^{n} x_i g\left(\left\|\frac{x - x_i}{h}\right\|^2\right)}{\sum_{i=1}^{n} g\left(\left\|\frac{x - x_i}{h}\right\|^2\right)} - x \qquad (9)$$

is called the *mean shift vector*. The expression (9) shows that at location x the weighted mean of the data points selected with kernel G is proportional to the normalized density gradient estimate obtained with kernel K. The mean shift vector thus points toward the direction of maximum increase in the density. The implication of the mean shift property is that the iterative procedure

$$y_{j+1} = \frac{\sum_{i=1}^{n} x_i g\left(\left\|\frac{y_j - x_i}{h}\right\|\right)}{\sum_{i=1}^{n} g\left(\left\|\frac{y_j - x_i}{h}\right\|\right)} \qquad j = 1, 2, \dots \qquad (10)$$

is a hill climbing technique to the nearest stationary point of the density, i.e., a point in which the density gradient vanishes. The initial position of the kernel, the starting point of the procedure y_1 can be chosen as one of the data points x_i. Most often the points of convergence of the iterative procedure are the modes (local maxima) of the density. All points which converge to the same mode are considered members of a cluster. The number of clusters is therefore the number of modes.

4   Mean Shift Computing Using The MD_E Distance

In our previous work in [1] we derived the MD_E distance and integrated it within the framework of the kNN and kMeans algorithms. In order to integrate the MD_E distance function within the framework of the mean shift algorithm, we will first compute the mean shift vector using the MD_E distance.

Using the MD_E distance the density estimator in (6) will written as

$$\hat{f}_{h,k}(x) = \frac{c_{k,d}}{nh^d} \sum_{i=1}^{n} k\left(\left\|\frac{x-x_i}{h}\right\|^2\right) = \frac{c_{k,d}}{nh^d} \sum_{i=1}^{n} k\left(\frac{\sum_{j=1}^{d} MD_E(x^j, x_i^j)^2}{h^2}\right). \quad (11)$$

Since each point x_i may contain missing attributes, $\hat{f}_{h,k}(x)$ will be:

$$\hat{f}_{h,k}(x) = \frac{c_{k,d}}{nh^d} \sum_{i=1}^{n} k\left(\underbrace{\frac{\sum_{j=1}^{kn_i} MD_E(x^j, x_i^j)^2}{h^2}}_{\text{each } x_i \text{ has } kn_i \text{ known attributes}} + \underbrace{\frac{\sum_{j=1}^{unkn_i} MD_E(x^j, x_i^j)^2}{h^2}}_{\text{each } x_i \text{ has } unkn_i \text{ missing attributes}} \right).$$

According to the definition of the MD_E distance, we obtain

$$\hat{f}_{h,k}(x) = \frac{c_{k,d}}{nh^d} \sum_{i=1}^{n} k\left(\frac{\sum_{j=1}^{kn_i} (x^j - x_i^j)^2}{h^2} + \frac{\sum_{j=1}^{unkn_i} (x^j - \mu^j)^2 + (\sigma^j)^2}{h^2}\right). \quad (12)$$

Now we will compute the gradient of the density estimator in (12)

$$\nabla \hat{f}_{h,k}(x) = \frac{c_{k,d}}{nh^{d+2}} \sum_{i=1}^{n} \left[\sum_{j=1}^{kn_i}(x^j - x_i^j)^2 + \sum_{j=1}^{unkn_i}(x^j - \mu^j)^2 + (\sigma^j)^2\right]'$$

$$\cdot k'\left(\frac{\sum_{j=1}^{kn_i}(x^j - x_i^j)^2}{h^2} + \frac{\sum_{j=1}^{unkn_i}(x^j - \mu^j)^2 + (\sigma^j)^2}{h^2}\right)$$

$$= \frac{c_{k,d}}{nh^{d+2}} \sum_{i=1}^{n} \left[\sum_{j=1}^{kn_i}(x^j - x_i^j)^2\right]' \cdot k'\left(\frac{\sum_{j=1}^{kn_i}(x^j - x_i^j)^2}{h^2} + \frac{\sum_{j=1}^{unkn_i}(x^j - \mu^j)^2 + (\sigma^j)^2}{h^2}\right)$$

$$+ \left[\sum_{j=1}^{unkn_i}(x^j - \mu^j)^2 + (\sigma^j)^2\right]' \cdot k'\left(\frac{\sum_{j=1}^{kn_i}(x^j - x_i^j)^2}{h^2} + \frac{\sum_{j=1}^{unkn_i}(x^j - \mu^j)^2 + (\sigma^j)^2}{h^2}\right).$$

In our computation we will first deal with one coordinate l and then we will generate the computation for all the other coordinates.

$$\Rightarrow f'_{x^l} = \frac{2c_{k,d}}{nh^{d+2}} \sum_{i=1}^{n_l} (x^l - x_i^l) \cdot k'\left(\frac{\sum_{j=1}^{kn_i}(x^j - x_i^j)^2}{h^2} + \frac{\sum_{j=1}^{unkn_i}(x^j - \mu^j)^2 + (\sigma^j)^2}{h^2}\right)$$

$$+ \frac{2c_{k,d}}{nh^{d+2}} \sum_{i=1}^{m_l} (x^l - \mu^l) \cdot k'\left(\frac{\sum_{j=1}^{kn_i}(x^j - x_i^j)^2}{h^2} + \frac{\sum_{j=1}^{unkn_i}(x^j - \mu^j)^2 + (\sigma^j)^2}{h^2}\right)$$

$$= \frac{2c_{k,d}}{nh^{d+2}} \left[x^l \cdot \sum_{i=1}^{n} k'\left(\frac{\sum_{j=1}^{kn_i}(x^j - x_i^j)^2}{h^2} + \frac{\sum_{j=1}^{unkn_i}(x^j - \mu^j)^2 + (\sigma^j)^2}{h^2}\right)\right.$$

$$- \sum_{i=1}^{n_l} x_i^l \cdot k'\left(\frac{\sum_{j=1}^{kn_i}(x^j - x_i^j)^2}{h^2} + \frac{\sum_{j=1}^{unkn_i}(x^j - \mu^j)^2 + (\sigma^j)^2}{h^2}\right)$$

$$\left. - \sum_{i=1}^{m_l} \mu^l \cdot k'\left(\frac{\sum_{j=1}^{kn_i}(x^j - x_i^j)^2}{h^2} + \frac{\sum_{j=1}^{unkn_i}(x^j - \mu^j)^2 + (\sigma^j)^2}{h^2}\right)\right],$$

where there are n_l points for which the x^l coordinate is known, and there are m_l points where it is missing.

$$f'_{x^l} = \frac{2c_{k,d}}{nh^{d+2}} \cdot \left[\sum_{i=1}^{n} g\left(\sum_{j=1}^{d} MD_E(x^j, x_i^j)^2 \right) \right]$$

$$\cdot \left[\frac{\sum_{i=1}^{n_l} x_i^l \cdot g\left(\sum_{j=1}^{d} MD_E(x^j, x_i^j)^2 \right) + \sum_{i=1}^{m_l} \mu^l \cdot g\left(\sum_{j=1}^{d} MD_E(x^j, x_i^j)^2 \right)}{\sum_{i=1}^{n} g\left(\sum_{j=1}^{d} MD_E(x^j, x_i^j)^2 \right)} - x^l \right].$$

As a result the mean shift vector using the MD_E distance is defined as:

$$m_{MD_E, G}(x) = \qquad\qquad\qquad\qquad\qquad\qquad\qquad\qquad (13)$$

$$\frac{\sum_{i=1}^{n_l} x_i^l \cdot g\left(\sum_{j=1}^{d} MD_E(x^j, x_i^j)^2 \right) + \sum_{i=1}^{m_l} \mu^l \cdot g\left(\sum_{j=1}^{d} MD_E(x^j, x_i^j)^2 \right)}{\sum_{i=1}^{n} g\left(\sum_{j=1}^{d} MD_E(x^j, x_i^j)^2 \right)} - x^l.$$

Now we can use this equation to run the mean shift procedure over datasets with missing values. Computing the mean shift vector using (13) seems like the computed mean shift vector using MA-method, where the mean of a specific attribute is selected to replace the unknown value of the attribute, except that in (13) the weights for the incomplete data points are lower than the computed mean shift using the MA-method, because the MD_E distance equals the Euclidian distance between the known value and the *mean* plus the *variance* of that attribute which is bigger than the Euclidian distance between the known value and the *mean* which is the distance when the MA-method is used.

The mean shift procedure starts the iterative process from each point in the dataset. It therefore also starts from incomplete points where in some cases the distance from a given incomplete point to all the other data points is larger than the bandwidth h. In that case we consider two cases, finite kernels or infinite kernels. If the kernel is infinite the computation will work as described above in (13). When the kernel is finite, there may be cases that there are no points within the radius h. In this case the mean shift will be the first nearest neighbor to the incomplete point, and here we again have two cases for the nearest neighbor. Where the nearest point is the incomplete point itself the mean shift iteration will stop and this point will also be the mode of the point. In the other case the nearest point will be another point from the data and then the next iteration of the mean shift procedure will start with this point, and the algorithm will iteratively continue until convergence. Formally, this is done by replacing the finite kernel g with an infinite kernel g_{inf} where $|g(x) - g_{inf}(x)| < \varepsilon$ for $0 \le x \le 1$ for an infinitesimal value ε.

5 Mean Shift Experiments on Numerical Datasets

In order to measure the ability to implement the mean shift algorithm over datasets with missing values we compare the performance of the mean shift clustering algorithm on complete data (i.e., without missing values) to its performance on data with missing values, using our distance measure ($MS - MD_E$) and then again using MS-(MCA, MA, MI), where each missing value in each attribute is replaced using the MCA, MA or MI method respectively and then a standard mean shift is run. We use the Rand index [11], which is a measure of similarity between two data clusterings, to compare how similar the results of the standard mean shift clustering algorithm were to the results of the other algorithms for datasets with missing values.

We ran our experiments on six standard numerical datasets from the Speech and Image Processing Unit [14] from different fields: the Flame dataset, the Jain dataset, the Path based dataset, the Spiral dataset, the Compound dataset, and the Aggregation dataset. The dataset characteristics are shown in Table 1.

Table 1. Speech and Image Processing Unit Dataset properties

Dataset	Dataset size	Clusters
Flame	240×2	2
Jain	373×2	2
Path based	300×2	3
Spiral	312×2	3
Compound	399×2	6
Aggregation	788×2	7

A set consisting of 10%-40% of the data was randomly drawn from each dataset. These randomly drawn sets serve as samples of missing data, where one coordinate from each instance was randomly selected to be missing.

The results are averaged over 10 different runs on each dataset. For all the cases the bandwidth $h = 4$ was used (The standard mean shift worked well for this value on all the data sets). A resulting curve was constructed for each dataset to evaluate how well the algorithm performed, by plotting the Rand Index.

As can be seen in Figure 1, for the Flame, Spiral, Path based,Compound, and Aggregation datasets, the curves show that our mean shift clustering algorithm outperformed the other methods for all missing value percentages, while for the Jain dataset its benefit became apparent when the percent of the missing values was large, as can be seen in Figure 1(b). Moreover, we can see from these curves that the $MS - MC$ method outperforms the $MS - MA$ method for the Flame and Path Based datasets and the $MS - MC$ outperforms $MS - MA$ for the other datasets. It means that we cannot decide unequivocally which algorithm is better. On the other hand we surely can state that the $MS - MD_E$ outperforms the other methods. If the percentage of the missing values further increases the performance of the algorithm degrades gracefully.

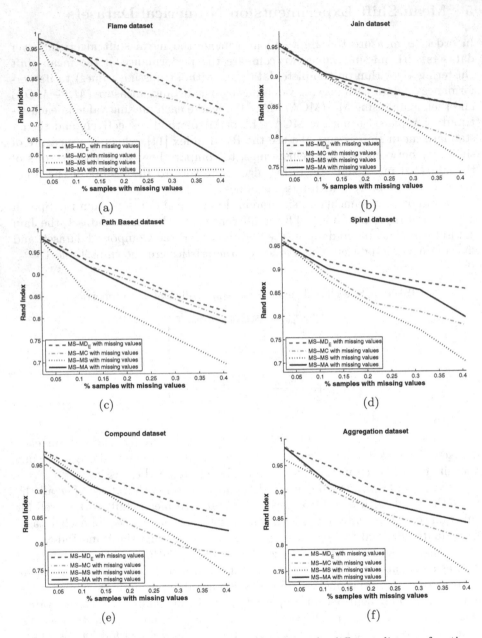

Fig. 1. Results of mean shift clustering algorithm using the different distance functions on the six datasets from the Speech and Image Processing Unit

6   Conclusions

Missing attribute values are very common in real-world datasets. Several meth ods have been proposed to measure the similarity between objects with missing values. In this work, we have proposed a new mean shift clustering algorithm over dataset with missing values using the MD_E distance that was presented in [1]. In order to do that we derived a formula for the mean shift vector for a given dataset when it contains points with missing values. The computational complexity for computing the mean shift vector using the MD_E distance is the same as that of the standard mean shift vector using the Euclidian distance.

From the experiments we conclude that our method is more appropriate for measuring the mean shift vectors for objects with missing values, especially when the percent of missing values is large.

References

1. AbdAllah, L., Shimshoni, I.: A distance function for data with missing values and its applications on $k$nn and $k$means algorithms. Submitted to Int. J. Advances in Data Analysis and Classification
2. Batista, G., Monard, M.C.: An analysis of four missing data treatment methods for supervised learning. Applied Artificial Intelligence 17(5-6), 519–533 (2003)
3. Cheng, Y.: Mean shift, mode seeking, and clustering. IEEE Trans. PAMI 17(8), 790–799 (1995)
4. Comaniciu, D., Meer, P.: Mean shift: A robust approach toward feature space analysis. IEEE Trans. PAMI 24(5), 603–619 (2002)
5. Comaniciu, D., Ramesh, V., Meer, P.: Kernel-based Object Tracking. IEEE Trans. PAMI 25(5), 564–577 (2003)
6. DeMenthon, D., Megret, R.: Spatio-temporal segmentation of video by hierarchical mean shift analysis. Computer Vision Laboratory, Center for Automation Research, University of Maryland (2002)
7. Fukunaga, K., Hostetler, L.: The estimation of the gradient of a density function, with applications in pattern recognition. IEEE Transactions on Information Theory 21(1), 32–40 (1975)
8. Georgescu, B., Shimshoni, I., Meer, P.: Mean shift based clustering in high dimensions: A texture classification example. In: Proceedings of the 9th International Conference on Computer Vision, pp. 456–463 (2003)
9. Grzymała-Busse, J.W., Hu, M.: A comparison of several approaches to missing attribute values in data mining. In: Ziarko, W.P., Yao, Y. (eds.) RSCTC 2000. LNCS (LNAI), vol. 2005, pp. 378–385. Springer, Heidelberg (2001)
10. Magnani, M.: Techniques for dealing with missing data in knowledge discovery tasks. Obtido 15(01), 2007 (2004),
http://magnanim.web.cs.unibo.it/index.html
11. Rand, W.M.: Objective criteria for the evaluation of clustering methods. Journal of the American Statistical Association 66(336), 846–850 (1971)
12. Suguna, N., Thanushkodi, K.G.: Predicting missing attribute values using k-means clustering. Journal of Computer Science 7(2), 216–224 (2011)
13. Tao, W., Jin, H., Zhang, Y.: Color image segmentation based on mean shift and normalized cuts. IEEE Trans. on Systems, Man, and Cybernetics, Part B 37(5), 1382–1389 (2007)

14. Speech University of Eastern Finland and Image Processing Unit. Clustering dataset, http://cs.joensuu.fi/sipu/datasets/
15. Zhang, S., Qin, Z., Ling, C.X., Sheng, S.: "Missing is useful": missing values in cost-sensitive decision trees. IEEE Trans. on Knowledge and Data Engineering 17(12), 1689–1693 (2005)
16. Zhang, S.: Shell-neighbor method and its application in missing data imputation. Applied Intelligence 35(1), 123–133 (2011)

Mining Recurrent Concepts in Data Streams
Using the Discrete Fourier Transform

Sakthithasan Sripirakas and Russel Pears

Auckland University of Technology, New Zealand
{ssakthit,rpears}@aut.ac.nz

Abstract. In this research we address the problem of capturing recurring concepts in a data stream environment. Recurrence capture enables the re-use of previously learned classifiers without the need for re-learning while providing for better accuracy during the concept recurrence interval. We capture concepts by applying the Discrete Fourier Transform (DFT) to Decision Tree classifiers to obtain highly compressed versions of the trees at concept drift points in the stream and store such trees in a repository for future use. Our empirical results on real world and synthetic data exhibiting varying degrees of recurrence show that the Fourier compressed trees are more robust to noise and are able to capture recurring concepts with higher precision than a meta learning approach that chooses to re-use classifiers in their originally occurring form.

Keywords: Data Stream Mining, Concept Drift Detection, Recurrent Concepts, Discrete Fourier Transform.

1 Introduction

Data stream mining has been the subject of extensive research over the last decade or so. One of the major issues with data stream mining is dealing with concept drift that causes models built by classifiers to degrade in accuracy over a period of time.

While data steam environments require that models are updated to reflect current concepts, the capture and storage of recurrent concepts allows a classifier to use an older version of the model that provides a better fit with newly arriving data in place of the current model. This approach removes the need to explicitly re-learn the model, thus improving both accuracy and computational cost. A number of methods have been proposed that deal with the capture and exploitation of recurring concepts [4], [5], [7], [1] and [12]. Although achieving higher accuracy as expected during phases of concept recurrence in the stream, a major issue with existing approaches is the setting of user defined parameters to determine whether a current concept matches with one from the past.

Such parameters are difficult to set, particularly due to the drifting nature of real world data streams. Our approach avoids this problem by applying the Discrete Fourier Transform (DFT) as a meta learner. The DFT, when applied on a concept (Decision Tree model) results in a spectral representation that captures the classification power of the original models. One very attractive property of the Fourier representation of Decision Tree is that most of the energy and classification power is contained within

L. Bellatreche and M.K. Mohania (Eds.): DaWaK 2014, LNCS 8646, pp. 439–451, 2014.
© Springer International Publishing Switzerland 2014

the low order coefficients [9]. The implication of this is that that when a concept C recurs as concept C* with relatively small differences caused by noise or concept drift, then such differences are likely to manifest in the high order coefficients of spectra S and S* (derived from C and C* respectively), thus increasing the likelihood of C* being recognized as a recurrence of C.

The DFT, apart from its use in meta learning, has a number of other desirable properties that make it attractive for mining high speed data streams. This includes the ability to classify directly from the spectra generated, thus eliminating the need for expensive traversal of a tree structure.

Our experimental results in section 5 clearly show the accuracy, processing speed and memory advantages of applying the DFT as opposed to the meta learning approach proposed by Gama and Kosina in [4].

The rest of the paper is as follows. In section 2 we review work done in the area of capturing recurrences. We describe the basics of applying the DFT to decision trees in section 3. In section 4 we discuss a novel approach to optimizing the computation of the Fourier spectrum from a Decision Tree. Our experimental results are presented in section 5 and we conclude the paper in section 6 where we draw conclusions on the research and discuss some directions for future research.

2   Related Research

While a vast literature on concept drift detection exists [14] only a small body of work exists so far on exploitation of recurrent concepts. The methods that exist fall into two broad categories. Firstly, methods that store past concepts as models and then use a meta learning mechanism to find the best match when a concept drift is triggered [4], [5]. Secondly, methods that store past concepts as an ensemble of classifiers.

Lazarescu in [10] proposes an evidence forgetting mechanism for data instances based on a multiple window approach and a prediction module to adapt classifiers based on an estimation of the future rate of change. Whenever the difference between the observed and estimated rates of change are above a user defined threshold a classifier that best represents the current concept is stored in a repository. Experimentation on the STAGGER dataset showed that the proposed approach outperformed the FLORA method on classification accuracy with re-emergence of previous concepts in the stream.

Ramamurthy and Bhatnagar [15] use an ensemble approach based on a set of classifiers in a global set G. An ensemble of classifiers is built dynamically from a collection of classifiers in G if none of the existing individual classifiers are able to meet a minimum accuracy threshold based on a user defined acceptance factor. Whenever the ensemble accuracy falls below the accuracy threshold, then the global set G is updated with a new classifier trained on the current chunk of data.

Another ensemble based approach by Katakis et al. is proposed in [8]. A mapping function is applied on data stream instances to form conceptual vectors which are then grouped together into a set of clusters. A classifier is incrementally built on each cluster and an ensemble is formed based on the set of classifiers. Experimentation on the Usenet dataset showed that the ensemble approach produced better accuracy than a simple incremental version of the Naive Bayes classifier.

Gomes et al [5] used a two layer approach with the first layer consisting of a set of classifiers trained on the current concept while the second contains classifiers created from past concepts. A concept drift detector is used to flag changes in concept and when a warning state is triggered incoming data instances are buffered in a window to prepare a new classifier. If the number of instances in the warning window is below a user defined threshold then the classifier in layer 1 is used instead of re-using classifiers in layer 2. One major issue with this method is validity of the assumption that explicit contextual information is available in the data stream.

Gama and Kosina also proposed a two layered system in [4] designed for delayed labeling, similar in some respects to the Gomes et al. [5] approach. In their approach Gama and Kosina pair a base classifier in the first layer with a referee in second layer. Referees learn regions of feature space which its corresponding base classifier predicts accurately and is thus able to express a level of confidence on its base classifier with respect to a newly generated concept. The base classifier which receives the highest confidence score is selected, provided that it is above a user defined hit ratio parameter; if not, a new classifier is learned.

3   Application of the Discrete Fourier Transform on Decision Trees

The Discrete Fourier Transform (DFT) has a vast area of application in very diverse domains such as time series analysis, signal processing, image processing and so on. It turns out as Park [13] and Kargupta [9] show that the DFT is very effective in terms of classification when applied on a decision tree model. Kargupta and Park in [9] explored the use of the DFT in a distributed environment but did not explore its usage in a data stream environment as this research sets out to do.

Kargupta and Park in [9] showed that the Fourier spectrum consisting of a set of Fourier coefficients fully captures a decision tree in algebraic form, meaning that the Fourier representation preserves the same classification power as the original decision tree.

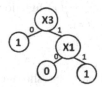

A decision tree can be represented in compact algebraic form by applying the DFT to the paths of the tree. We illustrate the process by considering a binary tree for simplicity but in practice the DFT can be applied to non binary trees as well [9]. For trees with a total of d binary valued features the jth Fourier coefficient ω_j is given by:

Fig. 1.   Decision Tree with 3 binary features

$$\omega_j = \frac{1}{2^d} \sum_x f(x)\psi_j(x)$$ (1)

where $f(x)$ is the classification outcome of path vector x and $\psi_j(x)$, the Fourier basis function is given by:

$$\psi_j(x) = (-1)^{(j.x)}$$ (2)

Figure 1 shows a simple example with 3 binary valued features x_1, x_2 and x_3, out of which only x_1 and x_3 are actually used in the classification.

As shown in [13] only coefficients for paths that are defined by attributes that actually appear in the tree need to be computed as all other coefficients are guaranteed to be zero in value. Thus any coefficient of the form ω_{*1*} will be zero since attribute x_2 does not appear in the tree.

With the wild card operator * in place we can use equations (1) and (2) to calculate non zero coefficients. Thus for example we can compute:

$$\omega_{000} = \frac{4}{8}f(**0)\psi_{000}(**0) + \frac{2}{8}f(0*1)\psi_{000}(0*1) + \frac{2}{8}f(1*1)\psi_{000}f(1*1) = \frac{3}{4}$$

$$\omega_{001} = \frac{4}{8}f(**0)\psi_{001}(**0) + \frac{2}{8}f(0*1)\psi_{001}(0*1) + \frac{2}{8}f(1*1)\psi_{001}f(1*1) = \frac{1}{4}$$

and so on. In addition to the properties discussed above, the Fourier spectrum of a given decision tree has two very useful properties that make it attractive as a tree compression technique [9]:

1. All coefficients corresponding to partitions not defined in the tree are zero.
2. The magnitudes of the Fourier coefficients decrease exponentially with their order, where the order is taken as the number of defining attributes in the partition.

Taken together these properties mean that the spectrum of a decision tree can be approximated by computing only a small number of low order coefficients, thus reducing storage overhead. With a suitable thresholding scheme in place, the Fourier spectrum consisting of the set of low order coefficients is thus an ideal mechanism for capturing past concepts.

Furthermore, classification of unlabeled data instances can be done directly in the Fourier domain as it is well known that the inverse of the DFT defined in expression (3) can be used to recover the classification vector, instead of the use of a tree traversal which can be expensive in the case of deep trees. Expression 3 uses the complex conjugate $\overline{\psi}_j(x)$ function for the inverse operation in place of the original basis function of $\psi_j(x)$.

$$f(x) = \sum_j \omega_j \overline{\psi}_j(x) \tag{3}$$

Due to thresholding and loss of some high order coefficient values the classification value $f(x)$ for a given data instance x may need to be rounded to the nearest integer in order to assign the class value. For example, with binary classes a value for f is rounded up to 1 if it is in the range $[0.5, 1)$ and rounded down to 0 in the range $(0, 0.5)$.

4   Exploitation of the Fourier Transform for Recurrent Concept Capture

We first present the basic algorithm used in section 5.1 and then go on to discuss an optimization that we used for energy thresholding in section 5.2.

4.1   The FCT Algorithm

We use CBDT [6] as the base algorithm which maintains a forest of trees. This forest of trees is dynamic in the sense that it can adapt to changing concepts at drift detection points. We thus define the memory consumed by this forest as *active*.

We integrate the basic CBDT algorithm with the ADWIN [2] drift detector to signal concept drift. At the first concept drift point the best performing tree (in terms of accuracy) is identified and the DFT is applied after energy thresholding after which the resulting spectrum is stored in the repository for future use if the current concept recurs. The spectra stored in the repository are fixed in nature as the intention is to capture past concepts. At each subsequent drift point a winner model is chosen by polling both the active memory and the repository. If the winner emerges from the active memory, two checks are made before the DFT is applied. First of all, we check whether the difference in accuracy between the winner tree in active memory (T) and the best performing model in the repository is greater than a tie threshold τ. If this check is passed then the DFT is applied to T and a further check is made to ensure that its Fourier representation is not already in the Repository. If the winner model at a drift point emerges from an already existing spectrum in the Repository then no Fourier conversion is applied on any of the trees in active memory. Whichever model is chosen as the winner it is applied to classify all unlabeled data instances until a new winner emerges at a subsequent drift point. The least performing model M having the lowest weighted accuracy function is deleted if the repository has no room for new models. The weighted accuracy of M is defined by: $weight(M) = winner\_tally(M) * accuracy(M)$, where $winner\_tally$ is the number of times that M was declared a winner since it was inserted into the repository.

Algorithm *FCT*
Input: Energy Threshold E_T , Accuracy Tie Threshold τ
Output: Best Performing model M that suits current concept
1.   read an instance I from the data stream
2.   **repeat**
3.   Call *Classify* to classify I using the best model M
4.   append 0 to ADWIN's window for M if classification is correct, else append 1
5.   **until** drift is detected by ADWIN
6.   **if** M is from active memory
7.   identify best performing model F in repository
8.   **if** (accuracy(M)-accuracy(F))> τ
9.   apply DFT on model M to produce F* using energy threshold E_T
10.   **if** F* is not already in repository
11.   insert F* into repository
12. Identify best performing model M by polling active memory and repository
13. GoTo 1

Algorithm *Classify*
Input: Instance I, Classifier M
Output: class value
1.   **if** M is a Decision Tree, route I to a leaf and return the class label of the leaf
2.   **else** using all coefficients (j) of M, Calculate $f(x)$ using $f(x) = \sum_j \omega_j \overline{\psi}_j(x)$ where
         $\overline{\psi}_j(x)$ is the the complex conjugate function of $\psi_j(x)$ and x is the instance I
3.   If $f(x)$ is greater than 0.5, return class1, otherwise class2

4.2 Optimizing the Energy Thresholding Process

In order to avoid unnecessary computation of higher order coefficients which yield increasingly low returns on classification accuracy, energy threshold is highly desirable. To threshold on energy a subset S of the (lower order) coefficients needs to be determined such that $\frac{E(S)}{E(T)} > \epsilon$, where $E(T)$ denotes the total energy across the spectrum and ϵ is the desired energy threshold value.

In our optimized thresholding, we first compute the cumulative energy CE_i at order i given by: $CE_i = \sum_{j=0}^{i} \sum_k (w_k{}^2 | order(k) = j)$.

Given an order i, an upper bound estimate for the cumulative energy across the rest of the spectrum is given by: $(d+1-(i+1)+1)CE_i$, as the exponential decay property ensures that the energy at each of the orders $i+1, i+2, \cdots, d$ is less than energy E_i at order i, where d is number of attributes in the dataset. Thus a lower bound estimate for the fraction of the cumulative energy CEF_i at order i to the total energy across all orders can then be expressed as:

$$CEF_i = \frac{CE_i}{CE_i + (d-i+1)E_i} \tag{4}$$

where E_i is actual (computed) energy at order i. The lower bound estimate allows the specification of a threshold ϵ based on the energy captured by a given order i which is more meaningful to set rather than an arbitrary threshold.

The scheme expressed by equation (4) enables the thresholding process to be done algorithmically. If the cumulative energy $CEF_i \geq \epsilon$, then we can guarantee that the actual energy captured is at least ϵ, since CEF_i is a lower bound estimate. On the other hand if $CEF_i < \epsilon$, then CEF_{i+1} can be expressed as:

$$CEF_{i+1} = \frac{CE_{i+1}}{CE_{i+1} + (d-i)E_{i+1}} = \frac{CE_i + E_{i+1}}{CE_i + dE_{i+1}} \tag{5}$$

Thus equation (5) enables the cumulative fraction to be easily updated incrementally for the next higher order (i+1) by simply computing the energy at that order while exploiting the exponential decay property of Fourier spectrum. The thresholding method guarantees that no early termination will take place. This is because CEF_i is a lower bound estimate and hence the order that it returns will never be less than the true order that captures a given fraction ϵ of the total actual energy in the spectrum.

5 Experimental Study

This section elaborates on our empirical study involving the following learning systems: CBDT, FCT (Fourier Concept Trees) and MetaCT. The FCT incorporates the Fourier compressed trees in a repository in addition to the forest of trees that standard CBDT maintains. We implement Gama's meta learning approach with CBDT as the base learner, namely MetaCT. The main focus of the study is to assess the extent to which recurrences are recognized using old models preserved in classifier pools.

5.1 Parameter Values

All experimentation was done with the following parameter values:

- Hoeffding Tree Parameters The desired probability of choosing the correct split attribute=0.99, Tie Threshold=0.01, Growth check interval=32
- Tree Forest Parameters Maximum Node Count=5000, Maximum Number of Fourier Trees=50, Accuracy Tie Threshold τ=0.01
- ADWIN Parameters drift significance value=0.01, warning significance value=0.3 (MetaCT only)

All experiments were done on the same software with C# .net runtime and hardware with Intel i5 CPU and 8GB RAM, clearning the memory in each run to have a fair comparison.

5.2 Datasets Used for the Experimental Study

We experimented with data generated from 3 data generators commonly used in drift detection and recurrent concept mining, namely SEA concept [16], RBF and Rotating hyperplane generators. In addition we used 2 real-world datasets, *Spam* and the *NSW electricity* which have also been commonly used in previous research.

For the synthetic datasets, each of the 4 concepts spanned 5,000 instances and reappeared 25 times in a data set, yielding a total of 500,000 instances with 100 true concept drift points.

In order to challenge the concept recognition process, we added a 10% noise level for all synthetic data sets to ensure that concepts recur in similar, but not exact form.

Synthetic Data Sets. We used MOA [3] as the tool to generate these datasets.

1. **SEA:** The concepts are defined by the function $feature1 + feature2 > threshold$. We ordered the concepts as concept1, concept2, concept3 and concept4 generated using threshold values 8,7,9 and 9.5 respectively on the first data segment of size 20,000. We generated 96 recurrences of a modified form of these concepts by using different seed values in MOA for each sequence of recurrence. Thus, our version of this dataset differed from the one used by Gama and Kosina [4]. who simply used 3 concepts with the third being an exact copy of the first.
2. **RBF:** The number of centroids parameter was adjusted to generate different concepts for the RBF dataset. Concept1, concept2, concept3 and concept4 were produced with the number of centroids set to 5,15, 25 and 35 respectively. Similar to the SEA dataset, the seed parameter helped in producing similar concepts for a given centroid count value. This dataset had 10 attributes.
3. **Rotating hyperplane:** The number of drifting attributes was adjusted to 2,4,6, and 8 in a 10 dimensional data set to create the four concepts. The concept ordering, generation of similar concepts and concatenation were exactly the same as in the other data sets mentioned above.

Real World Datasets

1. *Spam Data Set:* The Spam dataset was used in it original form[1] which encapsulates an evolution of Spam messages. There are 9,324 instances and 499 informative attributes, which was different from the one version used by Gama that had 850 attributes.
2. *Electricity Data Set:* NSW Electricity dataset is also used in its original form [2]. There are two classes *Up* and *Down* that indicate the change of price with respect to the moving average of the prices in last 24 hours.

5.3   Tuning MetaCT Key Parameters

In our preliminary experiments, we found optimal values for the two parameters, *delay* in receiving labels for the instances in short term memory, and *hit percentage threshold value* as 200 and 80%, respectively. The latter parameter reflects the estimated similarity of the current concept with one from the past and thus controls the degree of usage of classifiers from the pool.

5.4   Comparative Study: CBDT vs FCT vs MetaCT

Our focus in this series of experiments was to assess the models in terms of accuracy, memory consumption and processing times. None of the previous studies reported in the recurrent concept mining literature undertook a comparative study against other approaches and so we believe our empirical study to be the first such effort. Furthermore, all of the previous studies focused exclusively on accuracy without tracking memory and execution time overheads and so this study would also be the first of its kind.

Accuracy. A delay period of 200 was used with all three approaches in order to perform a fair comparison. Figure 2 clearly shows that overall, FCT significantly outperforms its two rivals with respect to classification accuracy. The major reason for FCT's superior performance was its ability to re-use previous classifiers as shown in the segment 20k-25k on the RBF dataset where the concept is similar to concept1 that occurred in interval 1-5K. This is in contrast to MetaCT which was unable to recognize the recurrence of concept1. A similar situation occurs in the interval 25k-35k where the concept is similar to the previously occurring concepts, which are concept2 and 3. As expected CBDT, operating on its own without support for concept recurrence had a relatively flat trajectory throughout the stream segment.

A similar trend to the RBF dataset was observed in Rotating Hyperplane and SEA datasets as well. It can be clearly seen that FCT was successful in reusing the models learned before on data segments from 20k to 25k and from 30k to 35k. Though a preserved model was reused on the data segment from 25k to 30k (corresponding to concept3), the accuracy was not as high as in the above two segments. On the segment from 35k to 40k, concept recurrence was not picked up by either FCT or MetaCT resulting in a new classifier being used.

---

[1] From http://www.liaad.up.pt/kdus/
products/datasets-for-concept-drift
[2] From http://moa.cms.waikato.ac.nz/datasets/

We omit the figure for the SEA dataset due to space constraints. In summary, FCT outperformed MetaCT over 90 recurring concepts whereas MetaCT did better in 6 occurrences, thus maintaining the same trend as with the other 2 synthetic datasets that we experimented with.

The next experiment was on the NSW Electricity data set. Figure 2 shows that overall, FCT was the winner here as well, outperforming MetaCT at 25 segments out of 35 that we tracked. Sudden fall in accuracy of MetaCT is occational but due to incorrect selection of winner which was a decision stump.

Memory. Our experimentation on accuracy has revealed, especially in the case of FCT, the key role that concept capture and re-use has played in improving accuracy. The question is, what price has to be paid in terms of memory overhead in storing these recurrent concepts? Table1 clearly shows that the Fourier transformed trees consume a small fraction of the memory used by the pool of trees kept in FCT's active memory, despite the fact that collectively these models outperform their decision tree counterparts at a greater number of points in the stream progression.

Comparison of overall memory consumption across FCT and MetaCT is complicated by the fact that the latter tended to have immature trees in its classifier pool that under fits concepts. Despite this, Table 1 reveals that FCT's memory consumption is competitive to that of MetaCT. The only case where MetaCT had a substantially lower consumption was with the Spam dataset with a lower overhead for active memory.

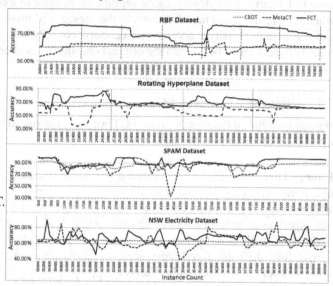

Fig. 2. Classification Accuracy for CBDT, FCT and MetaCT

Processing Speed. FCT and MetaCT have two very contrasting methods of classification. The former routes each unlabeled instance to a single tree, which is the best performing tree selected at the last concept drift point. In contrast MetaCT classifies by routing an unlabeled instance to all referees to obtain their assessment of their corresponding models and in general will have more processing overhead on a per instance basis. However, FCT has potentially more overhead at concept drift points if the winner tree is one that is selected from the active forest as this tree needs to be converted into its Fourier representation. Thus it is interesting to contrast the run time performances of the two approaches.

Table 1 shows that in general FCT has a higher processing speed (measured in instances processed per second); the only exception was with the Electricity dataset where MetaCT was faster. The electricity data contains a relatively larger number of drift points in comparison to the other datasets and this in turn required a greater number of DFT operations to be

Table 1. Average Memory Consumption (in KBs) and Processing Speed (Instances per second) Comparison

Datasets	Memory FCT		Memory MetaCT		Processing Speed	Processing Speed
	Tree Forest	Fourier Pool	Tree Forest	Pool	FCT	MetaCT
RBF	97.9	24.8	122.7	14.9	3540.6	2662.5
Rot. Hy/plane	187.4	59.7	148.7	43.4	2686.2	2180.1
SEA	29.3	34.8	28.0	18.1	11368.2	10125.8
Spam	1712.8	18.8	878.0	15.3	4.1	4.3
Electricity	48.4	39.9	19.8	18.9	5705.7	7191.42

performed, thus slowing down the processing. In our future research we will investigate methods of optimizing the DFT process.

Finally, we close this section with two general observations on FCT which hold across all experiments reported above. Firstly, we note that the Discrete Fourier Transform (DFT), as expected, was able to capture the *essence* of a concept to the extent that when it reappeared in a modified form in the presence of noise it was still recognizable and was able to classify it accurately. Secondly, not only was the DFT robust to noise, it actually performed better than the original decision trees at concept recurrence points due to its better generalization capability.

5.5   Sensitivity Analysis on FCT

Having established the superiority of FCT we were interested in exploring the sensitivity of FCT's accuracy on two key factors.

Energy Threshold. FCT's energy threshold parameter controls the extent to which it captures recurring contexts. We ran experiments with all datasets we experimented with and tracked accuracy across four different thresholds: 95%, 80%, 40% and 20%. The trends observed for all datasets were very similar and hence we display results for the SEA

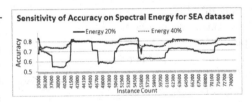

Fig. 3. Sensitivity of Accuracy on Spectral Energy

concepts dataset due to space constraints. Figure 3 clearly shows that very little difference in accuracy exists between the trajectories for 40% and 95%, showing the resilience of the DFT in capturing the classification power of concepts at low energy levels such as 40%. Thus the low order Fourier coefficients that survive the 40% threshold hold almost the same classification power of spectra at the 80% or 90% levels which contain more coefficients. Such higher energy spectra would represent larger decision trees in which some of the decision nodes would be split into leaf nodes, thus enabling them to reach a slightly higher level of accuracy.

Noise Level. In section 5.4 we observed that FCT outperformed MetaCT by recognizing concepts from the past even though the concepts did not recur exactly in their original form due partly to noise and partly due to different data instances being produced as a result of re-seeding of the concept generation functions. In this experiment we explicitly test the resilience of FCT to noise level by subjecting it to three different levels of noise - 10%, 20% and 30%. For reasons of completeness we also included MetaCT in he experimentation to aid in the interpretation of results.

Figure 4 reveals three interesting pieces of information. Firstly, FCT is still able to recognize recurring concepts at the 20% noise level even though the models it re-uses do not have quite the same classification power (when compared to the 10% noise level) on the current concept due to data instances being corrupted by a relatively higher level of noise.

Secondly, FCT's concept recurrence recognition is essentially disabled at the 30% noise level as shown by its flat trajectory, thus essentially performing at the level of the base CBDT system. It is able to avoid drops in accuracy on account of the forest of trees that is maintained and is able to switch quickly and seamlessly from one tree to another when concepts change occurs.

Thirdly, although MetaCT is not the focus of this experiment we see that MetaCT's ability to recognize recurring concepts is disabled at the 20% level, showing once gain the resilience of the DFT to noise. At the 30% level its accuracy drops quite sharply at certain points in the stream. This is due to the fact that it learns a single new classifier and relies on it to classify instances in the current concept. In contrast, FCT exploits the entire forest of trees and switches from one tree to another tree in its active forest in response to drift.

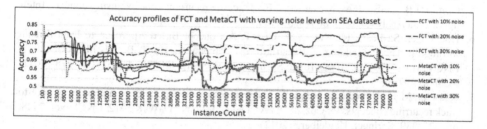

Fig. 4. Sensitivity of Accuracy for FCT and MetaCT on Noise Level

6 Conclusions and Future Work

In this research we proposed a novel mechanism for mining data streams by capturing and exploiting recurring concepts. Our experimentation showed that the Discrete Fourier Transform when applied on Decision Trees captures concepts very effectively, both in terms of information content and conciseness. The Fourier transformed trees were robust to noise and were thus able to recognize concepts that reappeared in modified form, thus contributing significantly to improving accuracy. Overall our proposed approach significantly outperformed the meta learning approach by Gama and Kosina [4] in terms of classification accuracy while being competitive in terms of memory and processing speed.

We were able to optimize the derivation of the Fourier spectrum by an efficient thresholding process but there is further scope for optimization in the computation of low order coefficients in streams exhibiting frequent drifts, as our experimentation with the NSW Electricity dataset reveals. Our future work will concentrate on two areas. Firstly we plan to investigate the use of multi-threading on a parallel processor platform to optimize the DFT operation. Allocating the DFT process to a thread while another thread processes the incoming stream will greatly speed up processing for FCT as the two processes are independent of each other and can be executed in parallel. Secondly, the computation of the Fourier basis function that requires a vector dot product computation can be optimized by using patterns in the two vectors involved.

References

1. Alippi, C., Boracchi, G., Roveri, M.: Just-In-Time Classifiers for Recurrent Concepts. IEEE Transactions on Neural Networks and Learning Systems 24(4), 620–634 (2013), doi:10.1109/tnnls.2013.2239309
2. Bifet, A., Gavaldà, R.: Learning from Time-Changing Data with Adaptive Windowing. In: Symposium Conducted at the Meeting of the 2007 SIAM International Conference on Data Mining (SDM 2007), Minneapolis, Minnesota (2007)
3. Bifet, A., Holmes, G., Kirkby, R., Pfahringer, B.: Moa: Massive online analysis. The Journal of Machine Learning Research 11, 1601–1604 (2010)
4. Gama, J., Kosina, P.: Tracking recurring concepts with meta-learners. In: Lopes, L.S., Lau, N., Mariano, P., Rocha, L.M. (eds.) EPIA 2009. LNCS, vol. 5816, pp. 423–434. Springer, Heidelberg (2009)
5. Gomes, J.B., Sousa, P.A., Menasalvas, E.: Tracking recurrent concepts using context. Intelligent Data Analysis 16(5), 803–825 (2012)
6. Hoeglinger, S., Pears, R., Koh, Y.S.: CBDT: A Concept Based Approach to Data Stream Mining. In: Theeramunkong, T., Kijsirikul, B., Cercone, N., Ho, T.-B. (eds.) PAKDD 2009. LNCS, vol. 5476, pp. 1006–1012. Springer, Heidelberg (2009), doi:10.1007/978-3-642-01307-2_107
7. Hosseini, M.J., Ahmadi, Z., Beigy, H.: New management operations on classifiers pool to track recurring concepts. In: Cuzzocrea, A., Dayal, U. (eds.) DaWaK 2012. LNCS, vol. 7448, pp. 327–339. Springer, Heidelberg (2012)
8. Katakis, I., Tsoumakas, G., Vlahavas, I.P.: An Ensemble of Classifiers for coping with Recurring Contexts in Data Streams. In: Symposium Conducted at the Meeting of the ECAI (2008)
9. Kargupta, H., Park, B.-H.: A Fourier Spectrum-Based Approach to Represent Decision Trees for Mining Data Streams in Mobile Environments. IEEE Trans. on Knowl. and Data Eng. 16(2), 216–229 (2004), doi:10.1109/tkde.2004.1269599
10. Lazarescu, M.: A Multi-Resolution Learning Approach to Tracking Concept Drift and Recurrent Concepts. In: Symposium Conducted at the Meeting of the PRIS (2005)
11. Linial, N., Mansour, Y., Nisan, N.: Constant depth circuits, Fourier transform, and learnability. Journal of the ACM 40(3), 607–620 (1993), doi:10.1145/174130.174138
12. Morshedlou, H., Barforoush, A.A.: A new history based method to handle the recurring concept shifts in data streams. World Acad. Sci. Eng. Technol. 58, 917–922 (2009)
13. Park, B.-H.: Knowledge discovery from heterogeneous data streams using fourier spectrum of decision trees. Washington State University (2001)
14. Pears, R., Sakthithasan, S., Koh, Y.: Detecting concept change in dynamic data streams. Machine Learning, 1–35 (2014), doi:10.1007/s10994-013-5433-9

15. Ramamurthy, S., Bhatnagar, R. (2007). Tracking recurrent concept drift in streaming data using ensemble classifiers. In: Symposium Conducted at the Meeting of the Sixth International Conference on Machine Learning and Applications (2007)
16. Street, W.N., Kim, Y.: A streaming ensemble algorithm (SEA) for large-scale classification. Presented at the Meeting of the Proceedings of the Seventh ACM SIGKDD International Conference on Knowledge Discovery and Data Mining, San Francisco, California (2001), doi:10.1145/502512.502568

Semi-Supervised Learning to Support the Exploration of Association Rules

Veronica Oliveira de Carvalho[1], Renan de Padua[2], and Solange Oliveira Rezende[2]

[1] Instituto de Geociências e Ciências Exatas,
UNESP - Univ Estadual Paulista, Rio Claro, Brazil
`veronica@rc.unesp.br`
[2] Instituto de Ciências Matemáticas e de Computação,
USP - Universidade de São Paulo, São Carlos, Brazil
`{padua,solange}@icmc.usp.br`

Abstract. In the last years, many approaches for post-processing association rules have been proposed. The automatics are simple to use, but they don't consider users' subjectivity. Unlike, the approaches that consider subjectivity need an explicit description of the users' knowledge and/or interests, requiring a considerable time from the user. Looking at the problem from another perspective, post-processing can be seen as a classification task, in which the user labels some rules as interesting [I] or not interesting [NI], for example, in order to propagate these labels to the other unlabeled rules. This work presents a framework for post-processing association rules that uses semi-supervised learning in which: (a) the user is constantly directed to the [I] patterns of the domain, minimizing his exploration effort by reducing the exploration space, since his knowledge and/or interests are iteratively propagated; (b) the users' subjectivity is considered without using any formalism, making the task simpler.

Keywords: Association Rules, Post-processing, Semi-supervised Learning (SSL).

1   Introduction

Association is a widely used task in data mining that has been applied in many domains due to its simplicity and comprehensibility. Since the task discovers strong correlations that may exist among the data set items, the problem is the number of patterns that are obtained. Even a small data set can generate a sufficient number of rules that can overload the user in the post-processing phase. In fact, finding interesting patterns in this huge exploration space becomes a challenge. It is infeasible for the user to explicitly explore all the obtained patterns in order to identify whether they are relevant or not.

Many post-processing approaches have been proposed to overcome the exposed problem (Section 2). The aim is to provide tools that allow users to find

L. Bellatreche and M.K. Mohania (Eds.): DaWaK 2014, LNCS 8646, pp. 452–464, 2014.

the interesting patterns of the domain so that their effort is minimized – the idea is that users don't need to explore all the rules in order to identify what is relevant or not. Post-processing approaches can be automatic or not, i.e., if it is necessary or not to provide information to achieve the desired answers. Although the automatics are simple to use, as objective evaluation measures, the users' subjectivity is not considered. Thereby, since the user is the person who will in fact validate the results, many approaches consider the user's domain knowledge and/or interests. In these cases, the user explicitly describes, through some formalism (ontologies, schemas, etc.), his knowledge and/or interests. However, providing such descriptions requires a considerable time from the user, which may lead to incomplete and/or incorrect specifications – sometimes known relations are forgotten and a previous knowledge which a user has another user may not have. Additionally, in most of the times, the user doesn't have an idea of what is probably interesting, nor from which relations to start the search, since the mining motivation is to support the user to find what he doesn't know.

Considering the post-processing phase from another perspective, the problem can be seen as a classification task, in which the user must label the rules as interesting [I] or not interesting [NI], for example (in fact, other classes may exist). As mentioned before, it is infeasible to explicitly explore all the obtained patterns in order to split the ones that are [I] from the patterns that are [NI]. However, if the user could label few rules and propagate these labels to the other unlabeled rules, the user would minimize his exploration effort. In this context, the semi-supervised learning (SSL) seems useful, since it is suitable when there are many unlabeled data and few labeled data. Besides, its use is also adequate when labeled data are expensive and/or scarce to obtain: it is expensive to discover the rules' labels, since the user must do the labeling process.

Based on the exposed, it would be relevant to develop a framework in which the user's knowledge and/or interests be implicitly obtained, through an iterative and interactive process, in such a way that this information be automatically propagated through the rule set, in order to direct the user to the [I] patterns of the domain. This work presents a framework for post-processing association rules that uses SSL to direct the users to the [I] patterns of the domain: starting from a subset of rules evaluated (labeled) by the user and suggested by the framework in order to implicitly capture the user's knowledge and/or interests, a SSL algorithm is applied to propagate the obtained labels to the rules which are not evaluated yet. Thereby, this paper contributes with current researches since: (a) the user is constantly directed to the potentially [I] patterns of the domain, which minimize his exploration effort through a reduction in the exploration space, once his knowledge and/or interests are iteratively propagated; (b) the user's subjectivity is considered, although his knowledge and/or interests be implicitly obtained, without using any formalism, making this specification task simpler. To the best of our knowledge, this is the first work that discusses a framework for post-processing association rules based on SSL.

This paper is organized as follows. Section 2 describes related researches as basic concepts. Section 3 presents the proposed framework. Section 4 describes some experiments that were carried out to analyze the framework. Section 5 discusses the results obtained in the experiments. Finally, conclusion is given in Section 6.

2   Background

In this section a brief introduction of the concepts related to the paper are presented, as well as the related works.

Association Rules Post-processing Approaches. The aim of the post-processing approaches is to provide tools that allow users to find the interesting patterns of the domain in order to minimize their effort. In the [*Filtering by Constraints*] approaches the user explores the rules through constraints imposed on the mined patterns (examples in [1,2]). To specify these constraints some formalism, as templates and/or schemas, are used. In the [*Evaluation Measures*] approaches the rules are evaluated according to their relevance (examples in [3]). These measures are usually classified as objective or subjective: the objectives depend on the pattern structure; the subjectives depend on the user's interests and/or needs. In the [*Summarization*] approaches the aim is to condense the discovered rules in general concepts to provide an overview of the extracted patterns (examples in [4,5]). The abstraction can be done, for example, through generalization processes through ontologies. In the [*Grouping*] approaches the aim is to provide groups of similar rules to organize the patterns (examples in [6,7]). Frequently, grouping is done through clustering algorithms. In the [*Pruning*] approaches the aim is to find what is interesting by removing what is not interesting (examples in [1,2,7]). Finally, in the [*Hybrid*] approaches two or more of the previously approaches are combined (examples in [6,1,2,7]). In this case, they are related with an interactive process, i.e., in which the user is needed – these interactions have been done through the codification of the user's knowledge. Thus, to the best of our knowledge, this is the first work that presents a framework for post-processing based on SSL. Besides, the framework contains some elements of the above approaches and, so, it can be categorized in some of them: "Filtering by Constraints", "Grouping", "Pruning" and "Hybrid".

Semi-supervised Learning [8]. SSL is between supervised and unsupervised learning and, so, it learns from both labeled and unlabeled data. SSL is useful when labeled data are scarce and/or expensive – it is difficult to obtain, in some tasks, a reasonable number of labeled data, since human annotators, special devices, etc. can be necessary. Thereby, the goal is to find a function f, both from labeled and unlabeled data, that will better map the domain in relation to a function f found only over a few number of labeled data. Distinct SSL methods exist, as self-training, co-training and graph-based models, in which each method considers a different assumption about the existing relation between the unlabeled data distribution and the target label. Thus, the performance of the SSL depends on the correctness of the assumptions made by these methods. However, selecting the best assumption for a given application is an open question in the field. One of the simplest SSL methods is self-training: in this case, the learning process uses its own predictions to teach itself. As advantages its simplicity and the fact of being a wrapper method can be cited. As a disadvantage the propagation of errors can be cited: an initial error through f can reinforce its own errors, leading to incorrect classifications. Finally, the self-training assumption considers that its own predictions, at least the more confident, tend to be correct.

3   Proposed Framework for Post-processing Association Rules

The proposed framework for post-processing association rules, seen in Figure 1, is presented in this section. Initially, the association rule set R, seen as the training set, contains all the non evaluated knowledge of the domain. In other words, the training set only contains unlabeled data, i.e., unlabeled rules. Therefore, in the beginning, a subset S of association rules (AR) is automatically selected to be classified, i.e., labeled, by the user (Step [A]). The user labels the rules in S, presented to him, based on some predefined classes (Step [B]). After this, the training set R contains both labeled and unlabeled rules. At this point, a SSL algorithm is applied in order to propagate the user's knowledge and/or interests (labels), implicitly obtained, for all the other rules (Step C). Finally, a stopping criterion is evaluated (Step [D]). If the stopping criterion is met, the rules, which are already classified, are shown to the user; otherwise (Step [E]), some rules are again specified as unlabeled, in order to re-start the process iteratively. This step allows some of the rules' labels to change during the process due to the user's knowledge and/or interests that are obtained through the iterations.

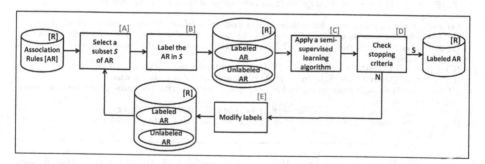

Fig. 1. The proposed framework

The proposed framework provides a post-processing strategy which has not been investigated yet in the literature (see Section 2) and which is flexible enough, since:

- many selection criteria, regarding Step [A], can be used. In this work, a combination of coverage, interest and rule size was used. A rule $r : a \Rightarrow c$ covers a rule $r' : b \Rightarrow d$ if $b \subseteq a$ and $d \subseteq c$; thus, the number of rules covered by r gives it coverage. The subset S is selected as described in Algorithm 1. Basically, n rules are chosen from R, in which $n/2$ are general and $n/2$ are specific. A rule is general if it has 2 or 3 items (rule size = 2 or 3) and specific if it has 4 or 5 items (rule size = 4 or 5) – we worked with rules composed of a maximum of 5 items (see Section 4). Since the rules in R form a lattice, the idea was to carry out a bidirectional search in this exploration space. First of all, the algorithm tries to pick up these n rules from the rules implicitly labeled as [I], also considering their coverage (the higher the coverage the better). Implicit means that the rule was automatically labeled by the SSL algorithm; explicit, on the other hand, means that the user saw the rule and labeled it[1]. If there are more than n implicit [I] rules, the non selected rules are set as unlabeled to ensure that the rules' labels can change through the iteration (details in Step [E]). If there aren't enough rules in this first subset (coverage+interest), the remaining rules to complete the amount of n are selected from the unlabeled ones, also considering their coverage. Specific details are presented in Algorithm 1 with comments. Other criteria will be explored in future works, since they correlate with the SSL methods in use – graphs based methods, for example, provide centrality measures (betweenness, closeness, etc.).

Algorithm 1. Step [A] procedure

Input: R: an association rule set, n: number of rules to be selected.
Output: S: subset of rules to be labeled by the user.
1: $S_1 :=$ rules in R implicitly labeled as [I] ordered by coverage (highest to lowest)
2: $S_2 :=$ unlabeled rules ordered by coverage
3: $nr := n/2$ {number of general and specific rules to be selected from S_1}
4: $S_g :=$ the first nr general rules from S_1
5: $S_e :=$ the first nr specific rules from S_1
6: {if there aren't enough rules in S_g and S_e, complete the sets with rules containing opposite sizes – ensures to select among the implicit [I] ones}
7: **if** $(|S_g| < nr)$ **then** $S_g := S_g \cup (nr - |S_g|)$ next specific rules from S_1 **end if**
8: **if** $(|S_e| < nr)$ **then** $S_e := S_e \cup (nr - |S_e|)$ next general rules from S_1 **end if**
9: $S_1 := S_1 - (S_g \cup S_e)$
10: Set the remaining rules in S_1 as unlabeled and update R {if more than n rules appear in S_1, set the non selected ones as unlabeled – ensures changes in the rules' labels during the iteration}
11: {if there aren't enough rules in S_g and S_e, complete the sets with the unlabeled ones containing the same sizes}
12: **if** $(|S_g| < nr)$ **then** $S_g := S_g \cup (nr - |S_g|)$ next general rules from S_2 **end if**
13: **if** $(|S_e| < nr)$ **then** $S_e := S_e \cup (nr - |S_e|)$ next specific rules from S_2 **end if**
14: {if there aren't enough rules in S_g and S_e, complete the sets with the unlabeled ones containing opposite sizes}
15: **if** $(|S_g| < nr)$ **then** $S_g := S_g \cup (nr - |S_g|)$ next specific rules from S_2 **end if**
16: **if** $(|S_e| < nr)$ **then** $S_e := S_e \cup (nr - |S_e|)$ next general rules from S_2 **end if**
17: $S := S_g \cup S_e$

- many classes, regarding Step [B], can be considered. In this work, the classes [I] and [NI] were used as labels. However, the task of specifying the classes could be allowed to the user, providing more flexibility.
- hypothetically, any SSL method, regarding Step [C], can be used, provided that the methods assumptions be correct. In this work, the self-training method based on kNN was initially used, as seen in Algorithm 2 – as mentioned before,

---

[1] In the first execution of step [A] S_1 is empty, since there are no implicity labeled rule in R – all rules in R start as unlabeled – and, therefore, Algorithm 1 starts in line 12.

the training set is the association rule set R composed with labeled and unlabeled rules. However, other methods will be explored in future works.

• many stopping criteria, regarding Step [D], can also be considered. In this work, the process is executed until a subset G of gold rules be found (details in Section 4). As in the items above, other stopping criteria will be explored in future works.

• the criterion related to the change of the labels, regarding Step [E], ensures the iterativity of the process and the update of the labels. Once the stopping criterion has been checked (Step [D]), some of the rules' labels can be modified in the next round of Step [C] due to the user's knowledge and/or interests that will be obtained during the next iteration (Steps [A]+[B]). This means that some of the labels can change depending on the next user information – these new information can direct the user to another subset of the exploration space. Thereby, this step modifies the implicit rules labeled as [NI] to unlabeled (the implicit rules labeled as [I] are modified to unlabeled in Step [A] due to the same reasons (see Step [A])).

Algorithm 2 Self-training with kNN

Input: R: training set with labeled and unlabeled data (i.e., rules); d: distance function; k: number of neighbors.
Output: R: training set with labeled data.
1: $L := $ labeled data from R; $U :=$ unlabeled data from R
2: **repeat**
3:     Select an unlabeled instance $u := \mathrm{argmin}_{u \in U} \min_{l \in L} d(u, l)$
4:     Set u's label with the same label of its nearest l neighbor
5:     Remove u from U; add u with its label to L
6: **until** U is empty
7: $R := L$

As seen, the proposed framework opens many researches possibilities, since it can be instantiated in many different manners. In this paper, the framework's ideas are explored through simple algorithms in order to demonstrate its feasibility (see Section 4). However, it can be noticed that: (a) the user is constantly directed to the potentially [I] patterns of the domain, which minimize his exploration effort through a reduction in the exploration space, since his knowledge and/or interests are iteratively propagated; (b) the user's subjectivity is considered, although his knowledge and/or interests be implicitly obtained, without using any formalism, making this specification task simpler. Besides, the framework can be seen as a hybrid approach since it contains characteristics of some other post-processing approaches described in Section 2. Finally, the framework also allows that some other post-processing approaches be used as an initial step in order to select and label some rules before starting Step [C] – in this case, the input is a rule set R containing both labeled and unlabeled rules (scenario obtained after applying Steps [A]+[B]).

4 Experiments

Some experiments were carried out in order to demonstrate the feasibility of the framework. Looking at Figure 1, it can be seen that, first of all, a rule set

R is needed. Thus, in order to evaluate the process, six data sets were used, which were divided in relational and transactional. In all of them the rules were extracted with an Apriori implementation[2] with a minimum of two items and a maximum of five items per rule.

The relational were Weather-Nominal (5;14), Contact-Lenses (5;24), Balloons (5;76) and Hayes-Roth (5;132). The numbers in parentheses indicate, respectively, number of attributes and number of instances. The first two are available in Weka[3]; the other two in the UCI Repository[4]. Before extracting the rules, the data sets were converted to a transactional format, where each transaction was composed by pairs of the form "attribute=value". Besides, in order to produce a suitable number of rules a minimum support (min-sup) of 0.0% and a minimum confidence (min-conf) of 0.0% were used in Weather-Nominal, Contact-Lenses and Balloons – in fact, all possible combinations were generated in each one. The values of min-sup=2.5% and min-conf=0.5% were used to Hayes-Roth set. 722 rules were obtained for Weather-Nominal, 890 for Contact-Lenses, 772 for Balloons and 889 for Hayes-Roth.

The transactional were Groceries (9835;169) and Sup (1716;1939). In this case, the numbers in parentheses indicate, respectively, number of transactions and number of distinct items. The first one is available in the R Project for Statistical Computing through the package "arules"[5]. The last one was donated by a supermarket located in São Carlos city, Brazil. With the Groceries data set 1092 rules were generated using a min-sup of 0.7% and a min-conf of 0.5% and with Sup 1149 rules considering a min-sup of 1.25% and a min-conf of 0.5%. The values, as in the cases above, were chosen experimentally.

Given a rule set R, Step [A] is executed. Thus, the number n of rules to be selected, in order to built S, was set to 4 (see Algorithm 1). In Step [B], the user must label the subset S of rules. As human evaluations in distinct data sets are difficult to obtain, a labeling process was simulated as presented in Algorithm 3. Consider that a set G of gold rules exists, i.e., a set containing the interesting rules of the domain that would be obtained by the user if he had evaluated all the rules in the set. For each rule to be labeled, it is computed its distance to each rule in G and the shortest distance is stored. After that, using this distance information, it is checked if the stored distance is $\leq$ a given threshold t, i.e., if the distance of the current rule in relation to a gold rule is small. If so, the rule is labeled as [I]; otherwise as [NI]. In the experiments, t was set to 0.3. The distance function used was the same as in the self-training method (see below). The set G was built as follows: (a) regarding the relational data sets, the C4.5[6] classifier was executed and the rules expressed through the decision trees were used to compose G – although the rules in G contain as consequent the classes of the

---

[2] Developed by Christian Borgelt: http://www.borgelt.net/apriori.html
[3] http://www.cs.waikato.ac.nz/ml/weka/
[4] http://archive.ics.uci.edu/ml/. In Ballons the four available sets were joined. In Hayes-Roth, the first attribute, regarding an ID number, was not considered.
[5] http://cran.r-project.org/web/packages/arules/index.html
[6] In fact, the J48 version available in Weka was executed with the default parameters.

sets, the rules in R contain any pair "attribute=value", since all possible relations were extracted; (b) regarding the transactional data sets, r rules were randomly selected to compose $G - r$ was set to a value representing less than 1% of the total of rules in R to maintain the same pattern as in the relational ones. The number of gold rules in G was: Weather-Nominal G=5 (0.69%), Contact-Lenses G=4 (0.45%), Balloons G=7 (0.91%), Hayes-Roth G=12 (1.35%)[7], Groceries G=7 (0.64%) and Sup G=9 (0.78%).

Algorithm 3 Algorithm used to simulate the labeling process.

Input: S: a subset S of unlabeled rules; G: a set of gold rules; t: threshold; d: distance function.
Output: S: a subset S of labeled rules.
1: **for all** $r \in S$ **do**
2:     $d(r) := \min_{g \in G} d(r, g)$
3: **end for**
4: **for all** $r \in S$ **do**
5:     **if** $d(r) \leq t$ **then** $r := [I]$ **end if**
6:     **if** $d(r) > t$ **then** $r := [NI]$ **end if**
7: **end for**

Table 1. Configurations used to apply the proposed framework

Data sets	Weather-Nominal; Contact-Lenses; Balloons; Hayes-Roth; Groceries; Sup
Step [A]	Algorithm 1; n=4
Step [B]	Algorithm 3; t=0.30; distance function: d_{jacc}
Step [C]	Algorithm 2; k=1; distance function: d_{jacc}
Step [D]	until set G is found

[1]Different values for n, t and k were tested, being the ones here presented the most suitable to the proposed framework.

In relation to kNN, used as the base to execute the self-training method (see Algorithm 2), k was set with 1. Regarding the distance function, an adaptation of Jaccard measure was used: $d_{jacc}(r_1, r_2) = 1 - \frac{Jacc + Jacc_A + Jacc_C}{3}$. The function considers the average similarity, using Jaccard, among all the items in the rules (Jacc $= \frac{|\{items\ in\ r_1\} \cap \{items\ in\ r_2\}|}{|\{items\ in\ r_1\} \cup \{items\ in\ r_2\}|}$), all the items appearing only in the antecedents of the rules (Jacc$_A = \frac{|\{items\ in\ r_1\ antecedent\} \cap \{items\ in\ r_2\ antecedent\}|}{|\{items\ in\ r_1\ antecedent\} \cup \{items\ in\ r_2\ antecedent\}|}$) and all the items appearing only in the consequents of the rules (Jacc$_C = \frac{|\{items\ in\ r_1\ consequent\} \cap \{items\ in\ r_2\ consequent\}|}{|\{items\ in\ r_1\ consequent\} \cup \{items\ in\ r_2\ consequent\}|}$). This strategy distinguishes the similarity between the rules concerning their items' position. Otherwise, only the itemset similarity is measured, not considering the rules' implication. Finally, in relation to the stopping criterion, the process was executed until all the set G of gold rules was found. Table 1 summarizes the configurations used in the experiments.

Lastly, for each data set, a comparison between the proposed framework and a traditional post-processing approach was done. In this case, the rules in R were ranked, considering the average rating obtained through 18 objective measures, as follows ([*Evaluation Measures*] approach (see Section 2)): (i) the value of 18 measures was computed for each rule; (ii) each rule received 18 ID's, each one corresponding to its position in one of the ranks related to a measure; (iii) the average was then calculated based on the ranks' position (ID's). Thereby, based on this rank (the higher the better), the number of rules the user would have to analyze, through this ordered list, to search for all $g \in G$, was computed.

---

[7] In fact, 19 rules were obtained through J.48. However, only 12 were considered, since not all of them were extracted through Apriori due to min-sup and min-conf constraints.

The measures used were Added Value, Certainty Factor, Collective Strength, Confidence, Conviction, IS, ϕ-coefficient, Gini Index, J-Measure, Kappa, Klosgen, λ, Laplace, Lift, Mutual Information (asymmetric), Novelty, Support and Odds Ratio. Details about the measures can be found on [3]. The choice regarding the post-processing approach, used to carry out the comparison, was based on its frequently application in many domains. Besides, it is simple to use and doesn't need any extra information to be processed. Others, like [*Summarization*] through ontologies, require domain specifications, which would imply on domain oriented experiments. However, other types of comparisons must be done in future works.

5   Results and Discussion

Considering the configurations in Table 1, the proposed framework was executed for each data set. Table 2 presents the obtained results. The columns of the table store:

- #rules: number of rules in the rule set R;
- #[I]: number of rules classified as [I] to be exhibited to the user in final of the process (after Step [D]). The pattern X/Y [Z%] indicates: X: number of rules classified as [I] in the end of the execution; Y: number of rules in X explicitly evaluated by the user as [I]; Z: exploration space reduction in relation to the number of rules in R. Looking at Table 2, regarding the Weather-Nominal data set, it can be noticed that: (a) 52 rules were labeled as [I] in the end of the process, in which 23 of them were explicitly classified by the user; (b) this set contains 7.20% of the rules in R and, therefore, the user would obtain a exploration space reduction of 92.80% (100-((52/722)*100));
- #rules evaluated: number of rules explicitly labeled by the user (Step [B]). The number in "[]" also indicates the exploration space reduction in relation to the number of rules in R. Looking at Table 2, regarding the Weather-Nominal data set, it can be noticed that: (a) 100 rules were explicitly labeled by the user; (b) the user would explore 13.85% of the rules in R and, therefore, would obtain a exploration space reduction of 86.15% (100-((100/722)*100));
- #gold rules: number of rules in G found in the end of the process (after Step [D]). In fact, since the stopping criterion occurs when all the set G is found, the pattern X/Y indicates: X: number of gold rules in G; Y: number of rules in X explicitly evaluated by the user. Looking at Table 2, regarding the Weather-Nominal data set, it can be noticed that the 5 rules in G were found, in which 3 of them were explicitly labeled by the user – this means that the other 2 were implicitly classified;
- #iterations: number of iterations executed;
- #explored rules: since the user explicitly labels x rules and, in the end, y rules are returned to him, the total of rules the user explores is: $((\#[I] - \#[I]_E) + \#rules\ evaluated)$. Looking at Table 2, regarding the Weather-Nominal data set, the user would explore the 52 rules in the [I] set, minus the 23 rules already seen, plus the 100 rules labeled during Step [B] ((52 - 23) + 100). This number leads to an exploration space reduction of 82.13% (100-((129/722)*100)).

For each data set, there are two lines: the first one regarding the framework results; the second one regarding the results of the traditional post-processing approach. Thereby, based on a rank list built through objective measures, as discussed in Section 4, Table 2 presents the number of rules the user would have to analyze to search for all the rules $g \in G$. Looking at Table 2, regarding the Weather-Nominal data set, it can be noticed that all the 5 gold rules were found after analyzing the first 120 best classified rules. Thereby, in this case, 120 iterations were made, since each iteration represents an explored rule. Besides, "#[I]" = "#rules evaluated" = "#explored rules" since all the seen rules were evaluated and, once they were on a rank list, all of them were considered as [I] until the stopping criterion was reached. The result set leads to an 83.38% (100-((120/722)*100)) exploration space reduction.

Table 2. Results obtained through the proposed framework, and through a traditional post-processing approach, considering Table 1 configurations

Data set	#rules	#[I]	#rules evaluated	#gold rules	#iterations	#explored rules
Weather-Nominal	722	52/23 [92.80%]	100 [86.15%]	5/3	25	129 [82.13%]
		120 [83.38%]		5/5	120	120 [83.38%] ▲
Contact-Lenses	890	38/31 [95.73%]	224 [74.83%]	4/3	56	231 [74.04%] ▲
		319 [64.16%]		4/4	319	319 [64.16%]
Balloons	772	154/1 [80.05%]	4 [99.48%]	7/0	1	157 [79.66%] ▲
		229 [70.34%]		7/7	229	229 [70.34%]
Hayes-Roth	889	102/64 [88.53%]	308 [65.35%]	12/9	77	346 [61.08%] ▲
		443 [50.17%]		12/12	443	443 [50.17%]
Groceries	1092	13/13 [98.81%]	488 [55.31%]	7/7	122	488 [55.31%] ▲
		1020 [6.59%]		7/7	1020	1020 [6.59%]
Sup	1149	27/27 [97.65%]	328 [71.45%]	9/9	82	328[71.45%] ▲
		1146 [0.26%]		9/9	1146	1146 [0.26%]

Evaluating the results, it can be noticed that:

Weather-Nominal. 52 of the 722 rules in R were labeled as [I], in which 23 of them were explicitly evaluated by the user, being 3 of the 23 in the set G. This [I] set leads to an exploration space reduction of 92.80%. In order to find all the 5 gold rules in G, 25 iterations were executed, enforcing the user to explicitly evaluate 100 rules, implying on an exploration space reduction of 86.15%. After labeling 100 rules and exploring 29 of the unseen [I] rules (29=52-23; 129=100+29), an exploration space reduction of 82.13% was obtained. Compared to the rank list, it would be necessary 120 iterations in order to find all the 5 gold rules in G, leading to an exploration space reduction of 83.38%. Thus, in this case, the traditional approach presents a better performance compared to the proposed framework (▲ sign), although near values have been obtained.

Contact-Lenses. 38 of the 890 rules in R were labeled as [I], in which 31 of them were explicitly evaluated by the user, being 3 of the 31 in the set G. This [I] set leads to an exploration space reduction of 95.73%. In order to find all the 4 gold rules in G, 56 iterations were executed, enforcing the user to explicitly evaluate 224 rules, implying on an exploration space reduction of 74.83%. After labeling 224 rules and exploring 7 of the unseen [I] rules (7=38-31; 231=224+7), an exploration space reduction of 74.04% was obtained. Compared to the rank list, it would be necessary 319 iterations in order to find all the 4 gold rules in G, leading to an exploration space reduction of 64.16%. Thus, in this case, the proposed framework presents a better performance compared to the traditional approach (▲ sign).

Balloons. 154 of the 772 rules in R were labeled as [I], in which 1 of them were explicitly evaluated by the user. This [I] set leads to an exploration space reduction of 80.05%. In order to find all the 7 gold rules in G, 1 iteration was executed, enforcing the user to explicitly evaluate 4 rules, implying on an exploration space reduction of 99.48%. After labeling 4 rules and exploring 153 of the unseen [I] rules (153=154-1; 157=4+153), an exploration space reduction of 79.66% was obtained. Compared to the rank list, it would be necessary 229 iterations in order to find all the 7 gold rules in G, leading to an exploration space reduction of 70.34%. Thus, in this case, the proposed framework presents a better performance compared to the traditional approach (▲ sign).

Hayes-Roth. 102 of the 889 rules in R were labeled as [I], in which 64 of them were explicitly evaluated by the user, being 9 of the 64 in the set G. This [I] set leads to an exploration space reduction of 88.53%. In order to find all the 12 gold rules in G, 77 iterations were executed, enforcing the user to explicitly evaluate 308 rules, implying on an exploration space reduction of 65.35%. After labeling 308 rules and exploring 38 of the unseen [I] rules (38=102-64; 346=308+38), an exploration space reduction of 61.08% was obtained. Compared to the rank list, it would be necessary 443 iterations in order to find all the 12 gold rules in G, leading to an exploration space reduction of 50.17%. Thus, in this case, the proposed framework presents a better performance compared to the traditional approach (▲ sign).

Groceries. 13 of the 1092 rules in R were labeled as [I], in which 13 of them were explicitly evaluated by the user, being 7 of the 13 in the set G. This [I] set leads to an exploration space reduction of 98.81%. In order to find all the 7 gold rules in G, 122 iterations were executed, enforcing the user to explicitly evaluate 488 rules, implying on an exploration space reduction of 55.31%. After labeling 488 rules, an exploration space reduction of 55.31% was obtained. Compared to the rank list, it would be necessary 1020 iterations in order to find all the 7 gold rules in G, leading to an exploration space reduction of 6.59%. Thus, in this case, the proposed framework presents a better performance compared to the traditional approach (▲ sign) – an expressive difference was obtained.

Sup. 27 of the 1149 rules in R were labeled as [I], in which 27 of them were explicitly evaluated by the user, being 9 of the 27 in the set G. This [I] set leads to an exploration space reduction of 97.65%. In order to find all the 9 gold rules in G, 82 iterations were executed, enforcing the user to explicitly evaluate 328 rules, implying on an exploration space reduction of 71.45%. After labeling 328 rules, an exploration space reduction of 71.45% was obtained. Compared to the rank list, it would be necessary 1146 iterations in order to find all the 9 gold rules in G, leading to an exploration space reduction of 0.26%. Thus, in this case, the proposed framework presents a better performance compared to the traditional approach (▲ sign) – an expressive difference was obtained.

Summarizing, it can be observed that: (a) the proposed framework presents excellent results regarding the transactional data sets compared to the objective measures approach, as well as good results regarding the relational data sets; (b) in almost all the cases (5 of 6 (83.33%)) the proposed framework presented better performance compared to the objective measures approach. Thereby, as seen through the experiments, since the user is constantly directed to the potentially [I] patterns of the domain, his exploration effort is minimized through a reduction in the exploration space, once his knowledge and/or interests are iteratively propagated.

6 Conclusion

In this paper a post-processing association rules framework, based on SSL, was proposed. The idea was to treat the post-processing phase as a classification task to: (a) automatically propagate the user's knowledge and/or interests over the rule set, minimizing his effort through a reduction in the exploration space; (b) implicitly obtain the user's knowledge and/or interests, through an iterative and interactive process, without using any formalism. Experiments were carried out in order to demonstrate the framework feasibility. It could be noticed that good results are obtained, using as baseline a traditional post-processing approach.

As seen, the proposed framework opens many researches possibilities, since it can be instantiated in many different manners. Many other configurations can be explored in the framework Steps, mainly regarding the SSL methods related to Step [C]. Furthermore, a case study, with real users, has to be done to analyze the process considering other aspects. Finally, other post-processing approaches could be used as baseline to complement the analysis here presented. However, we think this is the first step to a broad area to be explored.

Acknowledgments. We wish to thank CAPES and FAPESP (2013/12392-0) for the financial support.

References

1. Mansingh, G., Osei-Bryson, K., Reichgelt, H.: Using ontologies to facilitate post-processing of association rules by domain experts. Information Sciences 181(3), 419–434 (2011)
2. Marinica, C., Guillet, F.: Knowledge-based interactive postmining of association rules using ontologies. IEEE TKDE 22(6), 784–797 (2010)
3. Guillet, F., Hamilton, H.J.: Quality Measures in Data Mining. SCI, vol. 43. Springer, Heidelberg (2007)
4. Ayres, R.M.J., Santos, M.T.P.: Mining generalized association rules using fuzzy ontologies with context-based similarity. In: Proceedings of the 14th ICEIS, vol. 1, pp. 74–83 (2012)

5. Carvalho, V.O., Rezende, S.O., Castro, M.: Obtaining and evaluating generalized association rules. In: Proceedings of the 9th ICEIS, vol. 2, pp. 310–315 (2007)
6. de Carvalho, V.O., dos Santos, F.F., Rezende, S.O., de Padua, R.: PAR-COM: A new methodology for post-processing association rules. In: Zhang, R., Zhang, J., Zhang, Z., Filipe, J., Cordeiro, J. (eds.) ICEIS 2011. LNBIP, vol. 102, pp. 66–80. Springer, Heidelberg (2012)
7. Berrado, A., Runger, G.C.: Using metarules to organize and group discovered association rules. Data Mining and Knowledge Discovery 14(3), 409–431 (2007)
8. Zhu, X., Goldberg, A.B.: Introduction to Semi-Supervised Learning, vol. (6). Morgan & Claypool Publishers (2009)

Parameter-Free Extended Edit Distance

Muhammad Marwan Muhammad Fuad

Forskningsparken 3, Institutt for kjemi, NorStruct Universitetet i Tromsø,
NO-9037 Tromsø, Norway
mfu008@post.uit.no

Abstract. The edit distance is the most famous distance to compute the
similarity between two strings of characters. The main drawback of the edit
distance is that it is based on local procedures which reflect only a local view of
similarity. To remedy this problem we presented in a previous work the
extended edit distance, which adds a global view of similarity between two
strings. However, the extended edit distance includes a parameter whose
computation requires a long training time. In this paper we present a new
extension of the edit distance which is parameter-free. We compare the
performance of the new extension to that of the extended edit distance and we
show how they both perform very similarly.

Keywords: Edit Distance, Extended Edit Distance, Parameter-Free Extended
Edit Distance.

1   Introduction

Let U be a universe of objects. The similarity search problem is the process of finding
and retrieving the objects in U that are similar to a given object q; the query. This
problem comes in two flavors; *exact search*, i.e. a query q is given, and the algorithm
retrieves the data objects in U that exactly match q, and the other is *approximate
search* which is motivated by the fact that many exact similarity search methods are
time-consuming, that in some cases the response time becomes unacceptable. Besides,
in many applications, the overhead time necessary to achieve exact search is not
worth the importance of the results obtained.

There are several types of queries, the most famous of which is *range queries*,
which can be defined as: given a query q and a radius r, which represents a *threshold*,
tolerance, or *selectivity*. The range query problem can be specified as retrieving all
the data objects in U that are within a distance r of q. This can be represented as:

$$Range(q,r) = \{u \in U ; d(q,u) \le r\} \tag{1}$$

Another very important type of queries is the *k-nearest neighbor*. In this kind of
queries we look for the most similar, i.e. the closest, object in the database to a given
query. In the general case we look for the k most similar objects. Unlike the case with
range queries, the response set here is never empty. Moreover, its size is defined
beforehand by the user. Formally, this problem can be defined as:

L. Bellatreche and M.K. Mohania (Eds.): DaWaK 2014, LNCS 8646, pp. 465–475, 2014.

$$kNN(q) = \left\{ X \subseteq U, |X| = k \ \wedge \forall u \in X, v \in U - X : d(u,q) \leq d(v,q) \right\} \quad (2)$$

There are still other types of queries such as the *k- reverse nearest neighbor* and *similarity join*.

At the heart of the similarity search problem is the question of how this similarity can be depicted. One of the models that have been presented to tackle this problem is the *metric model*, which is based on the *distance metric*.

In this paper we present a new distance metric applied to sequential data. This new distance is an extension of the well-known *edit distance*. The particularity of the new distance compared with other extensions of the edit distance that we presented before is that it does not include any parameters, whose computing can be very time consuming, thus the new distance can be applied immediately.

The rest of the paper is organized as follows; the necessary background is presented in Section 2, and the new distance is presented in Section 3 with an analysis of its complexity in Section 4, we validate the new distance in Section 5, and related remarks are presented in Section 6. We conclude in Section 7.

2   Background

Let U be a collection of objects. A function d

$d : U \times U \to \mathbf{R}^+ \cup \{0\}$ is called a distance metric if the following holds:

(p1) $d(x,y) \geq 0$ (non-negativity)

(p2) $d(x,y) = d(y,x)$ (symmetry)

(p3) $x = y \Leftrightarrow d(x,y) = 0$ (identity)

(p4) $d(x,z) \leq d(x,y) + d(y,z)$ (triangle inequality)

$\forall x, y, z \in U$. We call (U, d) a metric space.   □

Of the distance metric properties, the triangle inequality is the key property for pruning the search space when processing queries [10].

Search in metric spaces has many advantages, the most famous of which is that a single indexing structure can be applied to several kinds of queries and data types that are so different in nature. This is mainly important in establishing unifying models for the search problem that are independent of the data type. This makes metric spaces a solid structure that is able to deal with several data types [12].

In metric spaces the only operation that can be performed on data objects is computing the distance between any two objects, which enables us to determine the relative location of the data objects to one another. This is different from the case of vector spaces; a special case of metric spaces, where data objects have k real-valued coordinates which makes it possible to perform operations that can not be performed in general metric spaces, like addition or subtraction, for instance. Vectors have certain geometric properties that can be exploited to construct indexing structures, but these properties can not be extended to general metric spaces [1].

A *string* is an ordered set of an alphabet Σ. Strings appear in a variety of domains in computer science and bioinformatics. The main distance used to compare two strings is the edit distance [11], also called the *Levenshtein distance* [3], which is defined as the minimum number of delete, insert, and substitute operations needed to transform string S into string R.

Formally, the edit distance is defined as follows: Let Σ be a finite alphabet, and let Σ^* be the set of strings on Σ. Given two strings $S = s_1 s_2 ... s_n$ and $R = r_1 r_2 ... r_m$ defined on Σ^*. An *elementary edit operation* is defined as a pair: $(a,b) \neq (\lambda, \lambda)$, where a and b are strings of lengths 0 and 1, respectively. The elementary edit operation is usually denoted $a \rightarrow b$ and the three elementary edit operations are $a \rightarrow \lambda$ (deletion) $\lambda \rightarrow b$ (insertion) and $a \rightarrow b$ (substitution). Those three operations can be weighted by a weighting function γ which assigns a nonnegative value to each of these operations. This function can be extended to edit transformations $T = T_1 T_2 .. T_m$.

The edit distance between S and R can then be defined as:

$$ED\ (S, R) = \{\gamma\ (T) \mid T \text{ is an edit transformation of } S \text{ into } R \ \} \tag{3}$$

ED is the main distance measure used to compare two strings and it is widely used in many applications. Fig. 1 shows the edit distance between the two strings $S_1 = \{M, A, R, W, A, N\}$ and $S_2 = \{F, U, A, D\}$

ED has a few drawbacks; the first is that it is a measure of local similarities in which matches between substrings are highly dependent on their positions in the strings. In fact, the edit distance is based on local procedures both in the way it is defined and also in the algorithms used to compute it.

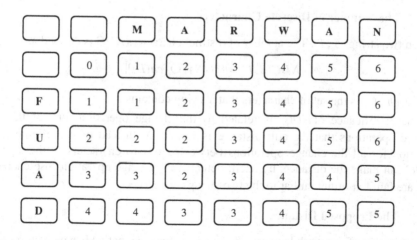

Fig. 1. The edit distance between two strings

Another drawback is that ED does not consider the length of the two strings.

Several modifications have been proposed to improve ED. One of them is the *extended edit distance* [7] [8], which adds a global level of similarity to that of ED.

The Extended Edit Distance: The extended edit distance (EED) is defined as follows:

Let Σ be a finite alphabet, and let Σ^* be the set of strings on Σ. Let $f_a^{(S)}, f_a^{(R)}$ be the frequency of the character a in S and R, respectively, where S, R are two strings in Σ^*. EED is defined as;

$$EED(S,R) = ED(S,R) + \lambda \left[|S| + |R| - 2\sum_{a \in \Sigma} min\left(f_a^{(S)}, f_a^{(R)}\right) \right] \qquad (4)$$

Where $|S|, |T|$ are the lengths of the two strings S, R respectively, and where $\lambda \geq 0$ ($\lambda \in \mathbf{R}^+$). We call λ the co-occurrence frequency factor.

EED is based on the principle that the ED distance does not take into account whether the change operation used a character that is more "familiar" to the two strings or not, because ED considers a local level of similarity only, while EED adds to this local level of similarity a global one, which makes EED more intuitive as we showed in [7] [8].

We also presented other modifications of the edit distance based on the frequencies of *n*-grams [5], [6], [9] but they all include parameters whose computing can be very time consuming. In this paper we try to introduce a new extension of the edit distance which is parameter-free, thus can be applied directly.

3   The Parameter-Free Extended Edit Distance

3.1   The Number of Discrete Characters

Given two strings S, T. The number of distinct characters NDC is defined as:

$$NDC(S,T) = |\{ch(S)\} \cup \{ch(T)\}| \qquad (5)$$

where ch ( ) is the set of characters that a string contains.

The significance of NDC is related to the change operation; one of the three atomic operations that the edit distance is based on. Instead of predefining a cost function for all the change operations between any two characters in the alphabet, NDC can make the distance, by itself, detect if the change operations use characters that are familiar or unfamiliar to the two strings concerned.

3.2   The Proposed Distance

The objective of our work is to introduce a new distance that adds new features to the edit distance to make it detect global similarity, all this is done by keeping the new distance metric, which is the same objective as that of EED, but unlike EED, our new distance is parameter-free, which makes it more generic.

Definition: The Parameter-Free Extended Edit Distance (PFEED) is defined as:

$$PFEED\ (S,T) = ED\ (S,T) + \left(1 - \frac{2\sum_i min\left(f_i^{(S)}, f_i^{(T)}\right)}{|S| + |T|} \right) \tag{6}$$

Where $|S|, |T|$ are the lengths of the two strings S, T, respectively.

3.3   Theorem

PFEED is a metric distance   □

3. 3. 1. Proof: Before we prove the theorem, we can easily notice that:

$$\left(1 - \frac{2\sum_i min(\ f_i^{(S)}, f_i^{(T)}\ )}{|S| + |T|} \right) \geq 0 \qquad \forall S, T \tag{7}$$

In order to prove the theorem we have to prove that:

i) $PFEED(S,T) = 0 \Leftrightarrow S = T$

    i. a) $PFEED(S,T) = 0 \Rightarrow S = T$

-Proof:

If $PFEED(S,T) = 0$, and taking into account (7), we get the following relations:

$$\left(1 - \frac{2\sum_i min(\ f_i^{(S)}, f_i^{(T)}\ )}{|S| + |T|} \right) = 0 \tag{8}$$

$$ED(S,T) = 0 \tag{9}$$

From (9), and since ED is metric we get: $S=T$.

    i. b) $S = T \Rightarrow PFEED(\ S,T\ ) = 0$ (obvious).

From i. a) and i. b) we get $PFEED(S,T) = 0 \Leftrightarrow S = T$

ii) $PFEED(S,T) = PFEED(T,S)$ (obvious).

iii) $PFEED(S,T) \leq PFEED(S,R) + PEEED(R,T)$

-Proof: $\forall S, T, R$, we have:

$$ED(S,T) \le ED(S,R) + ED(R,T) \tag{10}$$

(Valid since ED is metric).
We also have:

$$\left(1 - \frac{2\sum_i \min(f_i^{(S)}, f_i^{(T)})}{|S| + |T|}\right) \le \left(1 - \frac{2\sum_i \min(f_i^{(S)}, f_i^{(R)})}{|S| + |R|}\right) + \left(1 - \frac{2\sum_i \min(f_i^{(R)}, f_i^{(T)})}{|R| + |T|}\right) \tag{11}$$

See the appendix for a proof of (11))
Adding (10), (11) side to side we get: $PFEED(S,T) \le PFEED(S,R) + PFEED(R,T)$.
From i), ii), and iii) we conclude that the theorem is valid.

4   Complexity Analysis

The time complexity of PFEED is $O(m \times n)$, where m is the length of the first string and n is the length of the second string, or $O(n^2)$ if the two strings are of the same lengths. This is the same complexity as that of ED and EED.

5   Experiments

The objective of our experiments is to compare PFEED with EED and show that they have similar performance but one, PFEED, can be applied directly while the other needs a long training time.

A *time series TS* is an ordered collection of observations at different time points. Time series data mining handles several tasks such as classification, clustering, similarity search, motif discovery, anomaly detection, and others. Time series are high-dimensional data so they are usually processed by using representation methods that are used to extract features from these data and project them on lower-dimensional spaces.

The *Symbolic Aggregate approXimation* method (SAX) [4] is one of the most important representation methods of time series. SAX is applied as follows:

1-The time series are normalized.
2-The dimensionality of the time series is reduced using a dimensionality reduction technique called *piecewise aggregate approximation* (PAA).
3-The PAA representation of the time series is discretized by determining the number and location of the breakpoints. Their locations are determined using Gaussian lookup tables. The interval between two successive breakpoints is assigned to a symbol of the alphabet, and each segment of PAA that lies within that interval is discretized by that symbol.

The last step of SAX is using the following similarity measure:

$$MINDIST(\hat{S},\hat{R}) \equiv \sqrt{\frac{n}{N}} \sqrt{\sum_{i=1}^{N} (dist(\hat{s}_i, \hat{r}_i))^2} \qquad (12)$$

Where n is the length of the original time series, N is the length of the strings. $\hat{S}$ and $\hat{R}$ are the symbolic representations of the two time series S and R, respectively, and where the function $dist(\ )$ is implemented by using the appropriate lookup table. After reaching this last step, SAX converts the resulting strings into numeric values so that the MINDIST can be calculated.

Since EED is applied to strings of characters, in [7] and [8] EED was tested on symbolically represented time series using SAX. In this paper we proceed in the same manner and compare PFEED on symbolically represented time series using SAX as a representation method.

EED has two main drawbacks; the first is that the parameter λ does not have any semantics, so its choice is completely heuristic. Besides in all the applications in which EED should be applied directly, i.e. there is no training phase, it becomes difficult to choose and justify the numeric value assigned to the parameter λ to calculate EED. The second drawback is that for each dataset we have to train the training datasets for 5 times at least for parameter λ ($\lambda = 0,...,1$ step=0.25, and sometimes we have to go beyond $\lambda = 1$), which takes a long time.

Table 1. A comparison of the classification error between PFEED and EED on different datasets

Dataset	Distance	
	PFEED	EED
Synthetic Control	0.030 $\alpha^* = 7$	0.037 $\alpha = 7, \lambda = 0$
Gun-Point	0.067 $\alpha = 4$	0.060 $\alpha = 4, \lambda = 0.25$
CBF	0.010 $\alpha = 6$	0.026 $\alpha = 3, \lambda = 0.27$
Face (all)	0.323 $\alpha = 5$	0.324 $\alpha = 7=, \lambda = 0$
OSULeaf	0.310 $\alpha = 5$	0.293 $\alpha = 5, \lambda = 0.75$
50words	0.270 $\alpha = 7$	0.266 $\alpha = 7, \lambda = 0$
Adiac	0.650 $\alpha = 9$	0.642 $\alpha = 9, \lambda = 0.5$
Yoga	0.155 $\alpha = 8$	0.155 $\alpha = 7, \lambda = 0$

(*: α is the alphabet size).

We conducted the experiments on PFEED using the same datasets on which EED was tested. These datasets are available at UCR [2]. We used the same protocol used with EED; we used SAX to get a symbolic representation of the time series, and then we replaced MINDIST with PFEED (or EED, when testing EED). We also used the same compression ratio that was used to test EED (i.e. 1:4) and the same range of alphabet size [3,10]. The experiments are a time series classification task based on thefirst nearest-neighbor (1-NN) rule using leaving-one-out cross validation. This means that every time series is compared to the other time series in the dataset. If the 1-NN does not belong to the same class, the error counter is incremented by 1.

We varied the alphabet size on the training set to get the optimal value of the alphabet size; i.e. the value that minimizes the error rate, and then we utilized this optimal value of the alphabet size on the testing sets. We obtained the results shown in Table. 1.

We see from Table 1 that the performance of PFEED is very similar to that of EED, yet PFEED can be applied directly as it requires no training to get the value of λ, which is not the case with EED.

6   Remarks

1-In the experiments we conducted we had to use time series of equal lengths for comparison reasons only, since SAX can be applied only to strings of equal lengths. But PFEED and EED can both be applied to strings of different lengths

2- We did not conduct experiments for alphabet size=2 because SAX is not applicable in this case, but PFEED and EED.

7   Conclusion

In this paper we presented a new extension of the edit distance; the parameter-free extended edit distance (PFEED) and we compared it to another extension; the extended edit distance (EED). The experiments we conducted show that PFEED gives similar results on a classification task, yet the new distance does not include any parameters, thus can be applied directly, which is not the case with EED.

References

1. Chavez, E., Navarro, G., Baeza-Yates, R.A., Marroquin, J.L.: Searching in Metric Spaces. ACM Computing Surveys (2001)
2. Keogh, E., Zhu, Q., Hu, B., Hao, Y., Xi, X., Wei, L., Ratanamahatana, C.A.: The UCR Time Series Classification/Clustering Homepage (2011), http://www.cs.ucr.edu/~eamonn/time_series_data/
3. Levenshtein, V.I.: Binary Codes Capable of Correcting Spurious Insertions and Deletions of Ones. Problems of Information Transmission 1, 8–17 (1965)
4. Lin, J., Keogh, E., Lonardi, S., Chiu, B.Y.: A Symbolic Representation of Time Series, with Implications for Streaming Algorithms. In: DMKD 2003, pp. 2–11 (2003)

5. Muhammad Fuad, M.M.: ABC-SG: A New Artificial Bee Colony Algorithm-Based Distance of Sequential Data Using Sigma Grams. In: The Tenth Australasian Data Mining Conference - AusDM 2012, Sydney, Australia, December 5-7, vol. 134. Published in the CRPIT Series (2012)
6. Muhammad Fuad, M.M.: Towards Normalizing the Edit Distance Using a Genetic Algorithms–Based Scheme. In: Zhou, S., Zhang, S., Karypis, G. (eds.) ADMA 2012. LNCS, vol. 7713, pp. 477–487. Springer, Heidelberg (2012)
7. Muhammad Fuad, M.M., Marteau, P.-F.: Extending the Edit Distance Using Frequencies of Common Characters. In: Bhowmick, S.S., Küng, J., Wagner, R. (eds.) DEXA 2008. LNCS, vol. 5181, pp. 150–157. Springer, Heidelberg (2008)
8. Muhammad Fuad, M.M., Marteau, P.F.: The Extended Edit Distance Metric. In: Sixth International Workshop on Content-Based Multimedia Indexing (CBMI 2008), London, UK, June 18-20 (2008)
9. Muhammad Fuad, M.M., Marteau, P.F.: The Multi-resolution Extended Edit Distance. In: Third International ICST Conference on Scalable Information Systems, Infoscale 2008, Vico Equense, Italy, June 4-6. ACM Digital Library (2008)
10. Samet, H.: Foundations of Multidimensional and Metric Data Structures. Elsevier (2006)
11. Wagner, R.A., Fischer, M.J.: The String-to-String Correction Problem. Journal of the Association for Computing Machinery 21(I) (January 1974)
12. Zezula, et al.: Similarity Search - The Metric Space Approach. Springer (2005)

Appendix

We present a brief proof of the theorem presented in Section 3.3

Lemma

Let Σ be a finite alphabet, $f_a^{(S)}$ be the frequency of the character a in S, where $S \in \Sigma^*$. Then $\forall S_1, S_2, S_3$ we have:

$$\left(1 - \frac{2\sum_i \min(f_i^{(S_1)}, f_i^{(S_2)})}{|S_1| + |S_2|}\right) \leq \left(1 - \frac{2\sum_i \min(f_i^{(S_1)}, f_i^{(S_3)})}{|S_1| + |S_3|}\right) + \left(1 - \frac{2\sum_i \min(f_i^{(S_3)}, f_i^{(S_2)})}{|S_3| + |S_2|}\right) \quad (A1)$$

For all n, where n is the number of characters used to represent the strings

Proof

i) $n = 1$, this is a trivial case, where the strings are represented with one character . Given three strings S_1, S_2, S_3 represented by the same character a .

Let $f_a^{(S_1)}, f_a^{(S_2)}, f_a^{(S_3)}$ be the frequency of a in S_1, S_2, S_3, respectively. We have six configurations in this case:

1) $f_a^{(S_1)} \le f_a^{(S_2)} \le f_a^{(S_3)}$

2) $f_a^{(S_1)} \le f_a^{(S_3)} \le f_a^{(S_2)}$

3) $f_a^{(S_2)} \le f_a^{(S_1)} \le f_a^{(S_3)}$

4) $f_a^{(S_2)} \le f_a^{(S_3)} \le f_a^{(S_1)}$

5) $f_a^{(S_3)} \le f_a^{(S_1)} \le f_a^{(S_2)}$

6) $f_a^{(S_3)} \le f_a^{(S_2)} \le f_a^{(S_1)}$

We will prove that relation (A1) holds in these six configurations.

1) $f_a^{(S_1)} \le f_a^{(S_2)} \le f_a^{(S_3)}$

In this case we have:

$$min\left(f_a^{(S_1)}, f_a^{(S_2)}\right) = f_a^{(S_1)}, \; min\left(f_a^{(S_1)}, f_a^{(S_3)}\right) = f_a^{(S_1)}, \; min\left(f_a^{(S_2)}, f_a^{(S_3)}\right) = f_a^{(S_2)}$$

By substituting the above values in (A1) we get:

$$1 - \frac{2\,min\left(f_a^{(S_1)}, f_a^{(S_2)}\right)}{f_a^{(S_1)} + f_a^{(S_2)}} \le 1 - \frac{2\,min\left(f_a^{(S_1)}, f_a^{(S_3)}\right)}{f_a^{(S_1)} + f_a^{(S_3)}} + 1 - \frac{2\,min\left(f_a^{(S_3)}, f_a^{(S_2)}\right)}{f_a^{(S_3)} + f_a^{(S_2)}}$$

$$1 - \frac{2f_a^{(S_1)}}{f_a^{(S_1)} + f_a^{(S_2)}} \le 1 - \frac{2f_a^{(S_1)}}{f_a^{(S_1)} + f_a^{(S_3)}} + 1 - \frac{2f_a^{(S_2)}}{f_a^{(S_3)} + f_a^{(S_2)}}$$

$$\frac{2f_a^{(S_1)}}{f_a^{(S_1)} + f_a^{(S_2)}} \ge \frac{2f_a^{(S_1)}}{f_a^{(S_1)} + f_a^{(S_3)}} + \frac{2f_a^{(S_2)}}{f_a^{(S_3)} + f_a^{(S_2)}} - 1$$

If we substitute $f_a^{(S_2)}, f_a^{(S_1)}, f_a^{(S_2)}$ with $f_a^{(S_1)}, f_a^{(S_3)}, f_a^{(S_3)}$, respectively in the denominators of the last relation it still holds according to the stipulation of this configuration. We get:

$$\frac{2f_a^{(S_1)}}{f_a^{(S_1)} + f_a^{(S_1)}} \ge \frac{2f_a^{(S_1)}}{f_a^{(S_3)} + f_a^{(S_3)}} + \frac{2f_a^{(S_2)}}{f_a^{(S_3)} + f_a^{(S_3)}} - 1$$

$$\frac{2f_a^{(S_1)}}{2f_a^{(S_1)}} \ge \frac{2f_a^{(S_1)}}{2f_a^{(S_3)}} + \frac{2f_a^{(S_2)}}{2f_a^{(S_3)}} - 1$$

$$1 \geq \frac{f_a^{(S_1)} + f_a^{(S_2)}}{f_a^{(S_3)}} - 1$$

$$2 f_a^{(S_3)} \geq f_a^{(S_1)} + f_a^{(S_2)}$$

This is valid according to the stipulation of this configuration.

The proofs of cases 2), 3), 4), 5) and 6) are similar to that of case 1).

From 1)-6) we conclude that the lemma is valid for $n = 1$

ii) $n > 1$

This is a generalization of the case where $n = 1$.

$\forall i \in n$, then

$$\left(1 - \frac{2\min\left(f_i^{(S_1)}, f_i^{(S_2)}\right)}{|S_1| + |S_2|}\right) \leq \left(1 - \frac{2\min\left(f_i^{(S_1)}, f_i^{(S_3)}\right)}{|S_1| + |S_3|}\right) + \left(1 - \frac{2\min\left(f_i^{(S_3)}, f_i^{(S_2)}\right)}{|S_3| + |S_2|}\right)$$

holds, according to the first case i)

By summing over n we get

$$\left(1 - \frac{2\sum_i \min\left(f_i^{(S_1)}, f_i^{(S_2)}\right)}{|S_1| + |S_2|}\right) \leq \left(1 - \frac{2\sum_i \min\left(f_i^{(S_1)}, f_i^{(S_3)}\right)}{|S_1| + |S_3|}\right) + \left(1 - \frac{2\sum_i \min\left(f_i^{(S_3)}, f_i^{(S_2)}\right)}{|S_3| + |S_2|}\right)$$

VGEN: Fast Vertical Mining of Sequential Generator Patterns

Philippe Fournier-Viger[1], Antonio Gomariz[2], Michal Šebek[3],
and Martin Hlosta[3]

[1] Dept. of Computer Science, University of Moncton, Canada
[2] Dept. of Information and Communication Engineering, University of Murcia, Spain
[3] Faculty of Information Technology, Brno University of Technology, Czech Republic
philippe.fournier-viger@umoncton.ca, agomariz@um.es,
{isebek,ihlosta}@fit.vutbr.cz

Abstract. *Sequential pattern mining* is a popular data mining task with wide applications. However, the set of all sequential patterns can be very large. To discover fewer but more representative patterns, several compact representations of sequential patterns have been studied. The set of *sequential generators* is one the most popular representations. It was shown to provide higher accuracy for classification than using all or only closed sequential patterns. Furthermore, mining generators is a key step in several other data mining tasks such as sequential rule generation. However, mining generators is computationally expensive. To address this issue, we propose a novel mining algorithm named *VGEN* (*Vertical sequential GENerator miner*). An experimental study on five real datasets shows that VGEN is up to two orders of magnitude faster than the state-of-the-art algorithms for sequential generator mining.

Keywords: sequential patterns, generators, vertical mining, candidate pruning.

1 Introduction

Sequential pattern mining is a popular data mining task, which aims at discovering interesting patterns in sequences [10]. A subsequence is called *sequential pattern* or *frequent sequence* if it frequently appears in a sequence database, and its frequency is no less than a user-specified *minimum support threshold* called *minsup* [1]. Sequential pattern mining plays an important role in data mining and is essential to a wide range of applications such as the analysis of web clickstreams, program executions, medical data, biological data and e-learning data [10]. Several algorithms have been proposed for sequential pattern mining such as *PrefixSpan* [11], *SPAM* [2], *SPADE* [15] and *CM-SPADE* [5]. However, a critical drawback of these algorithms is that they may present too many sequential patterns to users. A very large number of sequential patterns makes it difficult for users to make a good analysis of results to gain insightful knowledge. It may also cause the algorithms to become inefficient in terms of time and memory because the more sequential patterns the algorithms produce, the more resources

L. Bellatreche and M.K. Mohania (Eds.): DaWaK 2014, LNCS 8646, pp. 476–488, 2014.

they consume. The problem becomes worse when the database contains long sequential patterns. For example, consider a sequence database containing a sequential pattern having 20 distinct items. A sequential pattern mining algorithm will present the sequential pattern as well as its $2^{20} - 1$ subsequences to the user. This will most likely make the algorithm fail to terminate in reasonable time and run out of memory. To reduce the computational cost of the mining task and present fewer but more representative patterns to users, many studies focus on developing concise representations of sequential patterns [3,4,6,13]. Two of those representations are *closed sequential patterns* [6,13] and *sequential generator patterns* [8,12] that allow us to keep all the information about all the frequent patterns that can be potentially generated.

Among the aforementioned representations, mining sequential generators is desirable for several reasons. First, if sequential generators are used with closed patterns, they can provide additional information that closed patterns alone cannot provide [12]. For example, a popular application of sequential generators is to generate sequential rules with a minimum antecedent and a maximum consequent [9]. When we focus on obtaining rules, by using generators as antecedents and closed patterns as consequents, we obtain rules which allows deriving the maximum amount of information based on the minimum amount of information [9]. Second, it was shown that generators provide higher classification accuracy and are more useful for model selection than using all patterns or only closed patterns [7,12]. Third, sequential generators are preferable according to the principles of the MDL (Minimum Description Length) as they represent the smallest members of equivalence classes rather than the largest ones [8,12]. Lastly, the set of sequential generators is compact. It has a similar size or can be smaller than the set of closed sequential patterns [8].

Although mining sequential generators is desirable, it remains computationally expensive and few algorithms have been proposed for this task. Most algorithms such as GenMiner [8], FEAT [7] and FSGP [14] employ a pattern-growth approach by extending the PrefixSpan algorithm [11]. These algorithms differ by how they store patterns, prune the search space, and whether they identify generators on-the-fly or by post-processing. Because these algorithms all adopts a pattern-growth approach, they also suffer from its main limitation which is to repeatedly perform database projections to grow patterns, which is an extremely costly operation (in the worst case, a pattern-growth algorithm will perform a database projection for each item of each frequent pattern). Recently, an algorithm named MSGPs was also proposed. But it only provides a marginal performance improvement over previous approaches (up to 10%)[12].

In this paper, we propose a novel algorithm for mining sequential generators named *VGEN* (*Vertical sequential GENerator miner*). VGEN performs a depth-first exploration of the search space using a vertical representation of the database. The algorithm incorporates three efficient strategies named ENG (Efficient filtering of Non-Generator patterns), BEC (Backward Extension checking) and CPC (Candidate Pruning by Co-occurrence map) to effectively identify generator patterns and prune the search space. VGEN can capture the complete set of sequential generators and requires a single database scan to build its vertical structure. An experimental study with five real-life datasets shows that VGEN is up to two orders of magnitude faster than the state-of-the-art algorithms for sequential generator mining and performs very well on dense datasets.

The rest of the paper is organized as follows. Section 2 formally defines the problem of sequential generator mining and its relationship to sequential pattern mining. Section 3 describes the VGEN algorithm. Section 4 presents the experimental study. Finally, Section 5 presents the conclusion.

2   Problem Definition

Definition 1 (sequence database). Let $I = \{i_1, i_2, ..., i_l\}$ be a set of items (symbols). An *itemset* $I_x = \{i_1, i_2, ..., i_m\} \subseteq I$ is an unordered set of distinct items. The *lexicographical order* $\succ_{lex}$ is defined as any total order on I. Without loss of generality, we assume that all itemsets are ordered according to $\succ_{lex}$. A *sequence* is an ordered list of itemsets $s = \langle I_1, I_2, ..., I_n \rangle$ such that $I_k \subseteq I$ ($1 \le k \le n$). A *sequence database* SDB is a list of sequences $SDB = \langle s_1, s_2, ..., s_p \rangle$ having sequence identifiers (SIDs) $1, 2...p$.

Running Example. Consider the following sequence database:

sequence 1: $\langle \{a, b\}, \{c\}, \{f, g\}, \{g\}, \{e\} \rangle$
sequence 2: $\langle \{a, d\}, \{c\}, \{b\}, \{a, b, e, f\} \rangle$
sequence 3: $\langle \{a\}, \{b\}, \{f, g\}, \{e\} \rangle$
sequence 4: $\langle \{b\}, \{f, g\} \rangle$

It contains four sequences having the SIDs 1, 2, 3 and 4. Each single letter represents an item. Items between curly brackets represent an itemset. The first sequence $\langle \{a, b\}, \{c\}, \{f, g\}, \{g\}, \{e\} \rangle$ contains five itemsets. It indicates that items a and b occurred at the same time, were followed by c, then f and g at the same time, followed by g and lastly e.

Definition 2 (sequence containment). A sequence $s_a = \langle A_1, A_2, ..., A_n \rangle$ is said to be *contained in* a sequence $s_b = \langle B_1, B_2, ..., B_m \rangle$ iff there exist integers $1 \le i_1 < i_2 < ... < i_n \le m$ such that $A_1 \subseteq B_{i1}, A_2 \subseteq B_{i2}, ..., A_n \subseteq B_{in}$ (denoted as $s_a \sqsubseteq s_b$). In such a case, s_a is said to be a *sub-pattern* of s_b, and s_b to be a *super-pattern* of s_a **Example.** Sequence 4 of the running example is contained in Sequence 1.

Definition 3 (prefix). A sequence $s_a = \langle A_1, A_2, ..., A_n \rangle$ is a *prefix* of a sequence $s_b = \langle B_1, B_2, ..., B_m \rangle$, $\forall n < m$, iff $A_1 = B_1, A_2 = B_2, ..., A_{n-1} = B_{n-1}$ and the first $|A_n|$ items of B_n according to $\succ_{lex}$ are equal to A_n.

Definition 4 (extensions). A sequence s_b is said to be an *s*-extension of a sequence $s_a = \langle I_1, I_2, ...I_h \rangle$ with an item x, iff $s_b = \langle I_1, I_2, ...I_h, \{x\} \rangle$, i.e. s_a is a prefix of s_b and the item x appears in an itemset later than all the itemsets of s_a. In the same way, the sequence s_c is said to be an *i*-extension of s_a with an item x, iff $s_c = \langle I_1, I_2, ...I_h \cup \{x\} \rangle$, i.e. s_a is a prefix of s_c and the item x occurs in the last itemset of s_a, and the item x is the last one in I_h, according to $\succ_{lex}$.

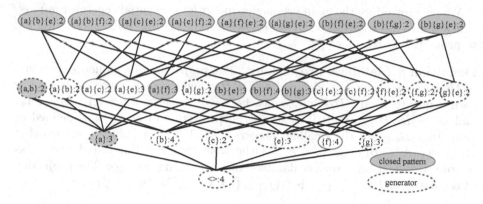

Fig. 1. All/closed/generator sequential patterns for the running example when $minsup = 2$

Definition 5 (support). The *support* of a sequence s_a in a sequence database SDB is defined as the number of sequences $s \in SDB$ such that $s_a \sqsubseteq s$ and is denoted by $sup_{SDB}(s_a)$.

Definition 6 (sequential pattern mining). Let $minsup$ be a minimum threshold set by the user and SDB be a sequence database. A sequence s is a *sequential pattern* and is deemed *frequent* iff $sup_{SDB}(s) \geq minsup$. The *problem of mining sequential patterns* is to discover all sequential patterns [1]. **Example.** The lattice shown in Fig. 1 presents the 30 sequential patterns found in the database of the running example for $minsup = 2$, and their support. For instance, the patterns $\langle\{a\}\rangle$, $\langle\{a\}, \{g\}\rangle$ and $\langle\rangle$ (the empty sequence) are frequent and have respectively a support of 3, 2 and 4 sequences.

Definition 7 (closed/generator sequential pattern mining). A sequential pattern s_a is said to be *closed* if there is no other sequential pattern s_b, such that $s_a \sqsubseteq s_b$, and their supports are equal. A sequential pattern s_a is said to be a *generator* if there is no other sequential pattern s_b, such that $s_b \sqsubseteq s_a$, and their supports are equal. An alternative and equivalent definition is the following. Let an *equivalence class* be the set of all patterns supported by the same set of sequences, partially-ordered by the $\sqsubseteq$ relation. Generator (closed) patterns are the minimal (maximal) members of each equivalence class [8]. The *problem of mining closed (generator) sequential patterns* is to discover the set of closed (generator) sequential patterns. **Example.** Consider the database of the running example and $minsup = 2$. There are 30 sequential patterns (shown in Fig. 1), such that 15 are closed (identified by a gray color) and only 11 are generators (identified by a dashed line). It can be observed that in this case, the number of generators is less than the number of closed patterns. Consider the equivalence class $\{\langle\{c\}\rangle, \langle\{a\}, \{c\}\rangle, \langle\{c\}, \{e\}\rangle, \langle\{a\}, \{c\}, \{e\}\rangle, \langle\{a\}, \{c\}, \{f\}\rangle\}$ supported by sequences 1 and 2. In this equivalence class, pattern $\langle\{c\}\rangle$ is the only generator and $\langle\{a\}, \{c\}, \{e\}\rangle$ and $\langle\{a\}, \{c\}, \{f\}\rangle$ are the closed patterns.

We next introduce the main pruning property for sequential generator mining, which is used in various forms by state-of-the-art algorithms for generator mining to prune the search space [8].

Definition 8 (database projection). The projection of a sequence database SDB by a sequence s_a is denoted as SDB_{s_a} and defined as the projection of each sequence from SDB by s_a. Let be two sequences $s_a = \langle A_1, A_2, ..., A_n \rangle$ and $s_b = \langle B_1, B_2, ..., B_m \rangle$. If $s_a \sqsubseteq s_b$, the projection of s_b by s_a is defined as $\langle B_{k1}, B_{k2}...B_{km} \rangle$ for the smallest integers $0 < k1 < k2 < ...km \leq m$ such that $A_1 \subseteq B_{k1}, A_2 \subseteq B_{k2}, ... A_n \subseteq B_{km}$. Otherwise, the projection is undefined. **Example.** Consider the sequence database of the running example. The projection of the database by $\langle \{a\}, \{c\} \rangle$ is $\langle \langle \{f, g\}, \{g\}, \{e\} \rangle, \langle \{b\}, \{a, b, e, f\} \rangle \rangle$.

Property 1. (non-generator pruning). Let SDB be a sequence database, and s_a and s_b be two distinct sequential patterns. If $s_b \sqsubseteq s_a$ and $SDB_{s_a} = SDB_{s_b}$, then s_a and any extensions of s_a are not generators [8]. **Example.** For the running example, the projections of $\langle \{f\} \rangle$ and $\langle \{f, g\} \rangle$ are identical. Therefore, $\langle \{f, g\} \rangle$ and any of its extensions are not generators.

Definition 9 (horizontal/vertical database format). A *sequence database in horizontal format* is a database where each entry is a sequence. A *sequence database in vertical format* is a database where each entry represents an item and indicates the list of sequences where the item appears and the position(s) where it appears [2]. **Example.** The sequence database of the running example was presented in horizontal format. Fig. 2 shows the same database but in vertical format.

a		b		c		e		f		g	
SID	**Itemsets**	**SID**	**Itemsets**	**SID**	**Itemsets**	**SID**	**Itemsets**	**SID**	**Itemsets**	**SID**	**Itemsets**
1	1	1	1	1	2	1	5	1	3	1	3,4
2	1,4	2	3,4	2	2	2	4	2	4	4	2
3	1	3	2			3	4	3	3		
		4	1	**d**				4	2		
				SID	**Itemsets**						
				2	1						

Fig. 2. The vertical representation of the database from our running example

Vertical mining algorithms associate a structure named *IdList* [15,2] to each pattern. IdLists allow calculating the support of a pattern quickly by making join operations with IdLists of smaller patterns. To discover sequential patterns, vertical mining algorithms perform a single database scan to create IdLists of patterns containing single items. Then, larger patterns are obtained by performing the join operation of IdLists of smaller patterns (see [15] for details). Several works proposed alternative representations for IdLists to save time in join operations, being the bitset representation the most efficient one [2].

3   The VGEN Algorithm

We present VGEN, our novel algorithm for sequential generator mining. It adopts the IdList structure implemented as bitsets [2,15]. We first describe the general search procedure used by VGEN to explore the search space of sequential patterns [2]. Then, we describe how it is adapted to discover sequential generators efficiently.

The pseudocode of the search procedure is shown in Fig. 3. The procedure takes as input a sequence database SDB and the $minsup$ threshold. The procedure first scans SDB once to construct the vertical representation of the database $V(SDB)$ and the set of frequent items F_1. For each frequent item $s \in F_1$, the procedure calls the SEARCH procedure with $\langle s \rangle$, F_1, $\{e \in F_1 | e \succ_{lex} s\}$, and $minsup$. The SEARCH procedure outputs the pattern $\langle \{s\} \rangle$ and recursively explores candidate patterns starting with prefix $\langle \{s\} \rangle$. The SEARCH procedure takes as parameters a sequential pattern pat and two sets of items to be appended to pat to generate candidates. The first set S_n represents items to be appended to pat by s-extension. The second set S_i represents items to be appended to pat by i-extension. For each candidate pat' generated by an extension, the procedure calculate the support to determine if it is frequent. This is done by the IdList join operation (see [2,15] for details) and counting the number of sequences where the pattern appears. If the pattern pat' is frequent, it is then used in a recursive call to SEARCH to generate patterns starting with the prefix pat'. It can be easily seen that the above procedure is correct and complete to explore the search space of sequential patterns since it starts with frequent patterns containing single items and then extend them one item at a time while only pruning infrequent extensions of patterns using the anti-monotonicity property (any infrequent sequential pattern cannot be extended to form a frequent pattern)[1]. We now describe how the search procedure is adapted to discover only generator patterns. This is done by integrating three strategies to efficiently filter non-generator patterns and prune the search space. The result is the VGEN algorithm, which returns the set of generator patterns.

Strategy 1. Efficient filtering of Non-Generator patterns (ENG). The first strategy identifies generator patterns among patterns generated by the search procedure. This is performed using a novel structure named Z that stores the set of generator patterns found so far. The structure Z is initialized as a set containing the empty sequence $\langle \rangle$ with its support equal to $|SDB|$. Then, during the search for patterns, every time that a pattern s_a, is generated by the search procedure, two operations are performed to update Z.

- *Sub-pattern checking.* During this operation, s_a is compared with each pattern $s_b \in Z$ to determine if there exists a pattern s_b such that $s_b \sqsubseteq s_a$ and $sup(s_a) = sup(s_b)$. If yes, then s_a is not a generator (by Definition 7) and thus, s_a is not inserted into Z. Otherwise, s_a is a generator with respect to all patterns found until now and it is thus inserted into Z.

PATTERN-ENUMERATION(*SDB, minsup*)
1. Scan *SDB* to create *V*(*SDB*) and identify S_{init}, the list of frequent items.
2. **FOR** each item s ∈ S_{init},
3.   **SEARCH**(⟨*s*⟩, S_{init}, the set of items from S_{init} that are lexically larger than *s*, *minsup*).

SEARCH(*pat, S_n, I_n, minsup*)
1. Output pattern *pat*.
2. $S_{temp} := I_{temp} := \emptyset$
3. **FOR** each item *j* ∈ S_n,
4.   **IF** the s-extension of *pat* is frequent **THEN** $S_{temp} := S_{temp} \cup \{i\}$.
5. **FOR** each item *j* ∈ S_{temp},
6.   **SEARCH**(the s-extension of *pat* with *j*, S_{temp}, elements in S_{temp} greater than *j*, *minsup*).
7. **FOR** each item *j* ∈ I_n,
8.   **IF** the i-extension of *pat* is frequent **THEN** $I_{temp} := I_{temp} \cup \{i\}$.
9. **FOR** each item *j* ∈ I_{temp},
10.   **SEARCH**(i-extension of *pat* with *j*, S_{temp}, all elements in I_{temp} greater than *j*, *minsup*).

Fig. 3. The search procedure

- *Super-pattern checking.* If s_a is determined to be a generator according to sub-pattern checking, the pattern s_a is compared with each pattern $s_b \in Z$. If there exists a pattern s_b such that $s_a \sqsubseteq s_b$ and $sup(s_a) = sup(s_b)$, then s_b is not a generator (by Definition 7) and s_b is removed from Z.

By using the above strategy, it is obvious that when the search procedure terminates, Z contains the set of sequential generator patterns. However, to make this strategy efficient, we need to reduce the number of pattern comparisons and containment checks ($\sqsubseteq$). We propose four optimizations.

1. *Size check optimization.* Let n be the number of items in the largest pattern found until now. The structure Z is implemented as a list of maps $Z = \{M_1, M_2, ...M_n\}$, where M_x contains all generator patterns found until now having x items ($1 \leq x \leq n$). To perform sub-pattern checking (super-pattern checking) for a pattern s containing w items, an optimization is to only compare s with patterns in $M_1, M_2...M_{w-1}$ (in $M_{w+1}, M_{w+2}...M_n$) because a pattern can only contain (be contained) in smaller (larger) patterns.
2. *SID count optimization.* To verify the pruning property 1, it is required to compare pairs of patterns s_a and s_b to see if their projected databases are identical, which will be presented in the BEC strategy. A necessary condition to have identical projected databases is that the sum of SIDs (Sequence IDs) containing s_a and s_b are the same. To check this condition efficiently, the sum of SIDs is computed for each pattern and each map M_k contains mappings of the form (l, S_k) where S_k is the set of all patterns in Z having l as sum of SIDS (Sequence IDs).
3. *Sum of items optimization.* In our implementation, each item is represented by an integer. For each pattern s, the *sum of the items* appearing in the pattern is computed, denoted as $sum(s)$. This allows the following optimization.

Consider super-pattern checking for pattern s_a and s_b. If $sum(s_a) > sum(s_b)$ for a pattern s_b, then we don't need to check $s_a \sqsubseteq s_b$. A similar optimization is done for sub-pattern checking. Consider sub-pattern checking for a pattern s_a and a pattern s_b. If $sum(s_b) > sum(s_a)$ for a pattern s_b, then we don't need to check $s_b \sqsubseteq s_a$.

4. *Support check optimization.* This optimization uses the support to avoid containment checks ($\sqsubseteq$). If the support of a pattern s_a is less than the support of another pattern s_b (greater), then we skip checking $s_a \sqsubseteq s_b$ ($s_b \sqsubseteq s_a$).

Strategy 2. Backward Extension Checking (BEC). The second strategy aims at avoiding sub-pattern checks. The search procedure discovers patterns by growing a pattern by appending one item at a time by s-extension or i-extension. Consider a pattern x' that is generated by extension of a pattern x. An optimization is to not perform sub-pattern checking if x' has the same support as x (because this pattern would have x as prefix, thus indicating that x is not a generator).

This optimization allows to avoid some sub-pattern checks. However it does allows the algorithm to prune the search space of frequent patterns to avoid considering patterns that are non generators. To prune the search space, we add a pruning condition based on Property 1. During sub-pattern checking for a pattern x, if a smaller pattern y can be found in Z such that the projected database is identical, then x is not a generator as well as any extension of x. Therefore, extensions of x should not be explored (by Property 1). Checking if projected databases are identical is done by comparing the IdLists of x and y.

Strategy 3. Candidate Pruning with Co-occurrence Map (CPC). The last strategy aims at pruning the search space of patterns by exploiting item co-occurrence information. We introduce a structure named *Co-occurrence MAP* (CMAP) [5] defined as follows: an item k is said to *succeed by i-extension* to an item j in a sequence $\langle I_1, I_2, ..., I_n \rangle$ iff $j, k \in I_x$ for an integer x such that $1 \le x \le n$ and $k \succ_{lex} j$. In the same way, an item k is said to *succeed by s-extension* to an item j in a sequence $\langle I_1, I_2, ..., I_n \rangle$ iff $j \in I_v$ and $k \in I_w$ for some integers v and w such that $1 \le v < w \le n$. A CMAP is a structure mapping each item $k \in I$ to a set of items succeeding it.

We define two CMAPs named $CMAP_i$ and $CMAP_s$. $CMAP_i$ maps each item k to the set $cm_i(k)$ of all items $j \in I$ succeeding k by i-extension in no less than *minsup* sequences of SDB. $CMAP_s$ maps each item k to the set $cm_s(k)$ of all items $j \in I$ succeeding k by s-extension in no less than *minsup* sequences of SDB. For example, the $CMAP_i$ and $CMAP_s$ structures built for the sequence database of the running example are shown in Table 1. Both tables have been created considering a *minsup* of two sequences. For instance, for the item f, we can see that it is associated with an item, $cm_i(f) = \{g\}$, in $CMAP_i$, whereas it is associated with two items, $cm_s(f) = \{e, g\}$, in $CMAP_s$. This indicates that both items e and g succeed to f by s-extension and only item g does the same for i-extension, being all of them in no less than *minsup* sequences.

VGEN uses CMAPs to prune the search space as follows. Let a sequential pattern pat being considered for s-extension (i-extension) with an item $x \in S_n$ by the SEARCH procedure (line 3). If the last item a in pat does not have an item $x \in cm_s(a)$ ($x \in cm_i$), then clearly the pattern resulting from the extension of pat with x will be infrequent and thus the join operation of x with pat to count the support of the resulting pattern does not need to be performed. Furthermore, the item x is not considered for generating any pattern by s-extension (i-extension) having pat as prefix, by not adding x to the variable S_{temp} (I_{temp}) that is passed to the recursive call to the SEARCH procedure. Moreover, note that we only have to check the extension of pat with x for the last item in pat, since other items have already been checked for extension in previous steps.

CMAPs are easily maintained and are built with a single database scan. With regards to their implementation, we define each one as a hash table of hashsets, where an hashset corresponding to an item k only contains the items that succeed to k in at least $minsup$ sequences.

Table 1. $CMAP_i$ and $CMAP_s$ for the database of Fig. 1 and $minsup = 2$

$CMAP_i$		$CMAP_s$	
item	is succeeded by (i-extension)	item	is succeeded by (s-extension)
a	$\{b\}$	a	$\{b, c, e, f\}$
b	$\emptyset$	b	$\{e, f, g\}$
c	$\emptyset$	c	$\{e, f\}$
e	$\emptyset$	e	$\emptyset$
f	$\{g\}$	f	$\{e, g\}$
g	$\emptyset$	g	$\emptyset$

4 Experimental Evaluation

We performed several experiments to assess the performance of the proposed algorithm. Experiments were performed on a computer with a third generation Core i5 64 bit processor running Windows 7 and 5 GB of free RAM. We compared the performance of VGEN with FSGP [14] and FEAT [7] for mining sequential generators and BIDE [13] for closed sequential pattern mining. Note that we do not compare with MSGPs for generator mining since it only provide a marginal speed improvement over FSGP (up to 10%) [12]. Furthermore, we do not compare with GenMiner, since authors of GenMiner reported that it is slower than BIDE [8]. All memory measurements were done using the Java API. Experiments were carried on five real-life datasets commonly used in the data mining literature, having varied characteristics and representing three different types of data (web click stream, text from a book and protein sequences). Table 2 summarizes the datasets' characteristics. The source code of all compared algorithms and datasets can be downloaded from http://goo.gl/xat4k.

Table 2. Dataset characteristics

dataset	sequence count	item count	avg. seq. length (items)	type of data
Leviathan	5834	9025	33.81 (std= 18.6)	book
Snake	163	20	60 (std = 0.59)	protein sequences
FIFA	20450	2990	34.74 (std = 24.08)	web click stream
BMS	59601	497	2.51 (std = 4.85)	web click stream
Kosarak10k	10000	10094	8.14 (std = 22)	web click stream

Experiment 1. Influence of the *minsup* **Parameter.** The first experiment consisted of running all the algorithms on each dataset while decreasing the *minsup* threshold until an algorithm became too long to execute, ran out of memory or a clear winner was observed. For each dataset, we recorded the execution time, memory usage and pattern count. Note that the execution time of VGEN includes the creation of ID-Lists.

In terms of execution time, results (see Fig. 4) show that VGEN is from one to two orders of magnitude faster than FEAT, FSGP and BIDE for all datasets. Furthermore, we observe that VGEN has excellent performance on dense datasets. For example, on Snake, VGEN is 127 times faster than FEAT, 149 times faster than FSGP, and 263 times faster than BIDE.

In terms of pattern count (see Fig. 4), as expected, we observe that the set of sequential generators can be much smaller than the set of all sequential patterns, and that there is about as much sequential generators as closed patterns.

In terms of memory consumption the maximum memory usage in megabytes of VGEN/FSGP/FEAT/BIDE for the Kosarak, BMS, Leviathan, Snake and FIFA datasets for the lowest *minsup* value were respectively 820/840/381/427, 1641/759/*/383, 3422/1561/*/*, 135/795/776/329 and 1815/1922/1896/2102, where * indicates that an algorithm has no result because it ran out of memory or failed to execute within the time limit of 1000s. We note that VGEN has the lowest memory consumption for Snake and FIFA, the second lowest for Leviathan and the third lowest for BMS and Kosarak.

Experiment 2. Influence of the Strategies. We next evaluated the benefit of using strategies in VGEN. We compared VGEN with a version of VGEN without strategy CPC (VGEN_WC) and a version without strategy BEC (VGEN_WB). Results for the Kosarak and Leviathan datasets are shown in Fig. 5. Results for other datasets are similar and are not shown due to space limitation. As a whole, strategies improved execution time by up to to 8 times, CPC being the most effective strategy.

We also measured the memory footprint used by the CPC strategy to build the CMAPs data structure. We found that the required amount memory is very small. For the BMS, Kosarak, Leviathan, Snake and FIFA datasets, the memory footprint is respectively 0.5 MB, 33.1 MB, 15 MB, 64 KB and 0.4 MB.

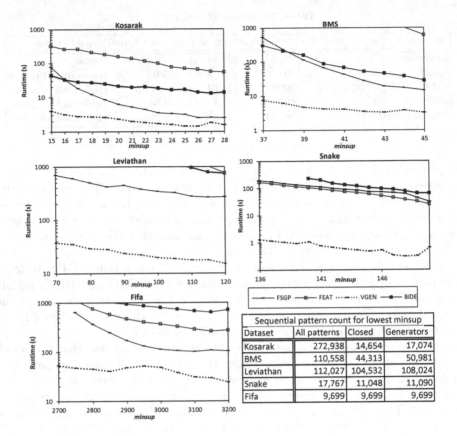

Fig. 4. Execution times and pattern count

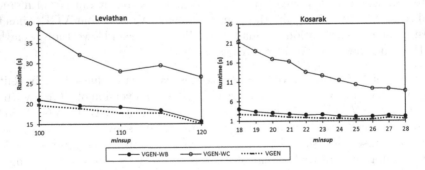

Fig. 5. Influence of optimizations for Leviathan (left) and Kosarak (right)

5   Conclusion

In this paper, we presented a depth-first search algorithm for mining sequential generators named VGEN (Vertical sequential GENerator miner). It relies on a vertical representation of the database and includes three novel strategies named ENG (Efficient filtering of Non-Generator patterns), BEC (Backward Extension checking) and CPC (Candidate Pruning by Co-occurrence map) to efficiently identify generators and prune the search space. We performed an experimental study with five real-life datasets to evaluate the performance of VGEN. Results show that VGEN is up to two orders of magnitude faster than the state-of-the-art algorithms for sequential generator mining. The source code of VGEN and all compared algorithms can be downloaded from http://goo.gl/xat4k.

Acknowledgement. This work is financed by a National Science and Engineering Research Council (NSERC) of Canada research grant.

References

1. Agrawal, R., Ramakrishnan, S.: Mining sequential patterns. In: Proc. 11th Intern. Conf. Data Engineering, pp. 3–14. IEEE (1995)
2. Ayres, J., Flannick, J., Gehrke, J., Yiu, T.: Sequential pattern mining using a bitmap representation. In: Proc. 8th ACM Intern. Conf. Knowl. Discov. Data Mining, pp. 429–435. ACM (2002)
3. Fournier-Viger, P., Wu, C.-W., Tseng, V.S.: Mining Maximal Sequential Patterns without Candidate Maintenance. In: Motoda, H., Wu, Z., Cao, L., Zaiane, O., Yao, M., Wang, W. (eds.) ADMA 2013, Part I. LNCS, vol. 8346, pp. 169–180. Springer, Heidelberg (2013)
4. Fournier-Viger, P., Wu, C.-W., Gomariz, A., Tseng, V.S.: VMSP: Efficient Vertical Mining of Maximal Sequential Patterns. In: Sokolova, M., van Beek, P. (eds.) Canadian AI 2014. LNCS, vol. 8436, pp. 83–94. Springer, Heidelberg (2014)
5. Fournier-Viger, P., Gomariz, A., Campos, M., Thomas, R.: Fast Vertical Mining of Sequential Patterns Using Co-occurrence Information. In: Tseng, V.S., Ho, T.B., Zhou, Z.-H., Chen, A.L.P., Kao, H.-Y. (eds.) PAKDD 2014, Part I. LNCS, vol. 8443, pp. 40–52. Springer, Heidelberg (2014)
6. Gomariz, A., Campos, M., Marin, R., Goethals, B.: ClaSP: An Efficient Algorithm for Mining Frequent Closed Sequences. In: Pei, J., Tseng, V.S., Cao, L., Motoda, H., Xu, G. (eds.) PAKDD 2013, Part I. LNCS, vol. 7818, pp. 50–61. Springer, Heidelberg (2013)
7. Gao, C., Wang, J., He, Y., Zhou, L.: Efficient mining of frequent sequence generators. In: Proc. 17th Intern. Conf. World Wide Web, pp. 1051–1052 (2008)
8. Lo, D., Khoo, S.-C., Li, J.: Mining and Ranking Generators of Sequential Patterns. In: Proc. SIAM Intern. Conf. Data Mining, pp. 553–564 (2008)
9. Lo, D., Khoo, S.-C., Wong, L.: Non-redundant sequential rules: Theory and algorithm. Information Systems 34(4), 438–453 (2011)
10. Mabroukeh, N.R., Ezeife, C.I.: A taxonomy of sequential pattern mining algorithms. ACM Computing Surveys 43(1), 1–41 (2010)
11. Pei, J., Han, J., Mortazavi-Asl, B., Wang, J., Pinto, H., Chen, Q., Dayal, U., Hsu, M.: Mining sequential patterns by pattern-growth: the PrefixSpan approach. IEEE Trans. Known. Data Engin. 16(11), 1424–1440 (2004)

12. Pham, T.-T., Luo, J., Hong, T.-P., Vo, B.: MSGPs: A novel algorithm for mining sequential generator patterns. In: Nguyen, N.-T., Hoang, K., Jędrzejowicz, P. (eds.) ICCCI 2012, Part II. LNCS, vol. 7654, pp. 393–401. Springer, Heidelberg (2012)
13. Wang, J., Han, J., Li, C.: Frequent closed sequence mining without candidate maintenance. IEEE Trans. on Knowledge Data Engineering 19(8), 1042–1056 (2007)
14. Yi, S., Zhao, T., Zhang, Y., Ma, S., Che, Z.: An effective algorithm for mining sequential generators. Procedia Engineering 15, 3653–3657 (2011)
15. Zaki, M.J.: SPADE: An efficient algorithm for mining frequent sequences. Machine Learning 42(1), 31–60 (2001)

Author Index